Robust Nonparametric Statistical Methods

KENDALL'S ADVANCED THEORY OF STATISTICS
and
KENDALL'S LIBRARY OF STATISTICS

Advisory Editorial Board: PJ Green, University of Bristol; RJ Little, University of Michigan; JK Ord, Pennsylvania State University

The development of statistical theory in the past fifty years is faithfully reflected in the history of the late Sir Maurice Kendall's volumes THE ADVANCED THEORY OF STATISTICS. The ADVANCED THEORY began life as a two volume work (Volume 1, 1943; Volume 2, 1946) and grew steadily, as a single authored work, until the late fifties. At this point, Alan Stuart became co-author and the ADVANCED THEORY was rewritten in three volumes. When Keith Ord joined in the early eighties, Volume 3 became the largest and plans were developed to expand, yet again, to a four-volume work. Even so, it became evident that there were gaps in the coverage and that it was becoming increasingly difficult to provide timely updates to all volumes, so a new strategy was devised.

In future, the ADVANCED THEORY will be in the form of three core volumes together with a series of related monographs called KENDALL'S LIBRARY OF STATISTICS. The three volumes of ADVANCED THEORY will be:

1   Distribution Theory
2A Classical Inference and Relationships
2B Bayesian Inference (a new companion volume by Anthony O'Hagan)

KENDALL'S LIBRARY OF STATISTICS will encompass the areas previously appearing in the old Volume 3, such as sample surveys, design of experiments, multivariate analysis and time series as well as nonparametrics and log-linear models, previously covered to some extent in Volume 2. In the preface to the first edition of THE ADVANCED THEORY Kendall declared that his aim was 'to develop a systematic treatment of [statistical] theory as it exists at the present time' while ensuring that the work remained 'a book on statistics, not on statistical mathematics'. These aims continue to hold true for KENDALL'S LIBRARY OF STATISTICS and the flexibility of the monograph format will enable the series to maintain comprehensive coverage over the whole of modern statistics.

Published volumes:
1. MULTIVARIATE ANALYSIS Part 1 Distributions, Ordination and Inference, WJ Krzanowski (University of Exeter) and FHC Marriott (University of Oxford)   1994   0 340 59326 1

2. MULTIVARIATE ANALYSIS Part 2 Classification, Covariance Structures and Repeated Measurements, WJ Krzanowski (University of Exeter) and FHC Marriott (University of Oxford)   1995   0 340 59325 3

3. MULTILEVEL STATISTICAL MODELS Second Edition, H Goldstein (University of London)   1995   0 340 59529 9

4. THE ANALYSIS OF PROXIMITY DATA, BS Everitt (Institute of Psychiatry) and S Rabe-Hesketh (Institute of Psychiatry)   1997   0 340 67776 7

5. ROBUST NONPARAMETRIC STATISTICAL METHODS, TP Hettmansperger (Penn State University) and JW McKean (Western Michigan University)   1998   0 340 54937 8

ROBUST NONPARAMETRIC STATISTICAL METHODS

T. P. Hettmansperger
Penn State University

and

J. W. McKean
Western Michigan University

A member of the Hodder Headline Group
LONDON • SYDNEY • AUCKLAND

Copublished in North, Central and South America by
John Wiley & Sons Inc., New York • Toronto

First published in Great Britain in 1998 by Arnold,
a member of the Hodder Headline Group
338 Euston Road, London NW1 3BH

http://www.arnoldpublishers.com

Copublished in North, Central and South America by
John Wiley & Sons Inc., 605 Third Avenue, New York, NY 10158-0012

© 1998 T. P. Hettmansperger and J. W. McKean

All rights reserved. No part of this publication may be reproduced or transmitted
in any form or by any means, electronically or mechanically, including
photocopying, recording or any information storage or retrieval system, without
either prior permission in writing from the publisher or a licence permitting
restricted copying. In the United Kingdom such licences are issued by the
Copyright Licensing Agency: 90 Tottenham Court Road, London W1P 9HE.

British Library Cataloguing in Publication Data
A catalogue record for this book is available from the British Library

Library of Congress Cataloging-in-Publication Data
A catalog record for this book in available from the Library of Congress

ISBN 0 340 54937 8

ISBN 0 471 19479 4

Publisher: Nicki Dennis
Production Editor: Wendy Rooke
Production Controller: Rose James

Typeset by AFS Image Setters Ltd, Glasgow
Printed in Great Britain by J W Arrowsmith, Bristol

To Ann and to Marge

Contents

Preface   **xiii**

1   One-Sample Problems   **1**
   1.1   Introduction   1
   1.2   Location Model   2
   1.3   Geometry and Inference in the Location Model   4
   1.4   Examples   12
   1.5   Properties of Norm-Based Inference   16
      1.5.1   Basic Properties of the Power Function $\gamma_S(\theta)$ 17
      1.5.2   Asymptotic Linearity and Pitman Regularity   19
      1.5.3   Asymptotic Theory and Efficiency Results for $\widehat{\theta}$ 22
      1.5.4   Asymptotic Power and Efficiency Results for the Test Based on $S(\theta)$ 24
      1.5.5   Efficiency Results for Confidence Intervals Based on $S(\theta)$ 25
   1.6   Robustness Properties of Norm-Based Inference   28
      1.6.1   Robustness Properties of $\widehat{\theta}$ 28
      1.6.2   Breakdown Properties of Tests   31
   1.7   Inference and the Wilcoxon Signed-Rank Norm   33
      1.7.1   Null Distribution Theory of $T(0)$ 34
      1.7.2   Statistical Properties   35
      1.7.3   Robustness Properties   40
   1.8   Inference Based on General Signed-Rank Norms   42
      1.8.1   Null Properties of the Test   44
      1.8.2   Efficiency and Robustness Properties   45
   1.9   Tests for Symmetry   49
   1.10   Ranked-Set Sampling   52
   1.11   Interpolated Confidence Intervals for the L_1 Inference   56
   1.12   Two-Sample Analysis   60
   1.13   Exercises   64

2 Two-Sample Problems — 69
- 2.1 Introduction — 69
- 2.2 Geometric Motivation — 69
  - 2.2.1 Least-Squares (LS) Analysis — 73
  - 2.2.2 Mann–Whitney–Wilcoxon (MWW) Analysis — 73
- 2.3 Examples — 75
- 2.4 Inference Based on the Mann–Whitney–Wilcoxon Statistic — 77
  - 2.4.1 Testing — 77
  - 2.4.2 Confidence Intervals — 86
  - 2.4.3 Statistical Properties of the Inference Based on the MWW — 86
  - 2.4.4 Estimation of Δ — 90
  - 2.4.5 Efficiency Results Based on Confidence Intervals — 92
- 2.5 General Rank Pseudo-Norms — 93
  - 2.5.1 Statistical Methods — 96
  - 2.5.2 Efficiency Results — 97
  - 2.5.3 Connection between One- and Two-Sample Scores — 100
- 2.6 L_1 Analyses — 101
  - 2.6.1 Analysis Based on the L_1 Pseudo-norm — 102
  - 2.6.2 Analysis Based on the L_1 Norm — 105
- 2.7 Robustness Properties — 108
  - 2.7.1 Breakdown Properties — 108
  - 2.7.2 Influence Functions — 109
- 2.8 Lehmann Alternatives and Proportional Hazards — 111
  - 2.8.1 The Log Exponential and the Savage Statistic — 112
  - 2.8.2 Efficiency Properties — 114
- 2.9 Two-Sample Ranked Set Sampling (RSS) — 116
- 2.10 Two-Sample Scale Problem — 118
  - 2.10.1 $\mathcal{L}(Y) = \mathcal{L}(\eta X)$ — 118
  - 2.10.2 Unknown Locations — 121
  - 2.10.3 Linear Rank Statistics of the Observations — 123
- 2.11 Behrens–Fisher Problem — 128
  - 2.11.1 Behavior of the Usual MWW Test — 128
  - 2.11.2 General Rank Tests — 130
  - 2.11.3 Modified Mathisen Test — 131
  - 2.11.4 Modified MWW Test — 133
  - 2.11.5 Efficiencies and Discussion — 134
- 2.12 Paired Designs — 136
  - 2.12.1 Behavior under Alternatives — 139
- 2.13 Exercises — 140

3 Linear Models — 145
- 3.1 Introduction — 145
- 3.2 Geometry of Estimation and Tests — 145
  - 3.2.1 Estimation — 146
  - 3.2.2 The Geometry of Testing — 149
- 3.3 Examples — 151
- 3.4 Assumptions for Asymptotic Theory — 157
- 3.5 Theory of Rank-Based Estimates — 159

		3.5.1 R-Estimators of the Regression Coefficients	159
		3.5.2 R-Estimates of the Intercept	164
	3.6	Theory of Rank-Based Tests	170
		3.6.1 Null Theory of Rank-Based Tests	170
		3.6.2 Theory of Rank-Based Tests under Alternatives	175
		3.6.3 Further Remarks on the Dispersion Function	179
	3.7	Implementation of the R Analysis	181
		3.7.1 Estimates of the Scale Parameter τ_φ	181
		3.7.2 Algorithms for Computing the R Analysis	184
		3.7.3 An Algorithm for a Linear Search	186
	3.8	L_1 Analysis	187
	3.9	Diagnostics	189
		3.9.1 Properties of R Residuals and Model Misspecification	190
		3.9.2 Standardization of R Residuals	197
		3.9.3 Measures of Influential Cases	202
	3.10	Survival Analysis	207
	3.11	Correlation Model	215
		3.11.1 Huber's Condition for the Correlation Model	215
		3.11.2 Traditional Measure of Association and its Estimate	217
		3.11.3 Robust Measure of Association and its Estimate	218
		3.11.4 Properties of R Coefficients of Multiple Determination	220
		3.11.5 Coefficients of Determination for Regression	225
	3.12	Exercises	226
4	**Experimental Designs**		**233**
	4.1	Introduction	233
	4.2	One-way Design	234
		4.2.1 R Fit of the One-way Design	235
		4.2.2 Rank-Based Tests of $H_0 : \mu_1 = \cdots = \mu_k$	239
		4.2.3 Tests of General Contrasts	241
		4.2.4 More on Estimation of Contrasts and Location	242
		4.2.5 Pseudo-observations	244
	4.3	Multiple Comparison Procedures	246
		4.3.1 Discussion	252
	4.4	Two-way Crossed Factorial	254
	4.5	Analysis of Covariance	258
	4.6	Further Examples	262
	4.7	Rank Transform	267
		4.7.1 Monte Carlo Study	269
	4.8	Exercises	273
5	**Bounded Influence and High-Breakdown Methods**		**279**
	5.1	Geometry of the GR-Estimates	279
	5.2	Asymptotic Theory	280
	5.3	Implementation	288
	5.4	Diagnostics	293
		5.4.1 Interpretation of GR Residuals	293
		5.4.2 Studies for Curvature Detection	298

		5.4.3	Standardization of GR Residuals	299
	5.5	Diagnostics that Detect Differences between Fits		300
	5.6	Coefficients of Multiple Determination		308
	5.7	Robustness Properties		311
		5.7.1	Influence Functions	311
		5.7.2	Breakdown	313
	5.8	High-Breakdown (HBR) Estimates		313
		5.8.1	Definition of the HBR-Estimates	314
		5.8.2	Asymptotic Normality of $\widehat{\boldsymbol{\beta}}_{HBR}$	315
		5.8.3	Robustness Properties of the HBR-Estimate	317
		5.8.4	Discussion	320
	5.9	Implementation and Examples of High-Breakdown Fits		322
		5.9.1	Standard Errors of $\widehat{\boldsymbol{\beta}}_{HBR}$ and Internal Studentized Residuals	323
		5.9.2	Examples	324
	5.10	Exercises		326

6 Multivariate Models — 329

	6.1	Bivariate Location Model		329
	6.2	Componentwise Estimating Equations		334
		6.2.1	Estimation	336
		6.2.2	Testing	339
		6.2.3	Componentwise Rank Methods	342
	6.3	Spatial Methods		344
		6.3.1	Spatial Rank Methods	350
	6.4	Affine Equivariant and Invariant L_1 Methods		356
		6.4.1	The Oja Criterion Function	360
		6.4.2	Affine Invariant Rank Methods	363
		6.4.3	Summary	366
	6.5	Robustness of Multivariate Estimates of Location		367
		6.5.1	Location and Scale Invariance: Componentwise Methods	368
		6.5.2	Rotation Invariance: Spatial Methods	368
		6.5.3	Spatial R-Estimate of Hössjer and Croux	369
		6.5.4	The Spatial Hodges–Lehmann Estimate	369
		6.5.5	Affine Equivariance: The Oja Median	369
	6.6	Linear Model		372
		6.6.1	Test for Regression Effect	374
		6.6.2	The Estimate of the Regression Effect	380
		6.6.3	Tests of General Hypotheses	381
	6.7	Experimental Designs		389
	6.8	Exercises		392

A Asymptotic Results — 397

	A.1	Central Limit Theorems		397
	A.2	Simple Linear Rank Statistics		398
		A.2.1	Null Asymptotic Distribution Theory	399
		A.2.2	Local Asymptotic Distribution Theory	400
		A.2.3	Signed-Rank Statistics	407

A.3		Results for Rank-Based Analysis of Linear Models	410
	A.3.1	Convex Functions	412
	A.3.2	Asymptotic Linearity and Quadraticity	413
	A.3.3	Asymptotic Distance Between $\widehat{\boldsymbol{\beta}}$ and $\overline{\boldsymbol{\beta}}$	416
	A.3.4	Consistency of the Test Statistic F_φ	417
	A.3.5	Proof of Lemma 3.5.8	418
A.4		Asymptotic Linearity for the L_1 Analysis	419
A.5		Influence Functions	422
	A.5.1	Influence Function for Estimates Based on Signed-Rank Statistics	423
	A.5.2	Influence Functions for Chapter 3	424
	A.5.3	Influence Function of $\widehat{\boldsymbol{\beta}}_{HBR}$ of Chapter 5	430
A.6		Asymptotic Theory for Chapter 5	432

Bibliography   441

Author Index   457

Subject Index   461

Preface

I don't believe I can really do without teaching. The reason is, I have to have something so that when I don't have any ideas and I'm not getting anywhere I can say to myself, "At least I'm living; at least I'm doing something; I'm making some contribution" – it's just psychological.

Richard Feynman

This book is based on the premise that nonparametric or rank-based statistical methods are a superior choice in many data-analytic situations. We cover location models, regression models including designed experiments, and multivariate models. Geometry provides a unifying theme throughout much of the development. We emphasize the similarity in interpretation with least-squares methods. Basically, we replace the Euclidean norm with a weighted L_1 norm. This results in rank-based methods or L_1 methods depending on the choice of weights. The rank-based methods proceed much like the traditional analysis. Using the norm, models are easily fitted. Diagnostics procedures can then be used to check the quality of fit (model criticism) and to locate outlying points and points of high influence. Upon satisfaction with the fit, rank-based inferential procedures can be used to conduct the statistical analysis. The benefits include significant gains in power and efficiency when the error distribution has tails heavier than those of a normal distribution and superior robustness properties in general.

The main text concentrates on Wilcoxon and L_1 methods. The theoretical development for general scores (weights) is contained in the Appendix. By restricting attention to Wilcoxon rank methods, we can recommend a unified approach to data analysis beginning with the simple location models and extending through complex regression models and designed experiments. All major methodology is illustrated on real data. The examples are intended as guides for the application of the rank and L_1 methods. Furthermore, all the data sets in this book can be obtained from the Web site

http://www.stat.wmich.edu/home.html.

Selected topics from the first four chapters provide a basic graduate course in rank-based methods. The prerequisites are an introductory course in mathematical statistics and some background in applied statistics. The first seven sections of Chapter 1 and the first four sections of Chapter 2 are fundamental for the development of Wilcoxon signed-rank and Mann–Whitney–Wilcoxon rank-sum methods in the one- and two-sample location models. In Chapter 3, on the linear model, sections one through seven and section nine present the basic material for estimation, testing and diagnostic procedures for model criticism. Sections two through four of Chapter 4 give extensive development of methods for the one- and two-way layouts. Then, depending on individual tastes, there are several more exotic topics in each chapter to choose from.

Chapters 5 and 6 contain more advanced material. In Chapter 5 we extend rank-based methods for a linear model to bounded influence, high-breakdown estimates and tests. In Chapter 6 we take up the concept of multidimensional rank. We then discuss various approaches to the development of rank-like procedures that satisfy various invariant/equivariant restrictions.

Computation of the procedures discussed in this book is very important. Minitab contains an undocumented RREG (rank regression) command. It contains various subcommands that allow for testing and estimation in the linear model. The reader can contact Minitab at (info@minitab.com) and request a technical report that describes the RREG command. In many of the examples of this book the package rglm is used to obtain the rank-based analyses. The basic algorithms behind this package are described in Chapter 3. Information (including online rglm analyses of examples) can be obtained from the Web site: http://www.stat.wmich.edu/home.html. Students can also be encouraged to write their own S-Plus functions for specific methods.

We are indebted to many of our students and colleagues for valuable discussions, stimulation, and motivation. In particular, the first author would like to express his sincere thanks for many stimulating hours of discussion with Steve Arnold, Bruce Brown, and Hannu Oja, while the second author wants to express his sincere thanks for discussions with John Kapenga, Joshua Naranjo, Jerry Sievers, and Tom Vidmar. We both would like to express our debt to Simon Sheather, our friend, colleague, and co-author on many papers. Finally, we would like to thank Jun Recta for assistance in creating several of the plots and Nicki Dennis, our editor at Arnold, who possesses an infinite store of patience and has given us encouragement and support at every opportunity.

Tom Hettmansperger
Joe McKean

June 1997
State College, PA
Kalamazoo, MI

1
One-Sample Problems

1.1 Introduction

Traditional statistical procedures are widely used because they offer the user a unified methodology with which to attack a multitude of problems, from simple location problems to highly complex experimental designs. These procedures are based on least-squares fitting. Once the problem has been cast into a model then least squares offers the user:

1. a way of fitting the model by minimizing the Euclidean normed distance between the responses and the conjectured model;
2. diagnostic techniques that check the adequacy of the fit of the model, explore the quality of fit, and detect outlying and/or influential cases;
3. inferential procedures, including confidence procedures, tests of hypotheses and multiple comparison procedures;
4. computational feasibility.

Procedures based on least squares, though, are easily impaired by outlying observations. Indeed one outlying observation is enough to spoil the least-squares fit, its associated diagnostics and inference procedures. Even though traditional inference procedures are exact when the errors in the model follow a normal distribution, they can be quite inefficient when the distribution of the errors has longer tails than the normal distribution.

For simple location problems, nonparametric methods were proposed by Wilcoxon (1945). These methods consist of test statistics based on the ranks of the data and associated estimates and confidence intervals for location parameters. The test statistics are distribution-free in the sense that their null distributions do not depend on the distribution of the errors. It was soon realized that these procedures are almost as efficient as the traditional methods when the errors follow a normal distribution and, furthermore, are often much more efficient relative to the traditional methods when the error distributions deviate from normality; see Hodges and Lehmann (1956). These procedures possess both robustness of validity and power. In recent years these nonparametric methods have been extended to linear and nonlinear models. In addition, from the perspective of modern robustness theory, contrary to least-squares estimates, these rank-based procedures have bounded influence functions and positive breakdown points.

Often these nonparametric procedures are thought of as disjoint methods that differ from one problem to another. In this text, we intend to show that this is not the case. Instead, these procedures present a unified methodology analogous to the traditional methods. The four items cited above for the traditional analysis hold for these procedures too. Indeed the only operational difference is that the Euclidean norm is replaced by another norm.

1.2 Location Model

In this chapter we will consider the one-sample location problem. This will allow us to explore some useful concepts such as distribution-freeness and robustness in a simple setting. We will extend many of these concepts to more complicated situations in later chapters. We need first to define a location parameter. For a random variable X we often subscript its distribution function by X to avoid confusion.

Definition 1.2.1 *Let $T(H)$ be a function defined on the set of distribution functions. We say $T(H)$ is a* **location functional** *if the following conditions hold:*

1. *if G is stochastically larger than F – i.e. $G(x) \leq F(x)$ – for all x, then $T(G) \geq T(F)$;*
2. $T(H_{aX+b}) = aT(H_X) + b, a > 0;$
3. $T(H_{-X}) = -T(H_X).$

Then, we will call $\theta = T(H)$ a **location parameter** *of H.*

Note that if X has location parameter θ it follows from the second item in the above definition that the random variable $e = X - \theta$ has location parameter 0. Suppose $X_1, \ldots, X_n$ is a random sample having the common distribution function $H(x)$ and $\theta = T(H)$ is a location parameter of interest. We express this by saying that X_i follows the **statistical location model**,

$$X_i = \theta + e_i, \quad i = 1, \ldots, n, \qquad (1.2.1)$$

where $e_1, \ldots, e_n$ are independent and identically distributed (iid) random variables with distribution function $F(x)$, density function $f(x)$, and location $T(F) = 0$. It follows that $H(x) = F(x - \theta)$ and that $T(H) = \theta$. We next discuss three examples of location parameters that we will use throughout this chapter. Other location parameters are discussed in Section 1.8. See Bickel and Lehmann (1975) for additional discussion of location functionals.

Example 1.2.1 *The Median Location Functional*

First define the inverse of the cdf $H(x)$ by $H^{-1}(u) = \inf\{x : H(x) \geq u\}$. Generally we will suppose that $H(x)$ is strictly increasing on its support and this will eliminate ambiguities on the selection of the parameter. Now define

$\theta_1 = T_1(H) = H^{-1}(\frac{1}{2})$. This is the median functional. Note that if $G(x) \leqslant F(x)$ for all x, then $G^{-1}(u) \geqslant F^{-1}(u)$ for all u; and, in particular, $G^{-1}(\frac{1}{2}) \geqslant F^{-1}(\frac{1}{2})$. Hence, $T_1(H)$ satisfies the first condition for a location functional. Next let $H^*(x) = P(aX + b \leqslant x) = H[a^{-1}(x-b)]$. Then it follows at once that $H^{*-1}(u) = aH^{-1}(u) + b$ and the second condition is satisfied. The third condition follows with an argument similar to the the one for the second condition.

Example 1.2.2 *The Mean Location Functional*

For the mean functional let $\theta_2 = T_2(H) = \int x\,dH(x)$, when the mean exists. Note that $\int x\,dH(x) = \int H^{-1}(u)\,du$. Now if $G(x) \leqslant F(x)$ for all x, then $x \leqslant G^{-1}(F(x))$. Let $x = F^{-1}(u)$ and we have $F^{-1}(u) \leqslant G^{-1}(F(F^{-1}(u))) \leqslant G^{-1}(u)$. Hence, $T_2(G) = \int G^{-1}(u)\,du \geqslant \int F^{-1}(u)\,du = T_2(F)$ and the first condition is satisfied. The other two conditions follow easily from the definition of the integral.

Example 1.2.3 *The Pseudo-Median Location Functional*

Assume that X_1 and X_2 are iid, with distribution function $H(x)$. Let $Y = (X_1 + X_2)/2$. Then Y has distribution function $H^*(y) = P(Y \leqslant y) = \int H(2y - x)h(x)\,dx$. Let $\theta_3 = T_3(H) = H^{*-1}(\frac{1}{2})$. To show that T_3 is a location functional, suppose $G(x) \leqslant F(x)$ for all x. Then

$$G^*(y) = \int G(2y-x)g(x)\,dx = \int\left[\int_{-\infty}^{2y-x} g(t)\,dt\right]g(x)\,dx \leqslant \int\left[\int_{-\infty}^{2y-x} f(t)\,dt\right]g(x)\,dx$$

$$= \int\left[\int_{-\infty}^{2y-t} g(x)\,dx\right]f(t)\,dt \leqslant \int\left[\int_{-\infty}^{2y-t} f(x)\,dx\right]f(t)\,dt = F^*(y);$$

hence, as in Example 1.2.1., it follows that $G^{*-1}(u) \geqslant F^{*-1}(u)$ and, hence, that $T_3(G) \geqslant T_3(F)$. For the second property, let $W = aX + b$, where X has distribution function H and $a > 0$. Then W has distribution function $F_W(t) = H((t-b)/a)$. Then by the change of variable $z = (x-b)/a$, we have

$$F_W^*(y) = \int H\left(\frac{2y - x - b}{a}\right)\frac{1}{a}h\left(\frac{x-b}{a}\right)dx = \int H\left(2\frac{y-b}{a} - z\right)h(z)\,dz.$$

Thus the defining equation for $T_3(F_W)$ is

$$\frac{1}{2} = \int H\left(2\frac{T_3(F_W) - b}{a} - z\right)h(z)\,dz,$$

which is satisfied for $T_3(F_W) = aT_3(H) + b$. For the third property, let $V = -X$, where X has distribution function H. Then V has distribution function $F_V(t) = 1 - H(-t)$. Hence, by the change in variable $z = -x$,

$$F_V^*(y) = \int(1 - H(-2y+x))h(-x)\,dx = 1 - \int H(-2y-z)h(z)\,dz.$$

Because the defining equation of $T_3(F_V)$ can be written as

$$\frac{1}{2} = \int H(2(-T_3(F_V)) - z)h(z)\,dz,$$

it follows that $T_3(F_V) = -T_3(H)$. Therefore, T_3 is a location functional. It has been called the **pseudo-median** by Høyland (1965) and is more appropriate for symmetric distributions.

The next theorem characterizes all the location functionals for a symmetric distribution.

Theorem 1.2.1 *Suppose that the pdf $h(x)$ is symmetric about some point a. If $T(H)$ is a location functional, then $T(H) = a$.*

Proof. Let the random variable X have pdf $h(x)$ symmetric about a. Let $Y = X - a$; then Y has pdf $g(y) = h(y + a)$ symmetric about 0. Hence Y and $-Y$ have the same distribution. By the third property of location functionals, this means that $T(G_Y) = T(G_{-Y}) = -T(G_Y)$; $T(G_Y) = 0$. But by the second property, $0 = T(G_Y) = T(H) - a$; that is, $a = T(H)$.

This theorem means that when we sample from a symmetric distribution we can unambiguously define location as the center of symmetry. Then all location functionals that we may wish to study will specify the same location parameter.

1.3 Geometry and Inference in the Location Model

Letting $\mathbf{X} = (X_1, \ldots, X_n)'$ and $\mathbf{e} = (e_1, \ldots, e_n)'$, we then write the statistical location model, (1.2.1), as,

$$\mathbf{X} = \mathbf{1}\theta + \mathbf{e}, \tag{1.3.1}$$

where **1** denotes the vector all of whose components are 1 and $T(F_e) = 0$. If Ω_F denotes the one-dimensional subspace spanned by **1**, then we can express the model more compactly as $\mathbf{X} = \boldsymbol{\eta} + \mathbf{e}$, where $\boldsymbol{\eta} \in \Omega_F$. The subscript F on Ω stands for **full model** in the context of hypothesis testing as discussed below.

Let **x** be a realization of **X**. Note that except for random error, **x** would lie in Ω_F. Hence an intuitive fitting criterion is to estimate θ by a value $\widehat{\theta}$ such that the vector $\mathbf{1}\widehat{\theta} \in \Omega_F$ lies 'closest' to **x**, where 'closest' is defined in terms of a norm. Furthermore, a norm, as the following general discussion shows, provides a complete inference for the parameter θ.

Recall that a **norm** is a nonnegative function, $\|\cdot\|$, defined on $\mathcal{R}^n$ such that $\|\mathbf{y}\| \geq 0$ for all **y**; $\|\mathbf{y}\| = 0$ if and only if $\mathbf{y} = \mathbf{0}$; $\|a\mathbf{y}\| = |a|\|\mathbf{y}\|$ for all real a; and $\|\mathbf{y} + \mathbf{z}\| \leq \|\mathbf{y}\| + \|\mathbf{z}\|$. The distance between two vectors is $d(\mathbf{z}, \mathbf{y}) = \|\mathbf{z} - \mathbf{y}\|$.

Given a location model, (1.3.1), and a specified a norm, $\|\cdot\|$, the **estimate of θ induced by the norm** is

$$\widehat{\theta} = \text{Argmin}\|\mathbf{x} - \mathbf{1}\theta\|, \tag{1.3.2}$$

i.e. the value which minimizes the distance between **x** and the space Ω_F. As discussed in Exercise 1.13.1, a minimizing value always exists. The **dispersion function induced by the norm** is given by

$$D(\theta) = \|\mathbf{x} - \mathbf{1}\theta\|. \tag{1.3.3}$$

The **minimum distance** between the vector of observations **x** and the space Ω_F is $D(\widehat{\theta})$. As Exercise 1.13.2 shows, $D(\theta)$ is a convex, continuous function of θ which is differentiable almost everywhere. Actually the norms discussed in this book are differentiable at all but at most a finite number of points. We define the **gradient process** by the function

$$S(\theta) = -\frac{d}{d\theta} D(\theta). \tag{1.3.4}$$

As Exercise 1.13.2, shows, $S(\theta)$ is a nonincreasing function. Its discontinuities are the points where $D(\theta)$ is nondifferentiable. Furthermore, the minimizing value is a value where $S(\theta)$ is 0 or, due to a discontinuity, steps through 0. We express this by saying that $\widehat{\theta}$ solves the equation

$$S(\widehat{\theta}) \doteq 0. \tag{1.3.5}$$

Suppose we can represent the above estimate by $\widehat{\theta} = \widehat{\theta}(\mathbf{x}) = \widehat{\theta}(H_n)$, where H_n denotes the empirical distribution function of the sample. The notation $\widehat{\theta}(H_n)$ is suggestive of the functional notation used in the previous section. This is as it should be, since it is easy to show that $\widehat{\theta}$ satisfies the sample analogs of properties (2) and (3) of Definition 1.2.1. For property (2), consider the estimating equation of the translated sample $\mathbf{y} = a\mathbf{x} + \mathbf{1}b$, for $a > 0$, given by

$$\widehat{\theta}(\mathbf{y}) = \text{Argmin}\|\mathbf{y} - \mathbf{1}\theta\| = a\,\text{Argmin}\left\|\mathbf{x} - \mathbf{1}\frac{\theta - b}{a}\right\|.$$

From this we immediately have that $\widehat{\theta}(\mathbf{y}) = a\widehat{\theta}(\mathbf{x}) + b$. For property (3), the defining equation for the sample $\mathbf{y} = -\mathbf{x}$ is

$$\widehat{\theta}(\mathbf{y}) = \text{Argmin}\|\mathbf{y} - \mathbf{1}\theta\| = \text{Argmin}\|\mathbf{x} - \mathbf{1}(-\theta)\|$$

from which we have $\widehat{\theta}(\mathbf{y}) = -\widehat{\theta}(\mathbf{x})$. Furthermore, for the norms considered in this book it is easy to check that $\widehat{\theta}(H_n) \geqslant \widehat{\theta}(G_n)$ when H_n and G_n are empirical cdfs for which H_n is stochastically larger than G_n. Hence, the norms generate location functionals on the set of empirical cdfs. The L_1 norm provides an easy example. We can think of $\widehat{\theta}(H_n) = H_n^{-1}(\frac{1}{2})$ as the restriction of $\theta(H) = H^{-1}(\frac{1}{2})$ to the class of discrete distributions which assign mass $1/n$ to n points. Generally we can think of $\widehat{\theta}(H_n)$ as the restriction of $\theta(H)$ or, conversely, we can think of $\theta(H)$ as the extension of $\widehat{\theta}(H_n)$. We let the norm determine the location. This is especially simple in the symmetric location model where all location functionals are equal to the point of symmetry.

Next consider the hypotheses,

$$H_0: \theta = \theta_0 \text{ versus } H_A: \theta \neq \theta_0, \tag{1.3.6}$$

for a specified θ_0. Because of the second property of location functionals in Definition 1.2.1, we can assume without loss of generality that $\theta_0 = 0$; otherwise we need only subtract θ_0 from each X_i. Based on the data, the most acceptable value of θ is the value at which the gradient $S(\theta)$ is zero. Hence large values of $|S(0)|$ favor H_A. Formally the level α **gradient test** or **score test** for the hypotheses (1.3.6) is given by

$$\text{Reject } H_0 \text{ in favor of } H_A \text{ if } |S(0)| \geqslant c, \tag{1.3.7}$$

6   One-Sample Problems

where c is such that $P_0[|S(0)| \geq c] = \alpha$. Typically, the null distribution of $S(0)$ is symmetric so there is no loss in generality in considering symmetric critical regions.

A second formulation of a test statistic is based on the difference in minimizing dispersions or the reduction in dispersion. Call model (1.2.1) the **full model**. As noted above, the distance between **x** and the subspace Ω_F is $D(\widehat{\theta})$. The **reduced model** is the full model subject to H_0. In this case the reduced model space is $\{0\}$. Hence the distance between **x** and the reduced model space is $D(0)$. Under H_0, **x** should be close to this space; therefore, the **reduction in dispersion** test is given by

$$\text{Reject } H_0 \text{ in favor of } H_A \text{ if } RD = D(0) - D(\widehat{\theta}) \geq m, \quad (1.3.8)$$

where m is determined by the null distribution of RD. This test will be used in Chapter 3 and subsequent chapters.

A third formulation is based on the standardized estimate:

$$\text{Reject } H_0 \text{ in favor of } H_A \text{ if } \frac{|\widehat{\theta}|}{\sqrt{\widehat{\text{Var}\,\theta}}} \geq \gamma, \quad (1.3.9)$$

where γ is determined by the null distribution of $\widehat{\theta}$. Tests based directly on the estimate are often referred to as **Wald-type tests**.

The following useful theorem allows us to shift between computing probabilities when $\theta = 0$ and for general θ. Its proof is a straightforward application of a change of variables. See Theorem A.2.4 of the Appendix for a more general result.

Theorem 1.3.1 *Suppose that we can write $S(\theta) = S(x_1 - \theta, \ldots, x_n - \theta)$. Then $P_\theta(S(0) \leq t) = P_0(S(-\theta) \leq t)$.*

We now turn to the problem of the construction of a $(1 - \alpha)100\%$ **confidence interval** for θ based on $S(\theta)$. Such an interval is easily obtained by inverting the acceptance region of the level α test given by (1.3.7). The acceptance region is $|S(0)| < c$. Define

$$\widehat{\theta}_L = \inf\{t : S(t) < c\} \text{ and } \widehat{\theta}_U = \sup\{t : S(t) > -c\}. \quad (1.3.10)$$

Then because $S(\theta)$ is nonincreasing,

$$\{\theta : |S(\theta)| < c\} = \{\theta : \widehat{\theta}_L \leq \theta \leq \widehat{\theta}_U\}. \quad (1.3.11)$$

Thus from Theorem 1.3.1,

$$P_\theta(\widehat{\theta}_L \leq \theta \leq \widehat{\theta}_U) = P_\theta(|S(\theta)| < c) = P_0(|S(0)| < c) = 1 - \alpha. \quad (1.3.12)$$

Hence, inverting a size α test results in the $(1 - \alpha)100\%$ confidence interval $(\widehat{\theta}_L, \widehat{\theta}_U)$.

Thus a norm provides not only a fitting criterion but also a complete inference. As with all statistical analyses, checks on the appropriateness of the model and the quality of fit are needed. Useful plots here include: stem-and-leaf plots and q–q plots to check shape and distributional assumptions, boxplots and dotplots to check for outlying observations, and a plot of X_i

versus i (or other appropriate variables) to check for dependence between observations. Some of these diagnostic checks are performed in the the next section of numerical examples.

In the next three examples, we discuss the inference for the norms associated with the location functionals presented in the last section. We state the results of their associated inference, which we will derive in later sections.

Example 1.3.1 L_1 Norm

Recall that the L_1 norm is defined as $\|\mathbf{x}\|_1 = \sum |x_i|$, hence the associated dispersion and negative gradient functions are given respectively by $D_1(\theta) = \sum |X_i - \theta|$ and $S_1(\theta) = \sum \text{sgn}(X_i - \theta)$. Letting H_n denote the empirical cdf, we can write the estimating equation as

$$0 = n^{-1} \sum \text{sgn}(x_i - \theta) = \int \text{sgn}(x - \theta) dH_n(x).$$

The solution, of course, is $\hat{\theta}$, the median of the observations. If we replace the empirical cdf H_n by the true underlying cdf H then the estimating equation becomes the defining equation for the parameter $\theta = T(H)$. In this case, we have

$$0 = \int \text{sgn}(x - T(H)) dH(x) = -\int_{-\infty}^{T(H)} dH(x) + \int_{T(H)}^{\infty} dH(x);$$

hence, $H(T(H)) = \frac{1}{2}$ and solving for $T(H)$ we find $T(H) = H^{-1}(\frac{1}{2})$ as expected.

As we show in Section 1.5,

$$\hat{\theta} \text{ has an asymptotic } N(\theta, \tau_S^2/n) \text{ distribution,} \qquad (1.3.13)$$

where $\tau_S = 1/(2h(\theta))$. Estimation of the standard deviation of $\hat{\theta}$ is discussed in Section 1.5.

Turning next to testing the hypotheses (1.3.6), the gradient test statistic is $S_1(0) = \sum \text{sgn}(X_i)$. But we can write $S_1(0) = S_1^+ - S_1^- + S_1^0$, where $S_1^+ = \sum I(X_i > 0)$, $S_1^- = \sum I(X_i < 0)$, and $S_1^0 = \sum I(X_i = 0) = 0$, with probability one (i.e. the probability of ties is zero) since we are sampling from a continuous distribution, and $I(\cdot)$ is the indicator function. In practice, we must deal with ties and this is usually done by setting aside those observations that are equal to the hypothesized value and carrying out the test with a reduced sample size. Now note that $n = S_1^+ + S_1^-$ so that we can write $S_1 = 2S_1^+ - n$ and the test can be based on S_1^+. The null distribution of S_1^+ is binomial with parameters n and $\frac{1}{2}$. Hence the level α **sign test** of the hypotheses (1.3.6) is

$$\text{Reject } H_0 \text{ in favor of } H_A \text{ if } S_1^+ \leqslant c_1 \text{ or } S_1^+ \geqslant n - c_1, \qquad (1.3.14)$$

and c_1 satisfies

$$P[\text{Bin}(n, \tfrac{1}{2}) \leqslant c_1] = \alpha/2, \qquad (1.3.15)$$

where $\text{Bin}(n, \frac{1}{2})$ denotes a binomial random variable based on n trials and with probability of success $\frac{1}{2}$. Note that the critical value of the test can be determined without specifying the shape of F. In this sense, the test based on S_1 is **distribution-free** or **nonparametric**. Using the asymptotic null distribution of S_1^+, c_1 can be approximated as $c_1 \doteq n/2 - n^{1/2}z_{\alpha/2}/2 - \frac{1}{2}$ where $\Phi(-z_{\alpha/2}) = \alpha/2$; $\Phi(\cdot)$ is the standard normal cdf, and $\frac{1}{2}$ is the continuity correction.

For the associated $(1-\alpha)100\%$ confidence interval, we follow the general development above, (1.3.12). Hence, we must find $\hat{\theta}_L = \inf\{t : S_1^+(t) < n - c_1\}$, where c_1 is given by (1.3.15). Note that $S_1^+(t) < n - c_1$ if and only if the number of X_i greater than t is less than $n - c_1$. But $\#\{i : X_i > X_{(c_1+1)}\} = n - c_1 - 1$ and $\#\{i : X_i > X_{(c_1+1)} - \varepsilon\} \geq n - c_1$ for any $\varepsilon > 0$, where $\#\{\cdot\}$ means the number of elements in the set described in the braces. Hence, $\hat{\theta}_L = X_{(c_1+1)}$. A similar argument shows that $\hat{\theta}_U = X_{(n-c_1)}$. We can summarize this by saying that the $(1-\alpha)100\%$ L_1 confidence interval is the half-closed, half-open interval

$$[X_{(c_1+1)}, X_{(n-c_1)}), \text{ where } \alpha/2 = P(S_1^+(0) \leq c_1) \text{ determines } c_1. \quad (1.3.16)$$

The critical value c_1 can be determined from the binomial$(n, \frac{1}{2})$ distribution or from the normal approximation cited above. The interval developed here is a distribution-free confidence interval since the confidence coefficient is determined from the binomial distribution without making any shape assumption on the underlying model distribution.

Example 1.3.2 L_2 *Norm*

Recall that the square of the L_2 norm is given by $\|\mathbf{x}\|_2^2 = \sum_{i=1}^n x_i^2$. As shown in Exercise 1.13.3, the estimate determined by this norm is the sample mean $\overline{X}$ and the functional parameter is $\mu = \int xh(x)\,dx$, provided it exists. Hence the L_2 norm is consistent for the mean location problem. The associated test statistic is equivalent to Student's t-test. The approximate distribution of $\overline{X}$ is $N(0, \sigma^2/n)$, provided the variance $\sigma^2 = \text{Var}\,X_1$ exists. Hence, the test statistic is not distribution-free. In practice, σ is replaced by its estimate $s = (\sum(X_i - \overline{X})^2/(n-1))^{1/2}$ and the test is based on the t-ratio, $t = \sqrt{n}\,\overline{X}/s$, which, under the null hypothesis, is asymptotically $N(0, 1)$. The usual confidence interval is $\overline{X} \pm t_{\alpha/2, n-1} s/\sqrt{n}$, where $t_{\alpha/2, n-1}$ is the $(1-\alpha/2)$ quantile of a t-distribution with $n-1$ degrees of freedom. This interval has the approximate confidence coefficient $(1-\alpha)100\%$, unless the errors, e_i, follow a normal distribution in which case it has exact confidence.

Example 1.3.3 *Weighted L_1 Norm*

Consider the function

$$\|\mathbf{x}\|_3 = \sum_{i=1}^n R(|x_i|)|x_i|, \quad (1.3.17)$$

where $R(|x_i|)$ denotes the rank of $|x_i|$ among $|x_1|, \ldots, |x_n|$. As the next theorem shows, this function is a norm on $\mathcal{R}^n$. See Section 1.8 for a general weighted L_1 norm.

Theorem 1.3.2 *The function* $\|\mathbf{x}\|_3 = \sum j|x|_{(j)} = \sum R(|x_j|)|x_j|$ *is a norm, where* $R(|x_j|)$ *is the rank of* $|x_j|$ *among* $|x_1|, \ldots, |x_n|$ *and* $|x|_{(1)} \leq \cdots \leq |x|_{(n)}$ *are the ordered absolute values.*

Proof. The equality relating $\|\mathbf{x}\|_3$ to the ranks is clear. To show that we have a norm, we first note that $\|\mathbf{x}\|_3 \geq 0$ and that $\|\mathbf{x}\|_3 = 0$ if and only if $\mathbf{x} = \mathbf{0}$. Also, clearly $\|a\mathbf{x}\|_3 = |a|\|\mathbf{x}\|_3$ for any real a. Hence, to finish the proof, we must verify the triangle inequality. Now

$$\|\mathbf{x} + \mathbf{y}\|_3 = \sum j |x+y|_{(j)}$$
$$= \sum R(|x_j + y_j|)|x_j + y_j|$$
$$\leq \sum R(|x_j + y_j|)|x_j| + \sum R(|x_j + y_j|)|y_j|. \quad (1.3.18)$$

Consider the first term on the right-hand side. By summing through another index we can write it as

$$\sum R(|x_j + y_j|)|x_j| = \sum b_j |x|_{(j)},$$

where $b_1, \ldots, b_n$ is a permutation on the integers $1, \ldots, n$. Suppose b_j is not in order, then there exist a t and an s such that $|x|_{(t)} \leq |x|_{(s)}$ but $b_t > b_s$. Whence,

$$[b_s|x|_{(t)} + b_t|x|_{(s)}] - [b_t|x|_{(t)} + b_s|x|_{(s)}] = (b_t - b_s)(|x|_{(s)} - |x|_{(t)}) \geq 0.$$

Hence such an interchange never decreases the sum. This leads to the result

$$\sum R(|x_j + y_j|)|x_j| \leq \sum j |x|_{(j)}.$$

A similar result holds for the second term on the right-hand side of (1.3.18). Therefore, $\|\mathbf{x} + \mathbf{y}\|_3 \leq \sum j|x|_{(j)} + \sum j|y|_{(j)} = \|\mathbf{x}\|_3 + \|\mathbf{y}\|_3$, and this completes the proof. The above argument is taken from Hardy, Littlewood, and Pólya (1952).

We shall call this norm the **weighted L_1 norm**. In the next theorem, we offer an interesting identity satisfied by this norm. First, though, we need another representation of it. For a random sample $X_1, \ldots, X_n$, define the **anti-ranks** to be the random variables $D_1, \ldots, D_n$ such that

$$Z_1 = |X_{D_1}| \leq \ldots \leq Z_n = |X_{D_n}|. \quad (1.3.19)$$

For example, if $D_1 = 2$ then $|X_2|$ is the smallest absolute value and Z_1 has rank 1. Note that the anti-rank function is just the inverse of the rank function. We can then write

$$\|\mathbf{x}\|_3 = \sum_{j=1}^{n} j|x|_{(j)} = \sum_{j=1}^{n} j|x_{D_j}|. \quad (1.3.20)$$

Theorem 1.3.3 *For any vector* $\mathbf{x}$,

$$\|\mathbf{x}\|_3 = \sum\sum_{i \leq j} \left|\frac{x_i + x_j}{2}\right| + \sum\sum_{i < j} \left|\frac{x_i - x_j}{2}\right|. \quad (1.3.21)$$

10 One-Sample Problems

Proof. Letting the index run through the anti-ranks, we have

$$\sum\sum_{i\leqslant j}\left|\frac{x_i+x_j}{2}\right|+\sum\sum_{i<j}\left|\frac{x_i-x_j}{2}\right|$$

$$=\sum_{i=1}^{n}|x_i|+\sum\sum_{i<j}\left\{\left|\frac{x_{D_i}+x_{D_j}}{2}\right|+\left|\frac{x_{D_j}-x_{D_i}}{2}\right|\right\}. \quad (1.3.22)$$

For $i < j$, hence $x_{D_i} \leqslant x_{D_j}$, consider the expression,

$$\left|\frac{x_{D_i}+x_{D_j}}{2}\right|+\left|\frac{x_{D_j}-x_{D_i}}{2}\right|.$$

There are four cases to consider: where x_{D_i} and x_{D_j} are both positive; where they are both negative; and the two cases where they have mixed signs. In all these cases, though, it is easy to show that

$$\left|\frac{x_{D_i}+x_{D_j}}{2}\right|+\left|\frac{x_{D_j}-x_{D_i}}{2}\right|=|x_{D_j}|.$$

Using this, we have that the right-hand side of expression (1.3.22) is equal to:

$$\sum_{i=1}^{n}|x_i|+\sum\sum_{i<j}|x_{D_j}|=\sum_{j=1}^{n}|x_{D_j}|+\sum_{j=1}^{n}(j-1)|x_{D_j}|=\sum_{j=1}^{n}j|x_{D_j}|=\|\mathbf{x}\|_3,$$
$$(1.3.23)$$

and we are finished.

The associated gradient function is

$$T(\theta)=\sum_{i=1}^{n}R(|X_i-\theta|)\text{sgn}(X_i-\theta)=\sum_{i\leqslant j}\text{sgn}\left(\frac{X_i+X_j}{2}-\theta\right). \quad (1.3.24)$$

The middle term is due to the fact that the ranks only change values at the finite number of points determined by $|X_i - \theta| = |X_j - \theta|$; otherwise $R(|X_i - \theta|)$ is constant. The third term is obtained immediately from the identity (1.3.21). The $n(n+1)/2$ pairwise averages $\{(X_i+X_j)/2 : 1 \leqslant i \leqslant j \leqslant n\}$ are called the **Walsh averages**. Hence, the estimate of θ is the median of the Walsh averages, which we shall denote as,

$$\widehat{\theta}_3=\text{med}_{i\leqslant j}\left\{\frac{X_i+X_j}{2}\right\}, \quad (1.3.25)$$

first discussed by Hodges and Lehmann (1963). Often $\widehat{\theta}_3$ is called the **Hodges–Lehmann estimate of location**. In order to obtain the corresponding location functional, note that

$$R(|X_i-\theta|)=\#\{|X_j-\theta|\leqslant|X_i-\theta|\}=\#\{\theta-|X_i-\theta|\leqslant X_j\leqslant\theta+|X_i-\theta|\}$$
$$=nH_n(\theta+|X_i-\theta|)-nH_n^-(\theta-|X_i-\theta|),$$

where H_n^- is the left limit of H_n. Hence (1.3.24) becomes

$$\int\{H_n(\theta+|x-\theta|)-H_n^-(\theta-|x-\theta|)\}\text{sgn}(x-\theta)\,dH_n(x)=0,$$

and in the limit we have

$$\int \{H(\theta + |x - \theta|) - H(\theta - |x - \theta|)\}\operatorname{sgn}(x - \theta)\, dH(x) = 0,$$

that is,

$$-\int_{-\infty}^{\theta} \{H(2\theta - x) - H(x)\}\, dH(x) + \int_{\theta}^{\infty} \{H(x) - H(2\theta - x)\}\, dH(x) = 0.$$

This simplifies to

$$\int_{-\infty}^{\infty} H(2\theta - x)\, dH(x) = \tfrac{1}{2}, \tag{1.3.26}$$

Hence, the functional is the pseudo-median defined in Example 1.2.3. If the density $h(x)$ is symmetric then from (1.7.11)

$$\widehat{\theta}_3 \text{ has an approximate } N(\theta_3, \tau^2/n) \text{ distribution}, \tag{1.3.27}$$

where $\tau = 1/(\sqrt{12} \int h^2(x)\, dx)$. Estimation of τ is discussed in Section 3.7.

The most convenient form of the gradient process is

$$T^+(\theta) = \sum\sum_{i \leq j} I\left(\frac{X_i + X_j}{2} > \theta\right) = \sum_{i=1}^{n} R(|X_i - \theta|)I(X_i > \theta). \tag{1.3.28}$$

The corresponding gradient test statistic for the hypotheses (1.3.6) is $T^+(0)$. In Section 1.7, provided that $h(x)$ is symmetric, it is shown that $T^+(0)$ is distribution-free under H_0 with null mean and variance $n(n+1)/4$ and $n(n+1)(2n+1)/24$, respectively. This test is often referred to as the **Wilcoxon signed-rank test**. Thus the test for the hypotheses (1.3.6) is

Reject H_0 in favor of H_A, if $T^+(0) \leq k$ or $T^+(0) \geq \dfrac{n(n+1)}{2} - k,$ \quad (1.3.29)

where $P(T^+(0) \leq k) = \alpha/2$. An approximation for k is given in the next paragraph.

Because of the similarity between the sign and signed-rank processes, the confidence interval based on $T^+(\theta)$ follows immediately from the argument given in Example 1.3.1 for the sign process. Instead of the order statistics which were used in the confidence interval based on the sign process, in this case we use the ordered Walsh averages, which we denote as $W_{(1)}, \ldots, W_{(n(n+1)/2)}$. Hence a $(1 - \alpha)100\%$ confidence interval for θ is given by

$$[W_{(k+1)}, W_{((n(n+1)/2)-k)}), \text{ where } k \text{ is such that } \alpha/2 = P(T^+(0) \leq k). \tag{1.3.30}$$

As with the sign process, k can be approximated using the asymptotic normal distribution of $T^+(0)$ by

$$k \doteq \frac{n(n+1)}{4} - z_{\alpha/2}\sqrt{\frac{n(n+1)(2n+1)}{24}} - \frac{1}{2},$$

where $z_{\alpha/2}$ is the $(1 - \alpha/2)$ quantile of the standard normal distribution. Provided that $h(x)$ is symmetric, this confidence interval is distribution free.

12 One-Sample Problems

Finally, note that the rank form of the Wilcoxon signed-rank test statistic, (1.3.27), is easy to compute. For estimation and confidence intervals either the Walsh averages need to be computed or a simple iterative search needs to be performed, as described in Section 3.7.3.

1.4 Examples

Example 1.4.1 *Cushney Peebles Data*

The data given in Table 1.4.1 give the average excess number of hours of sleep that each of 10 patients achieved from the use of two drugs. The third column gives the difference (Laevo − Dextro) in excesses across the two drugs. This is a famous data set; see Cushney and Peebles (1905). Gosset, writing under the pseudonym Student, published his landmark paper on the *t*-test in 1908 and used this data set for illustration. The differences, however, suggest that the L_2 methods may not be the methods of choice in this case. The normal quantile plot, of Figure 1.4.1(a), shows that the tails may be heavy and that there may be an outlier. A normal quantile plot has the data (differences) on the vertical axis and the expected values of the standard normal order statistics on the horizontal axis. When the data are consistent with a normal assumption, the plot should be roughly linear. The boxplot, with 95% L_1 confidence interval, Figure 1.4.1(b), further illustrates the presence of an outlier. The box is defined by the quartiles and the shaded notch represents the confidence interval.

For the sake of discussion and comparison of methods, we provide the *p*-values for the sign test, the Wilcoxon signed-rank test, and the *t*-test. In the following display, we show the Minitab output for the computation of the three tests. The Minitab commands are STEST, WTEST, and TTEST.

```
Sign test of median = 0.0 versus N.E. 0.0

           N    BELOW   EQUAL   ABOVE   P-VALUE   MEDIAN
Diff      10      0       1       9     0.0039    1.300
```

```
WILCOXON TEST OF MEDIAN = 0.0 VERSUS MEDIAN N.E. 0.0

              N FOR   WILCOXON              ESTIMATED
         N    TEST    STATISTIC   P-VALUE   MEDIAN
Diff    10     9        45.0       0.009    1.300
```

```
T-Test of mu = 0.0 vs mu not = 0.0

Variable   N    Mean    StDev   SE Mean     T     P-Value
Diff      10   1.580    1.230    0.389     4.06   0.0029
```

Table 1.4.1: Excess hours of sleep under the influence of two drugs and the difference in excesses.

Row	Dextro	Laevo	Diff(L−D)
1	−0.1	−0.1	0.0
2	0.8	1.6	0.8
3	3.4	4.4	1.0
4	0.7	1.9	1.2
5	−0.2	1.1	1.3
6	−1.2	0.1	1.3
7	2.0	3.4	1.4
8	3.7	5.5	1.8
9	−1.6	0.8	2.4
10	0.0	4.6	4.6

From the boxplot it is clear that the 95% L_1 confidence interval is shifted above 0. Note that the confidence interval is the interpolated interval discussed in Section 1.11. Hence, there is strong support for the alternative hypothesis that the location of the difference distribution is not equal to zero. That is, we reject $H_0 : \theta = 0$ in favor of $H_A : \theta \neq 0$ at $\alpha = 0.05$. All three tests support this conclusion. The estimates of location corresponding to the three tests are the median, the median of the Walsh averages (1.3.25), and the mean of the sample differences.

In order to see how sensitive the test statistics are to outliers, we change the value of the outlier (difference in the 10th row of Table 1.4.1) and plot the value of the test statistic against the value of the difference in the 10th row of Table 1.4.1; see Figure 1.4.1(c). Note that as the value of the 10th difference changes the t-test changes quite rapidly. In fact, the t-test can be pulled out of the rejection region by making the difference sufficiently small or large. However, the sign test, Figure 1.4.1(d), stays constant until the difference crosses zero and then only changes by 2. This illustrates the high sensitivity of the t-test to outliers and the relative resistance of the sign test. A similar plot can be prepared for the Wilcoxon signed-rank test; see Exercise 1.13.7. In addition, the corresponding p-values can be plotted to see how sensitive the decision to reject the null hypothesis is to outliers. Sensitivity plots are similar to influence functions. We discuss influence functions for estimates in Section 1.6.

Example 1.4.2 *Shoshoni Rectangles*

The golden rectangle is a rectangle in which the ratio of width (w) to length (l) is approximately 0.618. It can be characterized in various ways, for example, $w/l = l/(w + l)$. It is considered to be an aesthetic standard in Western civilization and appears in art and architecture going back to the ancient Greeks. It now appears even in such items as credit and business cards. In a cultural anthropology study, DuBois (1960) reports on a study of the Shoshoni beaded baskets. These baskets contain beaded rectangles and the question was whether the Shoshonis use the same aesthetic standard as the West. A sample of 20 width to length ratios from Shoshoni baskets is given in Table 1.4.2.

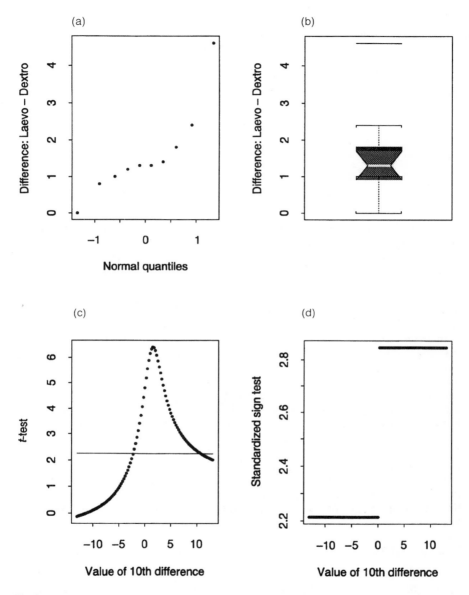

Figure 1.4.1: (a) Normal q−q plot of Cushney–Peebles data; (b) boxplot with 95% notched confidence interval; (c) sensitivity curve for *t*-test; (d) sensitivity curve for sign test

Table 1.4.2: Width to length ratios of rectangles

0.553	0.570	0.576	0.601	0.606	0.606	0.609	0.611	0.615	0.628
0.654	0.662	0.668	0.670	0.672	0.690	0.693	0.749	0.844	0.933

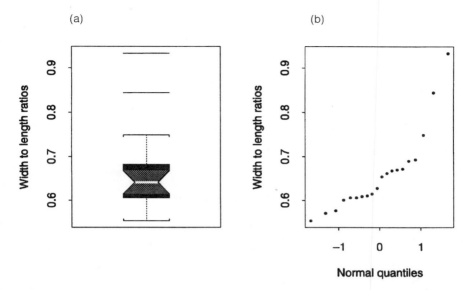

Figure 1.4.2: (a) Boxplot of width to length ratios of Shoshoni rectangles; (b) normal q–q plot

Figure 1.4.2(a) shows the notched boxplot containing the 95% L_1 confidence interval for the median of the population w/l ratios. In this example we do not want to assume a symmetric distribution. The normal quantile plot, Figure 1.4.2(b), is not linear and suggests a skewed sample. The boxplot shows two outliers.

This is an estimation problem rather than a testing problem. We want to know roughly what the median w/l ratio is for the population. Hence, a confidence interval is appropriate. Below is the 90% L_1 confidence interval. Again we use the interpolated confidence interval described in Section 1.11. The Minitab output is given below based on the commands SINT and TINT.

SIGN CONFIDENCE INTERVAL FOR MEDIAN

	N	MEDIAN	ACHIEVED CONFIDENCE	CONFIDENCE INTERVAL	POSITION
w/l	20	0.6410	0.8847	(0.6090, 0.6700)	7
			0.9000	(0.6087, 0.6702)	NLI
			0.9586	(0.6060, 0.6720)	6

We see that the golden rectangle 0.618 is contained in the confidence interval. This suggests that there is no evidence in these data that the Shoshonis are using a different standard.

The 90% t-interval in this case is

```
        N    MEAN    STDEV    SE MEAN    90.0 PERCENT C.I.
w/1    20   0.6605   0.0925   0.0207     (0.6247, 0.6963)
```

Note that the 90% t-interval does not contain the golden rectangle and the researcher might conclude that there is evidence that the Shoshonis are using a different standard. However, as pointed out above, there are two outliers in the data and they tend to pull the confidence interval upward. In this case we would have more faith in the simple sign test.

1.5 Properties of Norm-Based Inference

In this section, we establish statistical properties of the inference described in Section 1.3 for the norm fit of a location model. These properties describe the null and alternative distributions of the test, (1.3.7), and the asymptotic distribution of the estimate, (1.3.2). Furthermore, these properties allow us to derive relative efficiencies between competing procedures. While our discussion is general, we will illustrate the inference based on the L_1 and L_2 norms as we proceed. The inference based on the signed-rank norm will be considered in Section 1.7 and that based on norms of general signed-rank scores in Section 1.8.

We assume then that Model (1.2.1) holds for a random sample $X_1, \ldots, X_n$ with common distribution and density functions $H(x) = F(x - \theta)$ and $h(x) = f(x - \theta)$, respectively. Next a norm is specified to fit the model. We will assume that the induced functional is 0 at F, i.e. $T(F) = 0$. Let $S(\theta)$ be the gradient function induced by the norm. We establish the properties of the inference by considering the null and alternative behavior of the **gradient test**. For convenience we consider the one-sided hypothesis,

$$H_0 : \theta = 0 \text{ versus } H_A : \theta > 0. \quad (1.5.1)$$

Since $S(\theta)$ is nondecreasing, a level α test of these hypotheses based on $S(0)$ is

$$\text{Reject } H_0 \text{ in favor of } H_A \text{ if } S(0) \geq c, \quad (1.5.2)$$

where c is such that $P_0[S(0) \geq c] = \alpha$.

The power function of this test is given by,

$$\gamma_S(\theta) = P_\theta[S(0) \geq c] = P_0[S(-\theta) \geq c], \quad (1.5.3)$$

where the last equality follows from Theorem 1.3.1.

The power function forms a convenient summary of the test based on $S(0)$. The probability of a Type I error (level of the test) is given by $\gamma_S(0)$. The probability of a Type II error at the alternative θ is $\beta_S(\theta) = 1 - \gamma_S(\theta)$. For a given test of hypotheses (1.5.1) we want the power function to be increasing in θ with an upper limit of one. In the first subsection below, we establish these properties for the test (1.5.2). We can also compare level α tests of (1.5.1) by comparing their powers at alternative hypotheses. These are efficiency considerations and they are covered in later subsections.

1.5.1 Basic Properties of the Power Function $\gamma_S(\theta)$

As a first step we show that $\gamma_S(\theta)$ is nondecreasing:

Theorem 1.5.1 *Suppose the test of $H_0 : \theta = 0$ versus $H_A : \theta > 0$ rejects when $S(0) \geq c$. Then the power function is nondecreasing in θ.*

Proof. Recall that $S(\theta)$ is nonincreasing in θ since $D(\theta)$ is convex. By Theorem 1.3.1, $\gamma_S(\theta) = P_0[S(-\theta) \geq c]$. Now, if $\theta_1 \leq \theta_2$ then $S(-\theta_1) \leq S(-\theta_2)$ and, hence, $S(-\theta_1) \geq c$ implies that $S(-\theta_2) \geq c$. It then follows that $P_0(S(-\theta_1) \geq c) \leq P_0(S(-\theta_2) \geq c)$ and the power function is monotone in θ as required.

This theorem shows that the test of $H_0 : \theta = 0$ versus $H_A : \theta > 0$ based on $S(0)$ is **unbiased**, that is, $P_\theta(S(0) \geq c) \geq \alpha$ for positive θ, where α is the size of the test. At times it is convenient to consider the more general null hypothesis:

$$H_0^* : \theta \leq 0 \text{ versus } H_A : \theta > 0. \tag{1.5.4}$$

A test of H_0^* versus H_A with power function γ_S is said to have level α if

$$\sup_{\theta \leq 0} \gamma_S(\theta) = \alpha.$$

The proof of Theorem 1.5.1 shows that $\gamma_S(\theta)$ is nonincreasing in all $\theta \in \mathcal{R}$. Since the gradient test has level α for H_0, it follows immediately that it has level α for H_0^* also.

We next show that the power function of the gradient test converges to 1 as $\theta \to \infty$. We formally define this as:

Definition 1.5.1 *Consider a level α test for the hypotheses (1.5.1) which has power function $\gamma_S(\theta)$. We say the test is **resolving** if $\gamma_S(\theta) \to 1$ as $\theta \to \infty$.*

Theorem 1.5.2 *Suppose the test of $H_0 : \theta = 0$ versus $H_A : \theta > 0$ rejects when $S(0) \geq c$. Further, let $\eta = \sup_\theta S(\theta)$ and suppose that η is attained for some finite value of θ. Then the test is resolving, that is, $P_\theta(S(0) \geq c) \to 1$ as $\theta \to \infty$.*

Proof. Since $S(\theta)$ is nonincreasing, for any unbounded increasing sequence θ_m, $S(\theta_m) \geq S(\theta_{m+1})$. For fixed n and F, there is a real number a such that $P_0(|X_i| \leq a, i = 1, \ldots, n) > 1 - \varepsilon$ for any specified $\varepsilon > 0$. Let A_ε denote the event $\{|X_i| \leq a, i = 1, \ldots, n\}$. Now,

$$\begin{aligned} P_{\theta_m}(S(0) \geq c) &= P_0(S(-\theta_m) \geq c) \\ &= 1 - P_0(S(-\theta_m) < c) \\ &= 1 - P_0(\{S(-\theta_m) < c\} \cap A_\varepsilon) - P_0(\{S(-\theta_m) < c\} \cap A_\varepsilon^c). \end{aligned}$$

The hypothesis of the theorem implies that, for sufficiently large m, $\{S(-\theta_m) < c\} \cap A_\varepsilon$ is empty. Further, $P_0(\{S(-\theta_m) < c\} \cap A_\varepsilon^c) \leq P_0(A_\varepsilon^c) < \varepsilon$. Hence, for m sufficiently large, $P_{\theta_m}(S(0) \geq c) \geq 1 - \varepsilon$ and the proof is complete.

18 One-Sample Problems

The condition of boundedness imposed on $S(\theta)$ in the above theorem holds for almost all the nonparametric tests discussed in this book; hence, these nonparametric tests will be resolving. Thus they will be able to discern large alternative hypotheses with high power. What can be said at a fixed alternative? Recall the definition of a consistent test:

Definition 1.5.2 *We say that a test is* **consistent** *if the power tends to one for each fixed alternative as the sample size n increases. The alternatives consist in specific values of θ and a cdf F.*

Consistency implies that the test is behaving as expected when the sample size increases and the alternative hypothesis is true. To obtain consistency of the gradient test, we need to impose the following two assumptions on $S(\theta)$: first,

$$\overline{S}(\theta) = S(\theta)/n^{\gamma} \xrightarrow{P} \mu(-\theta) \text{ where } \mu(0) = 0 \text{ and } \mu(0) < \mu(\theta) \text{ for all } \theta > 0, \quad (1.5.5)$$

for some $\gamma > 0$; and second,

$$E_0 S(0) = 0 \text{ and } \sqrt{n}\,\overline{S}(0) \xrightarrow{D} N(0, \sigma^2(0)) \text{ under } H_0 \text{ for all } F, \quad (1.5.6)$$

for some positive constant $\sigma(0)$. The first assumption means that $\overline{S}(0)$ separates the null from the alternative hypothesis. Note, it is not crucial that $\mu(0) = 0$, since this can always be achieved by recentering. It will be useful to have the following result concerning the asymptotic null distribution of $S(0)$. Its proof follows readily from the definition of convergence in distribution.

Theorem 1.5.3 *Assume (1.5.6). The test defined by $\sqrt{n}\,\overline{S}(0) \geq z_\alpha \sigma(0)$, where z_α is the upper α percentile from the standard normal cdf, i.e. $1 - \Phi(z_\alpha) = \alpha$, is asymptotically size α. Hence, $P_0(\sqrt{n}\,\overline{S}(0) \geq z_\alpha \sigma(0)) \to \alpha$.*

It follows that a gradient test is consistent:

Theorem 1.5.4 *Assume conditions (1.5.5) and (1.5.6). Then the gradient test $\sqrt{n}\,\overline{S}(0) \geq z_\alpha \sigma(0)$ is consistent, i.e. the power at fixed alternatives tends to one as n increases.*

Proof. Fix $\theta^* > 0$ and F. For $\varepsilon > 0$ and for large n, we have $n^{-1/2} z_\alpha \sigma(0) < \mu(\theta^*) - \varepsilon$. This leads to the following string of inequalities:

$$P_{\theta^*,F}(\overline{S}(0) \geq n^{-1/2} z_\alpha \sigma(0)) \geq P_{\theta^*,F}(\overline{S}(0) \geq \mu(\theta^*) - \varepsilon)$$
$$\geq P_{\theta^*,F}(|\overline{S}(0) - \mu(\theta^*)| \leq \varepsilon) \to 1$$

which is the desired result.

Example 1.5.1 *The L_1 Case*

Assume that the model cdf F has the unique median 0. Consider the L_1 norm. The associated level α gradient test of (1.5.1) is equivalent to the sign test given by:

$$\text{Reject } H_0 \text{ in favor of } H_A \text{ if } S_1^+ = \sum I(X_i > 0) \geq c,$$

where c is such that $P[\text{Bin}(n, \frac{1}{2}) \geq c] = \alpha$. The test is nonparametric, i.e. it does not depend on F. From the above discussion its power function is nondecreasing in θ. Since $S_1^+(\theta)$ is bounded and attains its bound on a finite interval, the test is resolving. For consistency, take $\gamma = 1$ in expression (1.5.5). Then $E[n^{-1}S_1^+(0)] = P(X > 0) = 1 - F(-\theta) = \mu(\theta)$. An application of the weak law of large numbers shows that the limit in condition (1.5.5) holds. Further, $\mu(0) = \frac{1}{2} < \mu(\theta)$ for all $\theta > 0$ and all F. Finally, apply the central limit theorem to show that (1.5.6) holds. Hence, the sign test is consistent for location alternatives. Further, it is consistent for each pair θ, F such that $P(X > 0) > \frac{1}{2}$.

A discussion of these properties for the gradient test based on the L_2-norm can be found in Exercise 1.13.4.

1.5.2 Asymptotic Linearity and Pitman Regularity

In the previous section we discussed some of the basic properties of the power function for a gradient test. Next we establish some general results that will allow us to compare power functions for different level α tests. These results will also lead to the asymptotic distributions of the location estimators $\hat{\theta}$ based on norm fits. We will also make use of them in later sections and chapters.

Assume the setup found at the beginning of this section; i.e. we are considering the location model (1.3.1) and we have specified a norm with gradient function $S(\theta)$. We first define a Pitman regular process:

Definition 1.5.3 *We will say an estimating function $S(\theta)$ is **Pitman regular** if the following four conditions hold: first,*

$$S(\theta) \text{ is nonincreasing in } \theta; \quad (1.5.7)$$

second, letting $\overline{S}(\theta) = S(\theta)/n^\gamma$, for some $\gamma > 0$

there exists a function $\mu(\theta)$, such that $\mu(0) = 0$, $\mu'(\theta)$ is continuous at 0,

$$\mu'(0) > 0 \text{ and either } \overline{S}(0) \xrightarrow{P_\theta} \mu(\theta) \text{ or } E_\theta(\overline{S}(0)) = \mu(\theta); \quad (1.5.8)$$

third,

$$\sup_{|b| \leq B} \left| \sqrt{n}\,\overline{S}\left(\frac{b}{\sqrt{n}}\right) - \sqrt{n}\,\overline{S}(0) + \mu'(0)b \right| \xrightarrow{P} 0, \quad (1.5.9)$$

for any $B > 0$; and fourth, there is a constant $\sigma(0)$ such that

$$\sqrt{n}\left\{\frac{\overline{S}(0)}{\sigma(0)}\right\} \xrightarrow{D_0} N(0, 1). \quad (1.5.10)$$

Further, the quantity

$$c = \mu'(0)/\sigma(0) \quad (1.5.11)$$

*is called the **efficacy** of $S(\theta)$.*

20 One-Sample Problems

Condition (1.5.9) is called the **asymptotic linearity** of the process $S(\theta)$. Often we can compute c when we have the mean under general θ and the variance under $\theta = 0$. Thus

$$\mu'(0) = \frac{d}{d\theta} E_\theta[\overline{S}(0)]|_{\theta=0} \quad \text{and} \quad \sigma^2(0) = \lim\{n\mathrm{Var}_0(\overline{S}(0))\}. \tag{1.5.12}$$

Hence, another way of expressing the asymptotic linearity of $S(\theta)$ is

$$\sqrt{n}\left\{\frac{\overline{S}(b/\sqrt{n})}{\sigma(0)}\right\} = \sqrt{n}\left\{\frac{\overline{S}(0)}{\sigma(0)}\right\} - cb + o_p(1). \tag{1.5.13}$$

If we replace b by $\sqrt{n}\theta_n$ where, of course, $|\sqrt{n}\theta_n| \le B$ for $B > 0$, then we can write

$$\sqrt{n}\left\{\frac{\overline{S}(\theta_n)}{\sigma(0)}\right\} = \sqrt{n}\left\{\frac{\overline{S}(0)}{\sigma(0)}\right\} - c\sqrt{n}\theta_n + o_p(1). \tag{1.5.14}$$

We record one more result on limiting distributions whose proof follows from Theorems 1.3.1 and 1.5.6.

Theorem 1.5.5 *Suppose $S(\theta)$ is Pitman regular. Then*

$$\sqrt{n}\left\{\frac{\overline{S}(b/\sqrt{n})}{\sigma(0)}\right\} \xrightarrow{D_0} Z - cb \tag{1.5.15}$$

and

$$\sqrt{n}\left\{\frac{\overline{S}(0)}{\sigma(0)}\right\} \xrightarrow{D_{-b/\sqrt{n}}} Z - cb, \tag{1.5.16}$$

where $Z \sim N(0, 1)$ and, so, $Z - cb \sim N(-cb, 1)$.

The second part of this theorem says that the limiting distribution of $\overline{S}(0)$, when standardized by $\sigma(0)$, and computed along a sequence of alternatives $-b/n^{1/2}$, is still normal with the same variance of one but with a new mean, namely $-cb$. This result will be useful in approximating the power near the null hypothesis.

We will find asymptotic linearity to be useful in establishing statistical properties. Our next result provides sufficient conditions for linearity.

Theorem 1.5.6 *Let $\overline{S}(\theta) = (1/n^\gamma)S(\theta)$ for some $\gamma > 0$ such that the conditions (1.5.7), (1.5.8) and (1.5.10) of Definition 1.5.3 hold. Suppose for any $b \in \mathcal{R}$,*

$$n\mathrm{Var}_0(\overline{S}(n^{-1/2}b) - \overline{S}(0)) \to 0, \text{ as } n \to \infty. \tag{1.5.17}$$

Then

$$\sup_{|b| \le B}\left|\sqrt{n}\overline{S}\left(\frac{b}{\sqrt{n}}\right) - \sqrt{n}\overline{S}(0) + \mu'(0)b\right| \xrightarrow{P} 0, \tag{1.5.18}$$

for any $B > 0$.

Proof. First consider $U_n(b) = [\bar{S}(n^{-1/2}b) - \bar{S}(0)]/(b/\sqrt{n})$. By (1.5.8) we have

$$E_0(U_n(b)) = \frac{\sqrt{n}}{b}\mu\left(\frac{-b}{\sqrt{n}}\right) = \frac{\sqrt{n}}{b}\left[-\frac{b}{\sqrt{n}}\mu'(\xi_n)\right] \to -\mu'(0), \quad (1.5.19)$$

where ξ_n lies between 0 and $b/\sqrt{n}$. Furthermore,

$$\mathrm{Var}_0 U_n(b) = \frac{n}{b^2}\mathrm{Var}_0\left[\bar{S}\left(\frac{b}{\sqrt{n}}\right) - \bar{S}(0)\right] \to 0. \quad (1.5.20)$$

As Exercise 1.13.8 shows, (1.5.19) and (1.5.20) imply that $U_n(b)$ converges to $-\mu'(0)$ in probability, pointwise in b, i.e. $U_n(b) = -\mu'(0) + o_p(1)$.

For the second part of the proof, let $W_n(b) = \sqrt{n}[\bar{S}(b/\sqrt{n}) - \bar{S}(0)] + \mu'(0)b/\sqrt{n}]$. Further, let $\varepsilon > 0$ and $\gamma > 0$ and partition $[-B, B]$ into $-B = b_0 < b_1 < \ldots < b_m = B$ so that $b_i - b_{i-1} \leq \varepsilon/(2|\mu'(0)|)$ for all i. There exists N such that $n \geq N$ implies $P[\max_i |W_n(b_i)| > \varepsilon/2] < \gamma$.

Now suppose that $W_n(b) \geq 0$ (a similar argument can be given for $W_n(b) < 0$). Then

$$|W_n(b)| = \sqrt{n}\left[\bar{S}\left(\frac{b}{\sqrt{n}}\right) - \bar{S}(0)\right] + b\mu'(0) \leq \sqrt{n}\left[\bar{S}\left(\frac{b}{\sqrt{n}}\right) - \bar{S}(0)\right]$$

$$+ b_{i-1}\mu'(0) + (b - b_{i-1})\mu'(0)$$

$$\leq |W_n(b_{i-1})| + (b - b_{i-1})|\mu'(0)| \leq \max_i |W_n(b_i)| + \varepsilon/2.$$

Hence,

$$P_0\left(\sup_{|b| \leq B} |W_n(b)| > \varepsilon\right) \leq P_0(\max_i |W_n(b_i)| + \varepsilon/2 > \varepsilon) < \gamma,$$

and

$$\sup_{|b| \leq B} |W_n(b)| \xrightarrow{P} 0.$$

In the next three subsections we use these tools to handle the issues of power and efficiency for a general norm-based inference, but first we show that the L_1 gradient function is Pitman regular.

Example 1.5.2 *Pitman Regularity of the L_1 Process*

Assume that the model pdf satisfies $f(0) > 0$. Recall that the L_1 gradient function is

$$S_1(\theta) = \sum_{i=1}^{n} \mathrm{sgn}(X_i - \theta).$$

Take $\gamma = 1$ in Theorem 1.5.6; hence, the average of interest is $\bar{S}_1(\theta) = n^{-1}S_1(\theta)$. This is nonincreasing so condition (1.5.7) is satisfied. Next it is easy to check that $\mu(\theta) = E_\theta \bar{S}_1(0) = E_\theta \mathrm{sgn} X_i = E_0 \mathrm{sgn}(X_i + \theta) = 1 - 2F(-\theta)$.

Hence, $\mu'(0) = 2f(0)$. Then condition (1.5.8) is satisfied. We now consider condition (1.5.17). Consider the case $b > 0$, (similarly for $b < 0$),

$$\overline{S}_1(b/\sqrt{n}) - \overline{S}_1(0) = n^{-1} \sum_1^n [\text{sgn}(X_i - b/\sqrt{n}) - \text{sgn}(X_i)]$$

$$= -(2/n) \sum_1^n I(0 < X_i < b/n^{1/2}).$$

Because this is a sum of independent Bernoulli variables, we have

$$n\text{Var}_0[\overline{S}_1(b/n^{1/2}) - \overline{S}_1(0)] \leqslant 4P(0 < X_1 < b/\sqrt{n}) = 4[F(b/\sqrt{n}) - F(0)] \to 0.$$

The convergence to 0 occurs since F is continuous. Thus condition (1.5.17) is satisfied. Finally, note that $\sigma(0) = 1$ so $\sqrt{n}\overline{S}_1$ converges in distribution to $Z \sim N(0, 1)$ by the central limit theorem. Therefore the L_1 gradient process $S(\theta)$ is Pitman regular. It follows that the efficacy of the L_1 process is

$$c_{L_1} = 2f(0). \tag{1.5.21}$$

For future reference, we state the asymptotic linearity result for the L_1 process: if $|\sqrt{n}\theta_n| \leqslant B$ then

$$\sqrt{n}\overline{S}_1(\theta_n) = \sqrt{n}\overline{S}_1(0) - 2f(0)\sqrt{n}\theta_n + o_p(1). \tag{1.5.22}$$

Example 1.5.3 *Pitman Regularity of the L_2 Process*

In Exercise 1.13.5 it is shown that, provided X_i has finite variance, the L_2 gradient function is Pitman regular and that the efficacy is simply $c_{L_2} = 1/\sigma_f$.

We are now in a position to investigate the efficiency and power properties of the statistical methods based on the L_1 norm relative to the statistical methods based on the L_2 norm. As we will see in the next three subsections, these properties depend only on the efficacies.

1.5.3 Asymptotic Theory and Efficiency Results for $\hat{\theta}$

As at the beginning of this section, suppose we have the location model, (1.2.1), and that we have chosen a norm to fit the model with gradient function $S(\theta)$. In this part we will develop the asymptotic distribution of the estimate. The asymptotic variance will provide the basis for efficiency comparisons. We will use the asymptotic linearity that accompanies Pitman regularity. To do this, however, we first need to show that $\sqrt{n}\hat{\theta}$ is bounded in probability.

Lemma 1.5.7 *If the gradient function $S(\theta)$ is Pitman regular, then $\sqrt{n}(\hat{\theta} - \theta) = O_p(1)$.*

Proof. Assume without loss of generality that $\theta = 0$ and take $t > 0$. By the monotonicity of $S(\theta)$, if $S(t/\sqrt{n}) < 0$ then $\hat{\theta} \leqslant t/\sqrt{n}$. Hence, $P_0(S(t/\sqrt{n}) < 0) \leqslant P_0(\hat{\theta} \leqslant t)$. Theorem 1.5.5 implies that the first probability can be made as close to $\Phi(tc)$ as desired. This, in turn, can be made as close to 1 as desired. In a similar vein we note that if $S(-t/\sqrt{n}) > 0$, then $\hat{\theta} \geqslant -t/\sqrt{n}$

and $-\sqrt{n}\widehat{\theta} \leq t$. Again, the probability of this event can be made arbitrarily close to 1. Hence, $P_0(|\sqrt{n}\widehat{\theta}| \leq t)$ is arbitrarily close to 1 and we have boundedness in probability.

We are now in a position to exploit this boundedness in probability to determine the asymptotic distribution of the estimate.

Theorem 1.5.8 *Suppose $S(\theta)$ is Pitman regular with efficacy c. Then $\sqrt{n}(\widehat{\theta} - \theta)$ converges in distribution to $Z \sim N(0, c^{-2})$.*

Proof. As usual we assume, without loss of generality, that $\theta = 0$. First recall that $\widehat{\theta}$ is defined by $n^{-1/2}S(\widehat{\theta}) \doteq 0$. From Lemma 1.5.7, we know that $\sqrt{n}\widehat{\theta}$ is bounded in probability so that we can apply (1.5.13) to deduce

$$\frac{\sqrt{n}\,\overline{S}(\widehat{\theta})}{\sigma(0)} = \frac{\sqrt{n}\,\overline{S}(0)}{\sigma(0)} - c\sqrt{n}\widehat{\theta} + o_p(1).$$

Solving, we have

$$\sqrt{n}\widehat{\theta} = c^{-1}\sqrt{n}\,\overline{S}(0)/\sigma(0) + o_p(1);$$

hence, the result follows because $\sqrt{n}\,\overline{S}(0)/\sigma(0)$ is asymptotically $N(0, 1)$.

Definition 1.5.4 *If we have two Pitman regular estimates with efficacies c_1 and c_2, respectively, then the **efficiency** of $\widehat{\theta}_1$ with respect to $\widehat{\theta}_2$ is defined to be the reciprocal ratio of their asymptotic variances, namely, $e(\widehat{\theta}_1, \widehat{\theta}_2) = c_1^2/c_2^2$.*

The next example compares the L_1 estimate to the L_2 estimate.

Example 1.5.4 *Relative Efficiency between the L_1 and L_2 Estimates*

In this example we compare the L_1 and L_2 estimates, namely, the sample median and mean. We have seen that their respective efficacies are $2f(0)$ and σ_f^{-1}, and their asymptotic variances are $1/4f^2(0)n$ and σ_f^2/n, respectively. Hence, the relative efficiency of the median with respect to the mean is

$$e(\dot{X}, \bar{X}) = \text{asyvar}(\sqrt{n}\bar{X})/\text{asyvar}(\sqrt{n}\dot{X}) = c_{\dot{X}}^2/c_{\bar{X}}^2 = 4f^2(0)\sigma_f^2, \qquad (1.5.23)$$

where $\dot{X}$ is the sample median and $\bar{X}$ is the sample mean. The efficiency computation depends only on the Pitman efficacies. We illustrate the computation of the efficiency using the contaminated normal distribution. The pdf of the contaminated normal distribution consists of mixing the standard normal pdf with a normal pdf having mean zero and variance $\delta^2 > 1$. For ε between 0 and 1, the pdf can be written

$$f_\varepsilon(x) = (1 - \varepsilon)\phi(x) + \varepsilon\delta^{-1}\phi(\delta^{-1}x), \qquad (1.5.24)$$

with $\sigma_f^2 = 1 + \varepsilon(\delta^2 - 1)$. This distribution has tails heavier than the standard normal distribution and can be used to model data contamination; see Tukey (1960) for more discussion. We can think of ε as the fraction of the data contaminated. In Table 1.5.1 we provide values of the efficiencies for various values of contamination and with $\delta = 3$. Note that when we have 10% contam-

24 One-Sample Problems

Table 1.5.1: Efficiencies of the median relative to the mean for contaminated normal models

ε	$e(\dot{X}, \bar{X})$
0.00	0.637
0.03	0.758
0.05	0.833
0.10	1.000
0.15	1.134

ination that the efficiency is 1. This indicates that, for this distribution, the median and mean are equally effective. Finally, this example exhibits a distribution for which the median is superior to the mean as an estimate of the center. See Exercise 1.13.9 for other examples.

1.5.4 Asymptotic Power and Efficiency Results for the Test Based on $S(\theta)$

Consider the location model, (1.2.1), and assume that we have chosen a norm to fit the model with gradient function $S(\theta)$. Consider the gradient test (1.5.2) of the hypotheses (1.5.1). In Section 1.5.1, we showed that the power function of this test is nondecreasing with upper limit one and that it is typically resolving. Further, we showed that for a fixed alternative, the test is consistent. Thus the power will tend to one as the sample size increases. To offset this effect, we will let the alternative converge to the null value at a rate that will stabilize the power away from one. This will enable us to compare two tests along the same alternative sequence. Consider the null hypothesis $H_0 : \theta = 0$ versus $H_{An} : \theta = \theta_n$ where $\theta_n = \theta^*/\sqrt{n}$ and $\theta^* > 0$. Recall that the asymptotic size α test based on $S(0)$ rejects H_0 if $\sqrt{n}\,\bar{S}/\sigma(0) \geq z_\alpha$ where, $1 - \Phi(z_\alpha) = \alpha$.

The following theorem is called the **asymptotic power lemma**. Its proof follows immediately from expression (1.5.15).

Theorem 1.5.9 *Assume that $S(0)$ is Pitman regular with efficacy c, then the asymptotic local power along the sequence $\theta_n = \theta^*/\sqrt{n}$ is*

$$\gamma_S(\theta_n) = P_{\theta_n}\left[\sqrt{n}\,\bar{S}(0)/\sigma(0) \geq z_\alpha\right] = P_0\left[\sqrt{n}\,\bar{S}(-\theta_n)/\sigma(0) \geq z_\alpha\right] \to 1 - \Phi(z_\alpha - \theta^* c),$$

as $n \to \infty$.

Note that larger values of the efficacy imply larger values of the asymptotic local power.

Definition 1.5.5 *The Pitman **asymptotic relative efficiency** (ARE) of one test relative to another is defined to be $e(S_1, S_2) = c_1^2/c_2^2$.*

Note that this is the same formula as the efficiency of one estimate relative to another given in Definition 1.5.4. Therefore, the efficiency results discussed in Example 1.5.4 between the L_1 and L_2 estimates apply for the sign test and t-

test also. Hence, we have an example in which the simple sign test is asymptotically more powerful than the t-test.

We can also develop a sample size interpretation for the asymptotic power. Suppose we specify a power $\gamma < 1$. Further, let z_γ be defined by $1 - \Phi(z_\gamma) = \gamma$. Then $1 - \Phi(z_\alpha - cn^{1/2}\theta_n) = 1 - \Phi(z_\gamma)$ and $z_\alpha - cn^{1/2}\theta_n = z_\gamma$. Solving for n yields

$$n \doteq (z_\alpha - z_\gamma)^2/c^2\theta_n^2. \tag{1.5.25}$$

Typically we take $\theta_n = k_n \sigma$ with k_n small. Now if $S_1(0)$ and $S_2(0)$ are two Pitman regular asymptotically size α tests then the ratio of sample sizes required to achieve the same asymptotic power along the same sequence of alternatives is given by the approximation $n_2/n_1 \doteq c_1^2/c_2^2$. This provides additional motivation for the above definition of Pitman efficiency of two tests. The initial development of asymptotic efficiency was done by Pitman (1948) in an unpublished manuscript and later published by Noether (1955).

1.5.5 Efficiency Results for Confidence Intervals Based on $S(\theta)$

In this part we consider the length of the confidence interval as a measure of its efficiency. Suppose that we specify $\gamma = 1 - \alpha$ for the confidence coefficient. Then let $z_{\alpha/2}$ be defined by $1 - \Phi(z_{\alpha/2}) = \alpha/2$. Again we suppose throughout the discussion that the estimating functions are Pitman regular. Then the endpoints of the $100\gamma\%$ confidence interval are given asymptotically by $\widehat{\theta}_L$ and $\widehat{\theta}_U$ such that

$$\frac{\sqrt{n}\,\overline{S}(\widehat{\theta}_L)}{\sigma(0)} = z_{\alpha/2} \quad \text{and} \quad \frac{\sqrt{n}\,\overline{S}(\widehat{\theta}_U)}{\sigma(0)} = -z_{\alpha/2}; \tag{1.5.26}$$

see (1.3.10) for the exact versions of the endpoints.

The next theorem provides the asymptotic behavior of the length of this interval and, further, it shows that the standardized length of the confidence interval is a consistent estimate of the asymptotic standard deviation of $\sqrt{n}\widehat{\theta}$.

Theorem 1.5.10 *Suppose $S(\theta)$ is a Pitman regular estimating function with efficacy c. Let L be the length of the corresponding confidence interval. Then*

$$\frac{\sqrt{n}L}{2z_{\alpha/2}} \xrightarrow{P} \frac{1}{c}.$$

Proof. Using the same argument as in Lemma 1.5.7, we can show that $\widehat{\theta}_L$ and $\widehat{\theta}_U$ are bounded in probability when multiplied by $\sqrt{n}$. Hence, the above estimating equations can be linearized to obtain, for example:

$$z_{\alpha/2} = \sqrt{n}\,\overline{S}(\widehat{\theta}_L)/\sigma(0) = \sqrt{n}\,\overline{S}(0)/\sigma(0) - c\sqrt{n}\widehat{\theta}_L + o_P(1).$$

This can then be solved to find:

$$\sqrt{n}\widehat{\theta}_L = \sqrt{n}\,\overline{S}(0)/c\sigma(0) - z_{\alpha/2}/c + o_P(1).$$

When this is also done for $\widehat{\theta}_U$ and the difference is taken, we have:

$$n^{1/2}(\widehat{\theta}_U - \widehat{\theta}_L) = 2z_{\alpha/2}/c + o_P(1),$$

which concludes the argument.

If the ratio of squared asymptotic lengths is used as a measure of efficiency then the **efficiency of one confidence interval relative to another** is again the ratio of the squares of the efficacies.

The discussion of the properties of estimation, testing, and confidence interval construction shows that, asymptotically at least, the relative merit of a procedure is measured by its efficacy. This measure is the slope of the linear approximation of the standardized estimating function that determines these procedures. In the comparison of L_1 and L_2 methods, we have seen that the efficiency $e(L_1, L_2) = 4\sigma_f^2 f^2(0)$. There are other types of asymptotic efficiency that have been studied in the literature along with finite-sample versions of these asymptotic efficiencies. The conclusions drawn from these other efficiencies are consistent with the picture presented here. Finally, conclusions of simulation studies have also been consistent with the material presented here. Hence, we will not discuss these other measures; see Section 2.6 of Hettmansperger (1984a) for further references.

Example 1.5.5 *Estimation of the Standard Error of the Sample Median*

Recall that the sample median, when properly standardized, has a limiting normal distribution. Suppose we have a sample of size n from $H(x) = F(x - \theta)$, where θ is the unknown median. From Theorem 1.5.8, we know that the approximating distribution for $\widehat{\theta}$, the sample median, is normal with mean θ and variance $1/nh^2(\theta)$. We refer to this variance as the asymptotic variance. This normal distribution can be used to approximate probabilities concerning the sample median. When the underlying form of the distribution H is unknown, we must estimate this asymptotic variance. Theorem 1.5.10 provides one key to the estimation of the asymptotic variance. The square root of the asymptotic variance is sometimes called the asymptotic standard error of the sample median. We will discuss the estimation of this standard error rather than the asymptotic variance.

As a simple example, in Theorem 1.5.10, take $\alpha = 0.05, z_\alpha = 2$, and $k = n/2 - n^{1/2}$; then we have the following estimate of the asymptotic standard error of the median:

$$SE(median) \approx [X_{(n/2+n^{1/2})} - X_{(n/2-n^{1/2})}]/4. \tag{1.5.27}$$

This simple estimate of the asymptotic standard error is based on the length of the 95% confidence interval for the median. Sheather (1987) shows that the estimate can be improved by using the interpolated confidence intervals discussed in Section 1.11 below. Of course, other confidence intervals with different confidence coefficients can also be used. We recommend using 90% or 95%; again, see Sheather (1987).

There are other approaches to the estimation of this standard error. For example, we could estimate the density $h(x)$ directly and then use $h_n(\widehat{\theta})$, where h_n is the density estimate. Another possibility is to estimate the finite sample standard error of the sample median directly. Sheather (1987) surveys these approaches. We will discuss one further possibility here, namely the bootstrap. The bootstrap has gained wide attention recently because of its versatility in estimation and testing in nonstandard situations. See Efron and Tibshirani (1993) for a very readable account of the bootstrap.

If we know the underlying distribution $H(x)$, then we could estimate the standard error of the median by repeatedly drawing samples by computer from the distribution H. If we have B samples from H and have computed and stored the B values of the sample median, then our estimate of the standard error of the median is simply the sample standard deviation of these B values. When H is unknown we replace it by H_n, the empirical distribution function, and proceed with the simulation. Later in this chapter we will encounter an example where we want to compute a bootstrap p-value for a test; see Section 1.9. The bootstrap approach based on H_n is called the **nonparametric bootstrap** since nothing is assumed about the form of the underlying distribution H. In another version, called the parametric bootstrap, we suppose that we know the form of the underlying distribution H but there are some unknown parameters such as the mean and variance. We use the sample to estimate these unknown parameters, insert the values into H, and use this distribution to draw the B samples. In this book we will be concerned mainly with the nonparametric bootstrap and we will use the generic term bootstrap to refer to this approach. In either case, ready access to high-speed computing makes this method appealing. The following example illustrates the computations.

Example 1.5.6 *Generated Data*

Using Minitab, the 30 data points in Table 1.5.2 were generated from a normal distribution with mean 0 and variance 1. Thus, we know that the asymptotic standard error should be about $1/[30^{1/2}2f(0)] = 0.23$. We will use this to check what happens if we try to estimate the standard error from the data.

Using Minitab, the interpolated (see Section 1.11) 95% confidence interval is $[-0.7645, 0.3281]$. Hence, the length of confidence interval estimate, given in expression (1.5.27), is $(0.3281 + 0.7645)/4 = 0.27$.

A Minitab macro was written to bootstrap the sample; see Exercise 1.13.6. Using this macro, we obtained 300 bootstrap samples and the resulting standard deviation of the 300 bootstrap medians was 0.27. For this instance, the bootstrap procedure agrees to two decimal places with the length of confidence interval estimate.

Table 1.5.2: Generated N(0,1) variates (placed in order)

−1.79756	−1.66132	−1.46531	−1.45333	−1.21163	−0.92866	−0.86812
−0.84697	−0.81584	−0.78912	−0.68127	−0.37479	−0.33046	−0.22897
−0.02502	−0.00186	0.09666	0.13316	0.17747	0.31737	0.33125
0.80905	0.88860	0.90606	0.99640	1.26032	1.46174	1.52549
1.60306	1.90116					

Note that, from the data, the sample mean is −0.03575 and the sample standard deviation is 1.04769. If we assume the underlying distribution H is normal with unknown mean and variance, we would use the parametric bootstrap. Hence, instead of sampling from the empirical distribution function, we want to sample from a normal distribution with mean −0.03575 and standard deviation 1.04769. A Minitab macro was written that performs this parametric bootstrap. We used this to obtain 300 parametric bootstrapped samples. The sample standard deviation of the resulting medians was 0.23, just the value we would expect. You should not expect to get the precise value every time you bootstrap, either parametrically or nonparametrically. It is, however, a very versatile method to use to estimate such quantities as standard errors of estimates and p-values of tests.

An unusual aspect of this example is that the bootstrap distribution of the sample median can be found in closed form and does not have to be simulated as described above. The variance of the sample median computed from the bootstrap distribution can then be found. The result is another estimate of the variance of the sample median. This was discovered independently by Maritz and Jarrett (1978) and Efron (1979). We do not pursue this development here because in most cases we must simulate the bootstrap distribution and that is where the real strength of the bootstrap approach lies. For an interesting comparison of the various estimates of the variance of the sample median, see McKean and Schrader (1984).

1.6 Robustness Properties of Norm-Based Inference

We have just considered the statistical properties of the inference procedures. We have looked at ideas such as efficiency and power. We now turn to stability or robustness properties. By this we mean how the inference procedures are affected by outliers or corruption of portions of the data. Ideally, we would like procedures (tests and estimates) which do not respond too quickly to a single outlying value when it is introduced into the sample. Further, we would not like procedures that can be changed by arbitrary amounts by corrupting a small amount of the data. Response to outliers is measured by the influence curve and response to data corruption is measured by the breakdown value. We will introduce finite-sample versions of these concepts. They are easy to work with and, in the limit, they generally equal the more abstract versions based on the study of statistical functionals. We consider first the robustness properties of the estimates and then tests. As in the previous section, the discussion will be general but the L_1 and L_2 procedures will be discussed as we proceed. The robustness properties of the procedures based on the weighted L_1 norm will be covered in Sections 1.7 and 1.8. See Section A.5 of the Appendix for a development based on functionals.

1.6.1 Robustness Properties of $\widehat{\theta}$

We begin with the definition of breakdown for the estimator $\widehat{\theta}$.

Definition 1.6.1 *Estimation Breakdown. Let* $\mathbf{x} = (x_1, \ldots, x_n)$ *represent a realization of a sample and let*

$$\mathbf{x}^{(m)} = (x_1^*, \ldots, x_m^*, x_{m+1}, \ldots, x_n)'$$

represent the corruption of any m of the n observations. We define the **bias** *of an estimator* θ *to be* $\text{bias}(m; \theta, \mathbf{x}) = \sup |\theta(\mathbf{x}^{(m)}) - \theta(\mathbf{x})|$ *where the supremum is taken over all possible corrupted samples* $\mathbf{x}^{(m)}$. *Note that we change only* $x_1^*, \ldots, x_m^*$ *while* $x_{m+1}, \ldots, x_n$ *are fixed at their original values. If the bias is infinite, we say the estimate has broken down and the finite-sample breakdown value is given by*

$$\varepsilon_n^* = \min \{m/n : \text{bias}(m; \widehat{\theta}, \mathbf{x}) = \infty\}. \tag{1.6.1}$$

This approach to breakdown is called replacement breakdown because observations are replaced by corrupted values; see Donoho and Huber (1983) for more discussion of this approach. Often there exists an integer m such that $x_{(m)} \leq \theta \leq x_{(n-m+1)}$ and either θ tends to $-\infty$ as $x_{(m)}$ tends to $-\infty$ or θ tends to $+\infty$ as $x_{(n-m+1)}$ tends to $+\infty$. If m^* is the smallest such integer then $\varepsilon_n^* = m^*/n$. Hodges (1967) was the first to introduce these ideas.

To remove the effects of sample size, the limit, when it exists, can be computed. In this case we call $\lim \varepsilon_n^* = \varepsilon^*$ the **asymptotic breakdown value**.

Example 1.6.1 *Breakdown Values for the L_1 and L_2 Estimates*

The L_1 estimate is the sample median. If the sample size is $n = 2k$ then it is easy to see that when $x_{(k)}$ tends to $-\infty$, the median also tends to $-\infty$. Hence, the breakdown value of the sample median is k/n which tends to 0.5. By a similar argument, when the sample size is $n = 2k+1$, the breakdown value is $(k+1)/n$ and it also tends to 0.5 as the sample size increases. Hence, we say that the sample median is a 50% breakdown estimate. The L_2 estimate is the sample mean. A similar analysis shows that the breakdown value is $1/n$, which tends to zero. Hence, we say the sample mean is a zero breakdown estimate. This sharply contrasts the two estimates since we see that the median is the more resistant estimate and the sample mean is the less resistant estimate. In Exercise 1.13.10, the reader is asked to show that the pseudo-median induced by the signed-rank norm, (1.3.25), has breakdown 0.29.

We have just considered the effect of corrupting some of the observations. The estimate breaks down if we can force the estimate to change by an arbitrary amount by changing the observations over which we have control. Another important concept of stability entails measuring the effect of the introduction of a single outlier. An estimate is stable or resistant if it does not change by a large amount when the outlier is introduced. In particular, we want the change to be bounded no matter what the value of the outlier.

Suppose we have a sample of observations $x_1, \ldots, x_n$ from a distribution centered at 0 and an estimate $\widehat{\theta}_n$ based on these observations. By Pitman regularity, Definition 1.5.3, and Theorem 1.5.8, we have

$$n^{1/2}\widehat{\theta}_n = c^{-1}n^{-1/2}S(0)/\sigma(0) + o_P(1), \tag{1.6.2}$$

provided the true parameter is 0. Further, we often have a representation of $S(0)$ as a sum of independent random variables. We may have to make a projection of $S(0)$ to achieve this; see the next chapter for examples of projections. In any case, we then have the representation

$$c^{-1}n^{-1/2}S(0)/\sigma(0) = n^{-1/2}\sum_{i=1}^{n}\Omega(x_i) + o_P(1), \quad (1.6.3)$$

where $\Omega(\cdot)$ is the function needed in the representation. When we combine the above two statements we have

$$n^{1/2}\widehat{\theta}_n = n^{-1/2}\sum_{i=1}^{n}\Omega(x_i) + o_P(1). \quad (1.6.4)$$

Recall that the distribution that we are sampling is assumed to be centered at 0. The difference $(\widehat{\theta}_n - 0)$ is approximated by the average of n iid random variables. Since $\Omega(x_i)$ represents the effect of the ith observation on $\widehat{\theta}_n$ it is called the **influence function**.

The influence function approximates the rate of change of the estimate when an outlier is introduced. Let $x_{n+1} = x^*$ represent a new, outlying, observation. Since $\widehat{\theta}_n$ should be roughly 0, we have

$$(n+1)\widehat{\theta}_{n+1} - (n+1)\widehat{\theta}_n \doteq \Omega(x^*)$$

and

$$\frac{\widehat{\theta}_{n+1} - \widehat{\theta}_n}{1/(n+1)} \approx \Omega(x^*), \quad (1.6.5)$$

and this reveals the differential character of the influence function. Hampel (1974) developed the influence function from the theory of von Mises differentiable functions. In Sections A.5 and A.5.2 of the Appendix, we use his formulation to derive several influence functions for later situations. Here, though, we will identify influence functions for the estimates through the approximations described above. We now illustrate this approach.

Example 1.6.2 *Influence Function for the L_1 and L_2 Estimates*

We will briefly describe the influence functions for the sample median and the sample mean, the L_1 and L_2 estimates. From Example 1.5.2 we have immediately that, for the sample median,

$$n^{1/2}\widehat{\theta} \approx \frac{1}{\sqrt{n}}\sum_{i=1}^{n}\frac{\text{sgn}(X_i)}{2f(0)}$$

and

$$\Omega(x) = \frac{\text{sgn}(x)}{2f(0)}.$$

Note that the influence function is bounded but not continuous. Hence, outlying observations cannot have an arbitrarily large effect on the estimate. It is this feature, along with the 50% breakdown property, that makes the sample median the prototype of resistant estimates. The sample mean, on the other hand, has an unbounded influence function. It is easy to see that $\Omega(x) = x$, linear and unbounded. Hence, a single large outlier is sufficient to carry the sample mean beyond any bound. The unbounded influence is connected to the zero breakdown property. Hence, the L_2 estimate is the prototype of an estimate highly efficient for a specified model, the normal model in this case, but not resistant. This means that quite close to the model for which the estimate is optimal, the estimate may perform very poorly; recall Table 1.5.1.

1.6.2 Breakdown Properties of Tests

We now turn to the issue of breakdown in testing hypotheses. The problems are a bit different in this case since we typically want to move, by data corruption, a test statistic into or out of a critical region. It is not a matter of sending the statistic beyond any finite bound as it is in estimation breakdown.

Definition 1.6.2 *Suppose that V is a statistic for testing $H_0 : \theta = 0$ versus $H_0 : \theta > 0$ and we reject the null hypothesis when $V \geq k$, where $P_0(V \geq k) = \alpha$ determines k. The* **rejection breakdown** *of the test is defined by*

$$\varepsilon_n^*(reject) = \min\{m/n : \inf_{\mathbf{x}} \sup_{\mathbf{x}^{(m)}} V \geq k\}, \qquad (1.6.6)$$

where the supremum is taken over all possible corruptions of m data points. Likewise the **acceptance breakdown** *is defined to be*

$$\varepsilon_n^*(accept) = \min\{m/n : \sup_{\mathbf{x}} \inf_{\mathbf{x}^{(m)}} V < k\}. \qquad (1.6.7)$$

Rejection breakdown is the smallest portion of the data that can be corrupted to guarantee that the test will reject. Acceptance breakdown is interpreted as the smallest portion of the data that must be corrupted to guarantee that the test statistic will not be in the critical region; that is, the test is guaranteed to fail to reject the null hypothesis. We turn immediately to a comparison of the L_1 and L_2 tests.

Example 1.6.3 *Rejection Breakdown of the L_1 and L_2 tests*

We first consider the one-sided sign test for testing $H_0 : \theta = 0$ versus $H_A : \theta > 0$. The asymptotically size α test rejects the null hypothesis when $n^{-1/2}S_1(0) \geq z_\alpha$, the upper α quantile from a standard normal distribution. It is easier to see exactly what happens if we convert the test to $S_1^+(0) = \sum I(X_i > 0) \geq n/2 + (n^{1/2}z_\alpha)/2$. Now each time we make an observation positive it makes $S_1^+(0)$ increase by one. Hence, if we wish to guarantee that the test will reject, we make m^* observations positive, where $m^* = [n/2 + (n^{1/2}z_\alpha)/2] + 1$, $[\cdot]$ being the greatest integer function. Then the rejection breakdown is

$$\varepsilon_n^*(reject) = m^*/n \doteq \frac{1}{2} + \frac{z_\alpha}{2n^{1/2}}.$$

Likewise,

$$\varepsilon_n^*(accept) \doteq \frac{1}{2} - \frac{z_\alpha}{2n^{1/2}}.$$

Note that the rejection breakdown converges down to the estimation breakdown and the acceptance breakdown converges up to it.

We next turn to the one-sided Student t-test. Acceptance breakdown for the t-test is simple. By making a single observation approach $-\infty$, the t-statistic can be made negative, hence we can always guarantee acceptance with control of one observation. The rejection breakdown is more interesting. If we increase an observation both the sample mean and the sample standard deviation increase. Hence, it is not at all clear what will happen to the t-statistic. In fact it is not sufficient to increase a single observation in order to force the t-statistic to move into the critical region. We now show that the rejection breakdown for the t-statistic is

$$\varepsilon_n^*(reject) = \frac{t_\alpha^2}{n-1+t_\alpha^2} \to 0, \quad \text{as } n \to \infty,$$

where t_α is the upper α quantile from a t-distribution with $n-1$ degrees of freedom. The infimum part of the definition suggests that we set all observations at $-B < 0$ and then change m observations to $M > 0$. The result is

$$\bar{x} = \frac{mM - (n-m)B}{n} \quad \text{and} \quad s^2 = \frac{m(n-m)(M+B)^2}{(n-1)n}.$$

Putting these two quantities together, we have

$$\frac{n^{1/2}\bar{x}}{s} = [m - (n-m)B/M]\left(\frac{n-1}{m(n-m)(1+B/M)^2}\right)^{1/2} \to \frac{m(n-1)^{1/2}}{n-m},$$

as $M \to \infty$. We now equate the limit to t_α and solve for m to get $m = nt_\alpha^2/(n-1+t_\alpha^2)$, (actually we would take the greatest integer and add one). Then the rejection breakdown is m divided by n as stated. Table 1.6.1 compares rejection breakdown values for the sign and t-tests. We assume $\alpha = 0.05$ and the sample sizes are chosen so that the size of the sign test is quite close to 0.05. For further discussion, see Ylvisaker (1977).

These definitions of breakdown assume a worst-case scenario. They assume that the test statistic is as far away from the critical region (for rejection breakdown) as possible. In practice, however, it may be the case that

Table 1.6.1: Rejection breakdown values for size $\alpha = 0.05$ tests

n	Sign	t
10	0.71	0.27
13	0.70	0.21
18	0.67	0.15
30	0.63	0.09
100	0.58	0.03
∞	0.50	0

Table 1.6.2: Comparison of expected breakdown and worst-case breakdown for the size $\alpha = 0.05$ sign test.

n	$\text{Exp}_n^*(reject)$	$\varepsilon_n^*(reject)$
10	0.27	0.71
13	0.24	0.70
18	0.20	0.67
30	0.16	0.63
100	0.08	0.58
∞	0	0.50

a test statistic is quite near the edge of the critical region and only one observation is needed to change the decision from fail to reject to reject. An alternative form of breakdown considers the average number of observations that must be corrupted, conditional on the test statistic being in the acceptance region, to force a rejection.

Let M_R be the number of observations that must be corrupted to force a rejection; then, M_R is a random variable. The **expected rejection breakdown** is defined to be

$$\text{Exp}_n^*(reject) = E_{H_0}[M_R | M_R > 0]/n. \quad (1.6.8)$$

Note that we condition on $M_R > 0$ since $M_R = 0$ is equivalent to a rejection. It is left as Exercise 1.13.11 to show that the expected breakdown can be computed with unconditional expectation as

$$\text{Exp}_n^*(reject) = E_{H_0}[M_R]/(1 - \alpha). \quad (1.6.9)$$

In the following example we illustrate this computation on the sign test and show how it compares to the worst-case breakdown introduced earlier.

Example 1.6.4 *Expected Rejection Breakdown of the Sign Test*

Refer to Example 1.6.3. The one-sided sign test rejects when $\sum I(X_i > 0) \geq n/2 + n^{1/2} z_{\alpha/2}$. Hence, given that we fail to reject the null hypothesis, we will need to change (corrupt) $n/2 + n^{1/2} z_{\alpha/2} - \sum I(X_i > 0)$ negative observations into positive ones. This is precisely M_R and $E[M_R] = n^{1/2} z_{\alpha/2}$. It follows that $\text{Exp}_n^*(reject) = z_{\alpha/2} n^{1/2}(1 - \alpha) \to 0$ as $n \to \infty$ rather than 0.5 which happens in the worst-case breakdown. Table 1.6.2 compares the two types of rejection breakdown. This simple calculation clearly shows that even highly resistant tests such as the sign test may break down quite easily. This is contrary to what the worst-case breakdown analysis would suggest. For additional reading on test breakdown, see Coakley and Hettmansperger (1992). He, Simpson, and Portnoy (1990) discuss asymptotic test breakdown.

1.7 Inference and the Wilcoxon Signed-Rank Norm

In this section we develop the statistical properties for the procedures based on the Wilcoxon signed-rank norm, (1.3.17), that was defined in Example 1.3.3 of Section 1.3. Recall that the norm and its associated gradient function are given in expressions (1.3.17) and (1.3.24), respectively. Recall for a sample

$X_1, \ldots, X_n$ that the estimate of θ is the median of the Walsh averages given by (1.3.25). As in Section 1.3, our hypotheses of interest are

$$H_0 : \theta = 0 \text{ versus } H_0 : \theta \neq 0. \quad (1.7.1)$$

The level α test associated with the signed-rank norm is

$$\text{Reject } H_0 \text{ in favor of } H_A, \text{ if } |T(0)| \geq c, \quad (1.7.2)$$

where c is such that $P_0[|T(0)| \geq c]$. To complete the test we need to determine the null distribution of $T(0)$, which is given by Theorems 1.7.3 and 1.7.4.

In order to develop the statistical properties, in addition to (1.2.1), we assume that

$$h(x) \text{ is symmetrically distributed about } \theta. \quad (1.7.3)$$

We refer to this as the **symmetric location model**. Under symmetry, by Theorem 1.2.1, $T(H) = \theta$, for all location functionals T.

1.7.1 Null Distribution Theory of $T(0)$

In addition to expression (1.3.24), a third representation of $T(0)$ will be helpful in establishing its null distribution. Recall the definition of the anti-ranks, $D_1, \ldots, D_n$, given in expression (1.3.19). Using these anti-ranks, we can write

$$T(0) = \sum R(|X_i|)\text{sgn}(X_i) = \sum j\,\text{sgn}(X_{D_j}) = \sum jW_j,$$

where $W_j = \text{sgn}(X_{D_j})$.

Lemma 1.7.1 *Under H_0, $|X_1|, \ldots, |X_n|$ are independent of $\text{sgn}(X_1), \ldots, \text{sgn}(X_n)$.*

Proof. Since $X_1, \ldots, X_n$ is a random sample from $H(x)$, it suffices to show that $P[|X_i| \leq x, \text{sgn}(X_i) = 1] = P[|X_i| \leq x]P[\text{sgn}(X_i) = 1]$. But due to H_0 and the symmetry of $h(x)$ this follows from the following string of equalities:

$$P[|X_i| \leq x, \text{sgn}(X_i) = 1] = P[0 < X_i \leq x] = H(x) - \tfrac{1}{2}$$
$$= [2H(x) - 1]\tfrac{1}{2} = P[|X_i| \leq x]P[\text{sgn}(X_i) = 1].$$

Based on this lemma, the vector of ranks and, hence, the vector of anti-ranks $(D_1, \ldots, D_n)$, are independent of the vector $(\text{sgn}(X_1), \ldots, \text{sgn}(X_n))$. Based on these facts, we can obtain the distribution of $(W_1, \ldots, W_n)$, which we summarize in the following lemma; see Exercise 1.13.12 for its proof.

Lemma 1.7.2 *Under H_0 and the symmetry of $h(x)$, $W_1, \ldots, W_n$ are iid random variables with $P[W_i = 1] = P[W_i = -1] = \tfrac{1}{2}$.*

We can now easily derive the null distribution theory of $T(0)$ which we summarize in the following theorems. Details are given in Exercise 1.13.13.

Theorem 1.7.3 *Under H_0 and the symmetry of $h(x)$,*

$$T(0) \text{ is distribution-free and its distribution is symmetric} \quad (1.7.4)$$

$$E_0[T(0)] = 0 \quad (1.7.5)$$

$$\text{Var}_0(T(0)) = \frac{n(n+1)(2n+1)}{6} \quad (1.7.6)$$

$$\frac{T(0)}{\sqrt{\text{Var}_0(T(0))}} \text{ has an asymptotically } N(0,1) \text{ distribution.} \quad (1.7.7)$$

The exact distribution of $T(0)$ cannot be found in closed form. We do, however, have the following recursion formula; see Exercise 1.13.14.

Theorem 1.7.4 *Consider the version of the signed-rank test statistic given by T^+, (1.3.28). Let $p_n(k) = P[T^+ = k]$ for $k = 0, \ldots, n(n+1)/2$. Then*

$$p_n(k) = \tfrac{1}{2}[p_{n-1}(k) + p_{n-1}(k-n)], \quad (1.7.8)$$

where

$$p_0(0) = 1; \; p_0(k) = 0 \text{ for } k \neq 0; \text{ and } p_0(k) = 0 \text{ for } k < 0 \, .$$

Using this formula, algorithms can be developed which obtain the null distribution of the signed-rank test statistic. The moment generating function can also be inverted to find the null distribution; see Hettmansperger (1984a, Section 2.2). Software is now available which computes critical values and p-values of the null distribution; see, for example, STATXACT.

Theorem 1.7.3 justifies the confidence interval for θ given in (1.3.30); i.e. the $(1-\alpha)100\%$ confidence interval given by $[W_{(k+1)}, W_{((n(n+1))/2)-k)})$, where $W_{(i)}$ denotes the ith ordered Walsh average and $P(T^+(0) \leq k) = \alpha/2$. Based on (1.7.7), k can be approximated as $k \approx n(n+1)/4 - \tfrac{1}{2} - z_{\alpha/2}[n(n+1)(2n+1)/24]^{1/2}$. The computation of the estimate and confidence interval can be obtain by linear search algorithms discussed in Section 3.7.3.

1.7.2 Statistical Properties

From our earlier analysis of the statistical properties of the L_1 and L_2 methods we see that Pitman regularity is crucial. In particular, we need to compute the Pitman efficacy which determines the asymptotic variance of the estimate, the asymptotic local power of the test, and the asymptotic length of the confidence interval. In the following theorem we show that the weighted L_1 gradient function is Pitman regular and determine the efficacy. Then we make some preliminary efficiency comparisons with the L_1 and L_2 methods.

36   One-Sample Problems

Theorem 1.7.5 *Suppose that h is symmetric and that $\int h^2(x)dx < \infty$. Let*

$$\overline{T}(\theta) = \frac{2}{n(n+1)} \sum_{i \leq j} \text{sgn}\left(\frac{x_i + x_j}{2} - \theta\right).$$

Then the conditions of Definition 1.5.3 are satisfied and, thus, $T(\theta)$ is Pitman regular. Moreover, the Pitman efficacy is given by

$$c = \sqrt{12} \int_{-\infty}^{\infty} h^2(x)dx. \tag{1.7.9}$$

Proof. Since we have the L_1 norm applied to the Walsh averages, the estimating function is a nonincreasing step function with steps at the Walsh averages. Hence, (1.5.7) holds. Next note that $h(x) = h(-x)$ and, hence,

$$\mu(\theta) = E_\theta \overline{T}(0) = \frac{2}{n+1} E_\theta \text{sgn}(X_1) + \frac{n-1}{n+1} E_\theta \left(\text{sgn}\left\{\frac{X_1 + X_2}{2}\right\} \right).$$

Now

$$E_\theta \text{sgn} X_1 = \int \text{sgn}(x+\theta) h(x) dx = 1 - 2H(\theta),$$

and

$$E_\theta \text{sgn}(X_1 + X_2)/2 = \int \int \text{sgn}[(x+y)/2 + \theta] h(x) h(y) dx dy$$

$$= \int [1 - 2H(-2\theta - y)] h(y) dy.$$

Differentiate with respect to θ and set $\theta = 0$ to get

$$\mu'(0) = \frac{2h(0)}{n+1} + \frac{4(n-1)}{n+1} \int_{-\infty}^{\infty} h^2(y) dy \to 4 \int h^2(y) dy.$$

The finiteness of the integral is sufficient to ensure that the derivative can be passed through the integral; see Hodges and Lehmann (1961) or Olshen (1967). Hence, (1.5.8) also holds. We next establish condition (1.5.9). Since

$$\overline{T}(\theta) = \frac{2}{n(n+1)} \sum_{i=1}^{n} \text{sgn}(X_i - \theta) + \frac{2}{n(n+1)} \sum_{i<j} \text{sgn}\left(\frac{X_i + X_j}{2} - \theta\right),$$

the first term is of smaller order and we need only consider the second term. Now, for $b > 0$, let

$$V^* = \frac{2}{n(n+1)} \sum_{i<j} \left[\text{sgn}\left(\frac{X_i + X_j}{2} - n^{-1/2}b\right) - \text{sgn}\left(\frac{X_i + X_j}{2}\right) \right]$$

$$= \frac{-4}{n(n+1)} \sum \sum_{i<j} I\left(0 < \frac{X_i + X_j}{2} < n^{-1/2}b\right).$$

Hence,

$$n\text{Var}(V^*) = \frac{16n}{n^2(n+1)^2} E\left\{ \sum \sum_{i<j} \sum_{s<t} (I_{ij} I_{st} - EI_{ij} EI_{st}) \right\},$$

where $I_{ij} = I(0 < (x_i + x_j)/2 < n^{-1/2}b)$. This becomes

$$n\text{Var}(\overline{V}^*) = \frac{16n^2(n-1)}{2n^2(n+1)^2}\text{Var}(I_{12}) + \frac{16n^2(n-1)(n-2)}{2n^2(n+1)^2}[EI_{12}I_{13} - EI_{12}EI_{13}].$$

The first term tends to zero since it behaves like $1/n$. In the second term, consider $|EI_{12}I_{13} - EI_{12}EI_{13}| \leq EI_{12} + E^2I_{12} = EI_{12}(1 + EI_{12})$. Now, as $n \to \infty$,

$$EI_{12} = P\left(0 < \frac{X_i + X_j}{2} < n^{-1/2}b\right) = \int [H(2n^{-1/2}b - x) - H(-x)]h(x)dx \to 0.$$

Hence, by Theorem 1.5.6, condition (1.5.9) is true. Finally, asymptotic normality of the null distribution is established in Theorem 1.7.3 which also yields $n\text{Var}_0 \overline{T}(0) \to 4/3 = \sigma^2(0)$. It follows that the Pitman efficacy is

$$c = \frac{4\int h^2(y)dy}{\sqrt{4/3}} = \sqrt{12}\int h^2(y)dy,$$

and the proof is complete.

For future reference we display the asymptotic linearity result:

$$\frac{T(\theta)}{\sqrt{n(n+1)(2n+1)/6}} = \frac{T(0)}{\sqrt{n(n+1)(2n+1)/6}} - \sqrt{n}\theta\sqrt{12}\int_{-\infty}^{\infty} h^2(x)\,dx + o_p(1), \tag{1.7.10}$$

for $\sqrt{n}|\theta| \leq B$, where $B > 0$.

An immediate consequence of this theorem and Theorem 1.5.8 is that

$$\sqrt{n}(\widehat{\theta} - \theta) \xrightarrow{D} Z \sim N\left(0, 1/12\left[\int h^2(t)dt\right]^2\right), \tag{1.7.11}$$

and we thus have the limiting distribution of the median of the Walsh averages. Exercise 1.13.15 shows that $\int h^2(t)\,dt < \infty$, when h has finite Fisher information.

We will have more to say about this particular c in the next chapter where we will encounter it in the two-sample location model and later in the linear model.

From Example 1.5.3 and Definition 1.5.4, we have that the asymptotic relative efficiency between the signed-rank Wilcoxon process and the L_2 process is given by

$$e(\text{Wilcoxon}, L_2) = 12\sigma_h^2\left(\int h^2(x)\,dx\right)^2, \tag{1.7.12}$$

where h is the underlying density with variance σ_h^2.

In the following example, we consider the contaminated normal distribution and then find the efficiency of the rank methods relative to the L_1 and L_2 methods.

Example 1.7.1 *Asymptotic Relative Efficiency for Contaminated Normal Distributions*

Let $f_\varepsilon(x)$ denote the pdf of the contaminated normal distribution used in Example 1.5.4; the proportion of contamination is ε and the variance of the contaminated part is 9. A straightforward computation shows that

$$\int f_\varepsilon^2(y)dy = \frac{(1-\varepsilon)^2}{2\sqrt{\pi}} + \frac{\varepsilon^2}{6\sqrt{\pi}} + \frac{\varepsilon(1-\varepsilon)}{\sqrt{5}\sqrt{\pi}},$$

and we use this in the formula for c given above. The efficacies for the L_1 and L_2 methods are given in Example 1.5.4. We first consider the special case of $\varepsilon = 0$ corresponding to an underlying normal distribution. In this case we have for the rank methods $c_R^2 = 12/(4\pi) = 3/\pi = 0.955$, for the L_1 methods $c_1^2 = 2/\pi = 0.637$, and for the L_2 methods $c_2^2 = 1$. We have already seen that the efficiency $e_{\text{normal}}(L_1, L_2) = c_1^2/c_2^2 = 0.637$ from the first line of Table 1.5.1. We now have

$$e_{\text{normal}}(\text{Wilcoxon}, L_2) = 3/\pi \doteq 0.955 \quad \text{and} \quad e_{\text{normal}}(\text{Wilcoxon}, L_1) = 1.5. \tag{1.7.13}$$

The efficiency of the rank methods relative to the L_2 methods is extraordinary. It says that even at the distribution for which the t-test is uniformly most powerful, the Wilcoxon signed-rank test is almost as efficient. This means that replacing the values of the observations by their ranks (retaining only the order information) does not affect the statistical properties of the test. This was considered highly nonintuitive in the 1950s since nonparametric methods were thought of as quick and dirty. Now they must be considered highly efficient competitors of the optimal methods and, in addition, they are more robust than the optimal methods. This provides powerful motivation for the continued study of rank methods in other statistical models such as the two-sample location model and the linear model. The early work in the area of efficiency of rank methods is due largely to Lehmann and his students. See Hodges and Lehmann (1956; 1961) for two important early papers and Lehmann (1975, Appendix) for more discussion.

We complete this example with a table of efficiencies of the rank methods relative to the L_1 and L_2 methods for the contaminated normal model with $\sigma = 3$. Table 1.7.1 shows these efficiencies and extends Table 1.5.1. As ε increases, the weight in the tails of the distribution also increases. Note that

Table 1.7.1: Efficiencies of the rank, L_1, and L_2 methods for the contaminated normal distribution

ε	$e(L_1, L_2)$	$e(R, L_1)$	$e(R, L_2)$
0.00	0.637	1.500	0.955
0.01	0.678	1.488	1.009
0.03	0.758	1.462	1.108
0.05	0.833	1.436	1.196
0.10	1.000	1.373	1.373
0.15	1.134	1.320	1.497

the efficiencies of both the L_1 and rank methods relative to the L_2 methods increase with ε. On the other hand, the efficiency of the rank methods relative to the L_1 methods decreases slightly. The rank methods are still more efficient; however, this illustrates the fact that the L_1 methods are good for heavy-tailed distributions. The overall implication of this example is that the L_2 methods, such as the sample mean, the t-test and confidence interval, are not particularly efficient once the underlying distribution departs from the normal distribution. Further, the rank methods such as the Wilcoxon signed-rank test, confidence interval, and the median of the Walsh averages are surprisingly efficient, even for the normal distribution. Note that the rank methods are more efficient than the L_2 methods even for 1% contamination.

Finally, the following theorem shows that the Wilcoxon signed-rank statistic never loses much efficiency relative to the t-statistic. Let $\mathcal{F}_S$ denote the family of distributions which have symmetric densities and finite Fisher information; see Exercise 1.13.15.

Theorem 1.7.6 *Let $X_1, \ldots, X_n$ be a random sample from $H \in \mathcal{F}_S$. Then*

$$\inf_{\mathcal{F}_S} e(\text{Wilcoxon}, L_2) = 0.864. \tag{1.7.14}$$

Proof. By (1.7.12), $e(\text{Wilcoxon}, L_2) = 12\sigma_h^2 \left(\int h^2(x)\,dx\right)^2$. If $\sigma_h^2 = \infty$ then $e(\text{Wilcoxon}, L_2) > 0.864$; hence, we can restrict attention to $H \in \mathcal{F}_S$ such that $\sigma_h^2 < \infty$. As Exercise 1.13.16 indicates, $e(\text{Wilcoxon}, L_2)$ is location- and scale-invariant, so we can further assume that h is symmetric about 0 and $\sigma_h^2 = 1$. The problem, then, is to minimize $\int h^2$ subject to $\int h = \int x^2 h = 1$ and $\int xh = 0$. This is equivalent to minimizing

$$\int h^2 + 2b\int x^2 h - 2ba^2 \int h, \tag{1.7.15}$$

where a and b are positive constants to be determined later. We now write (1.7.15) as

$$\int [h^2 + 2b(x^2 - a^2)h] = \int_{|x| \leq a} [h^2 + 2b(x^2 - a^2)h]$$
$$+ \int_{|x| > a} [h^2 + 2b(x^2 - a^2)h]. \tag{1.7.16}$$

First complete the square of the first term on the right-hand side of (1.7.16) to get

$$\int_{|x| \leq a} [h + b(x^2 - a^2)]^2 - \int_{|x| \leq a} b^2(x^2 - a^2)^2. \tag{1.7.17}$$

Now (1.7.16) is equal to the two terms of (1.7.17) plus the second term on the right-hand side of (1.7.16). We can now write the density that minimizes (1.7.15).

If $|x| > a$ take $h(x) = 0$, since $x^2 > a^2$, and if $|x| \leq a$ take $h(x) = b(a^2 - x^2)$, since the integral in the first term of (1.7.17) is nonnegative. We can now determine the values of a and b from the side conditions. From $\int h = 1$ we have

$$\int_{-a}^{a} b(a^2 - x^2)\,dx = 1,$$

which implies that $a^3 b = \frac{3}{4}$. Further, from $\int x^2 h = 1$ we have

$$\int_{-a}^{a} x^2 b(a^2 - x^2)\,dx = 1,$$

from which $a^5 b = \frac{15}{4}$. Hence solving for a and b yields $a = \sqrt{5}$ and $b = 3\sqrt{5}/100$. Now

$$\int h^2 = \int_{-\sqrt{5}}^{\sqrt{5}} \left[\frac{3\sqrt{5}}{100}(5 - x^2)\right]^2 dx = \frac{3\sqrt{5}}{25},$$

which leads to the result,

$$\inf_{\mathcal{F}_s} e(\text{Wilcoxon}, L_2) = 12\left(\frac{3\sqrt{5}}{25}\right)^2 = \frac{108}{125} = 0.864.$$

1.7.3 Robustness Properties

We complete this section with a discussion of the breakdown point of the estimate and test and a heuristic derivation of the influence function of the estimate. In Example 1.6.1 we discussed the breakdown of the sample median and mean. In those cases we saw that the median is the most resistant, while the mean is the least resistant. In Exercise 1.13.10 the reader is asked to show that the breakdown point of the median of the Walsh averages, the R-estimate, is roughly 0.29. Our next result gives the influence function $\widehat{\theta}$.

Theorem 1.7.7 *The influence function of* $\widehat{\theta} = \text{med}_{i \leq j}(x_i + x_j)/2$ *is given by:*

$$\Omega(x) = \frac{H(x) - \frac{1}{2}}{\int_{-\infty}^{\infty} h^2(t)\,dt}.$$

We sketch a derivation of this result. A rigorous development is offered in Section A.5 of the Appendix. From Theorems 1.7.5 and 1.5.6 we have

$$n^{1/2} T(\theta)/\sigma(0) \approx n^{1/2} T(0)/\sigma(0) - c n^{1/2} \theta$$

and

$$\widehat{\theta}_n \approx T(0)/c\sigma(0),$$

where $\sigma(0) = (4/3)^{1/2}$ and $c = (12)^{1/2} \int h^2(t)\,dt$. Make these substitutions to get

$$\widehat{\theta}_n \doteq \frac{1}{n(n+1)2 \int h^2(t)\,dt} \sum_{i \leq j} \text{sgn}\left(\frac{X_i + X_j}{2}\right).$$

Now introduce an outlier $x_{n+1} = x^*$ and take the difference between $\widehat{\theta}_{n+1}$ and $\widehat{\theta}_n$. The result is

$$2\int h^2(t)dt[(n+2)\widehat{\theta}_{n+1} - n\widehat{\theta}_n] \doteq \frac{1}{(n+1)}\sum_{i=1}^{n+1}\text{sgn}\left(\frac{x_i + x^*}{2}\right).$$

We can replace $n+2$ and $n+1$ by n where convenient without affecting the asymptotics. Using the symmetry of the density of H, we have

$$\frac{1}{n}\sum_{i=1}^{n}\text{sgn}\left(\frac{x_i + x^*}{2}\right) \doteq 1 - 2H_n(-x^*) \to 1 - 2H(-x^*) = 2H(x^*) - 1.$$

It now follows that $(n+1)(\widehat{\theta}_{n+1} - \widehat{\theta}_n) \doteq \Omega(x^*)$, given in the statement of the theorem; see the discussion of the influence function in Section 1.6.

Note that we have a bounded influence function since the cdf H is a bounded function. Further, it is continuous, unlike the influence function of the median. Finally, as an additional check, note that $E\Omega^2(X) = 1/12[\int h^2(t)dt]^2 = 1/c^2$, the asymptotic variance of $n^{1/2}\widehat{\theta}$.

Let $\widehat{\theta}_c = \text{med}_{i,j}\{(X_i - cX_j)/(1-c)\}$ for $-1 \leq c < 1$. This extension of the Hodges–Lehmann estimate, (1.3.25), has some very interesting robustness properties for $c > 0$. The influence function of $\widehat{\theta}_c$ is not only bounded but also redescending, similar to the most robust M-estimates. In addition, $\widehat{\theta}_c$ has 50% breakdown. For a complete discussion of this estimate see Maritz, Wu, and Staudte (1977) and Brown and Hettmansperger (1994).

In the next theorem we develop the test breakdown for the Wilcoxon signed-rank test.

Theorem 1.7.8 *The rejection breakdown, Definition 1.6.2, for the Wilcoxon signed-rank test is*

$$\varepsilon_n^* \doteq 1 - \left(\frac{1}{2} - \frac{z_\alpha}{(3n)^{1/2}}\right)^{1/2} \to 1 - \frac{1}{2^{1/2}} \doteq 0.29.$$

Proof. Consider the form $T^+(0) = \sum\sum I[(x_i + x_j)/2 > 0]$, where the double sum is over all $i \leq j$. The asymptotically size α test rejects $H_0 : \theta = 0$ in favor of $H_A : \theta > 0$ when $T^+(0) \geq c \doteq n(n+1)/4 + z_\alpha[n(n+1)(2n+1)/24]^{1/2}$. Now we must guarantee that $T^+(0)$ is in the critical region. This requires at least c positive Walsh averages. Let $x_{(1)} \leq \ldots \leq x_{(n)}$ be the ordered observations. Then contamination of $x_{(n)}$ results in n contaminated Walsh averages, namely those Walsh averages that include $x_{(n)}$. Contamination of $x_{(n-1)}$ yields $n-1$ additional contaminated Walsh averages. When we proceed in this way, contamination of the b ordered values $x_{(n)}, \ldots, x_{(n-b+1)}$ yields $n + (n-1) + \ldots + (n-b+1) = [n(n+1)/2] - [(n-b)(n-b+1)/2]$ contaminated Walsh averages. We now set $[n(n+1)/2] - [(n-b)(n-b+1)/2] \doteq c$ and solve the resulting quadratic for b. We must solve $b^2 - (2n+1)b + 2c \doteq 0$. The appropriate root in this case is

$$b \doteq \frac{2n + 1 - [(2n+1)^2 - 8c]^{1/2}}{2}.$$

42 One-Sample Problems

Table 1.7.2: Rejection breakdown values for size $\alpha = 0.05$ tests

n	Sign	t	Signed-rank Wilcoxon
10	0.71	0.27	0.57
13	0.70	0.21	0.53
18	0.67	0.15	0.48
30	0.63	0.09	0.43
100	0.58	0.03	0.37
∞	0.50	0	0.29

Substituting the approximate critical value for c, dividing by n, and ignoring higher-order terms leads to the stated result.

Table 1.7.2 displays the finite rejection breakdowns of the Wilcoxon signed-rank test over the same sample sizes as the rejection breakdowns of the sign test given in Table 1.6.1. For convenience we have also reproduced the results for the sign and t-tests. The rejection breakdown for the Wilcoxon test converges from above to the estimation breakdown of 0.29. The Wilcoxon test is more resistant than the t-test but not as resistant as the simple sign test. It is interesting to note that from the discussion of efficiency, it is clear that we can now achieve high efficiency and not pay the price in lack of robustness. The rank-based methods seem to be a very attractive alternative to the highly resistant but relatively inefficient (for the normal model) L_1 methods and the highly efficient (for the normal model) but nonrobust L_2 methods.

1.8 Inference Based on General Signed-Rank Norms

In this section, we develop properties for a generalized signed-rank process. It includes the L_1 and the weighted L_1 as special cases. The development is similar to that of the weighted L_1 so a brief sketch suffices. For $\mathbf{x} \in \mathcal{R}^n$, consider the function,

$$\|\mathbf{x}\|_{\varphi^+} = \sum_{i=1}^n a^+(R|x_i|)|x_i|, \qquad (1.8.1)$$

where the **scores** $a^+(i)$ are generated as $a^+(i) = \varphi^+(i/(n+1))$ for a positive-valued, nondecreasing, square-integrable function $\varphi^+(u)$ defined on the interval $(0, 1)$. The proof that $\| \cdot \|_{\varphi^+}$ is a norm on $\mathcal{R}^n$ follows in the same way as in the weighted L_1 case; see the proof of Theorem 1.3.2 and Exercise 1.13.17. The gradient function associated with this norm is

$$T_{\varphi^+}(\theta) = \sum_{i=1}^n a^+(R|X_i - \theta|)\operatorname{sgn}(X_i - \theta). \qquad (1.8.2)$$

Note that it reduces to the L_1 norm if $\varphi^+(u) \equiv 1$ and the weighted L_1, Wilcoxon signed-rank, norm if $\varphi^+(u) = u$. A family of simple score functions between the weighted L_1 and the L_1 are of the form

$$\varphi_c^+ = (u) = \begin{cases} u & 0 < u < c \\ c & c \leqslant u < 1, \end{cases} \tag{1.8.3}$$

where the parameter c is between 0 and 1. These scores were proposed by Policello and Hettmansperger (1976); see also Hogg (1974). The frequently used normal scores are generated by the score function

$$\varphi_\Phi^+(u) = \Phi^{-1}\left(\frac{u+1}{2}\right), \tag{1.8.4}$$

where Φ is the standard normal distribution function. Note that $\varphi_\Phi^+(u)$ is the inverse cdf (or quantile function) of the absolute value of a standard normal random variable. The normal scores were originally proposed by Fraser (1957).

For the location model (1.2.1), the estimate of θ based on the norm (1.8.1) is the value of θ which minimizes the distance $\|\mathbf{X} - \mathbf{1}\theta\|_{\varphi^+}$ or equivalently solves the equation

$$T_{\varphi^+}(\theta) \doteq 0. \tag{1.8.5}$$

The linear searches discussed in Section 3.7.3, can be used to compute $\widehat{\theta}$.

To determine the corresponding functional, note that we can write $R|X_i - \theta| = \#_j\{\theta - |X_i - \theta| \leqslant X_j \leqslant |X_i - \theta| + \theta\}$. Let H_n denote the empirical distribution function of the sample $X_1, \ldots, X_n$ and let H_n^- denote the left limit of H_n. We can then write the defining equation of $\widehat{\theta}$ as

$$\int \varphi^+(H_n(|x - \theta| + \theta) - H_n^-(\theta - |x - \theta|))\operatorname{sgn}(x - \theta)\, dH_n(x) = 0,$$

which converges to

$$\delta(\theta) = \int_{-\infty}^{\infty} \varphi^+(H(|x - \theta| + \theta) - H(\theta - |x - \theta|))\operatorname{sgn}(x - \theta)\, dH(x) = 0. \tag{1.8.6}$$

For convenience, a second representation of $\delta(\theta)$ can be obtained if we extend $\varphi^+(u)$ to the interval $(-1, 0)$ as follows:

$$\varphi^+(t) = -\varphi^+(t), \quad \text{for } -1 < t < 0. \tag{1.8.7}$$

Using this extension, the functional $\theta = T(H)$ is the solution of

$$\delta(\theta) = \int_{-\infty}^{\infty} \varphi^+(H(x) - H(2\theta - x))\, dH(x). \tag{1.8.8}$$

Compare expressions (1.8.8) and (1.3.26).

The level α test of the hypotheses (1.3.6) based on $T_{\varphi^+}(0)$ is

Reject H_0 in favor of H_A, if $|T_{\varphi^+}(0)| \geqslant c$, (1.8.9)

where c solves $P_0[|T_{\varphi^+}(0)| \geqslant c] = \alpha$. We briefly develop the statistical and robustness properties of this test and the estimator $\widehat{\theta}_{\varphi^+}$ in the next two subsections.

1.8.1 Null Properties of the Test

For this subsection on null properties and the following subsection on efficiency properties of the test (1.8.9), we will assume that the sample $X_1, \ldots, X_n$ follows the symmetric location model, (1.7.3), with common symmetric density function $h(x) = f(x - \theta)$, where $f(x)$ is symmetric about 0. Let $H(x)$ denote the distribution function associated with $h(x)$.

As in Section 1.7.1, we can express $T_{\varphi^+}(0)$ in terms of the anti-ranks as

$$T_{\varphi^+}(0) = \sum a^+(R(|X_i|))\text{sgn}(X_i) = \sum a^+(j)\text{sgn}(X_{D_j}) = \sum a^+(j)W_j; \quad (1.8.10)$$

see the corresponding expression (1.3.20) for the weighted L_1 norm. Recall that under H_0 and the symmetry of $h(x)$, the variables $W_1, \ldots, W_n$ are iid with $P[W_i = 1] = P[W_i = -1] = \frac{1}{2}$, (Lemma 1.7.2). Thus we immediately have that $T_{\varphi^+}(0)$ is distribution-free under H_0 with mean and variance

$$E_0[T_{\varphi^+}(0)] = 0 \quad (1.8.11)$$

$$\text{Var}_0[T_{\varphi^+}(0)] = \sum_{i=1}^{n} a^{+2}(i). \quad (1.8.12)$$

Tables can be constructed for the null distribution of $T_{\varphi^+}(0)$ from which critical values, c, can be obtained to complete the test described in (1.8.9).

For the asymptotic null distribution of $T_{\varphi^+}(0)$, the following additional assumption on the scores will be sufficient:

$$\frac{\max_j a^{+2}(j)}{\sum a^{+2}(i)} \to 0. \quad (1.8.13)$$

Because φ^+ is square integrable, we have

$$\frac{1}{n}\sum a^{+2}(i) \to \sigma_{\varphi^+}^2 = \int_0^1 (\varphi^+(u))^2 \, du, \quad 0 < \sigma_{\varphi^+}^2 < \infty, \quad (1.8.14)$$

i.e. the left-hand side is a Riemann sum of the integral. Under these assumptions and the symmetric location model, Corollary A.1.2 of the Appendix can be used to show that the null distribution of $T_{\varphi^+}(0)$ is asymptotically normal; see also Exercise 1.13.13. Hence, an asymptotic level α test is

$$\text{Reject } H_0 \text{ in favor of } H_A, \text{ if } \left|\frac{T_{\varphi^+}(0)}{\sqrt{n}\sigma_{\varphi^+}}\right| \geq z_{\alpha/2}. \quad (1.8.15)$$

An approximate $(1 - \alpha)100\%$ confidence interval for θ based on the process $T_{\varphi^+}(\theta)$ is the interval $(\widehat{\theta}_{\varphi^+,L}, \widehat{\theta}_{\varphi^+,U})$ such that

$$T_{\varphi^+}(\widehat{\theta}_{\varphi^+,L}) = z_{\alpha/2}\sqrt{n}\sigma_{\varphi^+} \text{ and } T_{\varphi^+}(\widehat{\theta}_{\varphi^+,U}) = -z_{\alpha/2}\sqrt{n}\sigma_{\varphi^+}; \quad (1.8.16)$$

see (1.5.26). These equations can be solved by simple linear searches in the same way as the estimate $\widehat{\theta}_{\varphi^+}$ is obtained; see Section 3.7.3.

1.8.2 Efficiency and Robustness Properties

We derive the efficiency properties of the analysis described above by establishing the four conditions of Definition 1.5.3 to show that the process $T_{\varphi^+}(\theta)$ is Pitman regular. Assume that $\varphi^+(u)$ is differentiable. First define the quantity γ_h as

$$\gamma_h = \int_0^1 \varphi^+(u)\varphi_h^+(u)\,du, \qquad (1.8.17)$$

where

$$\varphi_h^+(u) = -\frac{h'\left(H^{-1}\left(\frac{u+1}{2}\right)\right)}{h\left(H^{-1}\left(\frac{u+1}{2}\right)\right)}. \qquad (1.8.18)$$

As discussed below, $\varphi_h^+(u)$ is called the **optimal score function**. We assume that our scores are such that $\gamma_h > 0$.

Since it is the negative of a gradient of a norm, $T_{\varphi^+}(\theta)$ is nondecreasing in θ; hence, the first condition, (1.5.7), holds. Let $\overline{T}_{\varphi^+}(0) = T_{\varphi^+}(0)/n$ and consider

$$\mu_{\varphi^+}(\theta) = E_\theta[\overline{T}_{\varphi^+}(0)] = E_0[\overline{T}_{\varphi^+}(-\theta)].$$

Note that $\overline{T}_{\varphi^+}(-\theta)$ converges in probability to $\delta(-\theta)$ in (1.8.8). Hence, $\mu_{\varphi^+}(\theta) = \delta(-\theta)$, where in (1.8.8) H is a distribution function with point of symmetry at 0, without loss of generality. If we differentiate $\delta(-\theta)$ and set $\theta = 0$, we get

$$\mu'_{\varphi^+}(0) = 2\int_{-\infty}^{\infty} \varphi^{+\prime}(2H(x)-1)h(x)\,dH(x)$$

$$= 4\int_0^{\infty} \varphi^{+\prime}(2H(x)-1)h^2(x)\,dx = \int_0^1 \varphi^+(u)\varphi_h^+(u)\,du > 0, \qquad (1.8.19)$$

where the third equality in (1.8.19) follows from an integration by parts. Hence, the second Pitman regularity condition holds.

For the third condition, (1.5.9), the asymptotic linearity for the process $T_{\varphi^+}(0)$ is given in Theorem A.2.11 of the Appendix. We restate the result here for reference:

$$P_0\left[\sup_{\sqrt{n}|\theta|\leqslant B}\left|\frac{1}{\sqrt{n}}T_{\varphi^+}(\theta) - \frac{1}{\sqrt{n}}T_{\varphi^+}(0) + \theta\gamma_h\right| \geqslant \varepsilon\right] \to 0, \qquad (1.8.20)$$

for all $\varepsilon > 0$ and all $B > 0$. Finally, the fourth condition, (1.5.10), concerns the asymptotic null distribution which was discussed above. The null variance of $T_{\varphi^+}(0)/\sqrt{n}$ is given by expression (1.8.12). Therefore the process $T_{\varphi^+}(\theta)$ is Pitman regular with efficacy given by

$$c_{\varphi^+} = \frac{\int_0^1 \varphi^+(u)\varphi_h^+(u)\,du}{\sqrt{\int_0^1 (\varphi^+(u))^2\,du}} = \frac{2\int_{-\infty}^{\infty} \varphi^{+\prime}(2H(x)-1)h^2(x)\,dx}{\sqrt{\int_0^1 (\varphi^+(u))^2\,du}}. \tag{1.8.21}$$

As our first result, we obtain the asymptotic power lemma for the process $T_{\varphi^+}(\theta)$. This, of course, follows immediately from Theorem 1.5.9 so we state it as a corollary.

Corollary 1.8.1 *Under the symmetric location model,*

$$P_{\theta_n}\left[\frac{T_{\varphi^+}(0)}{\sqrt{n}\sigma_{\varphi^+}} \geq z_\alpha\right] \to 1 - \Phi(z_\alpha - \theta^* c_{\varphi^+}), \tag{1.8.22}$$

for the sequence of hypotheses

$$H_0: \theta = 0 \text{ versus } H_{An}: \theta = \theta_n = \theta^*/\sqrt{n} \text{ for } \theta^* > 0.$$

Based on Pitman regularity, the asymptotic distribution of the estimate $\widehat{\theta}_{\varphi^+}$ is

$$\sqrt{n}(\widehat{\theta}_{\varphi^+} - \theta) \xrightarrow{D} N(0, \tau_{\varphi^+}^2), \tag{1.8.23}$$

where the scale parameter τ_{φ^+} is defined by the reciprocal of (1.8.21),

$$\tau_{\varphi^+} = c_{\varphi^+}^{-1} = \frac{\sigma_{\varphi^+}}{\int_0^1 \varphi^+(u)\varphi_h^+(u)\,du}. \tag{1.8.24}$$

The asymptotic relative efficiency between two estimates or two tests based on score functions $\varphi_1^+(u)$ and $\varphi_2^+(u)$ is the ratio

$$e(\varphi_1^+, \varphi_2^+) = \frac{c_{\varphi_1^+}^2}{c_{\varphi_2^+}^2} = \frac{\tau_{\varphi_2^+}^2}{\tau_{\varphi_1^+}^2}. \tag{1.8.25}$$

This can be used to compare different tests. For a specific distribution we can determine the optimum scores. Such a score should make the scale parameter τ_{φ^+} as small as possible. This scale parameter can be written as

$$c_{\varphi^+} = \tau_{\varphi^+}^{-1} = \left\{\frac{\int_0^1 \varphi^+(u)\varphi_h^+(u)\,du}{\sigma_{\varphi^+}\sqrt{\int_0^1 \varphi_h^{+2}(u)\,du}}\right\}\sqrt{\int_0^1 \varphi_h^{+2}(u)\,du}. \tag{1.8.26}$$

The quantity in braces is a correlation coefficient; hence, to minimize the scale parameter τ_{φ^+}, we need to maximize the correlation coefficient, which can be accomplished by selecting the **optimal score function** given by

$$\varphi^+(u) = \varphi_h^+(u),$$

where $\varphi_h^+(u)$ is given by expression (1.8.18). The quantity $\sqrt{\int_0^1 (\varphi_h^+(u))^2\,du}$ is the square root of the Fisher information; see Exercise 1.13.18. Therefore for this choice of scores the estimate $\widehat{\theta}_{\varphi^+_h}$ is **asymptotically efficient**. This is the reason for calling the score function φ^+_h the optimal score function.

It is shown in Exercise 1.13.19 that the optimal scores are the normal scores if $h(x)$ is a normal density, the Wilcoxon weighted L_1 scores if $h(x)$ is a logistic density, and the L_1 scores if $h(x)$ is a double exponential density. It is further shown that the scores generated by (1.8.3) are optimal for symmetric densities with a logistic center and exponential tails.

From Exercise 1.13.19, the efficiency of the normal scores methods relative to the least-squares methods is

$$e(\text{NS, LS}) = \left\{\int_{-\infty}^{\infty} \frac{f^2(x)}{\phi(\Phi^{-1}(F(x)))}\, dx\right\}^2, \qquad (1.8.27)$$

where $F \in \mathcal{F}_S$, the family of symmetric distributions with positive finite Fisher information and $\phi = \Phi'$ is the $N(0, 1)$ pdf.

We now prove a result similar to Theorem 1.7.6. We prove that the normal scores methods always have efficiency at least equal to one relative to the least-squares methods. Further, it is only equal to 1 for the normal distribution. The result was first proved by Chernoff and Savage (1958); however, the proof presented below is due to Gastwirth and Wolff (1968).

Theorem 1.8.2 *Let $X_1, \ldots, X_n$ be a random sample from $F \in \mathcal{F}_S$. Then*

$$\inf_{\mathcal{F}_S} e(\text{NS, LS}) = 1, \qquad (1.8.28)$$

and the value 1 is attained only for the normal distribution.

Proof. If $\sigma_f^2 = \infty$ then $e(\text{NS, LS}) > 1$; hence, we suppose that $\sigma_f^2 = 1$. Let $e = e(\text{NS, LS})$. Then from (1.8.27) we can write

$$\sqrt{e} = E\left[\frac{f(X)}{\phi(\Phi^{-1}(F(X)))}\right]$$

$$= E\left[\frac{1}{\phi(\Phi^{-1}(F(X)))/f(X)}\right].$$

Applying Jensen's inequality to the convex function $h(x) = 1/x$, we have

$$\sqrt{e} \geq \frac{1}{E[\phi(\Phi^{-1}(F(X)))/f(X)]}.$$

Hence,

$$\frac{1}{\sqrt{e}} \leq E\left[\frac{\phi(\Phi^{-1}(F(X)))}{f(X)}\right]$$

$$= \int \phi(\Phi^{-1}(F(x)))\, dx.$$

We now integrate by parts, using $u = \phi(\Phi^{-1}(F(x)))$, $du = \phi'(\Phi^{-1}(F(x)))f(x)\,dx/\phi(\Phi^{-1}(F(x))) = -\Phi^{-1}(F(x))f(x)\,dx$ since $\phi'(x)/\phi(x) = -x$. Hence, with $dv = dx$, we have

$$\int_{-\infty}^{\infty} \phi(\Phi^{-1}(F(x)))\,dx = x\phi(\Phi^{-1}(F(x)))\Big|_{-\infty}^{\infty} + \int_{-\infty}^{\infty} x\Phi^{-1}(F(x))f(x)\,dx. \quad (1.8.29)$$

Now transform $x\phi(\Phi^{-1}(F(x)))$ into $F^{-1}(\Phi(w))\phi(w)$ by first letting $t = F(x)$ and then $w = \Phi^{-1}(t)$. The integral $\int F^{-1}(\Phi(w))\phi(w)\,dw = \int xf(x)\,dx < \infty$, hence the limit of the integrand must be 0 as $x \to \pm\infty$. This implies that the first term on the right-hand side of (1.8.29) is 0. Hence, applying the Cauchy–Schwarz inequality,

$$\frac{1}{\sqrt{e}} \leq \int_{-\infty}^{\infty} x\Phi^{-1}(F(x))f(x)\,dx$$

$$= \int_{-\infty}^{\infty} x\sqrt{f(x)}\Phi^{-1}(F(x))\sqrt{f(x)}\,dx$$

$$\leq \left[\int_{-\infty}^{\infty} x^2 f(x)\,dx \int_{-\infty}^{\infty} \{\Phi^{-1}(F(x))\}^2 f(x)\,dx\right]^{1/2}$$

$$= 1,$$

since $\int x^2 f(x)\,dx = 1$ and $\int x^2 \phi(x)\,dx = 1$. Hence $e^{1/2} \geq 1$ and $e \geq 1$, which completes the proof. It should be noted that the inequality is strict except at the normal distribution. Hence the normal scores are strictly more efficient than the least-squares procedures except for the normal model where the asymptotic relative efficiency is 1.

The **influence function** for $\widehat{\theta}_{\varphi+}$ is derived in Section A.5 of the Appendix. It is given by

$$\Omega(t, \widehat{\theta}_{\varphi+}) = \frac{\varphi^+(2H(t) - 1)}{4\int_0^{\infty} \varphi^{+\prime}(2H(x) - 1)h^2(x)\,dx}. \quad (1.8.30)$$

Note, also, that $E[\Omega^2(X, \widehat{\theta}_{\varphi+})] = \tau_{\varphi+}^2$ as a check on the asymptotic distribution of $\widehat{\theta}_{\varphi+}$. Note that the influence function is bounded provided the score function is bounded. Thus the estimates based on the scores discussed in the previous paragraph are all robust except for the normal scores. In the case of the normal scores, when $H(t) = \Phi(t)$, the influence function is $\Omega(t) = \Phi^{-1}(t)$; see Exercise 1.13.20.

The **asymptotic breakdown** of the estimate $\widehat{\theta}_{\varphi+}$ is ε^* given by

$$\int_0^{1-\varepsilon^*} \varphi^+(u)\,du = \tfrac{1}{2}\int_0^1 \varphi^+(u)\,du. \quad (1.8.31)$$

We provide a heuristic argument for (1.8.31); for a rigorous development see Huber (1981). Recall Definition 1.6.1. The idea is to corrupt enough data so that the estimating equation, (1.8.5), no longer has a solution. Suppose that [εn] observations are corrupted, where [·] denotes the greatest integer function. Push the corrupted observations out towards $+\infty$ so that

$$\sum_{i=[(1-\varepsilon)n]+1}^{n} a^{+}(R(|X_i - \theta|))\text{sgn}(X_i - \theta) = \sum_{i=[(1-\varepsilon)n]+1}^{n} a^{+}(i).$$

This restrains the estimating function from crossing the horizontal axis provided

$$-\sum_{i=1}^{[(1-\varepsilon)n]} a^{+}(i) + \sum_{i=[(1-\varepsilon)n]+1}^{n} a^{+}(i) > 0.$$

Replacing the sums by integrals in the limit yields

$$\int_{0}^{1-\varepsilon} \varphi^{+}(u)\,du > \int_{1-\varepsilon}^{1} \varphi^{+}(u)\,du.$$

Now use the fact that

$$\int_{0}^{1-\varepsilon} \varphi^{+}(u)\,du + \int_{1-\varepsilon}^{1} \varphi^{+}(u)\,du = \int_{0}^{1} \varphi^{+}(u)\,du$$

and that we want the smallest possible ε to get (1.8.31).

Example 1.8.1 *Breakdowns of Estimates Based on Wilcoxon and Normal Scores*

For $\widehat{\theta} = \text{med}(X_i + X_j)/2$, $\varphi^{+}(u) = u$ and it follows at once that $\varepsilon^{*} = 1 - (1/\sqrt{2}) \doteq 0.293$. For the estimate based on the normal scores where $\varphi^{+}(u)$ is given by (1.8.4), expression (1.8.31) becomes

$$\exp\left\{-\frac{1}{2}\left[\Phi^{-1}\left(1 - \frac{\varepsilon}{2}\right)\right]^{2}\right\} = \frac{1}{2}$$

and $\varepsilon^{*} = 2(1 - \Phi(\sqrt{\log 4})) \doteq 0.239$. Hence we have the unusual situation that the estimate based on the normal scores has positive breakdown but an unbounded influence curve.

1.9 Tests for Symmetry

We have seen from the statistical properties of the Wilcoxon methods that when we are sampling from a symmetric distribution, we can achieve impressive efficiency gains over the L_1 methods provided that the tailweight of the distribution is not too far from that of a normal distribution. However, it is also clear that the Wilcoxon methods depend on the symmetry for the distribution-free property of the test and confidence interval. Hence, it makes sense to have some way to signal when there are indications in the data that the model may not be symmetric. We know that when the underlying distribution is symmetric the mean and median of the distribution coincide and, thus, when the mean and median are different the distribution is asymmetric. Gastwirth (1971) proposed a simple test based on the L_1 estimating function (the sign statistic) and the L_2 estimate (the sample mean). His test statistic is simply $S_1(\overline{X})$; see Example 1.3.1. We reject the null hypothesis of symmetry when $|S_1(\overline{X})|$ is too large. Certainly, if the underlying distribution is symmetric we would expect $S_1(\overline{X})$ to be close to zero. The following theorem shows that $S_1(\overline{X})$ is not distribution-free, even asymptotically, under the null hypothesis.

Theorem 1.9.1 *If $X_1, \ldots, X_n$ are iid $H(x) = F(x - \theta)$ and F has a density f symmetric about 0 with finite variance σ^2, then*

$$n^{-1/2} S_1(\overline{X}) = n^{-1/2} \sum_{i=1}^{n} \operatorname{sgn}(X_i - \overline{X})$$

$$= n^{-1/2} \sum_{i=1}^{n} \{\operatorname{sgn}(X_i - \theta) - 2f(0)(X_i - \theta)\} + o_P(1)$$

and $n^{-1/2} S_1(\overline{X})$ converges in distribution to $Z \sim N(0, \sigma_b^2)$, where

$$\sigma_b^2 = 1 - 4f(0) \int_{-\infty}^{\infty} |x| f(x) dx + 4f^2(0)\sigma^2.$$

Proof. The representation in the first line follows at once from expression (1.5.22) since $n^{1/2} \overline{X}$ is bounded in probability. Now the central limit theorem applies to establish the limiting normality. Finally, the asymptotic variance follows from computing $E\{\operatorname{sgn}(X - \theta) - 2f(0)(X - \theta)\}^2$.

Note that the asymptotic variance depends on the underlying distribution, and this is why the test is not distribution-free, even asymptotically. Gastwirth discusses the limiting distribution in his 1971 paper, shows that the true significance level of the test depends quite strongly on the form of the underlying distribution, and concludes that the test is disappointing from a practical point of view.

We consider using the bootstrap to assign a *p*-value to the test. This presents an interesting practical problem since we need to resample a symmetrized sample to approximate a *p*-value under the null hypothesis and the sample distribution may be very asymmetric.

The idea is this. To bootstrap a *p*-value for $S_1(\overline{X})$, we take B samples of size n, with replacement, from the $2n$ values $\pm(x_i - \overline{X})$, $i = 1, \ldots, n$. The bootstrap *p*-value is then the fraction of bootstrap values of $|S_1(\overline{X})|$ that exceed the observed value of $|S_1(\overline{X})|$.

Here is the rationale behind this suggestion. Denote by $H_n(x)$ the empirical cdf. If the true cdf $H(x)$ has a symmetric density then we can often improve H_n by using $H_n^*(x) = [H_n(x) + 1 - H_n(2\overline{X} - x)]/2$. See Schuster (1975) for a discussion of the properties of this symmetrized empirical cdf. Indeed, $H_n^*(x)$ is a consistent, asymptotically normal estimate of $H(x)$. To sample from $H_n^*(x)$, we simply take a sample of size n from the $2n$ values $\pm(x_i - \overline{X})$, $i = 1, \ldots, n$. Suppose, on the other hand, that $H(x)$ does not have a symmetric density. Then to bootstrap the *p*-value, we need to resample from an empirical cdf that estimates a cdf in the null hypothesis. Ideally, we would like to pick a least favorable cdf in the null hypothesis (one most difficult to distinguish from the asymmetric alternative) so that the *p*-value is as large as possible. Then we would have a test of the full composite null hypothesis with a controlled significance level. Our approach is not quite so ambitious but does provide a practical way to choose a representative of the null hypothesis that is close to the possibly asymmetric underlying distribution. We use $H_n^*(x)$ and note that it is a consistent estimate of $H^*(x) = [H(x) + 1 - H(2\theta - x)]/2$. The distribution $H^*(x)$ is symmetric about θ and is close to $H(x)$; see the next theorem. We then recommend rejection of the null hypothesis of symmetry when the bootstrap *p*-value, sampling from $H_n^*(x)$, is small, less than 0.05 for example. See also Schuster (1975; 1987) and Schuster and Barker (1987).

Theorem 1.9.2 *Given* $H(x)$, *let* $D(G, H) = \min\{\int [G(x) - H(x)]^2 dx : G \in \mathcal{S}\}$ *where* $\mathcal{S}$ *is the set of all symmetric distributions centered at 0. Then* $H_s(x) = [H(x) + 1 - H(-x)]/2$ *is in* $\mathcal{S}$ *and* $H_s(x)$ *minimizes* $D(G, H)$.

Proof. Consider $\int [H(x) - H_s(x) + H_s(x) - G(x)]^2 dx = \int [H(x) - H_s(x)]^2 dx + 2 \int [H(x) - H_s(x)][H_s(x) - G(x)] dx + \int [H_s(x) - G(x)]^2 dx$. If we can show the cross-product term is 0 then $G(x) = H_s(x)$ makes the third term 0 and must be the solution.

Let $D(x) = H(x) - H_s(x) = [H(x) - 1 - H(-x)]/2$. Then $D(x) = D(-x)$ and we will consider $\int D(x)[H_s(x) - G(x)] dx$. Now for any G in $\mathcal{S}$ we have:

$$\int_{-\infty}^{\infty} D(x)G(x)dx = \int_{-\infty}^{0} D(x)G(x)dx + \int_{0}^{\infty} D(x)G(x)dx$$

$$= \int_{0}^{\infty} D(-x)G(-x)dx + \int_{0}^{\infty} D(x)G(x)dx$$

$$= \int_{0}^{\infty} D(x)[1 - G(x)]dx + \int_{0}^{\infty} D(x)G(x)dx$$

$$= \int_{0}^{\infty} D(x)dx. \qquad (1.9.1)$$

Since this holds for any G in $\mathcal{S}$ and since H_s is also in $\mathcal{S}$, it follows at once that the cross product is 0 and this completes the proof.

As a partial check on this approach, we bootstrapped the test statistic from two symmetric distributions: the normal and the double exponential. Then we compared the bootstrap standard deviation of the test statistic to the asymptotic standard deviation suggested by Theorem 1.9.1. When sampling from the standard normal distribution, we have $\sigma_b^2 = 0.363$. If we sample 30 observations from the normal, the asymptotic variance is $(30)(0.363) = 10.89$ and the asymptotic standard deviation is 3.3. We took three samples of size 30, generated 200 bootstrap samples from each of the samples, computed $S_1(\overline{X})$ 200 times each, and then computed the standard deviation of the 200 values. The three results (to compare to 3.3) were 3.54, 3.23, 3.33, in pretty good agreement with that predicted by the asymptotic theory. Next, we repeated the computations with the double exponential distribution. In this case $\sigma_b^2 = 1$ and so the asymptotic variance is just the sample size 30. The asymptotic standard deviation is then 5.48. The results of three bootstraps (to compare to 5.48) were 6.59, 4.33, 5.42, again in pretty good agreement with the prediction of the asymptotic theory. We then bootstrapped from two asymmetric distributions: the exponential and a chi-square with 2 degrees of freedom, $\chi^2(2)$. We again took 200 bootstrap samples and computed the p-values. For the sample taken from the exponential distribution, we found $S_1(\overline{X}) = -10$ and, of the 200 bootstrap values of $S_1(\overline{X})$, we found only one value less than -10. This indicates a very small p-value and we would reject the hypothesis of symmetry. For the sample taken from the $\chi^2(2)$ distribution, we found $S_1(\overline{X}) = -12$ and there were no bootstrap values less than -12. Again, the p-value would be quite small and we easily reject the hypothesis of symmetry.

Boos (1982) proposed a test for asymmetry essentially based on comparing the two pieces of the weighted L_1 norm used to define the Wilcoxon methods,

$$\|\mathbf{x} - \mathbf{1}\theta\| = \sum_{i \leqslant j} \left| \frac{x_i + x_j}{2} - \theta \right| + \sum_{i \leqslant j} \left| \frac{x_i - x_j}{2} \right|;$$

see (1.3.21). He replaces θ by the Hodges–Lehmann estimate and his test statistic is a linear function of the ratio of the first to second terms given by

$$T_n = n \left(-1 + \frac{\sum\sum_{i<j} 0.5(X_i + X_j - 2\widehat{\theta})}{\sum\sum_{i<j} |X_i - X_j|} \right). \tag{1.9.2}$$

The idea is that if X_1 and X_2 are symmetrically distributed about θ then $X_1 + X_2 - 2\theta$ and $X_1 - X_2$ have the same distribution. The test statistic is not distribution-free, the asymptotic distribution is not easy to compute, and it is not robust (see Chapter 3). However, Boos shows that the asymptotic distribution does not change much if the logistic distribution is assumed. Table 1 in that reference provides estimated critical values for the test. For virtually any n and symmetric distributions, the 0.05 critical value is 0.90. Finally, the test is shown to have good power to detect asymmetry.

Randles et al. (1980) proposed an asymptotically distribution-free test of asymmetry. Their test is based on triples (X_i, X_j, X_k) and compares pairwise averages to the third observation through $\text{sgn}[(X_i + X_j)/2 - X_k]$ for all arrangements of i, j, and k. They show that their test is superior to tests based on sample skewness. Finally, an interesting approach based on rank-score functions which come from a complete orthonormal basis of $L_2(0, 1)$ is developed by Eubank, LaRiccia, and Rosenstein (1992).

1.10 Ranked-Set Sampling

In this section we discuss an alternative to simple random sampling (SRS) called ranked-set sampling (RSS). This method of data collection is useful when measurements are destructive or expensive while ranking of the data is relatively easy. Johnson et al. (1996) give an interesting application to environmental sampling. As a simple example consider the problem of estimating the mean volume of trees in a forest. To measure the volume, we must destroy the tree. On the other hand, an expert may well be able to rank the trees by volume in a small sample. The idea is to take a sample of size k of trees and ask the expert to pick the one with smallest volume. This tree is cut down and the volume measured and the other $k-1$ trees are returned to the population for possible future selection. Then a new sample of size k is taken and the expert identifies the second smallest which is then cut down and measured. This is repeated until we have k measurements, having looked at k^2 trees. This ends cycle 1. The measurements are represented as $x_{(1)1} \leq \ldots \leq x_{(k)1}$, where the number in parentheses indicates an order statistic and the second number indicates the cycle. We repeat the process for n cycles to get nk measurements:

$$x_{(1)1}, \ldots, x_{(1)n} \quad \text{iid} \quad h_{(1)}(t)$$
$$x_{(2)1}, \ldots, x_{(2)n} \quad \text{iid} \quad h_{(2)}(t)$$
$$\vdots \quad \vdots \quad \vdots$$
$$x_{(k)1}, \ldots, x_{(k)n} \quad \text{iid} \quad h_{(k)}(t).$$

It is important to note that all nk measurements are independent but are identically distributed only within each row. The density function $h_{(j)}(t)$ represents the pdf of the jth order statistic from a sample of size k and is given by:

$$h_{(j)}(t) = \frac{k!}{(j-1)!(k-j)!} H^{j-1}(t)[1-H(t)]^{k-j} h(t).$$

We suppose the measurements are distributed as $H(x) = F(x-\theta)$ and we wish to make a statistical inference concerning θ, such as an estimate, test, or confidence interval. We will illustrate the ideas on the L_1 methods since they are simple to work with. We also wish to compute the efficiency of the RSS L_1 methods relative to the SRS L_1 methods. We will see that there is a substantial increase in efficiency when using the RSS design. In particular, we will compare the RSS methods to SRS methods based on a sample of size nk. The RSS method was first applied by McIntyre (1952) in measuring mean pasture yields. See Hettmansperger (1995) for a development of the RSS L_1 methods. The most convenient form of the RSS sign statistic is the number of positive measurements, given by

$$S_{RSS}^+ = \sum_{j=1}^{k} \sum_{i=1}^{n} I(X_{(j)i} > 0). \quad (1.10.1)$$

Now note that S_{RSS}^+ can be written as $S_{RSS}^+ = \sum S_{(j)}^+$, where $S_{(j)}^+ = \sum_i I(X_{(j)i} > 0)$ has a binomial distribution with parameters n and $1 - H_{(j)}(0)$. Further, $S_{(j)}^+, j = 1, \ldots, k$, are stochastically independent. It follows at once that

$$ES_{RSS}^+ = n \sum_{j=1}^{k}(1 - H_{(j)}(0)) \quad (1.10.2)$$

$$\text{Var} S_{RSS}^+ = n \sum_{j=1}^{k}(1 - H_{(j)}(0))H_{(j)}(0).$$

54 One-Sample Problems

With k fixed and $n \to \infty$, it follows from the independence of $S_{(j)}^+, j = 1, \ldots, k$, that

$$(nk)^{-1/2} \left\{ S_{RSS}^+ - n \sum_{j=1}^{k} (1 - H_{(j)}(0)) \right\} \xrightarrow{D} Z \sim n(0, \xi^2), \tag{1.10.3}$$

and the asymptotic variance is given by

$$\xi^2 = k^{-1} \sum_{j=1}^{k} [1 - H_{(j)}(0)] H_{(j)}(0) = \tfrac{1}{4} - k^{-1} \sum_{j=1}^{k} (H_{(j)}(0) - \tfrac{1}{2})^2. \tag{1.10.4}$$

It is convenient to introduce a parameter $\delta^2 = 1 - (4/k) \sum (H_{(j)}(0) - 1/2)^2$; then $\xi^2 = \delta^2/4$. The reader is asked to prove the second equality above in Exercise 1.13.21. Using the formulas for the pdfs of the order statistics it is straightforward to verify that

$$h(t) = k^{-1} \sum_{j=1}^{k} h_{(j)}(t) \text{ and } H(t) = k^{-1} \sum_{j=1}^{k} H_{(j)}(t).$$

We now consider testing $H_0 : \theta = 0$ versus $H_A : \theta \neq 0$. The following theorem provides the mean and variance of the RSS sign statistic under the null hypothesis.

Theorem 1.10.1 *Under the assumption that $H_0 : \theta = 0$ is true, $F(0) = \tfrac{1}{2}$,*

$$F_{(j)}(0) = \frac{k!}{(j-1)!(k-j)!} \int_0^{1/2} u^{j-1}(1-u)^{k-j} du$$

and

$$ES_{RSS}^+ = nk/2, \text{ and } VarS_{RSS}^+ = \tfrac{1}{4} - k^{-1} \sum (F_{(j)}(0) - \tfrac{1}{2})^2.$$

Proof. Use the fact that $k^{-1} \sum F_{(j)}(0) = F(0) = \tfrac{1}{2}$, and the expectation formula follows at once. Note that

$$F_{(j)}(0) = \frac{k!}{(j-1)!(k-j)!} \int_{-\infty}^{0} F(t)^{j-1} (1 - F(t))^{k-j} f(t) dt,$$

and then make the change of variable $u = F(t)$.

The variance of S_{RSS}^+ does not depend on H, as expected; however, its computation requires the evaluation of the incomplete beta integral. Table 1.10.1 provides the values of $F_{(j)}(0)$, under $H_0 : \theta = 0$. The bottom line of the table provides the values of $\delta^2 = 1 - (4/k) \sum (F_{(j)}(0) - \tfrac{1}{2})^2$, an important parameter in assessing the gain of RSS over SRS.

We will compare the SRS sign statistic S_{SRS}^+ based on a sample of nk to the RSS sign statistic S_{RSS}^+. Note that the variance of S_{SRS}^+ is $nk/4$. Then the ratio of variances is $VarS_{RSS}^+/VarS_{SRS}^+ = \delta^2 = 1 - (4/k) \sum (F_{(j)}(0) - \tfrac{1}{2})^2$. The reduction in variance is given in the last row of Table 1.10.1 and can be quite large.

Table 1.10.1: Values of $F_{(j)}(0)$, $j = 1, \ldots, k$ and $\delta^2 = 1 - (4/k)\sum(F_{(j)}(0) - \tfrac{1}{2})^2$.

k:	2	3	4	5	6	7	8	9	10
1	0.750	0.875	0.938	0.969	0.984	0.992	0.996	0.998	0.999
2	0.250	0.500	0.688	0.813	0.891	0.938	0.965	0.981	0.989
3		0.125	0.313	0.500	0.656	0.773	0.856	0.910	0.945
4			0.063	0.188	0.344	0.500	0.637	0.746	0.828
5				0.031	0.109	0.227	0.363	0.500	0.623
6					0.016	0.063	0.145	0.254	0.377
7						0.008	0.035	0.090	0.172
8							0.004	0.020	0.055
9								0.002	0.011
10									0.001
δ^2	0.750	0.625	0.547	0.490	0.451	0.416	0.393	0.371	0.352

We next show that the parameter δ is an integral part of the efficacy of the RSS L_1 methods. It is straightforward using the methods of Section 1.5 and Example 1.5.2 to show that the RSS L_1 estimating function is Pitman regular. To compute the efficacy we first note that

$$\bar{S}_{RSS} = (nk)^{-1} \sum_{j=1}^{k} \sum_{i=1}^{n} \text{sgn}(X_{(j)i}) = (nk)^{-1}[2S_{RSS}^+ - nk].$$

We then have at once that

$$(nk)^{-1/2}\bar{S}_{RSS} \xrightarrow{D_0} Z \sim N(0, \delta^2), \qquad (1.10.5)$$

and $\mu'(0) = 2f(0)$; see Exercise 1.13.22. See Babu and Koti (1996) for a development of the exact distribution. Hence, the efficacy of the RSS L_1 methods is given by

$$c_{RSS} = \frac{2f(0)}{\delta} = \frac{2f(0)}{\left\{1 - (4/k)\sum_{j=1}^{k}(F_{(j)}(0) - \tfrac{1}{2})^2\right\}^{1/2}}.$$

We now summarize the inference methods and their efficiency in the following:

1. *The test.* Reject $H_0: \theta = 0$ in favor of $H_A: \theta > 0$ at significance level α if $S_{SRS}^+ > (nk/2) - z_\alpha \delta(nk/4)^{1/2}$ where, as usual, $1 - \Phi(z_\alpha) = \alpha$.
2. *The estimate.* $(nk)^{1/2}\{\text{med}X_{(j)i} - \theta\} \xrightarrow{D} Z \sim N(0, \delta^2/4f^2(0))$.
3. *The confidence interval.* Let $X_{(1)}^*, \ldots, X_{(nk)}^*$ be the ordered values of $X_{(j)i}$, $j = 1, \ldots, k$ and $i = 1, \ldots, n$. Then $[X_{(m+1)}^*, X_{(nk-m)}^*]$ is a $(1-\alpha)100\%$ confidence interval for θ where $P(S_{SRS}^+ \leq m) = \alpha/2$. Using the normal approximation we have $m \doteq (nk/2) - z_{\alpha/2}\delta(nk/4)^{1/2}$.
4. *Efficiency.* The efficiency of the RSS methods with respect to the SRS methods is given by $e(\text{RSS}, \text{SRS}) = c_{RSS}^2/c_{SRS}^2 = \delta^{-2}$. Hence, the reciprocal of the last line of Table 1.10.1 provides the efficiency values, which can be

quite substantial. Recall from the discussion following Definition 1.5.5 that efficiency can be interpreted as the ratio of sample sizes needed to achieve the same approximate variances, the same approximate local power, and the same confidence interval length. Hence, we write $(nk)_{RSS} \doteq \delta^2 (nk)_{SRS}$. This is really the point of the RSS design. Returning to the example of estimating the volume of wood in a forest, if we let $k = 5$, then, from Table 1.10.1, we would need to destroy and measure only about half as many trees using the RSS method compared to the SRS method.

As a final note, we mention the problem of assessing the effect of imperfect ranking. Suppose that the expert makes a mistake when asked to identify the jth ordered value in a set of k observations. As expected, there is less gain from using the RSS method. The interesting point is that if the expert simply identifies the supposed jth ordered value by random guess then $\delta^2 = 1$ and the two sign tests have the same information; see Hettmansperger (1995) for more detail.

1.11 Interpolated Confidence Intervals for the L_1 Inference

When we construct L_1 confidence intervals, we are limited in our choice of confidence coefficients because of the discreteness of the binomial distribution. The effect does not wear off very quickly as the sample size increases. For example, with a sample of size 50, we can have either a 93.5% or a 96.7% confidence interval, and that is as close as we can come to 95%. In the following discussion we provide a method to interpolate between confidence intervals. The method is nonlinear and seems to be essentially distribution-free. We will begin by presenting and illustrating the method and then derive its properties.

Suppose γ is the desired confidence coefficient. Further, suppose the following intervals are available from the binomial table: interval $(x_{(k)}, x_{(n-k+1)})$ with confidence coefficient γ_k and interval $(x_{(k+1)}, x_{(n-k)})$ with confidence coefficient γ_{k+1}, where $\gamma_{k+1} \leq \gamma \leq \gamma_k$. Then the interpolated interval is $[\hat{\theta}_L, \hat{\theta}_U]$,

$$\hat{\theta}_L = (1-\lambda)x_{(k)} + \lambda x_{(k+1)} \text{ and } \hat{\theta}_U = (1-\lambda)x_{(n-k+1)} + \lambda x_{(n-k)}, \quad (1.11.1)$$

where

$$\lambda = \frac{(n-k)I}{k + (n-2k)I} \text{ and } I = \frac{\gamma_k - \gamma}{\gamma_k - \gamma_{k+1}}. \quad (1.11.2)$$

We call I the interpolation factor and note that if we were using linear interpolation then $\lambda = I$. Hence, we see that the interpolation is distinctly nonlinear.

As a simple example we take $n = 10$ and ask for a 95% confidence interval. For $k = 2$ we find $\gamma_k = 0.9786$ and $\gamma_{k+1} = 0.8907$. Then $I = 0.325$ and $\lambda = 0.685$. Hence, $\hat{\theta}_L = 0.342x_{(2)} + 0.658x_{(3)}$ and $\hat{\theta}_U = 0.342x_{(9)} + 0.658x_{(8)}$. Note that linear interpolation is almost the reverse of the recommended mixtures, namely $\lambda = I = 0.325$, and this can make a substantial difference in small samples.

The method is based on the following theorem. This theorem highlights the nonlinear relationship between the interpolation factor and λ. After proving the theorem we will need to develop an approximate solution and then show that it works in practice.

Theorem 1.11.1 *The interpolation factor I is given by*

$$I = \frac{\gamma_k - \gamma}{\gamma_k - \gamma_{k+1}} = 1 - (n-k)2^n \int_0^\infty F^k\left(\frac{-\lambda}{1-\lambda}y\right)(1 - F(y))^{n-k-1}f(y)dy.$$

Proof. Without loss of generality, we will assume that θ is 0. Then we can write:

$$\gamma_k = P_0(x_k \leqslant 0 \leqslant x_{n-k+1}) = P_0(k-1 < S_1^+(0) < n-k-1)$$

and

$$\gamma_{k+1} = P_0(x_{k+1} \leqslant 0 \leqslant x_{n-k}) = P_0(k < S_1^+(0) < n-k).$$

Taking the difference, we have, using $\binom{n}{k}$ to denote the binomial coefficient,

$$\gamma_k - \gamma_{k+1} = P_0(S_1^+(0) = k) + P_0(S_1^+(0) = n-k) = \binom{n}{k}(1/2)^{n-1}. \quad (1.11.3)$$

We now consider the lower tail probability associated with the confidence interval. First consider

$$P_0(X_{k+1} > 0) = \frac{1 - \gamma_{k+1}}{2} = \int_0^\infty \frac{n!}{k!(n-k-1)!} F^k(t)(1 - F(t))^{n-k-1} dF(t) \quad (1.11.4)$$

$$= P_0(S_1^+(0) \geqslant n-k) = P_0(S_1^+(0) \leqslant k).$$

We next consider the lower end of the interpolated interval:

$$\frac{1-\gamma}{2} = P_0((1-\gamma)X_k + \lambda X_{k+1} > 0)$$

$$= \int_0^\infty \int_{(-\lambda/(1-\lambda))y}^y \frac{n!}{(k-1)!(n-k-1)!} F^{k-1}(x)(1 - F(y))^{n-k-1} f(x)f(y)dxdy$$

$$= \int_0^\infty \frac{n!}{(k-1)!(n-k-1)!} \frac{1}{k}\left[F^k(y) - F^k\left(\frac{-\lambda y}{1-\lambda}\right)\right](1 - F(y))^{n-k-1}f(y)dy$$

$$= \frac{1 - \gamma_{k+1}}{2} - \int_0^\infty \frac{n!}{k!(n-k-1)!} F^k\left(\frac{-\lambda y}{1-\lambda}\right)(1 - F(y))^{n-k-1}f(y)dy. \quad (1.11.5)$$

Use (1.11.4) in the last line above. Now with (1.11.3), substitute into the formula for the interpolation factor and the result follows.

Clearly, the relationship between I and λ is not only nonlinear but also depends on the underlying distribution F. Hence, the interpolated interval is not distribution-free. There is one interesting case in which we have a distribution-free interval given in the following corollary.

Corollary 1.11.2 *Suppose F is the cdf of a symmetric distribution. Then $I(\frac{1}{2}) = k/n$, where we write $I(\lambda)$ to denote the dependence of the interpolation factor on λ.*

This shows that when we sample from a symmetric distribution, the interval that lies halfway between the available intervals does not depend on the underlying distribution. Other interpolated intervals are not distribution-free. Our next theorem shows how to approximate the solution, and the solution is essentially distribution-free. We show by example that the approximate solution works in many cases.

Theorem 1.11.3

$$I(\lambda) \doteq \lambda k/(\lambda(2k - n) + n - k).$$

Proof. We consider the integral

$$\int_0^\infty F^k\left(\frac{-\lambda}{1-\lambda}y\right)(1 - F(y))^{n-k-1}f(y)dy.$$

The integrand decreases rapidly for moderate powers; hence, we expand the integrand around $y = 0$. First take logarithms; then

$$k \log F\left(\frac{-\lambda}{1-\lambda}y\right) = k \log F(0) - \frac{\lambda}{1-\lambda} k \frac{f(0)}{F(0)} y + o(y)$$

and

$$(n-k-1)\log(1-F(y)) = (n-k-1)\log(1-F(0)) - (n-k-1)\frac{f(0)}{1-F(0)} y + o(y).$$

Substitute $r = \lambda k/(1 - \lambda)$ and $F(0) = 1 - F(0) = \frac{1}{2}$ into the above equations, and add the two equations together. Add and subtract $r \log(\frac{1}{2})$, and group terms so the right-hand side of the second equation appears on the right-hand side along with $k \log(\frac{1}{2}) - r \log(\frac{1}{2})$. Hence, we have

$$k \log F\left(\frac{-\lambda}{1-\lambda}y\right) + (n - k - 1)\log(1 - F(y)) = k \log(\frac{1}{2}) - r \log(\frac{1}{2})$$
$$+ (n - r - k - 1)\log(1 - F(y)) + o(y),$$

and, hence,

$$\int_0^\infty F^k\left(\frac{-\lambda}{1-\lambda}y\right)(1 - F(y))^{n-k-1}f(y)dy \doteq \int_0^\infty 2^{-(k-r)}(1 - F(y))^{n+r-k-1}f(y)dy$$

$$= \frac{1}{2^n(n + r - k)}. \qquad (1.11.6)$$

Substitute this approximation into the formula for $I(\lambda)$, use $r = \lambda k/(1 - \lambda)$ and the result follows.

Note that the approximation agrees with Corollary 1.11.2 In addition Exercise 1.13.23 shows that the approximation formula is exact for the double

Table 1.11.1: Confidence coefficients for interpolated confidence intervals in Example 1.11.1. DE(Approx) = double exponential and the approximation in Theorem 1.11.3, U = Uniform, N = Normal, C = Cauchy, Linear = linear interpolation

λ	DE(Approx)	U	N	C	Linear
0.1	0.976	0.977	0.976	0.976	0.970
0.2	0.973	0.974	0.974	0.974	0.961
0.3	0.970	0.971	0.971	0.970	0.952
0.4	0.966	0.967	0.966	0.966	0.943
0.5	0.961	0.961	0.961	0.961	0.935
0.6	0.955	0.954	0.954	0.954	0.926
0.7	0.946	0.944	0.944	0.946	0.917
0.8	0.935	0.930	0.931	0.934	0.908
0.9	0.918	0.912	0.914	0.918	0.899

exponential (Laplace) distribution. In Table 1.11.1 we show how well the approximation works for several other distributions. The exact results were obtained by numerical integration of the integral in Theorem 1.11.1. Similar close results were found for asymmetric examples. For further reading see Hettmansperger and Sheather (1986) and Nyblom (1992).

Example 1.11.1 *Cushney–Peebles Example 1.4.1, continued*

We now return to the Cushney and Peebles data in order to illustrate the sign test and the L_1 interpolated confidence interval. Again we use Minitab for the computations. We take as our location model $X_1, \ldots, X_{10}$ iid from $H(x) = F(x - \theta)$, F and θ both unknown, along with the L_1 norm. We have already seen that the estimate of θ is the sample median equal to 1.3. Suppose we wish to test $H_0 : \theta = 0$ versus $H_A : \theta \neq 0$ and accompany the test with a confidence interval. The Minitab commands are STEST and SINT. The output for a test and a 95% interpolated confidence interval is:

```
SIGN TEST OF MEDIAN = 0.00000 VERSUS N.E. 0.00000

          N     BELOW    EQUAL    ABOVE    P-VALUE    MEDIAN
Diff     10       0        1        9       0.0039     1.300

SIGN CONFIDENCE INTERVAL FOR MEDIAN

                         ACHIEVED
         N     MEDIAN   CONFIDENCE    CONFIDENCE INTERVAL    POSITION
C3      10     1.300      0.8906      (1.000,   1.800)           3
                          0.9500      (0.932,   2.005)         NLI
                          0.9785      (0.800,   2.400)           2
```

Note the *p*-value of the test is 0.0039 and we would easily reject the null hypothesis at any reasonable level of significance. The interpolated 95% confidence interval for θ shows the reasonable set of values of θ to be between 0.932 and 2.005, given the level of confidence.

1.12 Two-Sample Analysis

We now propose a simple way to extend our one-sample methods to the comparison of two samples. Suppose $X_1, \ldots, X_m$ are iid $F(x - \theta_x)$ and $Y_1, \ldots, Y_n$ are iid $F(y - \theta_y)$ and the two samples are independent. Let $\Delta = \theta_y - \theta_x$, and we wish to test the null hypothesis $H_0 : \Delta = 0$ versus the alternative hypothesis $H_A : \Delta \neq 0$. Without loss of generality, we can consider $\theta_x = 0$ so that the X sample is from a distribution with cdf $F(x)$ and the Y sample is from a distribution with cdf $F(y - \Delta)$.

The hypothesis testing rule that we propose is:

1. Construct L_1 confidence intervals $[X_L, X_U]$ and $[Y_L, Y_U]$.
2. Reject H_0 if the intervals are disjoint.

If we consider the confidence interval as a set of reasonable values for the parameter, given the confidence coefficient, then we reject the null hypothesis when the respective reasonable values are disjoint. We must determine the significance level for the test. In particular, for given γ_x and γ_y, what is the value of α_c, the significance level for the comparison? Perhaps more pertinent: given α_c, what values should we choose for γ_x and γ_y? Below we show that for a broad range of sample sizes,

comparing two 84% CIs yields a 5% test of $H_0 : \Delta = 0$ versus $H_A : \Delta \neq 0$,

(1.12.1)

where CI denotes confidence interval. In the following theorem we provide the relationship between α_c and the pair γ_x, γ_y. Define z_x by $\gamma_x = 2\Phi(z_x) - 1$ and likewise z_y by $\gamma_y = 2\Phi(z_y) - 1$.

Theorem 1.12.1 *Suppose $m, n \to \infty$ so that $m/N \to \lambda$, $0 < \lambda < 1$, $N = m + n$. Then under the null hypothesis $H_0 : \Delta = 0$,*

$$\alpha_c = P(X_L > Y_U) + P(Y_L > X_U) \to 2\Phi[-(1-\lambda)^{1/2} z_x - \lambda^{1/2} z_y].$$

Proof. We will consider $\alpha_c/2 = P(X_L > Y_U)$. From (1.5.22) we have

$$X_L \doteq \frac{S_x(0)}{m 2 f(0)} - \frac{z_x}{m^{1/2} 2 f(0)} \quad \text{and} \quad Y_U \doteq \frac{S_y(0)}{n 2 f(0)} - \frac{z_y}{n^{1/2} 2 f(0)}.$$

Since $m/N \to \lambda$,

$$N^{1/2} X_L \xrightarrow{D} \lambda^{-1/2} Z_1, \quad Z_1 \sim N(-z_x/2f(0), 1/4f^2(0)),$$

and

$$N^{1/2} Y_U \xrightarrow{D} (1-\lambda)^{-1/2} Z_2, \quad Z_2 \sim N(-z_y/2f(0), 1/4f^2(0)).$$

Now $\alpha_c/2 = P(X_L > Y_U) = P(N^{1/2}(Y_U - X_L) < 0)$ and X_L, Y_U are independent, hence

$$N^{1/2}(Y_U - X_L) \xrightarrow{D} \lambda^{-1/2} Z_1 - (1-\lambda)^{-1/2} Z,$$

and

$$\lambda^{-1/2}Z_1 - (1-\lambda)^{-1/2}Z_2 \sim N\left(\frac{1}{2f(0)}\left\{\frac{z_x}{(1-\lambda)^{1/2}} + \frac{z_y}{\lambda^{1/2}}\right\}, \frac{1}{4f^2(0)}\left\{\frac{1}{\lambda} + \frac{1}{1-\lambda}\right\}\right).$$

It then follows that

$$P(N^{1/2}(Y_U - X_L) < 0) \to \Phi\left(-\left\{\frac{z_x}{(1-\lambda)^{1/2}} + \frac{z_y}{\lambda^{1/2}}\right\}\bigg/\left\{\frac{1}{\lambda(1-\lambda)}\right\}^{1/2}\right),$$

which, when simplified, yields the result in the statement of the theorem.

To illustrate, we take equal sample sizes so that $\lambda = \frac{1}{2}$ and we take $z_x = z_y = 2$. Then we have two 95% confidence intervals and we will reject the null hypothesis $H_0 : \Delta = 0$ if the two intervals are disjoint. The above theorem says that the significance level is approximately equal to $\alpha_c = 2\Phi(-2.83) = 0.0046$. This is a very small level and it will be difficult to reject the null hypothesis. We might prefer a significance level of say $\alpha_c = 0.05$. We then must find z_x and z_y so that $0.05 = 2\Phi(-(0.5)^{1/2}(z_x + z_y))$. Note that now we have an infinite number of solutions. If we impose the reasonable condition that the two confidence coefficients are the same then we require that $z_x = z_y = z$. Then we have the equation $0.025 = \Phi(-(2)^{1/2}z)$ and hence $-2 = -(2)^{1/2}z$. So $z = 2^{1/2} = 1.41$ and the confidence coefficient for the two intervals is $\gamma = \gamma_x = \gamma_y = 2\Phi(1.41) - 1 = 0.84$. Hence, if we have equal sample sizes and we use two 84% confidence intervals then we have a 5% two-sided comparison of the two samples.

If we set $\alpha_c = 0.10$, this would correspond to a 5% one-sided test. This means that we compare the two confidence intervals in the direction specified by the alternative hypothesis. For example, if we specify $\Delta = \theta_y - \theta_x > 0$, then we would reject the null hypothesis if the X interval is completely below the Y interval. To determine which confidence intervals to use we again assume that the two intervals will have the same confidence coefficient. Then we must find z such that $0.05 = \Phi(-(2)^{1/2}z)$, and this leads to $-1.645 = -(2)^{1/2}z$ and $z = 1.16$. Hence, the confidence coefficient for the two intervals is $\gamma = \gamma_x = \gamma_y = 2\Phi(1.16) - 1 = 0.75$. Hence, for a one-sided 5% test or a 10% two-sided test, when you have equal sample sizes, use two 75% confidence intervals.

We must now consider what to do if the sample sizes are not equal. Let z_c be determined by $\alpha_c/2 = \Phi(-z_c)$; then, again if we use the same confidence coefficient for the two intervals, $z = z_x = z_y = z_c/(\lambda^{1/2} + (1-\lambda)^{1/2})$. When $m = n$ so that $\lambda = 1 - \lambda = 0.5$, we have $z = z_c/2^{1/2} = 0.707z_c$ and so $z = 1.39$ when $\alpha_c = 0.05$. We now show by example that when $\alpha_c = 0.05$, z is not sensitive to the value of λ. Table 1.12.1 gives the relevant information. Hence,

Table 1.12.1: Confidence coefficients for 5% comparison

$\lambda = m/N$	0.500	0.550	0.600	0.650	0.750
m/n	1.00	1.22	1.50	1.86	3.00
$z_x = z_y$	1.39	1.39	1.39	1.40	1.43
$\gamma_x = \gamma_y$	0.84	0.84	0.84	0.85	0.86

if we use 84% confidence intervals, then the significance level will be roughly 5% for the comparison for a broad range of ratios of sample sizes. Likewise, we would use 75% intervals for a 10% comparison. See Hettmansperger (1984b) for additional discussion.

Next suppose that we want a confidence interval for $\Delta = \theta_y - \theta_x$. In the following simple theorem we show that the proposed test based on comparing two confidence intervals is equivalent to checking to see if zero is contained in a different confidence interval. This new interval will be a confidence interval for Δ.

Theorem 1.12.2 $[X_L, X_U]$ and $[Y_L, Y_U]$ are disjoint if and only if 0 is not contained in $[Y_L - X_U, Y_U - X_L]$.

If we specify our significance level to be α_c then we have immediately that

$$1 - \alpha_c = P_\Delta(Y_L - X_U \leq \Delta \leq Y_U - X_L)$$

and $[Y_L - X_U, Y_U - X_L]$ is a $\gamma_c = 1 - \alpha_c$ **confidence interval** for Δ.

This theorem simply points out that the hypothesis test can be equivalently based on a single confidence interval. Hence, two 84% intervals produce a roughly 95% confidence interval for Δ. The confidence interval is easy to construct since we need only find the least and greatest differences of the endpoints between the respective Y and X intervals.

Recall that one way to measure the efficiency of a confidence interval is to find its asymptotic length. This is directly related to the Pitman efficacy of the procedure; see Section 1.5.5. This would seem to be the most natural way to study the efficiency of the test based on confidence intervals. In the following theorem we determine the asymptotic length of the interval for Δ.

Theorem 1.12.3 Suppose $m, n \to \infty$ in such a way that $m/N \to \lambda$, $0 < \lambda < 1$, $N = m + n$. Further, suppose that $\gamma_c = 2\Phi(z_c) - 1$. Let Λ be the length of $[Y_L - X_U, Y_U - X_L]$. Then

$$\frac{N^{1/2}\Lambda}{2z_c} \to \frac{1}{[\lambda(1-\lambda)]^{1/2} 2f(0)}.$$

Proof. First note that $\Lambda = \Lambda_x + \Lambda_y$, the sum of the two lengths of the X and Y intervals, respectively. Further,

$$N^{1/2}\Lambda = \frac{N^{1/2}}{n^{1/2}} n^{1/2}\Lambda_y + \frac{N^{1/2}}{m^{1/2}} m^{1/2}\Lambda_x.$$

But by Theorem 1.5.10 this converges in probability to $z_x/\lambda^{1/2} + z_y/(1-\lambda)^{1/2}$. Now note that $(1-\lambda)^{1/2} z_x + \lambda^{1/2} z_y = z_c$ and the result follows.

The interesting point about this theorem is that the efficiency of the interval does not depend on how z_x and z_y are chosen so long as they satisfy $(1-\lambda)^{1/2} z_x + \lambda^{1/2} z_y = z_c$. In addition, this interval has inherited the efficacy of the L_1 interval in the one-sample location model. We will discuss the two-sample location model in detail in the next chapter. In Hettmansperger (1984b) other choices for z_x and z_y are discussed; for example, we could choose

z_x and z_y so that the asymptotic standardized lengths are equal. The corresponding confidence coefficients for this choice are more sensitive to unequal sample sizes than the method proposed here.

Example 1.12.1 *Hendy and Charles Coin Data*

Hendy and Charles (1970) study the change in silver content in Byzantine coins. During the reign of Manuel I (1143–1180) there were several mintings. We consider the research hypothesis that the silver content changed from the first to the fourth coinage. The data consists in 9 coins identified from the first coinage and 7 coins from the fourth. We suppose that they are realizations of random samples of coins from the two populations. Let $\Delta = \theta_1 - \theta_4$ where the 1 and 4 indicate the coinage. To test the null hypothesis $H_0 : \Delta = 0$ versus $H_A : \Delta \neq 0$ at $\alpha = 0.05$, we construct two 84% L_1 confidence intervals and reject the null hypothesis if they are disjoint. The confidence intervals can be found using the Minitab command SINT and specifying an 84% confidence coefficient. The percentage of silver in each coin is given in Table 1.12.2. The interpolated confidence intervals are:

SIGN CONFIDENCE INTERVAL FOR MEDIAN

	N	MEDIAN	ACHIEVED CONFIDENCE	CONFIDENCE INTERVAL	POSITION
First	9	6.800	0.8203	(6.400, 7.000)	3
			0.8400	(6.391, 7.009)	NLI
			0.9609	(6.200, 7.200)	2
Fourth	7	5.600	0.5469	(5.500, 5.800)	3
			0.8400	(5.346, 5.800)	NLI
			0.8750	(5.300, 5.800)	2

Clearly, the 84% confidence intervals are disjoint, hence, we reject the null hypothesis at a 5% significance level and claim that the emperor apparently held back a little on the fourth coinage. A 95% confidence interval for $\Delta = \theta_1 - \theta_4$ is found by taking the differences in the ends of the confidence intervals: $(6.391 - 5.8, 7.009 - 5.346) = (0.591, 1.663)$. Hence, this analysis suggests that the difference in median percentages is someplace between 0.6% and 1.7%, with a point estimate of $6.8 - 5.6 = 1.2\%$.

A more effective presentation provides a plot of the 84% confidence intervals shown in the boxplots. The plots were produced by the BOXPLOT command in Minitab.

```
                              ---------------
First                --------(I       + I)-------------
                              ---------------

            ----------
Fourth      ----(-I   +   )--------
            ----------
      +--------+--------+--------+--------+--------+-----Silver Pct.
     5.00    5.50     6.00     6.50     7.00     7.50
```

Table 1.12.2: Silver percentage in two mintings

First	5.9	6.8	6.4	7.0	6.6	7.7	7.2	6.9	6.2
Fourth	5.3	5.6	5.5	5.1	6.2	5.8	5.8		

Note the confidence intervals are clearly disjoint so we easily reject the null hypothesis at the 5% level and it is easy to see the sharp reduction in silver content from first to fourth coinage. In addition, the box for the fourth coinage is a bit narrower than the box for the first coinage, indicating that there may be less variation (as measured by the interquartile range) in the fourth coinage. There are no apparent outliers as indicated by the whiskers on the boxplot. Larson and Stroup (1976) analyze this example with a two-sample t-test.

1.13 Exercises

1.13.1 Show that if $\|\cdot\|$ is a norm, then there always exists a value of θ which minimizes $\|\mathbf{x} - \theta \mathbf{1}\|$ for any $x_1, \ldots, x_n$.

1.13.2 Show that $D(\theta)$, (1.3.3), is convex and continuous as a function of θ. Further, argue that $D(\theta)$ is differentiable almost everywhere. Let $S(\theta)$ be a function such that $S(\theta) = -D'(\theta)$ where the derivative exists. Then show that $S(\theta)$ is a nonincreasing function.

1.13.3 Consider the L_2 norm. Show that $\widehat{\theta} = \bar{x}$ and that $S_2(0) = \sqrt{nt}/\sqrt{n-1+t^2}$, where $t = \sqrt{n}\bar{x}/s$ and s is the sample standard deviation. Further, show $S_2(0)$ is an increasing function of t so the test based on t is equivalent to $S_2(0)$.

1.13.4 Discuss the consistency of the t-test. Is the t-test resolving?

1.13.5 Discuss the Pitman regularity in the L_2 case.

1.13.6 Using languages such as Minitab or S-Plus, write a computer program to bootstrap a single sample and obtain the bootstrap estimate of the sampling distribution of the median. Further, have it return estimates of the standard error of the median. Compare this estimate with the estimate based on the length of the confidence interval for the Shoshoni data, Example 1.4.2.

1.13.7 Using languages such as Minitab or S-Plus, obtain a plot of the test sensitivity curves based on the signed-rank Wilcoxon statistic for the Cushney–Peebles data, Example 1.4.1, similar to the sensitivity curves based on the t-test and the sign test as shown in Figure 1.4.1.

1.13.8 In the proof of Theorem 1.5.6, show that (1.5.19) and (1.5.20) imply that $U_n(b)$ converges to $-\mu'(0)$ in probability, pointwise in b, i.e. $U_n(b) = -\mu'(0) + o_p(1)$.

1.13.9 Show that (1.5.23) is scale-invariant. Hence the efficiency does not change if X is multiplied by a positive constant. Let

$$f(x, \delta) = \delta \exp(-|x|^\delta)/2\Gamma(\delta^{-1}), \quad -\infty < x < \infty, 1 \leqslant \delta \leqslant 2.$$

When $\delta = 2$, f is a normal distribution, and when $\delta = 1$, f is a Laplace distribution. Compute and plot as a function of δ the efficiency (1.5.23).

1.13.10 Show that the finite-sample breakdown of the Hodges–Lehmann estimate (1.3.25) is $\varepsilon_n^* = m/n$, where m is the solution to the quadratic inequality $2m^2 - (4n+2)m^* + n^2 + n \leq 0$. Table ε_n^* as a function of n and show that ε_n^* converges to $1 - 1\sqrt{2} \doteq 0.29$.

1.13.11 Derive (1.6.9).

1.13.12 Prove Lemma 1.7.2.

1.13.13 Prove Theorem 1.7.3. In particular, check the conditions of the Lindeberg central limit theorem to verify (1.7.7).

1.13.14 Prove Theorem 1.7.4.

1.13.15 Suppose $h(x)$ has finite Fisher information:

$$I(h) = \int \frac{(h'(x))^2}{h(x)} dx < \infty.$$

Prove that $h(x)$ is bounded and that $\int h^2(x) dx < \infty$.

Hint: Write

$$h(x) = \int_{-\infty}^{x} h'(t) dt \leq \int_{-\infty}^{x} |h'(t)| dt.$$

1.13.16 Repeat Exercise 1.13.9 for (1.7.12).

1.13.17 Show that (1.8.1) is a norm.

1.13.18 Show that $\int \varphi_h^{+2}(u) du$, $\varphi_h^+(u)$ given by (1.8.18), is equal to the Fisher information,

$$\int \frac{(h'(x))^2}{h(x)} dx.$$

1.13.19 Find (1.8.18) when h is normal, logistic, Laplace (double exponential) density, respectively.

1.13.20 Verify that the influence function of the normal score estimate is unbounded when the underlying distribution is normal.

1.13.21 Verify (1.10.4).

1.13.22 Derive the limit distribution in expression (1.10.5).

1.13.23 Show that approximation (1.11.6) is exact for the double exponential (Laplace) distribution.

1.13.24 Suppose $X_1, \ldots, X_{2n}$ are independent observations such that X_i has cdf $F(x - \theta_i)$, where F is symmetric about 0. For testing $H_0 : \theta_1 = \ldots = \theta_{2n}$ versus $H_A : \theta_1 \leq \ldots \leq \theta_{2n}$ with at least one strict inequality, consider the test statistic

$$S = \sum_{i=1}^{n} I(X_{n+i} > X_i).$$

66 One-Sample Problems

(a) Discuss the small sample and asymptotic distribution of S under H_0.

(b) Determine the alternative distribution of S under the alternative $\theta_{n+i} - \theta_i = \Delta, \Delta > 0$, for all $i = 1, \ldots, n$. Show that the test is consistent for this alternative. This test is called **Mann's (1945) test for trend**.

1.13.25 The data in Table 1.13.1 constitute a sample of size 59 of information on professional baseball players. The data were recorded from the back of a deck of baseball cards (complements of Carrie McKean).

Table 1.13.1: Data for professional baseball players (Exercise 1.13.25). The variables are: (H) height in inches; (W) weight in pounds; (B) side of plate from which the player bats (1, right-handed; 2, left-handed; 3, switch hitter); (A) throwing arm (0, right; 1, left); (P) pitch-hit indicator (0, pitcher; 1, hitter); and (Ave) average (earned run average (ERA) if pitcher; batting average if hitter)

H	W	B	A	P	Ave	H	W	B	A	P	Ave
74	218	1	1	0	3.330	79	232	2	1	0	3.100
75	185	1	0	1	0.286	72	190	1	0	1	0.238
77	219	2	1	0	3.040	75	200	2	0	0	3.180
73	185	1	0	1	0.271	70	175	2	0	1	0.279
69	160	3	0	1	0.242	75	200	1	0	1	0.274
73	222	1	0	0	3.920	78	220	1	0	0	3.880
78	225	1	0	0	3.460	73	195	1	0	0	4.570
76	205	1	0	0	3.420	75	205	2	1	1	0.284
77	230	2	0	1	0.303	74	185	1	0	1	0.286
78	225	1	0	0	3.460	71	185	3	0	1	0.218
76	190	1	0	0	3.750	73	210	1	0	1	0.282
72	180	3	0	1	0.236	76	210	2	1	0	3.280
73	185	1	0	1	0.245	73	195	1	0	1	0.243
73	200	2	1	0	4.800	75	205	1	0	0	3.700
74	195	1	0	1	0.276	73	175	1	1	0	4.650
75	195	1	0	0	3.660	73	190	2	1	1	0.238
72	185	2	1	1	0.300	74	185	3	1	0	4.070
75	190	1	0	1	0.239	72	190	3	0	1	0.254
76	200	1	0	0	3.380	73	210	1	0	0	3.290
76	180	2	1	0	3.290	71	195	1	0	1	0.244
72	175	2	1	1	0.290	71	166	1	0	1	0.274
76	195	2	1	0	4.990	71	185	1	1	0	3.730
68	175	2	0	1	0.283	73	160	1	0	0	4.760
73	185	1	0	1	0.271	74	170	2	1	1	0.271
69	160	1	0	1	0.225	76	185	1	0	0	2.840
76	211	3	0	1	0.282	71	155	3	0	1	0.251
77	190	3	0	1	0.212	76	190	1	0	0	3.280
74	195	1	0	1	0.262	71	160	3	0	1	0.270
75	200	1	0	0	3.940	70	155	3	0	1	0.261
73	207	3	0	1	0.251						

(a) Obtain dotplots of the weights and heights of the baseball players.

(b) Assume the weight of a typical adult male is 175 pounds. Use the Wilcoxon test statistic to test the hypotheses

$$H_0 : \theta_W = 175 \text{ versus } H_0 : \theta_W \neq 175,$$

where θ_W is the median weight of a professional baseball player. Compute the p-value. Next obtain a 95% confidence interval for θ_W using the confidence interval procedure based on the Wilcoxon. Use the dotplot in (a) to comment on the assumption of symmetry.

(c) Let θ_H be the median height of a baseball player. Repeat the analysis of (b) for the hypotheses

$$H_0 : \theta_H = 70 \text{ versus } H_0 : \theta_H \neq 70.$$

2

Two-Sample Problems

2.1 Introduction

Let $X_1, \ldots, X_{n_1}$ be a random sample with common distribution function $F(x)$ and density function $f(x)$. Let $Y_1, \ldots, Y_{n_2}$ be another random sample, independent of the first, with common distribution function $G(x)$ and density $g(x)$. We will call this the general model throughout this chapter. A natural null hypothesis is $H_0 : F(x) = G(x)$. In this chapter we will consider rank and sign tests of this hypothesis. A general alternative to H_0 is $H_A : F(x) \neq G(x)$ for some x. Except for Section 2.10 on the scale model, we will be generally concerned with the alternative models where one distribution is stochastically larger than the other; for example, the alternative that G is stochastically larger than F which can be expressed as $H_A : G(x) \leq F(x)$ with a strict inequality for some x. This family of alternatives includes the location model, described next, and the Lehmann alternative models discussed in Section 2.7, which are used in survival analysis.

As in Chapter 1, the location models will be of primary interest. For these models $G(x) = F(x - \Delta)$ for some parameter Δ. Thus the parameter Δ represents a shift in location between the two distributions. It can be expressed as $\Delta = \theta_Y - \theta_X$, where θ_Y and θ_X are the medians of the distributions of G and F, or equivalently as $\Delta = \mu_Y - \mu_X$, where, provided they exist, μ_Y and μ_X are the means of G and F. In the location problem the null hypothesis becomes $H_0 : \Delta = 0$. In addition to tests of this hypothesis we will develop estimates and confidence intervals for Δ. We will call this the location model throughout this chapter and we will show that this is a generalization of the location problem defined in Chapter 1.

2.2 Geometric Motivation

In this section, we work with the location model described above. As in Chapter 1, we will derive sign and rank-based tests and estimates from a geometric point of view. As we shall show, their development is analogous to

70 Two-Sample Problems

that of least-squares procedures in that other norms are used in place of the least-squares Euclidean norm. In order to do this we place the problem into the context of a linear model. This will facilitate our geometric development and will also serve as an introduction to Chapter 3.

Let $\mathbf{Z}' = (X_1, \ldots, X_{n_1}, Y_1, \ldots, Y_{n_2})$ denote the vector of all observations; let $n = n_1 + n_2$ denote the total sample size; and let

$$c_i = \begin{cases} 0 & \text{if } 1 \leq i \leq n_1 \\ 1 & \text{if } n_1 + 1 \leq i \leq n. \end{cases} \quad (2.2.1)$$

Then we can write the location model as

$$Z_i = \Delta c_i + e_i, \quad 1 \leq i \leq n, \quad (2.2.2)$$

where $e_1, \ldots, e_n$ are iid with distribution function $F(x)$. Let $\mathbf{C} = [c_i]$ denote the $n \times 1$ design matrix and let Ω_{FULL} denote the column space of $\mathbf{C}$. We can express the location model as

$$\mathbf{Z} = \mathbf{C}\Delta + \mathbf{e}, \quad (2.2.3)$$

where $\mathbf{e}' = (e_1, \ldots, e_n)$ is the $n \times 1$ vector of errors. Note that except for random error, the observations $\mathbf{Z}$ would lie in Ω_{FULL}. Thus given a norm, we estimate Δ so that $\mathbf{C}\hat{\Delta}$ minimizes the distance between $\mathbf{Z}$ and the subspace Ω_{FULL}; i.e. $\mathbf{C}\hat{\Delta}$ is the vector in Ω_{FULL} closest to $\mathbf{Z}$.

Before turning our attention to Δ, however, we write the problem in terms of the geometry discussed in Chapter 1. Consider any location functional T of the distribution of e. Let $\theta = T(F)$. Define the random variable $e^* = e - \theta$. Then the distribution function of e^* is $F^*(x) = F(x + \theta)$ and its functional is $T(F^*) = 0$. Thus model (2.2.3) can be expressed as

$$\mathbf{Z} = \mathbf{1}\theta + \mathbf{C}\Delta + \mathbf{e}^*. \quad (2.2.4)$$

Note that this is a generalization of the location problem discussed in Chapter 1. From the last paragraph, the distribution function of X_i can be expressed as $F(x) = F^*(x - \theta)$; hence, $T(F) = \theta$ is a location functional of X_i. Further, the distribution function of Y_j can be written as $G(x) = F^*(x - (\Delta + \theta))$. Thus $T(G) = \Delta + \theta$ is a location functional of Y_j. Therefore, Δ is precisely the difference in location functionals between X_i and Y_j. Furthermore, Δ does not depend on which location functional is used and will be called the **shift parameter**.

Let $\mathbf{b} = (\theta, \Delta)'$. Given a norm, we want to choose as our estimate of $\mathbf{b}$ a value $\hat{\mathbf{b}}$ such that $[\mathbf{1} \ \mathbf{C}]\hat{\mathbf{b}}$ minimizes the distance between the vector of observations $\mathbf{Z}$ and the column space V of the matrix $[\mathbf{1} \ \mathbf{C}]$. Thus we can use the norms defined in Chapter 1 to estimate $\mathbf{b}$.

If, as an example, we select the L_1 norm, then our estimate of $\mathbf{b}$ minimizes

$$D(\mathbf{b}) = \sum_{i=1}^{n} |Z_i - \theta - c_i \Delta|. \quad (2.2.5)$$

Differentiating D with respect to θ and Δ, respectively, and setting the resulting equations to 0, we obtain the equations

$$\sum_{i=1}^{n_1} \text{sgn}\,(X_i - \theta) + \sum_{j=1}^{n_2} \text{sgn}\,(Y_j - \theta - \Delta) \doteq 0 \qquad (2.2.6)$$

$$\sum_{j=1}^{n_2} \text{sgn}\,(Y_j - \theta - \Delta) \doteq 0. \qquad (2.2.7)$$

Subtracting the second equation from the first, we get $\sum_{i=1}^{n_1} \text{sgn}\,(X_i - \theta) \doteq 0$; hence, $\widehat{\theta} = \text{med}\,\{X_i\}$. Substituting this into the second equation, we get $\widehat{\Delta} = \text{med}\,\{Y_j - \widehat{\theta}\} = \text{med}\,\{Y_j\} - \text{med}\,\{X_i\}$; hence, $\widehat{\mathbf{b}} = (\text{med}\,\{X_i\}, \text{med}\,\{Y_j - \widehat{\theta}\} - \text{med}\,\{X_i\})$. We will obtain inference based on the L_1 norm in Sections 2.6.1 and 2.6.2.

If we select the L_2 norm, then, as shown in Exercise 2.13.1, the LS estimate $\widehat{\mathbf{b}} = (\overline{X}, \overline{Y} - \overline{X})'$. Another norm discussed in Chapter 1 was the weighted L_1 norm. In this case $\mathbf{b}$ is estimated by minimizing

$$D(\mathbf{b}) = \sum_{i=1}^{n} R(|Z_i - \theta - c_i \Delta|)|Z_i - \theta - c_i \Delta|. \qquad (2.2.8)$$

This estimate cannot be obtained in closed form; however, fast minimization algorithms for such problems are discussed later in Chapter 3.

In the initial statement of the problem, though, θ is a nuisance parameter and we are really interested in Δ, the shift in location between the populations. Hence, we want to define distance in terms of norms which are invariant to θ. The type of norm that is invariant to θ is a **pseudo-norm**, which we define next.

Definition 2.2.1 *An operator $\|\cdot\|_*$ is called a* **pseudo-norm** *if it satisfies the following four conditions:*

$$\|\mathbf{u} + \mathbf{v}\|_* \leq \|\mathbf{u}\|_* + \|\mathbf{v}\|_* \text{ for all } \mathbf{u}, \mathbf{v} \in \mathcal{R}^n$$

$$\|\alpha \mathbf{u}\|_* = |\alpha| \|\mathbf{u}\|_* \text{ for all } \alpha \in \mathcal{R}, \mathbf{u} \in \mathcal{R}^n$$

$$\|\mathbf{u}\|_* \geq 0 \text{ for all } \mathbf{u} \in \mathcal{R}^n$$

$$\|\mathbf{u}\|_* = 0 \text{ if and only if } u_1 = \cdots = u_n.$$

Note that a regular norm satisfies the first three properties but in lieu of the fourth property, the norm of a vector is 0 if and only if the vector is $\mathbf{0}$. The following inequalities establish the invariance of pseudo-norms to the parameter θ:

$$\|\mathbf{Z} - \theta \mathbf{1} - \mathbf{C}\Delta\|_* \leq \|\mathbf{Z} - \mathbf{C}\Delta\|_* + \|\theta \mathbf{1}\|_*$$
$$= \|\mathbf{Z} - \mathbf{C}\Delta\|_* = \|\mathbf{Z} - \theta \mathbf{1} - \mathbf{C}\Delta + \theta \mathbf{1}\|_*$$
$$\leq \|\mathbf{Z} - \theta \mathbf{1} - \mathbf{C}\Delta\|_*.$$

Hence, $\|\mathbf{Z} - \theta \mathbf{1} - \mathbf{C}\Delta\|_* = \|\mathbf{Z} - \mathbf{C}\Delta\|_*$.

Given a pseudo-norm, denote the associated dispersion function by $D_*(\Delta) = \|\mathbf{Z} - \mathbf{C}\Delta\|_*$. It follows from the above properties of a pseudo-norm that $D_*(\Delta)$ is a nonnegative, continuous, and convex function of Δ.

72 Two-Sample Problems

We next develop an inference which includes estimation of Δ and tests of hypotheses concerning Δ for a general pseudo-norm. As an estimate of the shift parameter Δ, we choose a value $\widehat{\Delta}$ which solves

$$\widehat{\Delta} = \text{Argmin} D_*(\Delta) = \text{Argmin} \|Z - C\Delta\|_*; \tag{2.2.9}$$

i.e. $C\widehat{\Delta}$ minimizes the distance between Z and Ω_{FULL}. Another way of defining $\widehat{\Delta}$ is as the stationary point of the gradient of the pseudo-norm. Define the function S_* by

$$S_*(\Delta) = -\nabla \|Z - C\Delta\|_*, \tag{2.2.10}$$

where ∇ denotes the gradient of $\|Z - C\Delta\|_*$ with respect to Δ. Because $D_*(\Delta)$ is convex, it follows immediately that

$$S_*(\Delta) \text{ is nonincreasing in } \Delta. \tag{2.2.11}$$

Hence $\widehat{\Delta}$ is such that

$$S_*(\widehat{\Delta}) \doteq 0. \tag{2.2.12}$$

Given a location functional $\theta = T(F)$, i.e. model (2.2.4), once Δ has been estimated we can base an estimate of θ on the residuals $Z_i - \widehat{\Delta} c_i$. For example, if we choose the median as our location functional then we could use the median of the residuals to estimate it. We will discuss this in more detail for general linear models in Chapter 3.

Next consider the hypotheses

$$H_0: \Delta = 0 \text{ versus } H_A: \Delta \neq 0. \tag{2.2.13}$$

The closer $S_*(0)$ is to 0 the more plausible is the hypothesis H_0. More formally, we define the **gradient test** of H_0 versus H_A by the rejection rule,

Reject H_0 in favor of H_A if $S_*(0) \leq k$ or $S_*(0) \geq l$,

where the critical values k and l depend on the null distribution of $S_*(0)$. Typically, the null distribution of $S_*(0)$ is symmetric about 0 and $k = -l$. The **reduction in dispersion** test is given by

Reject H_0 in favor of H_A if $D_*(0) - D_*(\widehat{\Delta}) \geq m$,

where the critical value m is determined by the null distribution of the test statistic. In this chapter, as in Chapter 1, we will be concerned with the gradient test while in Chapter 3 we will use the reduction in dispersion test. A **confidence interval** for Δ of confidence $(1-\alpha)100\%$ is the interval $\{\Delta : k < S_*(\Delta) < l\}$ and

$$1 - \alpha = P_\Delta[k < S_*(\Delta) < l]. \tag{2.2.14}$$

Since $D_*(\Delta)$ is convex, $S_*(\Delta)$ is nonincreasing and we have

$$\widehat{\Delta}_L = \inf\{\Delta : S_*(\Delta) < l\} \text{ and } \widehat{\Delta}_U = \sup\{\Delta : S_*(\Delta) > k\}; \tag{2.2.15}$$

compare (1.3.10). Often we will be able to invert $k < S_*(\Delta) < l$ to find an explicit formula for the upper and lower endpoints.

We will discuss a large class of general pseudo-norms in Section 2.5, but now we present the pseudo-norms that yield the pooled t-test and the Mann–Whitney–Wilcoxon test.

2.2.1 Least-Squares (LS) Analysis

The traditional analysis is based on the squared pseudo-norm given by

$$\|\mathbf{u}\|_{LS}^2 = \sum_{i=1}^{n}\sum_{j=1}^{n}(u_i - u_j)^2, \quad \mathbf{u} \in \mathcal{R}^n. \tag{2.2.16}$$

It follows (see Exercise 2.13.1) that

$$\nabla\|\mathbf{Z} - \mathbf{C}\Delta\|_{LS}^2 = -4n_1n_2(\overline{Y} - \overline{X} - \Delta);$$

hence the classical estimate is $\hat{\Delta}_{LS} = \overline{Y} - \overline{X}$. Eliminating the constant factor $4n_1n_2$, the classical test is based on the statistic

$$S_{LS}(0) = \overline{Y} - \overline{X}.$$

As shown in Exercise 2.13.1, standardizing S_{LS} results in the two-sample pooled t-statistic. An approximate confidence interval for Δ is given by

$$\overline{Y} - \overline{X} \pm t_{(\alpha/2, n_1+n_2-2)}\hat{\sigma}\sqrt{\frac{1}{n_1} + \frac{1}{n_2}},$$

where $\hat{\sigma}$ is the usual pooled estimate of the common standard deviation. This confidence interval is exact if e_i has a normal distribution. Asymptotically, we replace $t_{(\alpha/2, n_1+n_2-2)}$ by $z_{\alpha/2}$. The test is asymptotically distribution-free.

2.2.2 Mann–Whitney–Wilcoxon (MWW) Analysis

The rank-based analysis is based on the pseudo-norm defined by

$$\|\mathbf{u}\|_R = \sum_{i=1}^{n}\sum_{j=1}^{n}|u_i - u_j|, \quad \mathbf{u} \in \mathcal{R}^n. \tag{2.2.17}$$

Note that this pseudo-norm is the L_1-norm based on the differences between the components and that it is the second term of expression (1.3.20), which defines the norm of the signed-rank analysis of Chapter 1. Note, further, that this pseudo-norm differs from the LS pseudo-norm in that the square root is taken inside the double summation. In Exercise 2.13.2 the reader is asked to show that this indeed is a pseudo-norm and, further, that it can be written in terms of ranks as

$$\|\mathbf{u}\|_R = 4\sum_{i=1}^{n}\left(R(u_i) - \frac{n+1}{2}\right)u_i.$$

From (2.2.17), it follows that the **MWW gradient** is

$$\nabla\|\mathbf{Z} - \mathbf{C}\Delta\|_R = -2\sum_{i=1}^{n_1}\sum_{j=1}^{n_2}\operatorname{sgn}(Y_j - X_i - \Delta).$$

Our estimate of Δ is a value which makes the gradient zero; that is, makes half of the differences positive and the other half negative. Thus the rank-based estimate of Δ is

$$\widehat{\Delta}_R = \text{med}\,\{Y_j - X_i\}. \tag{2.2.18}$$

This pseudo-norm estimate is often called the **Hodges–Lehmann** estimate of **shift** for the two-sample problem (Hodges and Lehmann, 1963). As we show in Section 2.4.4, $\widehat{\Delta}_R$ has an approximate normal distribution with mean Δ and standard deviation $\tau\sqrt{(1/n_1) + (1/n_2)}$, where the scale parameter τ is given in (2.4.22).

From the gradient we define

$$S_R(\Delta) = \sum_{i=1}^{n_1} \sum_{j=1}^{n_2} \text{sgn}\,(Y_j - X_i - \Delta). \tag{2.2.19}$$

Next define

$$S_R^+(\Delta) = \#(Y_j - X_i > \Delta). \tag{2.2.20}$$

Note we have (with probability one) that $S_R(\Delta) = 2S_R^+(\Delta) - n$. The statistic $S_R^+ = S_R^+(0)$, originally proposed by Mann and Whitney (1947), will be more convenient to use. The gradient test for the hypotheses (2.2.13) is

Reject H_0 in favor of H_A if $S_R^+ \leq k$ or $S_R^+ \geq n_1 n_2 - k$,

where k is chosen by $P_0(S_R^+ \leq k) = \alpha/2$. We show in Section 2.4 that the test statistic is distribution-free under H_0 and, that further, it has an asymptotic normal distribution with mean $n_1 n_2 / 2$ and standard deviation $\sqrt{n_1 n_2 (n_1 + n_2 + 1)/12}$ under H_0. Hence, an asymptotic level α test rejects H_0 in favor of H_A if

$$|z| > z_{\alpha/2}, \text{ where } z = \frac{S_R^+ - (n_1 n_2 / 2)}{\sqrt{n_1 n_2 (n_1 + n_2 + 1)/12}}. \tag{2.2.21}$$

As shown in Section 2.4.2, the $(1 - \alpha)100\%$ **MWW confidence interval** for Δ is given by

$$[D_{(k+1)}, D_{(n_1 n_2 - k)}), \tag{2.2.22}$$

where k is such that $P_0[S_R^+ \leq k] = \alpha/2$ and $D_{(1)} \leq \cdots \leq D_{(n_1 n_2)}$ denotes the ordered $n_1 n_2$ differences $Y_j - X_i$. It follows from the asymptotic null distribution of S_R^+ that k can be approximated as $n_1 n_2 / 2 - \tfrac{1}{2} - z_{\alpha/2}\sqrt{(n_1 n_2 (n+1))/12}$.

A **rank formulation** of the MWW test statistic $S_R^+(\Delta)$ will also prove useful. Letting $R(u_i)$ denote the rank of u_i among $u_1, \ldots, u_n$, we can write

$$\sum_{j=1}^{n_2} R(Y_j - \Delta) = \sum_{j=1}^{n_2} \{\#_i(X_i < Y_j - \Delta) + \#_i(Y_i - \Delta \leq Y_j - \Delta)\}$$

$$= \#(Y_j - X_i > \Delta) + \frac{n_2(n_2 + 1)}{2}.$$

Defining

$$W(\Delta) = \sum_{i=1}^{n} R(Y_i - \Delta), \qquad (2.2.23)$$

we thus have the relationship that

$$S_R^+(\Delta) = W(\Delta) - \frac{n_2(n_2 + 1)}{2}. \qquad (2.2.24)$$

The test statistic $W(0)$ was proposed by Wilcoxon (1945). Since it is a linear function of the Mann–Whitney test statistic it has identical statistical properties. We will refer to the statistic, S_R^+, as the Mann–Whitney–Wilcoxon statistic and will label it as MWW.

As a final note on the geometry of the rank-based analysis, reconsider the model with the location functional θ in it, i.e. (2.2.4). Suppose we obtain the R-estimate of Δ, (2.2.18). Let $\widehat{\mathbf{e}}_R = \mathbf{Z} - \mathbf{C}\widehat{\Delta}_R$ denote the residuals. Next suppose we want to estimate the location parameter θ by using the weighted L_1 norm which was discussed for estimation of location in Section 1.7. Let $\|\mathbf{u}\|_{SR} = \sum_{j=1}^{n} j|u|_{(j)}$ denote this norm. For the residual vector $\widehat{\mathbf{e}}_R$, expression (1.3.10) is given by

$$\|\widehat{\mathbf{e}} - \theta \mathbf{1}\|_{SR} = \sum \sum_{i \leq j} \left| \frac{\widehat{e}_i + \widehat{e}_j}{2} - \theta \right| + \tfrac{1}{4} \|\widehat{\mathbf{e}}_R\|_R. \qquad (2.2.25)$$

Hence the estimate of θ determined by this geometry is the Hodges–Lehmann estimate based on the residuals; i.e.

$$\widehat{\theta}_R = \operatorname{med}_{i \leq j} \left\{ \frac{\widehat{e}_i + \widehat{e}_j}{2} \right\}. \qquad (2.2.26)$$

Asymptotic theory for the joint distribution of the random vector $(\widehat{\theta}_R, \widehat{\Delta}_R)'$ will be discussed in Chapter 3.

2.3 Examples

In this section we present two examples which illustrate the methods discussed in the previous section.

Example 2.3.1 *Quail Data*

The data for this problem are drawn from a high-volume drug screen designed to find compounds which reduce low-density lipoprotein (LDL) cholesterol in quail; see McKean, Vidmar and Sievers (1989) for a discussion of this screen. For the purposes of the present example, we have taken the plasma LDL levels of one group of quail who were fed over a specified period of time a special diet mixed with a drug compound and the LDL levels of a second group of quail who were fed the same special diet but without the drug compound over the same length of time. A completely randomized design was employed. We will refer to the first group as the treatment group and the second group as the control group. The data are displayed in Table 2.3.1. Let

76 Two-Sample Problems

Table 2.3.1: Data for quail example

Control	64	49	54	64	97	66	76	44	71	89
	70	72	71	55	60	62	46	77	86	71
Treated	40	31	50	48	152	44	74	38	81	64

θ_C and θ_T denote the true median levels of LDL for the control and treatment populations, respectively. The parameter of interest is $\Delta = \theta_C - \theta_T$. We are interested in the alternative hypothesis that the treatment has been effective; hence the hypotheses are

$$H_0 : \Delta = 0 \text{ versus } H_A : \Delta > 0.$$

Minitab comparison dotplots for the data are:

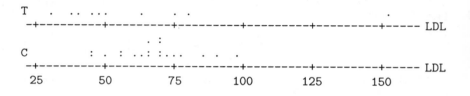

Note that there is one outlier, the 5th observation of the treated group, which has the value 152. Outliers such as this were typical with most of the data in this study; see McKean et al. (1989). For the data at hand, the treated group appears to have lower LDL levels.

The sum of the ranks of the control subjects is 344.5, which leads to a standardized test value of $z = 1.52$ with an observed significance level of 0.064. The two-sample t-statistic has the value $t = 0.56$ with a p-value of 0.29. The MWW indicates with marginal significance that the treatment performed better than the placebo. The two-sample t analysis was impaired by the outlier.

Based on the data in Table 2.3.1, the Hodges–Lehmann estimate of Δ, (2.2.18), is 14 and a 90% confidence interval, (2.2.22), obtained from the Minitab command MANN, is $(-2.01, 24.00)$. In contrast, the least squares estimate of shift is 5 and the corresponding 90% confidence interval is $(-10.25, 20.25)$.

Example 2.3.2 *Hendy–Charles Coin Data, continuation of Example 1.12.1*

Recall that the 84% L_1 confidence intervals for the data are disjoint. Thus we reject the null hypothesis that the silver content is the same for the two mintings at the 5% level. We now apply the MWW test and confidence interval to data and find the Hodges–Lehmann estimate of shift. If the tailweights of the underlying distributions are moderate, the MWW methods are more efficient.

The output from the minitab command MANN is:

```
First      N = 9      Median = 6.8000
Fourth     N = 7      Median = 5.6000
Point estimate for ETA1-ETA2 is 1.1000
95.6 pct c.i. for ETA1-ETA2 is (0.6002,1.6999)
W = 106.5
Test of ETA1 = ETA2 vs. ETA1 n.e. ETA2 is significant at 0.0018
The test is significant at 0.0018 (adjusted for ties)
```

Note that there is strong statistical evidence that the mintings are different. The Hodges–Lehmann estimate (2.2.18) is 1.1, which suggests that there is roughly a 1.1% decrease in the silver content from the first to the fourth mintings. The 95.6% confidence interval, (2.2.22), is (0.6, 1.7). Half the length of the confidence is 0.45 and this could be reported as the margin of error in estimating Δ, the change in median silver content from the first to the fourth mintings. Hence we could report $1.1\% \pm 0.45\%$.

2.4 Inference Based on the Mann–Whitney–Wilcoxon Statistic

We next develop the theory for inference based on the Mann–Whitney–Wilcoxon statistic, including the test, the estimate, and the confidence interval. Although much of the development is for the location model the general model will also be considered. We begin with testing.

2.4.1 Testing

Although the geometric motivation of the test statistic S_R^+ was derived under the location model, the test can be used for more general models. Recall that the general model is comprised of a random sample $X_1, \ldots, X_{n_1}$ with cdf $F(x)$ and a random sample $Y_1, \ldots, Y_{n_2}$ with cdf $G(x)$. For the discussion we select the hypotheses

$$H_0 : F(x) = G(x), \text{ for all } x \text{ versus } H_A : F(x) \geq G(x),$$

$$\text{with strict inequality for some } x. \quad (2.4.1)$$

Under this stochastically ordered alternative Y tends to dominate X; i.e. $P(Y > X) > \frac{1}{2}$. Our rank-based decision rule is to reject H_0 in favor of H_A if S_R^+ is too large, where $S_R^+ = \#(Y_j - X_i > 0)$. Our immediate goal is to make this precise. What we discuss will of course hold for the other one-sided alternative $F(x) \leq G(x)$ and the two-sided alternative $F(x) \leq G(x)$ or $F(x) \geq G(x)$ as well. Furthermore, since the location model is a submodel of the general model, what holds for the general model will hold for it also. It will always be clear which set of hypotheses is being considered.

Under H_0, we show first that S_R^+ is distribution-free and then that it is symmetrically distributed about $(n_1 n_2)/2$.

Theorem 2.4.1 *Under the general null hypothesis in (2.4.1), S_R^+ is distribution-free.*

Proof. Under the null hypothesis, the combined samples $X_1, \ldots, X_{n_1}$, $Y_1, \ldots, Y_{n_2}$ constitute a random sample of size n from the distribution function $F(x)$. Hence any assignment of n_2 ranks from the set of integers $\{1, \ldots, n\}$ to $Y_1, \ldots, Y_{n_2}$ is equilikely; i.e. has probability $\binom{n}{n_2}^{-1}$ independent of F.

Theorem 2.4.2 *Under H_0 in (2.4.1), the distribution of S_R^+ is symmetric about $(n_1 n_2)/2$.*

Proof. Under H_0 in (2.4.1), $\mathcal{L}(Y_j - X_i) = \mathcal{L}(X_i - Y_j)$ for all i, j; see Exercise 2.13.3. Thus if $S_R^- = \#(X_i - Y_j > 0)$ then, under H_0, $\mathcal{L}(S_R^+) = \mathcal{L}(S_R^-)$. Since $S_R^- = n_1 n_2 - S_R^+$ we have the following string of equalities which proves the result:

$$P\left[S_R^+ \geq \frac{n_1 n_2}{2} + x\right] = P\left[n_1 n_2 - S_R^- \geq \frac{n_1 n_2}{2} + x\right]$$
$$= P\left[S_R^- \leq \frac{n_1 n_2}{2} - x\right] = P\left[S_R^+ \leq \frac{n_1 n_2}{2} - x\right].$$

Hence for the hypotheses (2.4.1), a level α test based on S_R^+ would reject H_0 if $S_R^+ \geq c_{\alpha, n_1, n_2}$ where $P_{H_0}[S_R^+ \geq c_{\alpha, n_1, n_2}] = \alpha$. From the symmetry, note that the lower α critical point is given by $n_1 n_2 - c_{\alpha, n_1, n_2}$.

Although S_R^+ is distribution-free under the null hypothesis, its distribution cannot be obtained in closed form. The next theorem gives a recursive formula for its distribution. The proof can be found in Exercise 2.13.4; see also Hettmansperger (1984a, pp. 136–137).

Theorem 2.4.3 *Under the general null hypothesis in (2.4.1), let $P_{n_1, n_2}(k) = P_{H_0}[S_R^+ = k]$. Then*

$$P_{n_1, n_2}(k) = \frac{n_2}{n_1 + n_2} P_{n_1, n_2 - 1}(k - n_1) + \frac{n_1}{n_1 + n_2} P_{n_1 - 1, n_2}(k),$$

where $P_{n_1, n_2}(k)$ satisfies the boundary conditions $P_{i,j}(k) = 0$ if $k < 0$, $P_{i,0}(k)$ and $P_{0,j}(k)$ are 1 or 0 as $k = 0$ or $k \neq 0$.

Based on these recursion formulas, tables of the null distribution can be obtained readily, which then can be used to obtain the critical values for the rank-based test. Alternatively, the asymptotic null distribution of S_R^+ can be used to determine approximate critical values. This asymptotic test will be discussed later; see Theorem 2.4.10.

We next derive the mean and variance of S_R^+ under the three models:

(a) the general model where X has distribution function $F(x)$ and Y has distribution function $G(x)$;
(b) the location model where $G(x) = F(x - \Delta)$;
(c) and the null model in which $F(x) = G(x)$.

Of course, from Theorem 2.4.2, the null mean of S_R^+ is $(n_1 n_2)/2$. In our derivation we repeatedly make use of the fact that if H is the distribution function of a random variable Z then the random variable $H(Z)$ has a uniform distribution over the interval $(0, 1)$; see Exercise 2.13.5.

Theorem 2.4.4 *Assuming that $X_1, \ldots, X_{n_1}$ are iid $F(x)$ and $Y_1, \ldots, Y_{n_2}$ are iid $G(x)$ and that these two samples are independent of one another, the means of S_R^+ under the three models (a)–(c) are:*

(a) $E[S_R^+] = n_1 n_2 [1 - E[G(X)]] = n_1 n_2 E[F(Y)]$

(b) $E[S_R^+] = n_1 n_2 [1 - E[F(X - \Delta)]] = n_1 n_2 E[F(X + \Delta)]$

(c) $E[S_R^+] = \dfrac{n_1 n_2}{2}$.

Proof. We shall prove only (a), since results (b) and (c) follow directly from it. We can write S_R^+ in terms of indicator functions as

$$S_R^+ = \sum_{i=1}^{n_1} \sum_{j=1}^{n_2} I(Y_j - X_i > 0), \qquad (2.4.2)$$

where $I(t > 0)$ is 1 or 0 for $t > 0$ or $t \leqslant 0$, respectively. Let Y have distribution function G, let X have distribution function F, and let X and Y be independent. Then

$$E[I(Y - X > 0)] = E[P[Y > X|X]]$$
$$= E[1 - G(X)] = E[F(Y)],$$

where the second equality follows from the independence of X and Y. The results then follow.

Theorem 2.4.5 *The variances of S_R^+ under the models (a)–(c) are:*

(a) $\mathrm{Var}[S_R^+] = n_1 n_2 (E[G(X)] - E^2[G(X)])$
$\qquad + n_1 n_2 (n_1 - 1)\mathrm{Var}[F(Y)] + n_1 n_2 (n_2 - 1)\mathrm{Var}[G(X)]$

(b) $\mathrm{Var}[S_R^+] = n_1 n_2 (E[F(X - \Delta)] - E^2[F(X - \Delta)])$
$\qquad + n_1 n_2 (n_1 - 1)\mathrm{Var}[F(Y)] + n_1 n_2 (n_2 - 1)\mathrm{Var}[F(X - \Delta)]$

(c) $\mathrm{Var}[S_R^+] = \dfrac{n_1 n_2 (n + 1)}{12}$.

Proof. Again only the result (a) will be obtained. Using the indicator formulation of S_R^+, (2.4.2), we have

$$\mathrm{Var}[S_R^+] = \sum_{i=1}^{n_1} \sum_{j=1}^{n_2} \mathrm{Var}[I(Y_j - X_i > 0)]$$
$$+ \sum_{i=1}^{n_1} \sum_{j=1}^{n_2} \sum_{l=1}^{n_1} \sum_{k=1}^{n_2} \mathrm{Cov}[I(Y_j - X_i > 0), I(Y_k - X_l > 0)],$$

where the sums for the covariance terms are over all possible combinations except $(i,j) = (l,k)$. For the first term, note that the variance of $I(Y - X > 0)$ is

$$\text{Var}[I(Y > X)] = E[I(Y > X)] - E^2[I(Y > X)]$$
$$= E[1 - G(X)] - E^2[1 - G(X)]$$
$$= E[G(X)] - E^2[G(X)].$$

This yields the first term in (a). For the covariance terms, note that a covariance is 0 unless either $j = k$ or $i = l$. This leads to the following two cases:

Case(i) For the covariance terms with $j = k$ and $i \neq l$, we need $E[I(Y > X_1)I(Y > X_2)]$, which is

$$E[I(Y > X_1)I(Y > X_2)] = P[Y > X_1, Y > X_2]$$
$$= E[P[Y > X_1, Y > X_2 | Y]]$$
$$= E[P[Y > X_1 | Y]P[Y > X_2 | Y]]$$
$$= E[F(Y)^2].$$

There are n_2 ways to get a j and $n_1(n_1 - 1)$ ways to get $i \neq l$; hence there are $n_1 n_2 (n_1 - 1)$ covariances of this form. This leads to the second term of (a).

Case(ii) The terms for the covariances where $i = l$ and $j \neq k$ follow similarly to case (i). This leads to the third and final term of (a).

The last two theorems suggest that the random variable $Z = (S_R^+ - n_1 n_2/2)/(n_1 n_2 (n+1)/12)^{1/2}$ has an approximate $N(0, 1)$ distribution under H_0. This follows from the next results which yield the asymptotic distribution of S_R^+ under general alternatives as well as under the null hypothesis. We will obtain these results by projecting our statistic S_R^+ down onto a set of linear combinations of independent random variables. Then we can use central limit theory on the projection. See Hájek and Šidák (1967) for a discussion of this technique.

Let $T = T(Z_1, \ldots, Z_n)$ be a random variable based on a sample $Z_1, \ldots, Z_n$ such that $E[T] = 0$. Let

$$p_k^*(x) = E[T | Z_k = x], \quad k = 1, \ldots, n.$$

Next define the random variable T_p to be

$$T_p = \sum_{k=1}^{n} p_k^*(Z_k). \tag{2.4.3}$$

In the next theorem we show that T_p is the projection of T onto the space of linear functions of $Z_1, \ldots, Z_n$. Note that, unlike T, T_p is a linear combination of independent random variables; hence, its asymptotic distribution is often easier to obtain than that of T. As the following **projection theorem** shows, it is in a sense the 'closest' linear function of the form $\sum p_i(Z_i)$ to T.

Theorem 2.4.6 *If $W = \sum_{i=1}^{n} p_i(Z_i)$ then $E[(T - W)^2]$ is minimized by taking $p_i(x) = p_i^*(x)$. Furthermore, $E[(T - T_p)^2] = \text{Var}[T] - \text{Var}[T_p]$.*

Proof. First note that $E[p_k^*(Z_k)] = 0$. We have

$$E[(T - W)^2] = E\left[[(T - T_p) - (W - T_p)]^2\right]$$

$$= E[(T - T_p)^2] + E[(W - T_p)^2] - 2E[(T - T_p)(W - T_p)]. \quad (2.4.4)$$

We can write one-half the cross-product term as

$$\sum_{i=1}^{n} E[(T - T_p)(p_i(Z_i) - p_i^*(Z_i))] = \sum_{i=1}^{n} E[E[(T - T_p)(p_i(Z_i) - p_i^*(Z_i)) \mid Z_i]]$$

$$= \sum_{i=1}^{n} E\left[(p_i(Z_i) - p_i^*(Z_i))E\left[T - \sum_{j=1}^{n} p_j^*(Z_j) \mid Z_i\right]\right].$$

The conditional expectation can be written as

$$(E[T \mid Z_i] - p_i^*(Z_i)) - \sum_{j \neq i} E[p_j^*(Z_j)] = 0 - 0 = 0.$$

Hence the cross-product term is zero, and therefore the left-hand side of expression (2.4.4) is minimized with respect to W by taking $W = T_p$. Also since this holds, in particular, for $W = 0$ we get

$$E[T^2] = E[(T - T_p)^2] + E[T_p^2].$$

Since both T and T_p have zero means the second result of the theorem also follows.

From these results a strategy for obtaining the asymptotic distribution of T is apparent. Namely, find the asymptotic distribution of its projection, T_p and then show $\text{Var}[T] - \text{Var}[T_p] \to 0$ as $n \to \infty$. This implies that T and T_p have the same asymptotic distribution; see Exercise 2.13.7. We shall apply this strategy to get the asymptotic distribution of the rank-based methods. As a first step we obtain the projection of $S_R^+ - E[S_R^+]$ under the general model.

Theorem 2.4.7 *Under the general model the projection of the random variable $S_R^+ - E[S_R^+]$ is*

$$T_p = n_1 \sum_{j=1}^{n_2}(F(Y_j) - E[F(Y_j)]) - n_2 \sum_{i=1}^{n_1}(G(X_i) - E[G(X_i)]). \quad (2.4.5)$$

Proof. Define the n random variables $Z_1, \ldots, Z_n$ by

$$Z_i = \begin{cases} X_i & \text{if } 1 \leq i \leq n_1 \\ Y_{i-n_1} & \text{if } n_1 + 1 \leq i \leq n. \end{cases}$$

82 Two-Sample Problems

We have
$$p_k^*(x) = E[S_R^+ \mid Z_k = x] - E[S_R^+]$$
$$= \sum_{i=1}^{n_1} \sum_{j=1}^{n_2} E[I(Y_j > X_i) \mid Z_k = x] - E[S_R^+]. \qquad (2.4.6)$$

There are two cases, depending on whether $1 \leq k \leq n_1$ or $n_1 + 1 \leq k \leq n_1 + n_2 = n$.

Case 1. Suppose $1 \leq k \leq n_1$; then the conditional expectation in (2.4.6), depending on the value of i, becomes

(a) $i \neq k$, $E[I(Y_j > X_i) \mid X_k = x] = E[I(Y_j > X_i)]$
$$= P[Y > X]$$
(b) $i = k$, $E[I(Y_j > X_i) \mid X_i = x]$
$$= P[Y > X \mid X = x]$$
$$= 1 - G(x).$$

Hence, in this case,
$$p_k^*(x) = n_2(n_1 - 1)P[Y > X] + n_2(1 - G(x)) - E[S_R^+].$$

Case 2. Next suppose that $n_1 + 1 \leq k \leq n$; then the conditional expectation in (2.4.6), depending on the value of j, becomes

(a) $j \neq k$, $E[I(Y_j > X_i) \mid Y_k = x] = P[Y > X]$
(b) $j = k$, $E[I(Y_j > X_i) \mid Y_j = x] = F(x)$

Hence, in this case,
$$p_k^*(x) = n_1(n_2 - 1)P[Y > X] + n_1 F(x) - E[S_R^+].$$

Combining these results, we get
$$T_p = \sum_{i=1}^{n_1} p_i^*(X_i) + \sum_{j=1}^{n_2} p_j^*(Y_j)$$
$$= n_1 n_2(n_1 - 1)P[Y > X] + n_2 \sum_{i=1}^{n_1}(1 - G(X_i))$$
$$+ n_1 n_2(n_2 - 1)P[Y > X] + n_1 \sum_{j=1}^{n_2} F(Y_j) - nE[S_R^+].$$

This can be simplified by noting that
$$P(Y > X) = E[P(Y > X \mid X)] = E[1 - G(X)],$$
or similarly
$$P(Y > X) = E[F(Y)].$$

From (a) of Theorem 2.4.4,
$$E[S_R^+] = n_1 n_2 (1 - E[G(X)]) = n_1 n_2 P(Y > X).$$
Substituting these three results into (2.4.6), we get the desired result.

An immediate outcome is the following:

Corollary 2.4.8 *Under the general model, if T_p is given by (2.4.5) then*
$$\text{Var}(T_p) = n_1^2 n_2 \text{Var}(F(Y)) + n_1 n_2^2 \text{Var}(G(X)).$$
From this it follows that T_p should be standardized as
$$T_p^* = \frac{1}{\sqrt{n n_1 n_2}} T_p.$$

In order to obtain the asymptotic distribution of T_p and subsequently S_R^+, we need the following assumption on the design (sample sizes):

$$(D.1): \quad \frac{n_i}{n} \to \lambda_i, \quad 0 < \lambda_i < 1. \tag{2.4.7}$$

This says that the sample sizes go to ∞ at the same rate. Note that $\lambda_1 + \lambda_2 = 1$. The asymptotic variance of T_p^* is thus
$$\text{Var}(T_p^*) \to \lambda_1 \text{Var}(F(Y)) + \lambda_2 \text{Var}(G(X)).$$

We first want to obtain the asymptotic distribution under general alternatives. In order to do this we need an assumption concerning the ranges of X and Y. The support of a continuous random variable with distribution function H and density h is defined to be the set $\{x : h(x) > 0\}$, which is denoted by $\mathcal{S}(H)$.

Our second assumption states that the intersection of the supports of F and G has a nonempty interior; that is

$$(E.3): \quad \text{There is an open interval } I \text{ such that } I \subset \mathcal{S}(F) \cap \mathcal{S}(G). \tag{2.4.8}$$

Note that the asymptotic variance of T_p^* is not zero under (E.3).

We are now in the position to find the asymptotic distribution of T_p^*.

Theorem 2.4.9 *Under the general model and the assumptions (D.1) and (E.3), T_p^* has an asymptotic $N(0, \lambda_1 \text{Var}(F(Y)) + \lambda_2 \text{Var}(G(X)))$ distribution.*

Proof. By (2.4.5) we can write
$$T_p^* = \sqrt{\frac{n_1}{n n_2}} \sum_{j=1}^{n_2} (F(Y_j) - E[F(Y_j)]) - \sqrt{\frac{n_2}{n n_1}} \sum_{i=1}^{n_1} (G(X_i) - E[G(X_i)]). \tag{2.4.9}$$

Note that both sums on the right-hand side of expression (2.4.9) are composed of iid random variables and that the sums are independent of one another. The result then follows immediately by applying the simple central limit theorem to each sum.

84 Two-Sample Problems

This is the key result we need in order to obtain the asymptotic distribution of our test statistic S_R^+. We obtain the result first under the general model and then under the null hypothesis. As we will see, both results are immediate.

Theorem 2.4.10 *Under the general model and the conditions (E.3) and (D.1), the random variable $(S_R^+ - E[S_R^+])/\sqrt{\operatorname{Var}(S_R^+)}$ has a limiting $N(0, 1)$ distribution.*

Proof. By Theorems 2.4.9 and 2.4.6, we need only show that the difference in the variances of $S_R^+/\sqrt{nn_1n_2}$ and T_p^* goes to 0 as $n \to \infty$. Note that

$$\operatorname{Var}\left(\frac{1}{\sqrt{nn_1n_2}} S_R^+\right) = \frac{n_1 n_2}{nn_1 n_2}\left(E[G(X)] - E[G(X)]^2\right)$$

$$+ \frac{n_1 n_2 (n_1 - 1)}{nn_1 n_2}\operatorname{Var}(F(Y)) + \frac{n_1 n_2 (n_2 - 1)}{nn_1 n_2}\operatorname{Var}(G(X));$$

hence, $\operatorname{Var}(T_p^*) - \operatorname{Var}(S_R^+/\sqrt{nn_1n_2}) \to 0$ and the result follows from Exercise 2.13.7.

The asymptotic distribution of the test statistic under the null hypothesis follows immediately from this theorem. We record it in the next corollary.

Corollary 2.4.11 *Under $H_0 : F(x) = G(x)$ and (D.1) only, the test statistic S_R^+ is approximately $N(n_1 n_2/2, n_1 n_2(n+1)/12)$.*

Therefore an asymptotic size α test for $H_0 : F(x) = G(x)$ versus $H_A : F(x) \neq G(x)$ is to reject H_0 if $|z| \geqslant z_{\alpha/2}$, where

$$z = \frac{S_R^+ - \dfrac{n_1 n_2}{2}}{\sqrt{\dfrac{n_1 n_2 (n+1)}{12}}}$$

and

$$1 - \Phi(z_{\alpha/2}) = \alpha/2.$$

Since we approximate a discrete random variable with a continuous one, we think it is advisable in cases of small samples to use a continuity correction. Fix and Hodges (1955) give an Edgeworth approximation to the distribution of S_R^+ and Bickel (1974) discusses the error of this approximation.

Since the standard normal distribution function, Φ, is continuous on the entire real line, we can strengthen the convergence in Theorem 2.4.10 to uniform convergence; that is, the distribution function of the standardized MWW converges uniformly to Φ. Using this, it is not hard to show that the standardized critical values of the MWW converge to their standard normal counterparts. Thus if $c_{\alpha,n}$ is the MWW critical value defined by $\alpha = P_{H_0}[S_R^+ \geq c_{\alpha,n}]$, then

$$\frac{c_{\alpha,n} - \frac{n_1 n_2}{2}}{\sqrt{\frac{n_1 n_2 (n+1)}{12}}} \to z_\alpha, \qquad (2.4.10)$$

where $1 - \alpha = \Phi(z_\alpha)$; see Exercise 2.13.8 for details. This result will prove useful in the next section.

We now consider when the test based on S_R^+ is **consistent**. Consider the general setup; i.e. $X_1, \ldots, X_{n_1}$ is a random sample with distribution function $F(x)$ and $Y_1, \ldots, Y_{n_2}$ is a random sample with distribution function $G(x)$. Consider the hypotheses

$$H_0 : F = G \text{ versus } H_{A1} : F(x) \geq G(x) \text{ with } F(x_0) > G(x_0)$$
$$\text{for some } x_0 \in \text{Int}(\mathcal{S}(F) \cap \mathcal{S}(G)), \quad (2.4.11)$$

where Int (A) denotes the interior of the set A. Such an alternative is called a stochastically ordered alternative. The next theorem shows that the MWW test statistic is consistent for this alternative. Likewise, it is consistent for the other one-sided stochastically ordered alternative with F and G interchanged, H_{A2}, and also for the two-sided alternative which consists of the union of H_{A1} and H_{A2}. These results imply that the MWW test is consistent for location alternatives, provided F and G have overlapping support. As Exercise 2.13.9 shows, it will also be consistent when one support is shifted to the right of the other support.

Theorem 2.4.12 *Suppose that the assumptions (D.1) and (E.3) hold. Under the stochastic ordering alternatives given above, S_R^+ is a consistent test.*

Proof. Assume the stochastic ordering alternative H_{A1}, (2.4.11). For an arbitrary level α, select the critical level c_α such that the test that rejects H_0 if $S_R^+ \geq c_\alpha$ has asymptotic level α. We want to show that the power of the test goes to 1 as $n \to \infty$. Since $F(x_0) > G(x_0)$ for some point x_0 in the interior of $\mathcal{S}(F) \cap \mathcal{S}(G)$, there exists an interval N such that $F(x) > G(x)$ on N. Hence

$$E_{H_A}[G(X)] = \int_N G(y)f(y)dy + \int_{N^c} G(y)f(y)dy$$
$$< \int_N F(y)f(y)dy + \int_{N^c} F(y)f(y)dy = \tfrac{1}{2} \qquad (2.4.12)$$

The power of the test is given by

$$P_{H_A}[S_R^+ \geq c_\alpha] = P_{H_A}\left[\frac{S_R^+ - E_{H_A}(S_R^+)}{\sqrt{\text{Var}_{H_A}(S_R^+)}} \geq \frac{c_\alpha - (n_1 n_2/2)}{\sqrt{\text{Var}_{H_A}(S_R^+)}} + \frac{(n_1 n_2/2) - E_{H_A}(S_R^+)}{\sqrt{\text{Var}_{H_A}(S_R^+)}}\right].$$

Note by (2.4.10) that

$$\frac{c_\alpha - (n_1 n_2/2)}{\sqrt{\text{Var}_{H_A}(S_R^+)}} = \frac{c_\alpha - (n_1 n_2/2)}{\sqrt{\text{Var}_{H_0}(S_R^+)}} \frac{\sqrt{\text{Var}_{H_0}(S_R^+)}}{\sqrt{\text{Var}_{H_A}(S_R^+)}} \to z_\alpha K,$$

86 Two-Sample Problems

where κ is a real number (since the variances are of the same order). But by (2.4.12)

$$\frac{(n_1n_2/2) - E_{H_A}(S_R^+)}{\sqrt{\text{Var}_{H_A}(S_R^+)}} = \frac{(n_1n_2/2) - n_1n_2[1 - E_{H_A}(G(X))]}{\sqrt{\text{Var}_{H_A}(S_R^+)}}$$

$$= \frac{n_1n_2[-\frac{1}{2} + E_{H_A}(G(X))]}{\sqrt{\text{Var}_{H_A}(S_R^+)}} \to -\infty.$$

By Theorem 2.4.10, under H_A the random variable

$$\frac{S_R^+ - E_{H_A}(S_R^+)}{\sqrt{\text{Var}_{H_A}(S_R^+)}}$$

converges in distribution to a standard normal variate. Since the convergence is uniform, it follows from the above limits that the power converges to 1. Hence the MWW test is consistent.

2.4.2 Confidence Intervals

Consider the location model (2.2.4). We next obtain a distribution-free confidence interval for Δ by inverting the MWW test. As a first step we have the following result on the function $S_R^+(\Delta)$, defined in (2.2.20):

Lemma 2.4.13 $S_R^+(\Delta)$ *is a decreasing step function of Δ which steps down by 1 at each difference $Y_j - X_i$. Its maximum is n_1n_2 and its minimum is 0.*

Proof. Let $D_{(1)} \leq \cdots \leq D_{(n_1n_2)}$ denote the ordered n_1n_2 differences $Y_j - X_i$. The results follow immediately by writing $S_R^+(\Delta) = \#(D_{(i)} > \Delta)$.

Let α be given and choose $c_{\alpha/2}$ to be the lower $\alpha/2$ critical point of the MWW distribution; i.e., $P_\Delta[S_R^+(\Delta) \leq c_{\alpha/2}] = \alpha/2$. By the above lemma we have

$$1 - \alpha = P_\Delta[c_{\alpha/2} < S_R^+(\Delta) < n_1n_2 - c_{\alpha/2}]$$
$$= P_\Delta[D_{(c_{\alpha/2}+1)} \leq \Delta < D_{(n_1n_2-c_{\alpha/2})}].$$

Thus $[D_{(c_{\alpha/2}+1)}, D_{(n_1n_2-c_{\alpha/2})})$ is a $(1-\alpha)100\%$ confidence interval for Δ; compare (1.3.30). From the asymptotic null distribution theory for S_R^+, Corollary 2.4.11, we can approximate $c_{\alpha/2}$ as

$$c_{\alpha/2} \doteq \frac{n_1n_2}{2} - z_{\alpha/2}\sqrt{\frac{n_1n_2(n+1)}{12}} - 0.5. \quad (2.4.13)$$

2.4.3 Statistical Properties of the Inference Based on the MWW

In this subsection we derive the efficiency properties of the MWW test statistic and properties of its power function under the location model (2.2.4).

We begin with an investigation of the power function of the MWW test. For definiteness we will consider the one-sided alternative,

$$H_0: \Delta = 0 \text{ versus } H_A: \Delta > 0. \tag{2.4.14}$$

Results similar to those given below can be obtained for the power function of the other one-sided and the two-sided alternatives. Given a level α, let c_{α,n_1,n_2} denote the upper critical value for the MWW test of this hypothesis; hence, the test rejects H_0 if $S_R^+ \geq c_{\alpha,n_1,n_2}$. The power function of this test is given by

$$\gamma(\Delta) = P_\Delta[S_R^+ \geq c_{\alpha,n_1,n_2}], \tag{2.4.15}$$

where the subscript Δ on P denotes that the probability is determined when the true parameter is Δ. Recall that $S_R^+(\Delta) = \#\{Y_j - X_i > \Delta\}$.

The following theorem will prove useful, its proof is similar to that of Theorem 1.3.1 of Chapter 1 and the more general result Theorem A.2.4 of the Appendix.

Theorem 2.4.14 *For all t, $P_\Delta[S_R^+(0) \geq t] = P_0[S_R^+(-\Delta) \geq t]$.*

From Lemma 2.4.13 and Theorem 2.4.14 we have our first important result on the power function of the MWW test; namely, that it is monotone.

Theorem 2.4.15 *For the above hypotheses (2.4.14), the function $\gamma(\Delta)$ is monotonically increasing in Δ.*

Proof. Let $\Delta_1 < \Delta_2$. Then $-\Delta_2 < -\Delta_1$ and, hence, from Lemma 2.4.13, we have $S_R^+(-\Delta_2) \geq S_R^+(-\Delta_1)$. By applying Theorem 2.4.14, the desired result, $\gamma(\Delta_2) \geq \gamma(\Delta_1)$, follows from the following:

$$\begin{aligned} 1 - \gamma(\Delta_2) &= P_{\Delta_2}[S_R^+(0) < c_{\alpha,n_1,n_2}] \\ &= P_0[S_R^+(-\Delta_2) < c_{\alpha,n_1,n_2}] \\ &\leq P_0[S_R^+(-\Delta_1) < c_{\alpha,n_1,n_2}] \\ &= P_{\Delta_1}[S_R^+(0) < c_{\alpha,n_1,n_2}] \\ &= 1 - \gamma(\Delta_1). \end{aligned}$$

From this we immediately have that the MWW test is **unbiased**; that is, its power function evaluated at an alternative is always at least as large as its level of significance. We state it as a corollary.

Corollary 2.4.16 *For the above hypotheses (2.4.14), $\gamma(\Delta) \geq \alpha$ for all $\Delta > 0$.*

A more general null hypothesis is given by

$$H_0^*: \Delta \leq 0 \text{ versus } H_A: \Delta > 0.$$

If T is any test for these hypotheses with critical region C then we say T is a size α test provided

$$\sup_{\Delta \leq 0} P_\Delta[T \in C] = \alpha.$$

88   Two-Sample Problems

For selected α, it follows from the monotonicity of the MWW power function that the MWW test has size α for this more general null hypothesis.

From the above theorems, we have that the MWW power function is monotonically increasing in Δ. Since $S_R^+(\Delta)$ achieves its maximum for Δ finite, we have by Theorem 1.5.2 of Chapter 1 that the MWW test is resolving; hence, its power function approaches one as $\Delta \to \infty$. Even for the location model, though, we cannot get the power function of the MWW test in closed form. For local alternatives, however, we can obtain an asymptotic expression for the power function. Applications of this result include sample size determination for the MWW test and efficiency comparisons of the MWW with other tests, both of which we consider.

We will need the assumption that the density $f(x)$ has finite **Fisher information**, i.e.,

(E.1) f is absolutely continuous, $0 < I(f) = \int_0^1 \varphi_f^2(u)\,du < \infty,$ (2.4.16)

where

$$\varphi_f(u) = -\frac{f'(F^{-1}(u))}{f(F^{-1}(u))}. \qquad (2.4.17)$$

As discussed in Section 3.4, assumption (E.1) implies that f is uniformly bounded.

Once again we will consider the one-sided alternative, (2.4.14), (similar results hold for the other one-sided and two-sided alternatives). Consider a sequence of local alternatives of the form

$$H_{A_n} : \Delta_n = \frac{\delta}{\sqrt{n}}, \qquad (2.4.18)$$

where $\delta > 0$ is arbitrary but fixed.

As a first step, we need to show that $S_R^+(\Delta)$ is Pitman regular as discussed in Chapter 1. Let $\overline{S}_R^+(\Delta) = S_R^+(\Delta)/(n_1 n_2)$. We need to verify the four conditions of Definition 1.5.3. The first condition is true by Lemma 2.4.13 and the fourth condition follows from Corollary 2.4.11. By (b) of Theorem 2.4.4, we have

$$\mu(\Delta) = E_\Delta[\overline{S}_R^+(0)] = 1 - E[F(X - \Delta)]. \qquad (2.4.19)$$

By assumption (E.1), $\int f^2(x)\,dx \leq \sup f \int f(x)\,dx < \infty$. Hence, differentiating (2.4.19), we obtain $\mu'(0) = \int f^2(x)\,dx > 0$, and thus the second condition is true. Hence, we need only show that the third condition, asymptotic linearity of $\overline{S}_R^+(\Delta)$, is true. This will follow provided we can show the variance condition (1.5.17) of Theorem 1.5.6 is true. Note that

$$\overline{S}_R^+(\delta/\sqrt{n}) - \overline{S}_R^+(0) = (n_1 n_2)^{-1} \#(0 < Y_j - X_i \leq \delta/\sqrt{n}).$$

This is similar to the MWW statistic itself. Using essentially the same argument as that for the variance of the MWW statistic, Theorem 2.4.5, we get

$$n\text{Var}_0[\overline{S}_R^+(\delta/\sqrt{n}) - \overline{S}_R^+(0)] = \frac{n}{n_1 n_2}(a_n - a_n^2) + \frac{n(n_1 - 1)}{n_1 n_2}(b_n - c_n^2)$$
$$+ \frac{n(n_2 - 1)}{n_1 n_2}(d_n - a_n^2),$$

where $a_n = E_0[F(X + \delta/\sqrt{n}) - F(X)]$, $b_n = E_0[(F(Y) - F(Y - \delta/\sqrt{n}))^2]$, $c_n = E_0[(F(Y) - F(Y - \delta/\sqrt{n}))]$, and $d_n = E_0[(F(X + \delta/\sqrt{n}) - F(X))^2]$. Using the Lebesgue dominated convergence theorem, it is easy to see that a_n, b_n, c_n, and d_n all converge to 0. Therefore condition (1.5.17) of Theorem 1.5.6 holds and we have thus established the asymptotic linearity result given by

$$\sup_{|\delta| \leqslant B} \left\{ \left| n^{1/2}\overline{S}_R^+(\delta/\sqrt{n}) - n^{1/2}\overline{S}_R^+(0) + \delta \int f^2(x)\,dx \right| \right\} \xrightarrow{P} 0, \qquad (2.4.20)$$

for any $B > 0$. Therefore, it follows that $S_R^+(\Delta)$ is Pitman regular.

In order to get the efficacy of the MWW test, we need the quantity $\sigma^2(0)$ defined by

$$\sigma^2(0) = \lim_{n \to 0} n\text{Var}_0(\overline{S}_R(0))$$
$$= \lim_{n \to 0} \frac{nn_1 n_2(n+1)}{n_1^2 n_2^2 12} = (12\lambda_1 \lambda_2)^{-1};$$

see expression (1.5.12). Therefore by (1.5.11) the **efficacy of the MWW test** is

$$c_{MWW} = \mu'(0)/\sigma(0) = \sqrt{\lambda_1 \lambda_2}\sqrt{12}\int f^2(x)\,dx = \sqrt{\lambda_1 \lambda_2}\tau^{-1}, \qquad (2.4.21)$$

where τ is the scale parameter given by

$$\tau = \left(\sqrt{12}\int f^2(x)dx\right)^{-1}. \qquad (2.4.22)$$

In Exercise 2.13.10 it is shown that the efficacy of the two-sample pooled t-test is $\sqrt{\lambda_1 \lambda_2}\sigma^{-1}$, where σ^2 is the common variance of X and Y. Hence the efficiency of the MWW test relative to the two-sample t test is the ratio σ^2/τ^2. This of course is the same efficiency as that of the signed-rank Wilcoxon test relative to the one-sample t test; see (1.7.12). In particular if the distribution of X is normal then the efficiency of the MWW test relative to the two-sample t-test is 0.955. For heavier tailed distributions, this efficiency is usually larger than 1; see Example 1.7.1.

As in Chapter 1, it is convenient to summarize the asymptotic linearity result as follows:

$$\sqrt{n}\left\{\frac{\overline{S}_R^+(\delta/\sqrt{n}) - \mu(0)}{\sigma(0)}\right\} = \sqrt{n}\left\{\frac{\overline{S}_R^+(0) - \mu(0)}{\sigma(0)}\right\} - c_{MWW}\delta + o_p(1), \qquad (2.4.23)$$

uniformly for $|\delta| \leqslant B$ and any $B > 0$.

The next theorem is the asymptotic power lemma for the MWW test. As in Chapter 1 (see Theorem 1.5.9), its proof follows from the Pitman regularity of the MWW test.

Theorem 2.4.17 *Under the sequence of local alternatives, (2.4.18),*

$$\lim_{n \to \infty} \gamma(\Delta_n) = P_0[Z \geq z_\alpha - c_{MWW}\delta] = 1 - \Phi\left(z_\alpha - \sqrt{12\lambda_1\lambda_2}\int f^2(t)\,dt\,\delta\right),$$

where Z is $N(0, 1)$.

In Exercise 2.13.10, it is shown that if $\gamma_{LS}(\Delta)$ denotes the power function of the usual two-sample t-test then

$$\lim_{n \to \infty} \gamma_{LS}(\Delta_n) = 1 - \Phi\left[z_\alpha - \sqrt{\lambda_1\lambda_2}\frac{\delta}{\sigma}\right], \tag{2.4.24}$$

where σ^2 is the common variance of X and Y. By comparing these two power functions, it is seen that the Wilcoxon test is asymptotically more powerful if $\tau < \sigma$, i.e. if $e = c^2_{MWW}/c^2_t > 1$.

As an application of the asymptotic power lemma, we consider **sample size determination**. Consider the MWW test for the one-sided hypothesis (2.4.14). Suppose the level, α, and the power, β, for a particular alternative Δ_A are specified. For convenience, assume equal sample sizes, i.e. $n_1 = n_2 = n^*$, where n^* denotes the common sample size; hence, $\lambda_1 = \lambda_2 = 2^{-1}$. Express Δ_A as $\sqrt{2n^*}\Delta_A/\sqrt{2n^*}$. Then by Theorem 2.4.17 we have

$$\beta \doteq 1 - \Phi\left[z_\alpha - \sqrt{\frac{1}{4}}\frac{\sqrt{2n^*}\Delta_A}{\tau}\right].$$

But this implies

$$z_\beta = z_\alpha - \tau^{-1}\sqrt{n^*}\Delta_A/\sqrt{2},$$

$$n^* = \left(\frac{z_\alpha - z_\gamma}{\Delta_A}\right)^2 2\tau^2. \tag{2.4.25}$$

The above value of n^* is the approximate sample size. Note that it does depend on τ which, in applications, would have to be guessed or estimated in a pilot study; see the discussion in Section 2.4.5 (estimates of τ are discussed in Sections 2.4.5 and 3.7.1). For a specified distribution it can be evaluated; for instance, if the underlying density is assumed to be normal with standard deviation σ then $\tau = \sqrt{\pi/3}\sigma$.

Using (2.4.24), a similar derivation can be obtained for the usual two-sample t-test, resulting in an approximate sample size of

$$n^*_{LS} = \left(\frac{z_\alpha - z_\gamma}{\Delta_A}\right)^2 2\sigma^2.$$

The ratio of the sample size needed by the MWW test to that of the two-sample t-test is τ^2/σ^2. This provides additional motivation for the definition of efficiency.

2.4.4 Estimation of Δ

Recall from the geometry earlier in this chapter that the estimate of Δ based on the rank pseudo-norm is $\widehat{\Delta}_R = \text{med}_{i,j}\{Y_j - X_i\}$; see (2.2.18). We now obtain several properties of this estimate, including its asymptotic distribution. This will lead again to the efficiency properties of the rank-based methods discussed in the previous section.

For convenience, we note some equivariances of $\widehat{\Delta}_R = \widehat{\Delta}(Y, X)$, which are established in Exercise 2.13.11. First, $\widehat{\Delta}_R$ is translation-equivariant; i.e.

$$\widehat{\Delta}_R(Y + \Delta + \theta, X + \theta) = \widehat{\Delta}_R(Y, X) + \Delta,$$

for any Δ and θ. Second, $\widehat{\Delta}_R$ is scale-equivariant; i.e.

$$\widehat{\Delta}_R(aY, aX) = a\widehat{\Delta}_R(Y, X),$$

for any a. Based on these we next show that $\widehat{\Delta}_R$ is an unbiased estimate of Δ under certain conditions.

Theorem 2.4.18 *If the errors, e_i^*, in the location model (2.2.4) are symmetrically distributed about 0, then $\widehat{\Delta}_R$ is symmetrically distributed about Δ.*

Proof. Due to translation equivariance there is no loss of generality in assuming that Δ and θ are 0. Then Y and X are symmetrically distributed about 0; hence, $\mathcal{L}(Y) = \mathcal{L}(-Y)$ and $\mathcal{L}(X) = \mathcal{L}(-X)$. Thus, from the above equivariance properties, we have

$$\mathcal{L}(-\widehat{\Delta}(Y, X)) = \mathcal{L}(\widehat{\Delta}(-Y, -X)) = \mathcal{L}(\widehat{\Delta}(Y, X)).$$

Therefore $\widehat{\Delta}_R$ is symmetrically distributed about 0, and, in general, it is symmetrically distributed about Δ.

Theorem 2.4.19 *Under model (2.2.4), if $n_1 = n_2$ then $\widehat{\Delta}_R$ is symmetrically distributed about Δ.*

The reader is asked to prove this in Exercise 2.13.12. In general, $\widehat{\Delta}_R$ may be biased if the error distribution is not symmetrically distributed, but, as the following result shows, $\widehat{\Delta}_R$ is always asymptotically unbiased. Since the MWW process $S_R^+(\Delta)$ was shown to be Pitman regular the asymptotic distribution of $\sqrt{n}(\widehat{\Delta} - \Delta)$ is $N(0, c_{MWW}^{-2})$. In practice, we say

$\widehat{\Delta}_R$ has an approximate $N(\Delta, \tau^2(n_1^{-1} + n_2^{-1}))$ distribution,

where τ was defined in (2.4.22).

Recall from Definition 1.5.4 of Chapter 1 that the asymptotic relative efficiency of two Pitman regular estimators is the reciprocal of the ratio of their asymptotic variances. As Exercise 2.13.10 shows, the least-squares estimate $\widehat{\Delta}_{LS} = \overline{Y} - \overline{X}$ of Δ is approximately $N(\Delta, \sigma^2(1/n_1 + 1/n_2))$; hence,

$$e(\widehat{\Delta}_R, \widehat{\Delta}_{LS}) = \frac{\sigma^2}{\tau^2} = 12\sigma_f^2 \left(\int f^2(x)\, dx \right)^2.$$

This agrees with the asymptotic relative efficiency results for the MWW test relative to the t-test and (1.7.12).

2.4.5 Efficiency Results Based on Confidence Intervals

Let $L_{1-\alpha}$ be the length of the $(1-\alpha)100\%$ distribution-free confidence interval based on the MWW statistic discussed in Section 2.4.2. Since this interval is based on the Pitman regular process $S_R^+(\Delta)$, it follows from Theorem 1.5.10 of Chapter 1 that

$$\sqrt{\frac{n_1 n_2}{n}} \frac{L_{1-\alpha}}{2z_{\alpha/2}} \xrightarrow{P} \tau; \qquad (2.4.26)$$

that is, the standardized length of a distribution-free confidence interval is a consistent estimate of the scale parameter τ. It further follows from (2.4.26) that, as in Chapter 1, if efficiency is based on the relative squared asymptotic lengths of confidence intervals then we obtain the same efficiency results as quoted above for tests and estimates.

The distribution-free confidence interval is not symmetric about $\widehat{\Delta}_R$. Often in practice symmetric intervals are desired. Based on the asymptotic distribution of $\widehat{\Delta}_R$ we can formulate the approximate interval

$$\widehat{\Delta}_R \pm z_{\alpha/2} \hat{\tau} \sqrt{\frac{1}{n_1} + \frac{1}{n_2}}, \qquad (2.4.27)$$

where $\hat{\tau}$ is a consistent estimate of τ. If we use (2.4.26) as our estimate of τ with level α, then the confidence interval simplifies to

$$\widehat{\Delta}_R \pm \frac{L_{1-\alpha}}{2}. \qquad (2.4.28)$$

Besides the estimate given in (2.4.26), a consistent estimate of τ was proposed by by Koul, Sievers, and McKean (1987) and will be discussed in Section 3.7. Using this estimate, small-sample studies indicate that $z_{\alpha/2}$ should be replaced by the t critical value $t_{(\alpha/2, n-1)}$; see McKean and Sheather (1991) for a review of small-sample studies on R-estimates. In this case, the symmetric confidence interval based on $\widehat{\Delta}_R$ is directly analogous to the usual t-interval based on least squares in that the only difference is that $\hat{\sigma}$ is replaced by $\hat{\tau}$.

Example 2.4.1 *Hendy and Charles Coin Data, continued from Examples 1.12.1 and 2.3.2*

Recall from Chapter 1 that this example concerned the silver content in two coinages (the first and the fourth) minted during the reign of Manuel I. The data are given in Chapter 1. The Hodges–Lehmann estimate of the difference between the first and the fourth coinage is 1.10% of silver and a 95% confidence interval for the difference is (0.60, 1.70). The length of this confidence interval is 1.10; hence, the estimate of τ given in expression (2.4.26) is 0.54. The symmetrized confidence interval (2.4.27) based on the t upper 0.025 critical value is (0.50, 1.70). Both of these intervals are in agreement with the confidence interval obtained in Example 1.12.1 based on two L_1 confidence intervals.

Another estimate of τ can be obtained from a similar consideration of the distribution-free confidence intervals based on the signed-rank statistic discussed in Chapter 1; see Exercise 2.13.13. Note in this case for consistency, though, we would have to assume that f is symmetric.

2.5 General Rank Pseudo-Norms

In this section we will be concerned with the location model; i.e. $X_1, \ldots, X_{n_1}$ are iid $F(x)$, $Y_1, \ldots, Y_{n_2}$ are iid $G(x) = F(x - \Delta)$, and the samples are independent of one another. We will present an analysis for this problem based on general rank scores. In this terminology, the Mann–Whitney–Wilcoxon procedures are based on a linear score function. We will present the results for the hypotheses

$$H_0 : \Delta = 0 \text{ versus } H_0 : \Delta > 0. \tag{2.5.1}$$

The results for the other one-sided and two-sided alternatives are similar. We will also be concerned with estimation and confidence intervals for Δ. As in the preceeding sections, we will first present the geometry.

Recall that the pseudo-norm which generated the MWW analysis could be written as a linear combination of ranks times residuals. This is easily generalized. Consider the function

$$\|\mathbf{u}\|_* = \sum_{i=1}^n a(R(u_i))u_i, \tag{2.5.2}$$

where $a(i)$ are scores such that $a(1) \leq \cdots \leq a(n)$ and $\sum a(i) = 0$. For the next theorem, we will also assume that $a(i) = -a(n+1-i)$, although this is only used to show the scalar multiplicative property.

Theorem 2.5.1 *Suppose that* $a(1) \leq \cdots \leq a(n)$, $\sum a(i) = 0$, *and* $a(i) = -a(n+1-i)$. *Then the function* $\|\cdot\|_*$ *is a pseudo-norm.*

Proof. By the connection between ranks and order statistics we can write

$$\|\mathbf{u}\|_* = \sum_{i=1}^n a(i)u_{(i)}.$$

Next suppose that $u_{(j)}$ is the last order statistic with a negative score. Since the scores sum to 0, we can write

$$\|\mathbf{u}\|_* = \sum_{i=1}^n a(i)(u_{(i)} - u_{(j)})$$

$$= \sum_{i \leq j} a(i)(u_{(i)} - u_{(j)}) + \sum_{i \geq j} a(i)(u_{(i)} - u_{(j)}). \tag{2.5.3}$$

Both terms on the right-hand side are nonnegative; hence, $\|\mathbf{u}\|_* \geq 0$. Since all the terms in (2.5.3) are nonnegative, $\|\mathbf{u}\|_* = 0$ implies that all the terms are zero. But since the scores are not all 0, yet sum to zero, we must have $a(1) < 0$ and $a(n) > 0$. Hence, we must have $u_{(1)} = u_{(j)} = u_{(n)}$; i.e. $u_{(1)} = \cdots = u_{(n)}$. Conversely if $u_{(1)} = \cdots = u_{(n)}$ then $\|\mathbf{u}\|_* = 0$. By the condition $a(i) = -a(n+1-i)$ it follows that $\|\alpha\mathbf{u}\|_* = |\alpha|\|\mathbf{u}\|_*$; see Exercise 2.13.16.

In order to complete the proof we need to show the triangle inequality holds. This is established by the following string of inequalities:

$$\|\mathbf{u}+\mathbf{v}\|_* = \sum_{i=1}^{n} a(R(u_i+v_i))(u_i+v_i)$$

$$= \sum_{i=1}^{n} a(R(u_i+v_i))u_i + \sum_{i=1}^{n} a(R(u_i+v_i))v_i$$

$$\leq \sum_{i=1}^{n} a(i)u_{(i)} + \sum_{i=1}^{n} a(i)v_{(i)}$$

$$= \|\mathbf{u}\|_* + \|\mathbf{v}\|_*.$$

The proof of the above inequality is similar to that of Theorem 1.3.2 of Chapter 1.

Based on a set of scores satisfying the above assumptions, we can establish a rank inference for the two-sample problem similar to the MWW analysis. We shall do so for general rank scores of the form

$$a_\varphi(i) = \varphi(i/(n+1)), \tag{2.5.4}$$

where $\varphi(u)$ satisfies the following assumptions:

$$\begin{cases} \varphi(u) \text{ is a nondecreasing function defined on the interval } (0,1) \\ \int_0^1 \varphi(u)\,du = 0 \text{ and } \int_0^1 \varphi^2(u)\,du = 1; \end{cases} \tag{2.5.5}$$

see (S.1), (3.4.10) in Chapter 3, also. The last assumptions concerning standardization of the scores are for convenience. The Wilcoxon scores are generated in this way by the linear function $\varphi_R(u) = \sqrt{12}(u - \frac{1}{2})$ and the sign scores are generated by $\varphi_S(u) = \text{sgn}(2u - 1)$. We will denote the corresponding pseudo-norm for scores generated by $\varphi(u)$ as

$$\|\mathbf{u}\|_\varphi = \sum_{i=1}^{n} a_\varphi(R(u_i))u_i. \tag{2.5.6}$$

These two-sample sign and Wilcoxon scores are generalizations of the sign and Wilcoxon scores discussed in Chapter 1 for the one-sample problem. In Section 1.8 we presented one-sample analyses based on general score functions. Similar to the sign and Wilcoxon cases, we can generate a two-sample score function from any one-sample score function. For reference we establish this in the following theorem.

Theorem 2.5.2 *As discussed at the beginning of Section 1.8, let $\varphi^+(u)$ be a score function for the one-sample problem. For $u \in (-1, 0)$, let $\varphi^+(u) = -\varphi^+(-u)$. Define*

$$\varphi(u) = \varphi^+(2u - 1), \quad \text{for } u \in (0, 1). \tag{2.5.7}$$

and
$$\|\mathbf{x}\|_\varphi = \sum_{i=1}^n \varphi(R(x_i)/(n+1))x_i. \tag{2.5.8}$$

Then $\|\cdot\|_\varphi$ is a pseudo-norm on $\mathcal{R}^n$. Furthermore,
$$\varphi(u) = -\varphi(1-u), \tag{2.5.9}$$
and
$$\int_0^1 \varphi^2(u)\,du = \int_0^1 (\varphi^+(u))^2\,du. \tag{2.5.10}$$

Proof. As discussed at the beginning of Section 1.8, (see expression (1.8.1)), $\varphi^+(u)$ is a positive-valued and nondecreasing function defined on the interval $(0, 1)$. Based on these properties, it follows that $\varphi(u)$ is nondecreasing and that $\int_0^1 \varphi(u)\,du = 0$. Hence, $\|\cdot\|_\varphi$ is a pseudo-norm on $\mathcal{R}^n$. Properties (2.5.9) and (2.5.10) follow readily; see Exercise 2.13.17 for details.

The two-sample sign and Wilcoxon scores, cited above, are easily seen to be generated this way from their one-sample counterparts $\varphi^+(u) = 1$ and $\varphi^+(u) = \sqrt{3}u$, respectively. As discussed further in Section 2.5.3, properties such as efficiencies of the analysis based on the one-sample scores are the same for a two-sample analysis based on their corresponding two-sample scores.

In the notation of (2.2.3), the estimate of Δ is
$$\widehat{\Delta}_\varphi = \text{Argmin } \|\mathbf{Z} - \mathbf{C}\Delta\|_\varphi.$$

Denote the negative of the gradient of $\|\mathbf{Z} - \mathbf{C}\Delta\|_\varphi$ by $S_\varphi(\Delta)$. Then, based on (2.5.6),
$$S_\varphi(\Delta) = \sum_{j=1}^{n_2} a_\varphi(R(Y_j - \Delta)). \tag{2.5.11}$$

Hence $\widehat{\Delta}_\varphi$ equivalently solves the equation,
$$S_\varphi(\widehat{\Delta}_\varphi) \doteq 0. \tag{2.5.12}$$

As with pseudo-norms in general, the function $\|\mathbf{Z} - \mathbf{C}\Delta\|_\varphi$ is a convex function of Δ. The negative of its derivative, $S_\varphi(\Delta)$, is a decreasing step function of Δ which steps down at the differences $Y_j - X_i$; see Exercise 2.13.18. Unlike the MWW function $S_R(\Delta)$, the step sizes of $S_\varphi(\Delta)$ are not necessarily the same size. Simple iterative algorithms, however, are sufficient to obtain the estimates; details are discussed in Section 3.7.3.

The gradient rank test statistic for the hypotheses (2.5.1) is
$$S_\varphi = \sum_{j=1}^{n_2} a_\varphi(R(Y_j)). \tag{2.5.13}$$

96 Two-Sample Problems

Since the test statistic only depends on the ranks of the combined sample it is distribution-free under the null hypothesis. As shown in Exercise 2.13.18,

$$E_0[S_\varphi] = 0 \tag{2.5.14}$$

$$\sigma_\varphi^2 = \mathrm{Var}_0[S_\varphi] = \frac{n_1 n_2}{n(n-1)} \sum_{i=1}^n a^2(i). \tag{2.5.15}$$

Note that we can write the variance as

$$\sigma_\varphi^2 = \frac{n_1 n_2}{n-1} \left\{ \sum_{i=1}^n a^2(i) \frac{1}{n} \right\} \doteq \frac{n_1 n_2}{n-1}, \tag{2.5.16}$$

where the approximation is due to the fact that the term in braces is a Riemann sum of $\int \varphi^2(u) du = 1$ and, hence, converges to 1.

It will be convenient from time to time to use rank statistics based on unstandardized scores; i.e. a rank statistic of the form

$$S_a = \sum_{j=1}^{n_2} a(R(Y_j)), \tag{2.5.17}$$

where $a(i) = \varphi(i/(n+1))$, $i = 1, \ldots, n$ is a set of scores. As Exercise 2.13.18, shows the null mean μ_S and null variance σ_S^2 of S_a are given by

$$\mu_S = n_2 \bar{a} \quad \text{and} \quad \sigma_S^2 = \frac{n_1 n_2}{n(n-1)} \sum (a(i) - \bar{a})^2. \tag{2.5.18}$$

2.5.1 Statistical Methods

The asymptotic null distribution of the statistic S_φ, (2.5.13), easily follows from Theorem A.2.1 of the Appendix. To see this, note that we can use the notation (2.2.1) and (2.2.2) to write S_φ as a linear rank statistic; i.e.

$$S_\varphi = \sum_{i=1}^n c_i a(R(Z_i)) = \sum_{i=1}^n (c_i - \bar{c}) a\left(\frac{n}{n+1} F_n(Z_i) \right), \tag{2.5.19}$$

where F_n is the empirical distribution function of $Z_1, \ldots, Z_n$. Our score function φ is monotone and square integrable; hence, the conditions on scores in Section A.2 are satisfied. Also F is continuous so the distributional assumption is satisfied. Finally, we need only show that the constants c_i satisfy conditions, (D.2) and (D.3), (3.4.7) and (3.4.8). It is a simple exercise to show that

$$\sum_{i=1}^n (c_i - \bar{c})^2 = \frac{n_1 n_2}{n}$$

$$\max_{1 \leq i \leq n} (c_i - \bar{c})^2 = \max\left\{ \frac{n_2^2}{n^2}, \frac{n_1^2}{n^2} \right\}.$$

Under condition (D.1), (2.4.7), $0 < \lambda_i < 1$ where $\lim(n_i/n) = \lambda_i$ for $i = 1, 2$. Using this along with the last two expressions, it is immediate that Noether's condition, (3.4.9) holds for the c_is. Thus the assumptions of Section A.2 hold for the statistic S_φ.

As in expression (A.2.7) of Section A.2, define the random variable T_φ as

$$T_\varphi = \sum_{i=1}^{n}(c_i - \bar{c})\varphi(F(Z_i)). \tag{2.5.20}$$

By comparing expressions (2.5.19) and (2.5.20), it seems that the variable T_φ is an approximation of S_φ. This follows from Section A.2. Briefly, under H_0 the distribution of T_φ is approximately normal and $\text{Var}((T_\varphi - S_\varphi)/\sigma_\varphi) \to 0$; hence, S_φ is asymptotically normal with mean and variance given by expressions (2.5.14) and (2.5.15), respectively. Hence, an asymptotic level α test of the hypotheses (2.5.1) is

Reject H_0 in favor of H_A if $S_\varphi \geq z_\alpha \sigma_\varphi$,

where σ_φ is defined by (2.5.15).

As discussed above, the estimate $\widehat{\Delta}_\varphi$ of Δ solves the equation (2.5.12). The interval $(\widehat{\Delta}_L, \widehat{\Delta}_U)$ is a $(1 - \alpha)100\%$ confidence interval for Δ (based on the asymptotic distribution) provided $\widehat{\Delta}_L$ and $\widehat{\Delta}_U$ solve the equations

$$S_\varphi(\widehat{\Delta}_U) \doteq -z_{\alpha/2}\sqrt{\frac{n_1 n_2}{n}} \text{ and } S_\varphi(\widehat{\Delta}_L) \doteq z_{\alpha/2}\sqrt{\frac{n_1 n_2}{n}}, \tag{2.5.21}$$

where $1 - \Phi(z_{\alpha/2}) = \alpha/2$. As with the estimate of Δ, these equations can be easily solved with an iterative algorithm; see Section 3.7.3.

2.5.2 Efficiency Results

In order to obtain the efficiency results for these statistics, we first show that the process $S_\varphi(\Delta)$ is Pitman regular. For general scores we need to assume further that the density has finite Fisher information i.e. satisfies condition (E.1), (2.4.16). Recall that Fisher information is given by $I(f) = \int_0^1 \varphi_f^2(u)\,du$, where

$$\varphi_f(u) = -\frac{f'(F^{-1}(u))}{f(F^{-1}(u))}. \tag{2.5.22}$$

Below we will show that the score function φ_f is **optimal**. Define the parameter τ_φ as

$$\tau_\varphi^{-1} = \int \varphi(u)\varphi_f(u)du. \tag{2.5.23}$$

Estimation of τ_φ is dicussed in Section 3.7.

To show that the process $S_\varphi(\Delta)$ is Pitman regular, we show that the four conditions of Definition 1.5.3 are true. As noted after expression (2.5.12), $S_\varphi(\Delta)$ is nonincreasing; hence, the first condition holds. For the second condition, note that we can write

$$S_\varphi(\Delta) = \sum_{i=1}^{n_2} a(R(Y_i - \Delta)) = \sum_{i=1}^{n_2} \varphi\left(\frac{n_1}{n+1}F_{n_1}(Y_i - \Delta) + \frac{n_2}{n+1}F_{n_2}(Y_i)\right), \tag{2.5.24}$$

where F_{n_1} and F_{n_2} are the empirical cdfs of the samples $X_1, \ldots, X_{n_1}$ and $Y_1, \ldots, Y_{n_2}$, respectively. Hence, passing to the limit, we have

98 Two-Sample Problems

$$E_0\left[\frac{1}{n}S_\varphi(\Delta)\right] \to \lambda_2 \int_{-\infty}^{\infty} \varphi[\lambda_1 F(x) + \lambda_2 F(x-\Delta)]f(x-\Delta)\,dx$$

$$= \lambda_2 \int_{-\infty}^{\infty} \varphi[\lambda_1 F(x+\Delta) + \lambda_2 F(x)]f(x)\,dx = \mu_\varphi(\Delta); \qquad (2.5.25)$$

see Chernoff and Savage (1958) for a rigorous proof of the limit. Differentiating $\mu_\varphi(\Delta)$ and evaluating the derivative at 0, we obtain

$$\mu'_\varphi(0) = \lambda_1\lambda_2 \int_{-\infty}^{\infty} \varphi'[F(t)]f^2(t)\,dt$$

$$= \lambda_1\lambda_2 \int_{-\infty}^{\infty} \varphi[F(t)]\left(-\frac{f'(t)}{f(t)}\right)f(t)\,dt$$

$$= \lambda_1\lambda_2 \int_0^1 \varphi(u)\varphi_f(u)\,du = \lambda_1\lambda_2\tau_\varphi^{-1} > 0. \qquad (2.5.26)$$

Hence, the second condition is satisfied.

The null asymptotic distribution of $S_\varphi(0)$ was established in Section 2.5.1; hence the fourth condition is true. Hence, we need only establish asymptotic linearity. This result follows from the results for general rank regression statistics which are developed in Section A.2.2 of the Appendix. By Theorem A.2.8 of the Appendix, the asymptotic linearity result for $S_\varphi(\Delta)$ is given by

$$\frac{1}{\sqrt{n}}S_\varphi(\delta/\sqrt{n}) = \frac{1}{\sqrt{n}}S_\varphi(0) - \tau_\varphi^{-1}\lambda_1\lambda_2\delta + o_p(1), \qquad (2.5.27)$$

uniformly for $|\delta| \leq B$, where $B > 0$ and τ_φ is defined in (2.5.23).

Therefore, following Definition 1.5.3, the estimating function is Pitman regular.

By the discussion following (2.5.20), we have that $n^{-1/2}S_\varphi(0)/\sqrt{\lambda_1\lambda_2}$ is asymptotically $N(0, 1)$. The **efficacy** of the test based on S_φ is thus given by

$$c_\varphi = \frac{\tau_\varphi^{-1}\lambda_1\lambda_2}{\sqrt{\lambda_1\lambda_2}} = \tau_\varphi^{-1}\sqrt{\lambda_1\lambda_2}. \qquad (2.5.28)$$

As with the MWW analysis, several important items follow immediately from Pitman regularity. Consider first the behavior of S_φ under local alternatives. Specifically consider a level α test based on S_φ for the hypothesis (2.5.1) and the sequence of local alternatives $H_n : \Delta_n = \delta/\sqrt{n}$. As in Chapter 1, it is easy to show that the **asymptotic power** of the test based on S_φ is given by

$$\lim_{n\to\infty} P_{\delta/\sqrt{n}}[S_\varphi \geq z_\alpha \sigma_\varphi] = 1 - \Phi(z_\alpha - \delta c_\varphi). \qquad (2.5.29)$$

Based on this result, sample size determination for the test based on S_φ can be conducted similar to that based on the MWW test statistic; see (2.4.25).

Next consider the **asymptotic distribution of the estimator** $\widehat{\Delta}_\varphi$. Recall that the estimate $\widehat{\Delta}_\varphi$ solves the equation $S_\varphi(\widehat{\Delta}_\varphi) \doteq 0$. Based on Pitman regularity and Theorem 1.5.8, the asymptotic distribution $\widehat{\Delta}_\varphi$ is given by

$$\sqrt{n}(\widehat{\Delta}_\varphi - \Delta) \xrightarrow{\mathcal{D}} N(0, \tau_\varphi^2(\lambda_1\lambda_2)^{-1}); \qquad (2.5.30)$$

By using (2.5.27) and $T_\varphi(0)$ to approximate $S_\varphi(0)$, we have the following useful result:

$$\sqrt{n}\widehat{\Delta} = \frac{\tau_\varphi}{\lambda_1\lambda_2}\frac{1}{\sqrt{n}}T_\varphi(0) + o_p(1). \tag{2.5.31}$$

We want to select scores such that the efficacy c_φ, (2.5.28), is as large as possible, or equivalently such that the asymptotic variance of $\widehat{\Delta}_\varphi$ is as small as possible. How large can the efficacy be? Similar to (1.8.26), note that we can write

$$\tau_\varphi^{-1} = \int \varphi(u)\varphi_f(u)du$$

$$= \sqrt{\int \varphi_f^2(u)du} \frac{\int \varphi(u)\varphi_f(u)du}{\sqrt{\int \varphi_f^2(u)du}\sqrt{\int \varphi^2(u)du}}$$

$$= \rho\sqrt{\int \varphi_f^2(u)du}. \tag{2.5.32}$$

The second equation is true since the scores were standardized as above. In the third equation ρ is a correlation coefficient and $\int \varphi_f^2(u)du$ is Fisher location information, (2.4.16), which we denoted by $I(f)$. By the Cramér–Rao lower bound, the smallest asymptotic variance obtainable by an asymptotically unbiased estimate is $(\lambda_1\lambda_2 I(f))^{-1}$. Such an estimate is called **asymptotically efficient**. Choosing a score function to maximize (2.5.32) is equivalent to choosing a score function to make $\rho = 1$. This can be achieved by taking the score function to be $\varphi(u) = \varphi_f(u)$, (2.5.22). The resulting estimate, $\widehat{\Delta}_\varphi$, is asymptotically efficient. Of course this can be accomplished only provided that the form of f is known; see Exercise 2.13.19. Evidently, the closer the chosen score is to φ_f, the more powerful the rank analysis will be.

In Exercise 2.13.19, the reader is asked to show that the MWW analysis is asymptotically efficient if the errors have a logistic distribution. For normal errors, it follows in a few steps from expression (2.4.17) that the optimal scores are generated by the **normal scores** function,

$$\varphi_N(u) = \Phi^{-1}(u), \tag{2.5.33}$$

where $\Phi(u)$ is the distribution function of a standard normal random variable. Exercise 2.13.19 shows that this score function is standardized. These scores yield an asymptotically efficient analysis if the errors truly have a normal distribution and, further, $e(\varphi_N, L_2) \geq 1$; see Theorem 1.8.2. Also, unlike the MWW analysis, the estimate of the shift Δ based on the normal scores cannot be obtained in closed form. But as mentioned above for general scores, provided the score function is nondecreasing, simple iterative algorithms can be used to obtain the estimate and the corresponding confidence interval for Δ. In the next sections we will discuss analyses that are asymptotically efficient for other distributions.

Example 2.5.1 Quail Data, continued from Example 2.3.1

In the larger study, McKean et al. (1989), from which these data were drawn, the responses were positively skewed with long right tails; although outliers frequently occurred in the left tail also. McKean et al. conducted an investigation of estimates of the score functions for over 20 of these experiments. Classes of simple scores which seemed appropriate for such data were piecewise linear with one piece which is linear on the first part on the interval $(0, b)$ and with a second piece which is constant on the second part $(b, 1)$; i.e. scores of the form

$$\varphi_b(u) = \begin{cases} \dfrac{2}{b(2-b)} u - 1 & \text{if } 0 < u < b \\ \dfrac{b}{2-b} & \text{if } b \leq u < 1. \end{cases} \quad (2.5.34)$$

These scores are optimal for densities with left logistic and right exponential tails; see Exercise 2.13.19. A value of b which seemed appropriate for this type of data was $\tfrac{3}{4}$. For this data set, the resulting scores are

$$a_{3/4}(i) = \begin{cases} \tfrac{32}{15} \dfrac{i}{n+1} - 1 & \text{if } 1 \leq i \leq 22 \\ \tfrac{3}{5} & \text{if } 23 \leq i \leq 30. \end{cases}$$

Let $S_{3/4} = \sum a_{3/4}(R(Y_j))$ denote the test statistic based on these scores. It follows that $S_{3/4} = 3.7075$. An easy calculation shows that $E_{H_0}(S_{3/4}) = 0.14$ and $\text{Var}_{H_0}(S_{3/4}) = 0.185$. The resulting z-test statistic has the value 2.63 and a p-value of 0.004. Using an iterative routine, the corresponding point estimate of Δ has the value 16.

For another class of scores similar to (2.5.34), see the discussion around expression (3.10.6) in Chapter 3.

2.5.3 Connection between One- and Two-Sample Scores

In Theorem 2.5.2 we discussed how to obtain a corresponding two-sample score function given a one-sample score function. Here we reverse the problem, showing how to obtain a one-sample score function from a two-sample score function. This will provide a natural estimate of θ in (2.2.4). We also show the efficiencies and asymptotic properties are the same for such corresponding scores functions.

Consider the location model but further assume that X has a symmetric distribution. Then Y also has a symmetric distribution. For associated one-sample problems, we could then use the signed-rank methods developed in Chapter 1. What one-sample scores should we select?

First consider what two-sample scores would be suitable under symmetry. Assume without loss of generality that X is symmetrically distributed about 0. Recall that the optimal scores are given by expression (2.5.22). Using the fact that $F(x) = 1 - F(-x)$, it is easy to see (Exercise 2.13.20) that the optimal scores satisfy

$$\varphi_f(-u) = -\varphi_f(1-u), \text{ for } 0 < u < 1,$$

that is, the optimal score function is odd about $\frac{1}{2}$. Hence, for symmetric distributions, it makes sense to consider two-sample scores which are odd about $\frac{1}{2}$.

For this subsection, then, assume that the two-sample score generating function satisfies the property

$$\text{(S.3)} \quad \varphi(1-u) = -\varphi(u). \tag{2.5.35}$$

Note that such scores satisfy: $\varphi(\frac{1}{2}) = 0$ and $\varphi(u) \geq 0$ for $u \geq \frac{1}{2}$. Define a one-sample score generating function as

$$\varphi^+(u) = \varphi\left(\frac{u+1}{2}\right) \tag{2.5.36}$$

and the one-sample scores as

$$a^+(i) = \varphi^+\left(\frac{i}{n+1}\right). \tag{2.5.37}$$

It follows that these one-sample scores are nonnegative and nonincreasing.

For example, if we use Wilcoxon two-sample scores, that is, scores generated by the function $\varphi(u) = \sqrt{12}(u - \frac{1}{2})$, then the associated one-sample score generating function is $\varphi^+(u) = \sqrt{3}u$ and, hence, the one-sample scores are the Wilcoxon signed-rank scores. If instead we use the two-sample sign scores, $\varphi(u) = \text{sgn}(2u - 1)$, then the one-sample score function is $\varphi^+(u) = 1$. This results in the one-sample sign scores.

Suppose we use two-sample scores which satisfy (2.5.35) and use the associated one-sample scores. Then the corresponding one- and two-sample efficacies satisfy

$$c_\varphi = \sqrt{\lambda_1 \lambda_2} c_{\varphi^+}, \tag{2.5.38}$$

where the efficacies are given by expressions (2.5.28) and (1.8.21). Hence, the efficiency and asymptotic properties of the one- and two-sample analyses are the same. As a final remark, if we write the model as in expression (2.2.4), then we can use the rank statistic based on the two-sample analysis to estimate Δ. We next form the residuals $Z_i - \hat{\Delta} c_i$. Then using the one-sample scores statistic of Chapter 1, we can estimate θ based on these residuals, as discussed in Chapter 1. In terms of a regression problem, we are estimating the intercept parameter θ based on the residuals after fitting the regression coefficient Δ. This is discussed in some detail in Section 3.5.

2.6 L_1 **Analyses**

In this section, we present analyses based on the L_1 norm and pseudo-norm. We discuss the pseudo-norm first, showing that the corresponding test is the familiar Mood (1950) test. The test which corresponds to the norm is Mathisen's (1943) test.

2.6.1 Analysis Based on the L_1 Pseudo-norm

Consider the sign scores. These are the scores generated by the function $\varphi(u) = \text{sgn}(u - \frac{1}{2})$. The corresponding pseudo-norm is given by,

$$\|\mathbf{u}\|_\varphi = \sum_{i=1}^n \text{sgn}\left(R(u_i) - \frac{n+1}{2}\right) u_i. \tag{2.6.1}$$

This pseudo-norm is optimal for double exponential errors; see Exercise 2.13.19.

We have the following relationship between the L_1 pseudo-norm and the L_1 norm. Note that we can write

$$\|\mathbf{u}\|_\varphi = \sum_{i=1}^n \text{sgn}\left(i - \frac{n+1}{2}\right) u_{(i)}.$$

Next, consider

$$\sum_{i=1}^n |u_{(i)} - u_{(n-i+1)}| = \sum_{i=1}^n \text{sgn}(u_{(i)} - u_{(n-i+1)})(u_{(i)} - u_{(n-i+1)})$$

$$= 2 \sum_{i=1}^n \text{sgn}(u_{(i)} - u_{(n-i+1)}) u_{(i)}.$$

Finally, note that

$$\text{sgn}(u_{(i)} - u_{(n-i+1)}) = \text{sgn}(i - (n - i + 1)) = \text{sgn}\left(i - \frac{n+1}{2}\right).$$

Putting these results together, we have the relationship

$$\sum_{i=1}^n |u_{(i)} - u_{(n-i+1)}| = 2 \sum_{i=1}^n \text{sgn}\left(i - \frac{n+1}{2}\right) u_{(i)} = 2\|\mathbf{u}\|_\varphi. \tag{2.6.2}$$

Recall that the pseudo-norm-based Wilcoxon scores can be expressed as the sum of all absolute differences between the components; see (2.2.17). In contrast the pseudo-norm based on the sign scores only involves the n symmetric absolute differences $|u_{(i)} - u_{(n-i+1)}|$.

In the two-sample location model the corresponding R-estimate based on the pseudo-norm (2.6.1) is a value of Δ which solves the equation

$$S_\varphi(\Delta) = \sum_{j=1}^{n_2} \text{sgn}\left(R(Y_j - \Delta) - \frac{n+1}{2}\right) \doteq 0. \tag{2.6.3}$$

Note that we are ranking the set $\{X_1, \ldots, X_{n_1}, Y_1 - \Delta, \ldots, Y_{n_2} - \Delta\}$ which is equivalent to ranking the set $\{X_1 - \text{med } X_i, \ldots, X_{n_1} - \text{med } X_i, Y_1 - \Delta - \text{med } X_i, \ldots, Y_{n_2} - \Delta - \text{med } X_i\}$. We must choose Δ so that half of the ranks of the Y part of this set are above $(n + 1)/2$ and half are below. Note that in the X part of the second set, half of the X part is below 0 and half is above 0. Thus we need to choose Δ so that half of the Y part of this set is below 0 and half is above 0. This is achieved by taking

$$\widehat{\Delta} = \text{med } Y_j - \text{med } X_i. \tag{2.6.4}$$

This is the same estimate as produced by the L_1 norm, see the discussion following (2.2.5). We shall refer to the above pseudo-norm (2.6.1) as the L_1 pseudo-norm. Actually, as pointed out in Section 2.2, this equivalence between estimates based on the L_1 norm and the L_1 pseudo-norm is true for general regression problems in which the model includes an intercept, as it does here.

The corresponding test statistic for $H_0 : \Delta = 0$ is $\sum_{j=1}^{n_2} \text{sgn}(R(Y_j) - (n+1)/2)$. Note that the sgn function here is only counting the number of Y_js which are above the combined sample median $\widehat{M} = \text{med } \{X_1, \ldots, X_{n_1}, Y_1, \ldots, Y_{n_2}\}$ minus the number below $\widehat{M}$. Hence, a more convenient but equivalent test statistic is

$$M_0^+ = \#(Y_j > \widehat{M}), \tag{2.6.5}$$

which is called **Mood's median test** statistic; see Mood (1950).

Testing

Since this L_1 analysis is based on a rank-based pseudo-norm we could use the general theory discussed in Section 2.5 to handle the theory for estimation and testing. As we will point out, though, there are some interesting results pertaining to this analysis.

For the null distribution of M_0^+, first assume that n is even. Without loss of generality, assume that $n = 2r$ and $n_1 \geq n_2$. Consider the combined sample as a population of n items, where n_2 of the items are Ys and n_1 items are Xs. Think of the $n/2$ items which exceed $\widehat{M}$. Under H_0 each of these items is as likely to be an X as a Y. Hence, M_0^+, the number of Ys in the top half of the sample, follows the hypergeometric distribution, i.e.

$$P(M_0^+ = k) = \frac{\binom{n_2}{k}\binom{n_1}{r-k}}{\binom{n}{r}} \quad k = 0, \ldots, n_2,$$

where $r = n/2$. If n is odd the same result holds, except in this case $r = (n-1)/2$. Thus as a level α decision rule, we would reject $H_0 : \Delta = 0$ in favor of $H_A : \Delta > 0$, if $M_0^+ \geq c_\alpha$, where c_α could be determined from the hypergeometric distribution or approximated by the binomial distribution. From the properties of the hypergeometric distribution, $E_0[M_0^+] = r(n_2/n)$ and $\text{Var}_0[M_0^+] = (rn_1 n_2 (n-r))/(n^2(n-1))$. Under assumption (D.1), (2.4.7), it follows that the limiting distribution of M_0^+ is normal.

Confidence Intervals

Exercise 2.13.21 shows that, for $n = 2r$,

$$M_0^+(\Delta) = \#(Y_j - \Delta > \widehat{M}) = \sum_{i=1}^{n_2} I(Y_{(i)} - X_{(r-i+1)} - \Delta > 0), \tag{2.6.6}$$

and, furthermore, that the $n = 2r$ differences,

104 Two-Sample Problems

$$Y_{(1)} - X_{(r)} < Y_{(2)} - X_{(r-1)} < \cdots < Y_{(n_2)} - X_{(r-n_2+1)},$$

can be ordered only knowing the order statistics from the individual samples. It is further shown that if k is such that $P(M_0^+ \leq k) = \alpha/2$ then a $(1-\alpha)100\%$ confidence interval for Δ is given by

$$(Y_{(k+1)} - X_{(r-k)}, Y_{(n_2-k)} - X_{(r-n_2+k+1)}).$$

The above confidence interval simplifies when $n_1 = n_2 = m$, say. In this case the interval becomes

$$(Y_{(k+1)} - X_{(m-k)}, Y_{(m-k)} - X_{(k+1)}),$$

which is the difference in endpoints of the two simple L_1 confidence intervals $(X_{(k+1)}, X_{(m-k)})$ and $(Y_{(k+1)}, Y_{(m-k)})$ which were discussed in Section 1.12. Using the normal approximation to the hypergeometric distribution we have $k = m/2 - Z_{\alpha/2}\sqrt{m^2/(4(2m-1))} - 0.5$. Hence, the above two intervals have confidence coefficient

$$\gamma \doteq 1 - 2\Phi\left(\frac{k - m/2}{\sqrt{m/4}}\right) = 1 - 2\Phi\left(z_{\alpha/2}\sqrt{m/(2m-1)}\right)$$
$$\doteq 1 - 2\Phi(z_{\alpha/2}2^{-1/2}).$$

For example, for the equal sample size case, a 5% two-sided Mood test is equivalent to rejecting the null hypothesis if the 84% one-sample L_1 confidence intervals are disjoint. While this also could be done for the unequal sample sizes case, we recommend the direct approach of Section 1.12.

Efficiency Results

We will obtain the efficiency results from the asymptotic distribution of the estimate, $\hat{\Delta} = \text{med } Y_j - \text{med } X_i$, of Δ. Equivalently, we could obtain the results by asymptotic linearity that was derived for arbitrary scores in (2.5.27); see Exercise 2.13.22.

Theorem 2.6.1 *Under the conditions cited in Example 1.5.2 (L_1 Pitman regularity conditions), and (2.4.7), we have*

$$\sqrt{n}(\hat{\Delta} - \Delta) \xrightarrow{\mathcal{D}} N(0, (\lambda_1\lambda_2 4f^2(0))^{-1}). \qquad (2.6.7)$$

Proof. Without loss of generality assume that Δ and θ are 0. We can write

$$\sqrt{n}\hat{\Delta} = \sqrt{\frac{n}{n_2}}\sqrt{n_2}\text{med } Y_j - \sqrt{\frac{n}{n_1}}\sqrt{n_1}\text{med } X_i.$$

From Example 1.5.2, we have

$$\sqrt{n_2}\text{med } Y_j = \frac{1}{2f(0)}\frac{1}{\sqrt{n_2}}\sum_{j=1}^{n_2} \text{sgn } Y_j + o_p(1);$$

hence, $\sqrt{n_2}$med $Y_j \xrightarrow{\mathcal{D}} Z_2$, where Z_2 is $N(0,(4f^2(0))^{-1})$. Likewise $\sqrt{n_1}$med $X_i \xrightarrow{\mathcal{D}} Z_1$, where Z_1 is $N(0,(4f^2(0))^{-1})$. Since Z_1 and Z_2 are independent, we have that $\sqrt{n}\widehat{\Delta} \xrightarrow{\mathcal{D}} (\lambda_2)^{-1/2}Z_2 - (\lambda_1)^{-1/2}Z_1$ which yields the result.

The efficacy of Mood's test is thus $\sqrt{\lambda_1\lambda_2}2f(0)$. The asymptotic efficiency of Mood's test relative to the two-sample t-test is $4\sigma^2 f^2(0)$, while its asymptotic efficiency relative to the MWW test is $f^2(0)/(3(\int f^2)^2)$. These are the same as the efficiency results of the sign test relative to the t-test and to the Wilcoxon signed-rank test, respectively, that were obtained in Chapter 1; see Section 1.7.

Example 2.6.1 *Quail Data, continued, Example 2.3.1*

For the quail data the median of the combined samples is $\widehat{M} = 64$. For the subsequent test based on Mood's test we eliminated the three data points which had this value. Thus $n = 27$, $n_1 = 9$ and $n_2 = 18$. The value of Mood's test statistic is $M_0^+ = \#(P_j > 64) = 11$. Since $E_{H_0}(M_0^+) = 8.67$ and $\text{Var}_{H_0}(M_0^+) = 1.55$, the standardized value (using the continuity correction) is 1.47 with a p-value of 0.071. Using all the data, the point estimate corresponding to Mood's test is 19 while a 90% confidence interval, using the normal approximation, is $(-10, 31)$.

2.6.2 Analysis Based on the L_1 Norm

Another sign-type procedure is based on the L_1 norm. Reconsider expression (2.2.7), which is the partial derivative of the L_1 dispersion function with respect to Δ. We take the parameter θ as a nuisance parameter and we estimate it by med X_i. An **aligned sign test** procedure for Δ is then obtained by aligning the Y_js with respect to this estimate of θ. The process of interest, then, is

$$S(\Delta) = \sum_{j=1}^{n_2} \text{sgn}(Y_j - \text{med } X_i - \Delta).$$

A test of $H_0 : \Delta = 0$ is based on the statistic

$$M_a^+ = \#(Y_j > \text{med } X_i). \tag{2.6.8}$$

This statistic was proposed by Mathisen (1943) and is also referred to as the control median test; see Gastwirth (1968). The estimate of Δ obtained by solving $S(\Delta) \doteq 0$ is, of course, the L_1 estimate $\widetilde{\Delta} = \text{med } Y_j - \text{med } X_i$.

Testing

Mathisen's test statistic, similar to Mood's, has a hypergeometric distribution under H_0.

Theorem 2.6.2 *Suppose n_1 is odd and is written as $n_1 = 2n_1^* + 1$. Then under $H_0 : \Delta = 0$,*

$$P(M_a^+ = t) = \frac{\binom{n_1^* + t}{n_1^*}\binom{n_2 - t + n_1^*}{n_1^*}}{\binom{n}{n_1}}, \quad t = 0, 1, \ldots, n_2.$$

Proof. The proof will be based on a conditional argument. Given $X_{(n_1^*+1)} = x$, M_a^+ is binomial with n_2 trials and $1 - F(x)$ as the probability of success. The density of $X_{(n_1^*+1)}$ is

$$f^*(x) = \frac{n_1!}{(n_1^*!)^2}(1 - F(x))^{n_1^*} F(x)^{n_1^*} f(x).$$

Using this and the fact that the samples are independent, we get

$$P(M_a^+ = t) = \int \binom{n_2}{t}(1 - F(x))^t F(x)^{n_2 - t} f(x) dx$$

$$= \binom{n_2}{t} \frac{n_1!}{(n_1^*!)^2} \int (1 - F(x))^{t + n_1^*} F(x)^{n_1^* + n_2 - t} f(x) dx$$

$$= \binom{n_2}{t} \frac{n_1!}{(n_1^*!)^2} \int_0^1 (1 - u)^{t + n_1^*} u^{n_1^* + n_2 - t} du.$$

By properties of the β function this reduces to the result.

Once again using the conditional argument, we obtain the moments of M_a^+ as

$$E_0[M_a^+] = \frac{n_2}{2} \tag{2.6.9}$$

$$\text{Var}_0[M_a^+] = \frac{n_2(n+1)}{4(n_1+2)}; \tag{2.6.10}$$

see Exercise 2.13.23.

The result when n_1 is even is found in Exercise 2.13.23. For the asymptotic null distribution of M_a^+ we shall make use of the linearity result for the sign process derived in Chapter 1; see Example 1.5.2.

Theorem 2.6.3 *Under H_0 and (D.1), (2.4.7), M_a^+ has an approximate $N(n_2/2, n_2(n + 1)/4(n_1 + 2))$ distribution.*

Proof. Assume without loss of generality that the true median of X and Y is 0. Let $\widehat{\theta} = \text{med } X_i$. Note that

$$M_a^+ = \left(\sum_{j=1}^{n_2} \text{sgn}(Y_j - \widehat{\theta}) + n_2\right)/2. \tag{2.6.11}$$

Clearly under (D.1), $\sqrt{n_2}\widehat{\theta}$ is bounded in probability. Hence by the asymptotic linearity result for the L_1 analysis, obtained in Example 1.5.2, we have

$$n_2^{-1/2} \sum_{j=1}^{n_2} \text{sgn}(Y_j - \widehat{\theta}) = n_2^{-1/2} \sum_{j=1}^{n_2} \text{sgn}(Y_j) - 2f(0)\sqrt{n_2}\widehat{\theta} + o_p(1).$$

But we also have

$$\sqrt{n_1}\widehat{\theta} = (2f(0)\sqrt{n_1})^{-1} \sum_{i=1}^{n_1} \text{sgn}(X_i) + o_p(1).$$

Therefore

$$n_2^{-1/2} \sum_{j=1}^{n_2} \text{sgn}(Y_j - \widehat{\theta}) = n_2^{-1/2} \sum_{j=1}^{n_2} \text{sgn}(Y_j) - \sqrt{n_2/n_1} n_1^{-1/2} \sum_{i=1}^{n_1} \text{sgn}(X_i) + o_p(1).$$

Note that

$$n_2^{-1/2} \sum_{j=1}^{n_2} \text{sgn}(Y_j) \xrightarrow{D} N(0, \lambda_1^{-1})$$

and

$$\sqrt{n_2/n_1} n_1^{-1/2} \sum_{i=1}^{n_1} \text{sgn}(X_i) \xrightarrow{D} N(0, \lambda_2/\lambda_1).$$

The result follows from these asymptotic distributions, the independence of the samples, expression (2.6.11), and the fact that asymptotically the variance of M_a^+ satisfies

$$\frac{n_2(n+1)}{4(n_1+2)} \doteq n_2(4\lambda_1)^{-1}.$$

Confidence Intervals

Note that $M_a^+(\Delta) = \#(Y_j - \Delta > \widehat{\theta}) = \#(Y_j - \widehat{\theta} > \Delta)$; hence, if k is such that $P_0(M_a^+ \leq k) = \alpha/2$ then $(Y_{(k+1)} - \widehat{\theta}, Y_{(n_2-k)} - \widehat{\theta})$ is a $(1-\alpha)100\%$ confidence interval for Δ. For testing the two-sided hypothesis $H_0: \Delta = 0$ versus $H_A: \Delta \neq 0$ we would reject H_0 if 0 is not in the confidence interval. This is equivalent, however, to rejecting if $\widehat{\theta}$ is not in the interval $(Y_{(k+1)}, Y_{(n_2-k)})$.

Suppose we determine k by the normal approximation. Then

$$k \doteq \frac{n_2}{2} - z_{\alpha/2}\sqrt{\frac{n_2(n+1)}{4(n_1+2)}} - 0.5 \doteq \frac{n_2}{2} - z_{\alpha/2}\sqrt{\frac{n_2}{4\lambda_1}} - 0.5.$$

Written this way, the confidence interval $(Y_{(k+1)}, Y_{(n_2-k)})$, is a $\gamma 100\%$, $(\gamma = 1 - 2\Phi(-z_{\alpha/2}(\lambda_1)^{-1/2})$, confidence interval based on the sign procedure for the sample $Y_1, \ldots, Y_{n_2}$. Suppose we take $\alpha = 0.05$ and have the equal sample sizes case so that $\lambda_1 = 0.5$. Then $\gamma = 1 - 2\Phi(-2\sqrt{2})$. Hence, the two-sided 5% test rejects $H_0: \Delta = 0$ if $\widehat{\theta}$ is not in the confidence interval.

Remarks on Efficiency

Since the estimator of Δ based on the Mathisen procedure is the same as that of Mood's procedure, the asymptotic relative efficiency results for Mathisen's procedure are the same as for Mood's. Using another type of efficiency due to Bahadur (1967), Killeen, Hettmansperger, and Sievers (1972) show it is generally better to compute the median of the smaller sample.

Curtailed sampling on the Ys is one situation where Mathisen's test would be used instead of Mood's test since with Mathisen's test an early decision could be made; see Gastwirth (1968).

Example 2.6.2 *Quail Data, continued, Examples 2.3.1 and 2.6.1*

For these data, med $T_i = 49$. Since one of the placebo values was also 49, we eliminated it in the subsequent computation of Mathisen's test. The test statistic has the value $M_a^+ = \#(C_j > 49) = 17$. Using $n_2 = 19$ and $n_1 = 10$ the null mean and variance are 9.5 and 11.875, respectively. This leads to a standardized test statistic of 2.03 (using the continuity correction) with a p-value of 0.021. Utilizing all the data, the corresponding point estimate and confidence interval are 19 and (6, 27). These differ from MWW and Mood analyses; see Examples 2.3.1 and 2.6.1, respectively.

2.7 Robustness Properties

In this section we obtain the breakdown points and the influence functions of the L_1 and MWW estimates. We first consider the breakdown properties.

2.7.1 Breakdown Properties

We begin with the definition of an equivariant estimator of Δ. For convenience let the vectors **X** and **Y** denote the samples $\{X_1, \ldots, X_{n_1}\}$ and $\{Y_1, \ldots, Y_{n_2}\}$, respectively. Also let $\mathbf{X} + a\mathbf{1} = (X_1 + a, \ldots, X_{n_1} + a)'$.

Definition 2.7.1 *An estimator $\widehat{\Delta}(\mathbf{X}, \mathbf{Y})$ of Δ is said to be an* **equivariant estimator** *of Δ if $\widehat{\Delta}(\mathbf{X} + a\mathbf{1}, \mathbf{Y}) = \widehat{\Delta}(\mathbf{X}, \mathbf{Y}) - a$ and $\widehat{\Delta}(\mathbf{X}, \mathbf{Y} + a\mathbf{1}) = \widehat{\Delta}(\mathbf{X}, \mathbf{Y}) + a$.*

Note that the L_1 estimator and the Hodges–Lehmann estimator are both equivariant estimators of Δ. Indeed, as Exercise 2.13.24 shows, any estimator based on the rank pseudo-norms discussed in Section 2.5 is an equivariant estimator of Δ. As the following theorem shows, the breakdown point of an equivariant estimator is bounded above by 0.25.

Theorem 2.7.1 *Suppose $n_1 \leqslant n_2$. Then the breakdown point of an equivariant estimator satisfies $\varepsilon^* \leqslant \{[(n_1 + 1)/2] + 1\}/n$, where $[\cdot]$ denotes the greatest integer function.*

Proof. Let $m = [(n_1 + 1)/2] + 1$. Suppose $\widehat{\Delta}$ is an equivariant estimator such that $\varepsilon^* > m/n$. Then the estimator remains bounded if m points are corrupted.

Let $\mathbf{X}^* = (X_1 + a, \ldots, X_m + a, X_{m+1}, \ldots, X_{n_1})'$. Since we have corrupted m points there exists a $B > 0$ such that

$$|\widehat{\Delta}(\mathbf{X}^*, \mathbf{Y}) - \widehat{\Delta}(\mathbf{X}, \mathbf{Y})| \leq B. \tag{2.7.1}$$

Next let $\mathbf{X}^{**} = (X_1, \ldots, X_m, X_{m+1} - a, \ldots, X_{n_1} - a)'$. Then $\mathbf{X}^{**}$ contains $n_1 - m = [n_1/2] \leq m$ altered points. Therefore,

$$|\widehat{\Delta}(\mathbf{X}^{**}, \mathbf{Y}) - \widehat{\Delta}(\mathbf{X}, \mathbf{Y})| \leq B. \tag{2.7.2}$$

Equivariance implies that $\widehat{\Delta}(\mathbf{X}^{**}, \mathbf{Y}) = \widehat{\Delta}(\mathbf{X}^*, \mathbf{Y}) + a$. By (2.7.1) we have

$$\widehat{\Delta}(\mathbf{X}, \mathbf{Y}) - B \leq \widehat{\Delta}(\mathbf{X}^*, \mathbf{Y}) \leq \widehat{\Delta}(\mathbf{X}, \mathbf{Y}) + B \tag{2.7.3}$$

while from (2.7.2) we have

$$\widehat{\Delta}(\mathbf{X}, \mathbf{Y}) - B + a \leq \widehat{\Delta}(\mathbf{X}^{**}, \mathbf{Y}) \leq \widehat{\Delta}(\mathbf{X}, \mathbf{Y}) + B + a. \tag{2.7.4}$$

Taking $a = 3B$ leads to a contradiction between (2.7.2) and (2.7.4).

By this theorem the maximum breakdown point of any equivariant estimator is roughly half of the smaller sample proportion. If the sample sizes are equal then the best possible breakdown is $\frac{1}{4}$.

Example 2.7.1 *Breakdown of L_1 and MWW Estimates*

The L_1 estimator of Δ, $\widehat{\Delta} = \text{med } Y_j - \text{med } X_i$, achieves the maximal breakdown since med Y_j achieves the maximal breakdown in the one-sample problem.

The Hodges–Lehmann estimate $\widehat{\Delta}_R = \text{med } \{Y_j - X_i\}$ also achieves maximal breakdown. To see this, suppose we corrupt an X_i. Then n_2 differences $Y_j - X_i$ are corrupted. Hence between samples we maximize the corruption by corrupting the items in the smaller sample, so without loss of generality we can assume that $n_1 \leq n_2$. Suppose we corrupt m X_is. In order to corrupt med $\{Y_j - X_i\}$ we must corrupt $(n_1 n_2)/2$ differences. Therefore $m n_2 \geq (n_1 n_2)/2$; i.e. $m \geq n_1/2$. Hence med $\{Y_j - X_i\}$ has maximal breakdown. Based on Exercise 1.13.10 of Chapter 1, the one-sample estimate based on the Wilcoxon signed-rank statistic does not achieve the maximal breakdown value of $\frac{1}{2}$ in the one-sample problem.

2.7.2 Influence Functions

Recall from Section 1.6.1 that the influence function of a Pitman regular estimator based on a single sample $X_1, \ldots, X_n$ is the function $\Omega(z)$ when the estimator has the representation $n^{-1/2} \sum \Omega(X_i) + o_p(1)$. The estimators we are concerned with in this section are Pitman regular; hence, to determine their influence functions we need only obtain similar representations for them.

For the L_1 estimate we have from the proof of Theorem 2.6.1 that

$$\sqrt{n}\widehat{\Delta} = \text{med } Y_j - \text{med } X_i = \frac{1}{2f(0)} \frac{1}{\sqrt{n}} \left\{ \sum_{j=1}^{n_2} \frac{\text{sgn }(Y_j)}{\lambda_2} - \sum_{i=1}^{n_1} \frac{\text{sgn }(X_i)}{\lambda_1} \right\} + o_p(1).$$

110 Two-Sample Problems

Hence the influence function of the L_1 estimate is

$$\Omega(z) = \begin{cases} -(\lambda_1 2f(0))^{-1}\operatorname{sgn} z & \text{if } z \text{ is an } x \\ (\lambda_2 2f(0))^{-1}\operatorname{sgn} z & \text{if } z \text{ is a } y, \end{cases}$$

which is a bounded discontinuous function.

For the Hodges–Lehmann estimate (2.2.18), note that we can write the linearity result (2.4.23) as

$$\sqrt{n}(\overline{S}^+(\delta/\sqrt{n}) - \tfrac{1}{2}) = \sqrt{n}(\overline{S}^+(0) - \tfrac{1}{2}) - \delta \int f^2 + o_p(1),$$

which, upon substituting $\sqrt{n}\widehat{\Delta}_R$ for δ, leads to

$$\sqrt{n}\widehat{\Delta}_R = \left(\int f^2\right)^{-1} \sqrt{n}(\overline{S}^+(0) - \tfrac{1}{2}) + o_p(1).$$

Recall the projection of the statistic $\overline{S}_R(0) - \tfrac{1}{2}$ given in Theorem 2.4.7. Since the difference between it and this statistic goes to zero in probability we can, after some algebra, obtain the following representation for the Hodges–Lehmann estimator

$$\sqrt{n}\widehat{\Delta}_R = \left(\int f^2\right)^{-1} \frac{1}{\sqrt{n}} \left\{ \sum_{j=1}^{n_2} \frac{F(Y_j) - \tfrac{1}{2}}{\lambda_2} - \sum_{i=1}^{n_2} \frac{F(X_i) - \tfrac{1}{2}}{\lambda_1} \right\} + o_p(1).$$

Therefore the influence function for the Hodges–Lehmann estimate is

$$\Omega(z) = \begin{cases} -\left(\lambda_1 \int f^2\right)^{-1} (F(z) - \tfrac{1}{2}) & \text{if } z \text{ is an } x \\ \left(\lambda_2 \int f^2\right)^{-1} (F(z) - \tfrac{1}{2}) & \text{if } z \text{ is a } y \end{cases}$$

which is easily seen to be bounded and continuous.

For least squares, since the estimate is $\overline{Y} - \overline{X}$ the influence function is

$$\Omega(Z) = \begin{cases} -(\lambda_1)^{-1} z & \text{if } z \text{ is an } x \\ (\lambda_2)^{-1} z & \text{if } z \text{ is a } y \end{cases}$$

which is unbounded and continuous. The Hodges–Lehmann and L_1 estimates attain the maximal breakdown point and have bounded influence functions; hence, they are robust. On the other hand, the least-squares estimate has zero breakdown and an unbounded influence function. One bad point can destroy a least-squares analysis.

For a general score function $\varphi(u)$, by (2.5.31) we have the asymptotic representation

$$\widehat{\Delta} = \frac{1}{\sqrt{n}} \left[\sum_{i=1}^{n_1} \left(-\frac{\tau_\varphi}{\lambda_1}\right) \varphi(F(X_i)) + \sum_{i=1}^{n_2} \left(\frac{\tau_\varphi}{\lambda_2}\right) \varphi(F(Y_i)) \right].$$

Hence, the influence function of the R-estimate based on the score function φ is given by

$$\Omega(z) = \begin{cases} -\dfrac{\tau_\varphi}{\lambda_1} \varphi(F(z)) & \text{if } z \text{ is an } x \\ \dfrac{\tau_\varphi}{\lambda_2} \varphi(F(z)) & \text{if } z \text{ is a } y \end{cases}$$

where τ_φ is defined by expression (2.5.23). In particular, the influence function is bounded provided the score generating function is bounded. Note that the influence function for the R-estimate based on normal scores is unbounded; hence, this estimate is not robust. Recall Example 1.8.1 in which the one-sample normal scores estimate has an unbounded influence function (nonrobust) but has positive breakdown point (resistant). A rigorous derivation of these influence functions can be based on the influence function derived in Section A.5.2 of the Appendix.

2.8 Lehmann Alternatives and Proportional Hazards

Consider a two-sample problem where the responses are lifetimes of subjects. We shall continue to denote the independent samples by $X_1, \ldots, X_{n_1}$ and $Y_1, \ldots, Y_{n_2}$. Let X_i and Y_j have distribution functions $F(x)$ and $G(x)$, respectively. Since we are dealing with lifetimes both X_i and Y_j are positive-valued random variables. The **hazard function** for X_i is defined by

$$h_X(t) = \frac{f(t)}{1 - F(t)}$$

and represents the likelihood that a subject will die at time t given that he has survived until that time; see Exercise 2.13.25.

In this section, we will consider the class of lifetime models that are called **Lehmann alternative** models for which the distribution function G satisfies

$$1 - G(x) = (1 - F(x))^\alpha, \qquad (2.8.1)$$

where the parameter $\alpha > 0$. See Section 4.4 of Maritz (1981) for an overview of nonparametric methods for these models. The Lehmann model generalizes the exponential scale model $F(x) = 1 - \exp(-x)$ and $G(x) = 1 - (1 - F(x))^\alpha = 1 - \exp(-\alpha x)$. As shown in Exercise 2.13.25, the hazard function of Y_j is given by $h_Y(t) = \alpha h_X(t)$, i.e. the hazard function of Y_j is proportional to the hazard function of X_i. Hence, these models are also referred to as **proportional hazards models**; see also Section 3.10. The null hypothesis can be expressed as $H_{L0}: \alpha = 1$. The alternative we will consider is $H_{LA}: \alpha < 1$; Y is less hazardous than X, or, to put it another way, Y has more chance of long survival than X and is stochastically larger than X. Note that

$$P_\alpha(Y > X) = E_\alpha[P(Y > X \mid X)]$$
$$= E_\alpha[1 - G(X)]$$
$$= E_\alpha[(1 - F(X))^\alpha] = (\alpha + 1)^{-1}. \qquad (2.8.2)$$

The last equality holds since $1 - F(X)$ has a uniform $(0, 1)$ distribution. Under H_{LA}, then, $P_\alpha(Y > X) > \frac{1}{2}$; i.e. Y tends to dominate X.

The MWW test statistic $S_R^+ = \#(Y_j > X_i)$ is a consistent test statistic for H_{L0} versus H_{LA}, by Theorem 2.4.12. We reject H_{L0} in favor of H_{LA} for large values of S_R^+. Furthermore, by Theorem 2.4.4 and (2.8.2), we have that

$$E_\alpha[S_R^+] = n_1 n_2 E_\alpha[1 - G(X)] = \frac{n_1 n_2}{1 + \alpha}.$$

This suggests as an estimate of α the statistic,

$$\widehat{\alpha} = ((n_1 n_2)/S_R^+) - 1. \tag{2.8.3}$$

By Theorem 2.4.5 it can be shown that

$$\text{Var}_\alpha(S_R^+) = \frac{\alpha n_1 n_2}{(\alpha + 2)(\alpha + 1)^2} + \frac{n_1 n_2 (n_1 - 1)\alpha}{(\alpha + 2)(\alpha + 1)^2} + \frac{n_1 n_2 (n_2 - 1)\alpha^2}{(2\alpha + 1)(\alpha + 1)^2}; \tag{2.8.4}$$

see Exercise 2.13.27. Using this result and the asymptotic distribution of S_R^+ under general alternatives, Theorem 2.4.10, we can obtain, by the delta method, the asymptotic variance of $\widehat{\alpha}$ given by

$$\text{Var}\,\widehat{\alpha} \doteq \frac{(1 + \alpha)^2 \alpha}{n_1 n_2} \left\{ 1 + \frac{n_1 - 1}{\alpha + 2} + \frac{(n_2 - 1)\alpha}{2\alpha + 1} \right\}. \tag{2.8.5}$$

This can be used to obtain an asymptotic confidence interval for α; see Exercise 2.13.27 for details. As in the example below, the bootstrap could also be used to estimate $\text{Var}(\widehat{\alpha})$.

2.8.1 The Log Exponential and the Savage Statistic

Another rank test which is frequently used in this situation is the log-rank test proposed by Savage (1956). In order to obtain this test, first consider the special case where X has the exponential distribution function, $F(x) = 1 - e^{-x/\theta}$, for $\theta > 0$. In this case the hazard function of X is a constant function. Consider the random variable $\varepsilon = \log X - \log \theta$. In a few steps we can obtain its distribution function as

$$P[\varepsilon \leq t] = P[\log X - \log \theta \leq t]$$
$$= 1 - \exp(-e^t);$$

i.e. ε has an extreme value distribution. The density of ε is $f_\varepsilon(t) = \exp(t - e^t)$. Hence, we can model $\log X$ as the location model:

$$\log X = \log \theta + \varepsilon. \tag{2.8.6}$$

Next consider the distribution of the $\log Y$. Using expression (2.8.1) and a few steps of algebra we get

$$P[\log Y \leq t] = 1 - \exp\left(-\frac{\alpha}{\theta} e^t\right).$$

But from this it is easy to see that we can model Y as

$$\log Y = \log \theta + \log \frac{1}{\alpha} + \varepsilon, \tag{2.8.7}$$

where the error random variable has the above extreme value distribution. From (2.8.6) and (2.8.7) we see that the log transformation problem is simply a two-sample location problem with shift parameter $\Delta = -\log \alpha$. Here, H_{L0} is equivalent to $H_0: \Delta = 0$ and H_{LA} is equivalent to $H_A: \Delta > 0$. We shall refer to this model as the **log exponential model** for the remainder of this section.

Based on Section 2.5 and Exercise 2.13.19, the optimal scores for the extreme value distribution are generated by the function

$$\varphi_{f_\varepsilon}(u) = -(1 + \log(1-u)). \qquad (2.8.8)$$

Hence the optimal rank test in the log exponential model is given by

$$S_L = \sum_{j=1}^{n_2} \varphi_{f_\varepsilon}\left(\frac{R(Y_j)}{n+1}\right) = -\sum_{j=1}^{n_2}\left(1 + \log\left(1 - \frac{R(\log Y_j)}{n+1}\right)\right)$$

$$= -\sum_{j=1}^{n_2}\left(1 + \log\left(1 - \frac{R(Y_j)}{n+1}\right)\right). \qquad (2.8.9)$$

We reject H_{L0} in favor of H_{LA} for large values of S_L. By (2.5.14) the null mean of S_L is 0, while from (2.5.18) its null variance is given by

$$\sigma^2_{\varphi_{f_\varepsilon}} = \frac{n_1 n_2}{n(n-1)} \sum_{i=1}^{n}\left\{1 + \log\left(1 - \frac{i}{n+1}\right)\right\}^2. \qquad (2.8.10)$$

Then an asymptotic level α test rejects H_{L0} in favor of H_{LA} if $S_L \geq z_\alpha \sigma_{\varphi_{f_\varepsilon}}$.

Certainly the statistic S_L can be used in the general Lehmann alternative model described above, although it is not optimal if X does not have an exponential distribution. We shall discuss the efficiency of this test below. Let $\widehat{\Delta}$ be the estimate of Δ based on the optimal score function φ_{f_ε}; that is, $\widehat{\Delta}$ solves the equation

$$\sum_{j=1}^{n_2}\left(1 + \log\left(1 - \frac{R(\log(Y_j - \Delta))}{n+1}\right)\right) \doteq 0. \qquad (2.8.11)$$

Thus another estimate of α would be $\widehat{\alpha} = \exp\{-\widehat{\Delta}\}$. As discussed in Exercise 2.13.27, an asymptotic confidence interval for α can be formulated from this relationship. Keep in mind, though, that we are assuming that X is exponentially distributed.

As a further note, since $\varphi_{f_\varepsilon}(u)$ is an unbounded function it follows from Section 2.7.2 that the influence function of $\widehat{\Delta}$ is unbounded. Thus the estimate is not robust.

A frequently used test statistic equivalent to S_L was proposed by Savage. To derive it, denote $R(Y_j)$ by R_j. Then we can write

$$\log\left(1 - \frac{R_j}{n+1}\right) = \int_1^{1-R_j/(n+1)} \frac{1}{t} dt = \int_{R_j/(n+1)}^0 \frac{1}{1-t} dt.$$

We can approximate this last integral by the following Riemann sum:

$$\frac{1}{1-R_j/(n+1)}\frac{1}{n+1}+\frac{1}{1-(R_j-1)/(n+1)}\frac{1}{n+1}+\cdots$$
$$+\frac{1}{1-(R_j-(R_j-1))/(n+1)}\frac{1}{n+1}.$$

This simplifies to

$$\frac{1}{n+1-1}+\frac{1}{n+1-2}+\cdots+\frac{1}{n+1-R_j}=\sum_{i=n+1-R_j}^{n}\frac{1}{i}.$$

This suggests the rank statistic

$$\tilde{S}_L = -n_2 + \sum_{j=1}^{n_2}\sum_{i=n-R_j+1}^{n}\frac{1}{i}. \qquad (2.8.12)$$

This statistic was proposed by Savage (1956). Note that it is a rank statistic with scores defined by

$$a_j = -1 + \sum_{i=n-j+1}^{n}\frac{1}{i}. \qquad (2.8.13)$$

Exercise 2.13.28 shows that its null mean and variance are given by

$$E_{H_0}[\tilde{S}_L] = 0$$

$$\tilde{\sigma}^2 = \frac{n_1 n_2}{n-1}\left\{1 - \frac{1}{n}\sum_{j=1}^{n}\frac{1}{j}\right\}^2. \qquad (2.8.14)$$

Hence an asymptotic level α test is to reject H_{L0} in favor of H_{LA} if $\tilde{S}_L \geq \tilde{\sigma} z_\alpha$.

Based on the above Riemann sum it would seem that $\tilde{S}_L$ and S_L are close statistics. Indeed they are asymptotically equivalent and, hence, are both optimal when X is exponentially distributed; see Hájek and Šidák (1967) or Kalbfleisch and Prentice (1980) for details.

2.8.2 Efficiency Properties

We next derive the asymptotic relative efficiencies for the log exponential model with $f_\varepsilon(t) = \exp(t - e^t)$. The MWW statistic, S_R^+, is a consistent test for the log exponential model. By (2.4.21), the efficacy of the Wilcoxon test is

$$c_{MWW} = \sqrt{12}\int f_\varepsilon^2 \sqrt{\lambda_1\lambda_2} = \sqrt{\tfrac{3}{4}}\sqrt{\lambda_1\lambda_2}.$$

Since the Savage test is asymptotically optimal its efficacy is the square root of the Fisher information, i.e. $I^{1/2}(f_\varepsilon)$, discussed in Section 2.5. This efficacy is $\sqrt{\lambda_1\lambda_2}$. Hence the asymptotic efficiency of the Mann–Whitney–Wilcoxon test relative to the Savage test for the log exponential model is $\tfrac{3}{4}$; see Exercise 2.13.29.

Recall that the efficacy of the L_1 procedures, both Mood's and Mathisen's, is $2f_\varepsilon(\theta_\varepsilon)\sqrt{\lambda_1\lambda_2}$, where θ_ε denotes the median of the extreme value distribution.

This turns out to be $\theta_\varepsilon = \log(\log 2)$. Hence $f_\varepsilon(\theta_\varepsilon) = (\log 2)/2$, which leads to the efficacy $\sqrt{\lambda_1 \lambda_2} \log 2$ for the L_1 methods. Thus the asymptotic relative efficiency of the L_1 procedures with respect to the procedure based on Savage scores is $(\log 2)^2 = 0.480$. The asymptotic efficiency of the L_1 methods relative to the MWW for this model is 0.6406. Therefore there is a substantial loss of efficiency if L_1 methods are used for the log exponential model. This makes sense since the extreme value distribution has very light tails.

The variance of a random variable with density f_ε is $\pi^2/6$; hence the asymptotic efficiency of the t-test relative to the Savage test for the log exponential model is $6/\pi^2 = 0.608$. Hence, for the procedures analyzed in this chapter on the log exponential model the Savage test is optimal followed, in order, by the MWW, t, and L_1 tests.

Example 2.8.1 *Lifetimes of an Insulation Fluid*

The data below are drawn from an example from Lawless (1982, p. 3); see also Nelson (1982, p. 227). They consist of the breakdown times (in minutes) of an electrical insulating fluid when subject to two different levels of voltage stress, 30 and 32 kV. Suppose we are interested in testing to see if the lower level is less hazardous than the higher level.

Voltage Level	Times to Breakdown (Minutes)							
30 kV Y	17.05 194.90	22.66 47.30	21.02 7.74	175.88	139.07	144.12	20.46	43.40
32 kV X	0.40 3.91	82.85 0.27	9.88 0.69	89.29 100.58	215.10 27.80	2.75 13.95	0.79 53.24	15.93

Comparison boxplots, using 75% L_1 confidence intervals (Section 1.12), of these two samples are displayed next. Based on these plots the data appear to be positively skewed with the second sample having an outlier with value 215.1.

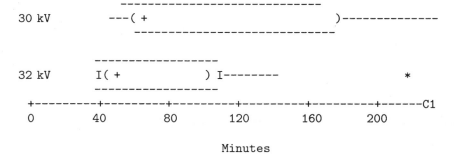

For these data the sum of the ranks of the 30 kV (Y) sample is 184. Using the continuity correction, this leads to the standardized test statistic $z = 2.57$ which has p-value 0.035. The estimate of α based on the MWW statistic is 0.40.

A 90% confidence interval for α based on the approximate (via the delta method) variance, (2.8.5), is $(0.06, 0.74)$; while a 90% bootstrap confidence interval based on 1000 bootstrap samples is $(0.15, 0.88)$. Hence, the MWW test, the corresponding estimate of α and the two confidence intervals indicate that the lower voltage level is less hazardous than the higher level.

The value of S_L, given by (2.8.9), is 2.0859, which yields the standardized test statistic value of $z = 1.30$ with a p-value of 0.096. Finally, the two sample t-test has value 1.34, with p-value 0.096.

Note that the outlier has somewhat impaired both the LS and the analysis based on the log-rank scores. Figure 2.8.1 displays the $q-q$ plots for the two samples. The population quantiles are drawn from an exponential distribution. On the basis of these plots, the data do not appear to be generated from an exponential distribution but from a distribution with heavier tails. Of the three procedures, the MWW procedure would be more appropriate for these data.

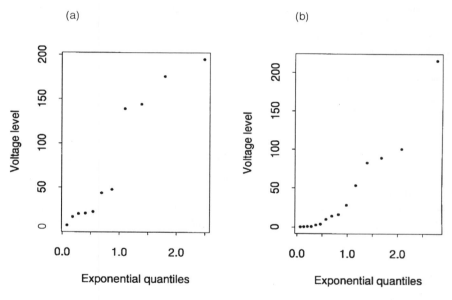

Figure 2.8.1: Exponential $q-q$ plot for: (a) 30 kV data; (b) 32 kV data

2.9 Two-Sample Ranked Set Sampling (RSS)

The basic background for ranked set sampling was discussed in Section 1.10. In this section we extend these ideas to the two-sample location problem. Suppose we have two samples in which $X_1, \ldots, X_{n_1}$ are iid $F(x)$ and $Y_1, \ldots, Y_{n_2}$ are iid $F(x - \Delta)$, and the two samples are independent of one another. In the corresponding RSS design, we take n_1 cycles of k samples for X and n_2 cycles of q samples for Y. Proceeding as in Section 1.10, we display the measured data as:

$$X_{(1)1}, \ldots, X_{(1)n_1} \text{ iid } f_{(1)}(t) \quad Y_{(1)1}, \ldots, Y_{(1)n_2} \text{ iid } f_{(1)}(t-\Delta)$$

$$\vdots \qquad \qquad \vdots \qquad \qquad \vdots \qquad \qquad \vdots$$

$$X_{(k)1}, \ldots, X_{(k)n_1} \text{ iid } f_{(k)}(t) \quad Y_{(q)1}, \ldots, Y_{(q)n_2} \text{ iid } f_{(q)}(t-\Delta).$$

To test $H_0 : \Delta = 0$ versus $H_A : \Delta > 0$ we compute the Mann–Whitney–Wilcoxon statistic with these ranked set samples. Letting $U_{si} = \sum_{t=1}^{n_2} \sum_{j=1}^{n_1} I(Y_{(s)t} > X_{(i)j})$, the test statistic is

$$U_{RSS} = \sum_{s=1}^{q} \sum_{i=1}^{k} U_{si}.$$

Note that U_{si} is the Mann–Whitney–Wilcoxon statistic computed on the sample of the sth Y order statistics and the ith X order statistics. Even under the null hypothesis $H_0 : \Delta = 0$, U_{si} is not based on identically distributed samples unless $s = i$. This complicates the null distribution of U_{RSS}.

Bohn and Wolfe (1992) present a thorough treatment of the distribution theory for U_{RSS}. We note that under $H_0 : \Delta = 0$, U_{RSS} is distribution-free and, further, using the same ideas as in Theorem 1.10.1, that $E_{H_0}(U_{RSS}) = qkn_1n_2/2$. For fixed k and q, provided assumption (D.1), (2.4.7), holds, Theorem 2.4.11 can be applied to show that $(U_{RSS} - qkn_1n_2/2)/\sqrt{\text{Var}_{H_0}(U_{RSS})}$ has a limiting $N(0, 1)$ distribution. The difficulty is in the calculation of $\text{Var}_{H_0}(U_{RSS})$; recall Theorem 1.10.1 for a similar calculation for the sign statistic. Bohn and Wolfe (1992) present a complex formula for the variance. They also give a table of the approximate null distribution of U_{RSS} for $q = k = 2$, $n_1 = 1, \ldots, 5$, $n_2 = 1, \ldots, 5$ and likewise for $q = k = 3$.

Another way to approximate the null distribution of U_{RSS} is to bootstrap it. Consider, for simplicity, the case $k = q = 3$ and $n_1 = n_2 = m$. Hence, the expert must rank three observations and each of the m cycles consists of three samples of size three for each of the X and Y measurements. In order to bootstrap the null distribution of U_{RSS}, first align the Y RSSs with $\widehat{\Delta}$, the Hodges–Lehmann estimate of shift computed across the two RSSs. Our bootstrap sampling is on the data with the indicated sampling distributions:

$$X_{(1)1}, \ldots, X_{(1)m} \text{ sample } \hat{F}_{(1)}(x) \quad Y_{(1)1}, \ldots, Y_{(1)m} \text{ sample } \hat{F}_{(1)}(y - \widehat{\Delta})$$

$$X_{(2)1}, \ldots, X_{(2)m} \text{ sample } \hat{F}_{(2)}(x) \quad Y_{(2)1}, \ldots, Y_{(2)m} \text{ sample } \hat{F}_{(2)}(y - \widehat{\Delta})$$

$$X_{(3)1}, \ldots, X_{(3)m} \text{ sample } \hat{F}_{(3)}(x) \quad Y_{(3)1}, \ldots, Y_{(3)m} \text{ sample } \hat{F}_{(3)}(y - \widehat{\Delta}).$$

In the bootstrap process, for each row $i = 1, 2, 3$, we take random samples $X^*_{(i)1}, \ldots, X^*_{(i)m}$ from $\hat{F}_{(i)}(x)$ and $Y^*_{(i)1}, \ldots, Y^*_{(i)m}$ from $\hat{F}_{(2)}(y - \widehat{\Delta})$. We then compute U^*_{RSS} on these samples. Repeating this B times, we obtain the sample of test statistics $U^*_{RSS,1}, \ldots, U^*_{RSS,B}$. Then the bootstrap p-value for our test is $\#(U^*_{RSS,j} \geq U_{RSS})/B$, where U_{RSS} is the value of the statistic based on the original data. Generally we take $B = 1000$ for a p-value. It is clear how to modify the above argument to allow for $k \neq q$ and $n_1 \neq n_2$.

2.10 Two-Sample Scale Problem

Frequently it is of interest to investigate whether or not one random variable is more dispersed than another. The most general case which we will consider is when the random variables differ in both location and scale. Suppose the distribution functions of X and Y are given by $F(x)$ and $G(y) = F((y - \Delta)/\eta)$, respectively; hence $\mathcal{L}(Y) = \mathcal{L}(\eta X + \Delta)$. For the discussion in this section, we will consider the hypotheses

$$H_0 : \eta = 1 \text{ versus } H_A : \eta > 1. \qquad (2.10.1)$$

The discussions would be similar for the other one-sided or two-sided hypotheses. Let $X_1, \ldots, X_{n_1}$ and $Y_1, \ldots, Y_{n_2}$ be samples drawn on the random variables X and Y, respectively.

In Section 2.10.2, we offer a practical test statistic for the above hypotheses based on the ranks of centered samples, where each sample is centered by subtracting its sample median from each observation. To fix ideas, though, we first consider the case of **known locations**. This implies the more restricted problem where $\mathcal{L}(Y) = \mathcal{L}(\eta X)$. An example of such a problem would be if X and Y have different exponential distributions, which was discussed in Section 2.8.

2.10.1 $\mathcal{L}(Y) = \mathcal{L}(\eta X)$

Because $\eta > 0$, in this case we also have $\mathcal{L}(|Y|) = \mathcal{L}(\eta|X|)$. If we use the linear model notation of Section 2.2 with $\mathbf{Z}' = (\log |X_1|, \ldots, \log |X_{n_1}|, \log |Y_1|, \ldots, \log |Y_{n_2}|)$ and c_i as defined in expression (2.2.1), then another equivalent formulation of this problem is

$$Z_i = \Delta c_i + e_i, \ 1 \leq i \leq n, \qquad (2.10.2)$$

where $\Delta = \log \eta$, $e_1, \ldots, e_n$ are iid with distribution function $F^*(x)$, and F^* is the distribution function of $\log |X|$. The hypotheses (2.10.1) are equivalent to

$$H_0 : \Delta = 0 \text{ versus } H_A : \Delta > 1. \qquad (2.10.3)$$

This is also an example of a simple log-linear model which is discussed in some detail in Section 3.10.

This, of course, is the two-sample location problem based on the logs of the absolute values of the observations. Following Section 2.5, the general scores process for this problem is given by

$$S_\varphi(\Delta) = \sum_{j=1}^{n_2} a_\varphi(R(\log|Y_j| - \Delta)), \qquad (2.10.4)$$

where the scores $a(i)$ are generated by $a(i) = \varphi(i/(n+1))$ and φ is a nondecreasing function following assumptions (S.2), (2.5.5).

A rank-test statistic for the hypotheses, (2.10.3) is given by

$$S_\varphi = S_\varphi(0) = \sum_{j=1}^{n_2} a_\varphi(R(\log|Y_j|)) = \sum_{j=1}^{n_2} a_\varphi(R(|Y_j|)), \qquad (2.10.5)$$

where the last equality holds because the log function is strictly increasing. It follows from (2.5.14)–(2.5.16) that the null mean of this statistic is 0 and its null variance is given by

$$\sigma_\varphi^2 = \frac{n_1 n_2}{n-1}\left\{\sum_{i=1}^{n} a^2(i)\frac{1}{n}\right\} \doteq \frac{n_1 n_2}{n-1}, \qquad (2.10.6)$$

where the approximation is due to the fact that the term in braces is a Riemann sum of $\int \varphi^2(u)du = 1$ and, hence, converges to 1. Hence, an asymptotically level α test is to reject H_0 in favor of H_A if $z \geq z_\alpha$, where

$$z = \frac{S_\varphi}{\sqrt{\frac{n_1 n_2}{n-1}}}. \qquad (2.10.7)$$

An estimate of Δ is a value $\widehat{\Delta}$ which solves equation (2.5.12); i.e.

$$S_\varphi(\widehat{\Delta}) \doteq 0. \qquad (2.10.8)$$

An estimate of η, the ratio of scale parameters, is then

$$\widehat{\eta} = e^{\widehat{\Delta}}. \qquad (2.10.9)$$

The interval $(\widehat{\Delta}_L, \widehat{\Delta}_U)$, where $\widehat{\Delta}_L$ and $\widehat{\Delta}_U$ solve equations (2.5.21), forms a $(1-\alpha)100\%$ confidence interval for Δ. The corresponding confidence interval for η is $(\exp\{\widehat{\Delta}_L\}, \exp\{\widehat{\Delta}_U\})$.

The efficacy of the test based on S_φ is given by expression (2.5.28); i.e.

$$c_\varphi = \tau_\varphi^{-1}\sqrt{\lambda_1 \lambda_2}, \qquad (2.10.10)$$

where τ_φ is given by

$$\tau_\varphi^{-1} = \int_0^1 \varphi(u)\varphi_{f^*}(u)\,du \qquad (2.10.11)$$

and

$$\varphi_{f^*}(u) = -\frac{f^{*\prime}(F^{*-1}(u))}{f^*(F^{*-1}(u))}. \qquad (2.10.12)$$

It is instructive to consider expression (2.10.12) since $\varphi_{f^*}(u)$ is an optimal score generating function. After some simplification (see Exercise 2.13.30), we have

$$-\frac{f^{*\prime}(x)}{f^*(x)} = \frac{e^x[f'(e^x) - f'(-e^x)]}{f(e^x) + f(-e^x)} + 1. \qquad (2.10.13)$$

In the following examples we will look at some specific situations. The first example is general but it will give the situation where the MWW statistic is

120 Two-Sample Problems

optimal. The last three examples focus on cases where f is a symmetric density.

Example 2.10.1 $\mathcal{L}(|X|)$ *Is a Member of the Generalized F family: MWW Statistic*

In Section 3.10 a discussion is devoted to a large family of commonly used distributions called the generalized F family for survival type data. In particular, as shown there, if $|X|$ follows an $F(2, 2)$ distribution, then it follows (Exercise 2.13.31) that the $\log |X|$ has a logistic distribution. Thus the MWW statistic is the optimal rank score statistic in this case. As discussed in Section 3.10, for distributions $F(v, v)$ as v approaches 0 the optimal rank scores approach the sign score function. These are, of course, distributions with heavy right tails. Larger values of v yield lighter tailed distributions. As noted in Section 3.10, as v approaches ∞, the score function approaches the normal scores function.

Expression (2.10.13) simplifies when f is symmetric. We state this result in the next example and follow it with the two subcases of interest, namely f normal and then f double exponential.

Example 2.10.2 $\mathcal{L}(X)$ *Is Symmetric*

Assuming f is symmetric about 0, expression (2.10.12) reduces to

$$\varphi_{f*}(u) = -F^{-1}\left(\frac{u+1}{2}\right) \frac{f'\left(F^{-1}\left(\frac{u+1}{2}\right)\right)}{f\left(F^{-1}\left(\frac{u+1}{2}\right)\right)} + 1; \qquad (2.10.14)$$

see Exercise 2.13.32. This is the optimal score function when X has symmetric density $f(x)$.

Example 2.10.3 $\mathcal{L}(X)$ *Is Normal*

Without loss of generality, assume that $f(x)$ is the standard normal density. In this case expression (2.10.14) simplifies to

$$\varphi_{f*}(u) = \left(\Phi^{-1}\left(\frac{u+1}{2}\right)\right)^2 - 1, \qquad (2.10.15)$$

where Φ is the standard normal distribution function; see Exercise 2.13.32. Hence, if we are sampling from a normal distribution this suggests the rank-test statistic

$$S_{FK} = \sum_{j=1}^{n_2} \left(\Phi^{-1}\left(\frac{R|Y_j|}{2(n+1)} + \tfrac{1}{2}\right)\right)^2, \qquad (2.10.16)$$

where the FK subscript indicates that the statistic is due to Fligner and Killeen (1976); see (2.10.21). This is not a standardized score function, but it follows from the discussion on general scores found in Section 2.5 and (2.5.18) that the null mean μ_{FK} and null variance σ_{FK}^2 of the statistic are given by

$$\mu_{FK} = n_2 \bar{a} \quad \text{and} \quad \sigma_{FK}^2 = \frac{n_1 n_2}{n(n-1)} \sum (a(i) - \bar{a})^2, \tag{2.10.17}$$

where $a(i) = (\Phi^{-1}((i/(2(n+1))) + \frac{1}{2}))^2$. The asymptotic version of this test statistic rejects H_0 at approximate level α if $z \geq z_\alpha$ where

$$z = \frac{S_{FK} - \mu_{FK}}{\sigma_{FK}}. \tag{2.10.18}$$

Example 2.10.4 $\mathcal{L}(X)$ *Is Double Exponential*

Suppose that the density of X is the double exponential, $f(x) = 2^{-1} \exp\{-|x|\}$, $-\infty < x < \infty$. Then, as Exercise 2.13.32 shows, the optimal rank-score function is given by

$$\varphi(u) = -(\log(1-u) + 1). \tag{2.10.19}$$

The corresponding rank-test statistic would have to be standardized as in the previous example. These scores are not surprising, because the distribution of $|X|$ is exponential. Hence, this is precisely the log-linear problem with exponentially distributed lifetime that was discussed in Section 2.8; see the discussion around expression (2.8.8).

2.10.2 Unknown Locations

Suppose we have the general problem where the distribution functions of X and Y are given by $F(x)$ and $G(y) = F((y - \Delta)/\eta)$, respectively; hence $\mathcal{L}(Y) = \mathcal{L}(\eta X + \Delta)$. Under the null hypothesis, $\eta = 1$, the distributions of $Y - \theta_Y$ and $X - \theta_X$ are the same. Our strategy is to align the observations first and then employ the rank statistics discussed above on these aligned samples. Let $X_i^* = X_i - \hat\theta_X$ and $Y_j^* = Y_j - \hat\theta_Y$ denote the aligned observations where $\hat\theta_X$ and $\hat\theta_Y$ are the sample medians of the X and Y samples, respectively. We then consider the linear rank statistic (2.10.5), where the ranking is performed on the folded-aligned observations; i.e.

$$S_\varphi^* = \sum_{j=1}^{n_2} a(R(|Y_j^*|)), \tag{2.10.20}$$

where the scores are generated as $a(i) = \varphi(i/(n+1))$ as discussed above. The statistic S^* is no longer distribution-free for finite samples. If we further assume that the distributions of X and Y are symmetric, then the test statistic S_φ^* is asymptotically distribution-free and has the same efficiency properties as S_φ; see Puri (1968) and Fligner and Hettmansperger (1979). The requirement that f is symmetric is discussed in detail by Fiigner and Hettmansperger (1979).

Conover, Johnson, and Johnson (1981) performed a large Monte Carlo study of tests of dispersion, including these aligned rank tests, over a wide variety of situations for the c-sample scale problem. The F-test (Bartlett's test) did poorly (as would be expected from our comments below about the lack of robustness of the classical F-test). In certain null situations its empirical α levels exceeded 0.80 when the nominal α level was 0.05. One rank test that performed very well was the aligned rank version of the test statistic S_{FK} (2.10.16), i.e.

$$S_{FK}^* = \sum_{j=1}^{n_2} \left(\Phi^{-1}\left(\frac{R|Y_j^*|}{2(n+1)} + \frac{1}{2} \right) \right)^2, \qquad (2.10.21)$$

where the rankings are over the absolute values of the aligned observations. In the study it was standardized using the mean and standard deviation (2.10.17) associated with these scores. This statistic for nonaligned samples is given by Hájek and Šidák (1967, p. 74). A version of it was also discussed by Fligner and Killeen (1976). This scale test possessed both robustness of validity and power in the study by Conover *et al.* and was one of the few tests of their study which they recommended for general use.

Example 2.10.5 *Laplace Data*

To illustrate the nonrobustness of the classical F-test, consider testing the hypotheses $H_0 : \sigma_1 = \sigma_2$ versus $H_A : \sigma_1 > \sigma_2$ for the following simulated data:

Sample 1	−0.38982	−2.17746	0.81368	−0.00072	0.76384	−0.57041	−2.56511	−1.73311
	−0.11032	−0.70976	0.45664	0.13583	0.40363	0.77812	−0.11548	
Sample 2	−1.06716	−0.57712	0.36138	−0.68037	−0.77576	−1.42159	−0.81898	0.32863
	−0.63445	−0.99624	−0.18128	0.23957	0.21390	1.42551	−0.16589	

Both samples of size 15 are generated data from the Laplace distribution with density $f(x) = 2^{-1} \exp(-|x|)$. In particular, $\sigma_1 = \sigma_2$. Note that the boxes of the boxplots given in Figure 2.10.1(a) indicate that the scales are roughly the same for the two samples.

The sample variances are $s_1^2 = 1.134$ and $s_2^2 = 0.540$. Notice from the comparison dotplots, Figure 2.10.1(b), that there are three outlying observations in the first sample and one in the second sample. The outlying observations in the first sample caused the disparity in these sample variances. The F-statistic has the value 2.10 with p-value 0.089, which, while not strongly significant, is indicative of more disparity between scale than is warranted by the boxplots. The standardized value of the aligned rank statistic S_{FK}^*, (2.10.21), is $z = −0.938$ with a p-value of 0.348, which indicates little disparity between scales.

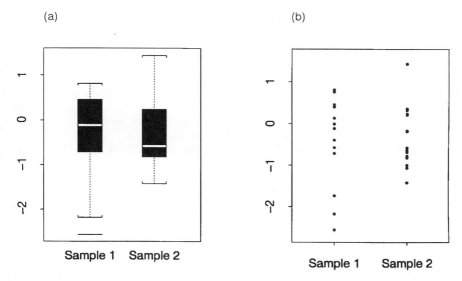

Figure 2.10.1: (a) Comparison boxplots of generated Laplace data; (b) comparison dotplots of generated Laplace data

2.10.3 Linear Rank Statistics of the Observations

Another class of linear rank statistics for the two-sample scale problem with **equal but possibly unknown location parameters** consists of simple linear rank statistics of the form

$$S = \sum_{j=1}^{n_2} a(R(Y_j)), \qquad (2.10.22)$$

where the scores are generated as $a(i) = \varphi(i/(n+1))$. The folded rank statistics discussed above suggest that φ is a convex (or concave) function. One popular score function is the quadratic function $\varphi(u) = (u - \frac{1}{2})^2$. The resulting statistic,

$$S_M = \sum_{j=1}^{n_2} \left(\frac{R(Y_j)}{n+1} - \frac{1}{2} \right)^2, \qquad (2.10.23)$$

was proposed by Mood (1954) as a test statistic for (2.10.1). We reject H_0 in favor of H_A for large values of S_M. Other popular scale tests include the Ansari and Bradley (1960) test where $\varphi(u) = |u - \frac{1}{2}|$ and the Klotz (1962) test where $\varphi(u) = \Phi^{-1}(u)^2$. Another convenient test is the Siegel and Tukey (1960) test which is discussed in Exercise 2.13.33.

As with the folded rank statistics, to fix ideas, we will assume that the locations of X and Y are the same and are at 0. Then the distribution function of Y is given by $G(y) = F(y/\eta)$, where F is the distribution function of X.

It follows from the discussion on general scores found in Section 2.5, that the test statistic given by (2.10.22) is distribution-free under H_0 with null mean μ_S and null variance σ_S^2 given by (2.5.18). Hence, tables of the null distribution of a given linear rank test statistic for scale can be constructed.

An approximate test can also be developed. Assume that F has density $f(x)$ and that the design condition (D.1), (2.4.7), holds. It then follows from Theorem A.2.1 of the Appendix that the statistic $z = (S - \mu_S)/\sigma_S$ is approximately $N(0, 1)$ under H_0. Hence, an asymptotic level α test rejects $H_0 : \eta = 1$ in favor of $H_A : \eta > 1$ if $z \geq z_\alpha$, where $1 - \Phi(z_\alpha) = \alpha$.

An observation that will be useful for Theorem 2.10.1 is to write the variance as

$$n^{-1}\sigma_S^2 = \frac{n_1 n_2}{n(n-1)}\left[\sum_{i=1}^{n}(a(i) - \bar{a})^2 \frac{1}{n}\right]. \quad (2.10.24)$$

Note that the term in brackets is a Riemann sum and, thus, converges to A_φ^2, defined by

$$A_\varphi^2 = \int (\varphi(u) - \bar{\varphi})^2 du, \quad (2.10.25)$$

where $\bar{\varphi} = \int \varphi(u) du$. Thus $n^{-1}\sigma_S^2$ converges to $\lambda_1 \lambda_2 A_\varphi^2$.

We next obtain efficiency results for the test statistic S, of (2.10.22), by deriving the associated asymptotic power lemma. As with the folded rank statistics, it is convenient to write the parameter as $\Delta = \log \eta$; hence, the hypotheses become

$$H_0 : \Delta = 0 \text{ versus } H_A : \Delta > 0. \quad (2.10.26)$$

A level α test for H_0 versus H_A is to reject H_0 if $S \geq c_\alpha$, where c_α is the critical value; i.e. $P_0(S \geq c_\alpha) = \alpha$. As in the location problem, based on the asymptotic distribution of S under H_0 we have

$$\frac{c_\alpha - \mu_S}{\sigma_S} \to z_\alpha, \quad (2.10.27)$$

where z_α is the upper α critical value of a standard normal distribution. Define the power function of this test by

$$\gamma(\Delta) = P_\Delta(S \geq c_\alpha).$$

Using the change of variable technique, we have

$$\gamma(\Delta) = P_0(S(-\Delta) \geq c_\alpha), \quad (2.10.28)$$

where $S(-\Delta) = \sum a(R(e^{-\Delta} Y_j))$ and the ranks are over the combined samples $X_1, \ldots, X_{n_1}$ and $e^{-\Delta} Y_1, \ldots, e^{-\Delta} Y_{n_2}$.

Next consider a sequence of alternatives given by

$$H_n : \Delta_n = \delta/\sqrt{n}, \quad \delta > 0. \quad (2.10.29)$$

Under this sequence of alternatives

$$\mathcal{L}(Y) = \mathcal{L}(e^{\delta/\sqrt{n}} X). \quad (2.10.30)$$

The asymptotic power lemma for linear rank scale tests is as follows:

Theorem 2.10.1 *Under the above model and (D.1), (2.4.7),*
$$\lim_{n \to \infty} \gamma(\Delta_n) = P(Z \geq z_\alpha - c_S \delta),$$
where Z has a standard normal distribution and

$$c_S = \frac{\gamma_1 \sqrt{\lambda_1 \lambda_2}}{A_\varphi}, \tag{2.10.31}$$

$$\gamma_1 = \int_0^1 \varphi(u) \varphi_{1,f}(u) du, \tag{2.10.32}$$

$$\varphi_{1,f}(u) = -1 - F^{-1}(u) \frac{f'(F^{-1}(u))}{f(F^{-1}(u))}, \tag{2.10.33}$$

where A_φ is defined in (2.10.25).

Proof. Using (2.10.28), we have the equalities:
$$\gamma(\Delta_n) = P_0(S(-\Delta_n) \geq c_\alpha)$$
$$= P_0\left(\frac{S(-\Delta_n) - \mu_S - \mu_n}{\sigma_S} \geq \frac{c_\alpha - \mu_S}{\sigma_S} - \frac{\mu_n}{\sigma_S}\right), \tag{2.10.34}$$

where
$$\mu_n = \sqrt{n} \delta \gamma_1 \frac{n_1 n_2}{n^2}.$$

We can establish that the random variable in expression (2.10.34) converges in distribution to a $N(0, 1)$ random variable by Theorem A.2.3 of the Appendix; see the discussion given by Hájek and Šidák (1967, p. 216). Also by (2.10.27) the first term on the right-hand side of the inequality converges to z_α and, finally, by the discussion around (2.10.24), the second term on the right-hand side converges to $c_S \delta$. This yields the result.

Hence the efficacy of the the scale test statistic S is c_S given by expression (2.10.31). Several items follow immediately from this result. Sample size determination based on a specified alternative, level, and power can be obtained as in the location problem; see Exercise 2.13.34. Secondly, to maximize power we want c_S as large as possible. As in the location problem we can write c_S as

$$c_S = \left[\frac{\int \varphi \varphi_{1,f}}{\sqrt{A_\varphi^2 I_1(f)}}\right] I_1(f) \sqrt{\lambda_1 \lambda_2}, \tag{2.10.35}$$

where $I_1(f) = \int \varphi_{1,f}^2$ which is Fisher information for a scale family of distributions; see Exercise 2.13.35. As shown in the same exercise, $\int \varphi_{1,f} = 0$; hence, the quantity in brackets is a correlation coefficient which takes its maximum value 1 when the **optimal score function** is $\varphi(u) = \varphi_{1,f}(u)$. The scale test statistic in this case is an **asymptotically most powerful rank test**; see Chapter 3 of Hájek and Šidák (1967). For example, if X has a normal distribution then it is easy to show that $\varphi_{1,f}(u) = \Phi^{-1}(u)^2 - 1$; that is, the Klotz (1962) test is an optimal rank test for scale if the observations are normally distributed. The efficacies for Mood's (1954) and Ansari and Bradley's (1960) scale test statistics are derived in Exercise 2.13.36.

We next obtain the efficacy of the **traditional F-test** for the ratio of scale parameters. Actually for our development we need not assume that X and Y have the same locations. Let σ_2^2 and σ_1^2 denote the variances of Y and X, respectively. Then in the notation in the first paragraph of this section, $\eta^2 = \sigma_2^2/\sigma_1^2$. The classical F-test of the hypotheses (2.10.1) is to reject H_0 if $F^* \geq F(\alpha, n_2 - 1, n_1 - 1)$, where

$$F^* = \widehat{\sigma}_2^2/\widehat{\sigma}_1^2,$$

and $\widehat{\sigma}_2^2$ and $\widehat{\sigma}_1^2$ are the sample variances of the samples $Y_1, \ldots, Y_{n_2}$ and $X_1, \ldots, X_{n_1}$, respectively. The F-test is of exact size α if f is a normal pdf. Also the test is invariant to differences in location.

We first need the asymptotic distribution of F^* under the null hypothesis. Instead of working with F^* it is more convenient mathematically to work with the equivalent test statistic $\sqrt{n}\log F^*$. We will assume that X has a finite fourth central moment; i.e. $\mu_{X,4} = E[(X - E(X))^4] < \infty$. Let $\xi = (\mu_{X,4}/\sigma_1^4) - 3$ denote the kurtosis of X. It easily follows that Y has the same kurtosis under the null and alternative hypotheses. A key result, established in Exercise 2.13.37, is that under these conditions

$$\sqrt{n_i}(\widehat{\sigma}_i^2 - \sigma_i^2) \xrightarrow{\mathcal{D}} N(0, \sigma_i^4(\xi + 2)), \text{ for } i = 1, 2. \qquad (2.10.36)$$

It follows immediately by the delta method that

$$\sqrt{n_i}(\log \widehat{\sigma}_i^2 - \log \sigma_i^2) \xrightarrow{\mathcal{D}} N(0, \xi + 2), \text{ for } i = 1, 2. \qquad (2.10.37)$$

Under H_0, $\sigma_i = \sigma$, say, and the last result,

$$\sqrt{n}\log F^* = \sqrt{\frac{n}{n_2}}\sqrt{n_2}(\log \widehat{\sigma}_2^2 - \log \sigma^2) - \sqrt{\frac{n}{n_1}}\sqrt{n_1}(\log \widehat{\sigma}_1^2 - \log \sigma^2)$$

$$\xrightarrow{\mathcal{D}} N(0, (\xi + 2)/(\lambda_1 \lambda_2)). \qquad (2.10.38)$$

The approximate test rejects H_0 if

$$\frac{\sqrt{n}\log F^*}{\sqrt{(\xi + 2)/(\lambda_1 \lambda_2)}} \geq z_\alpha. \qquad (2.10.39)$$

Note that $\xi = 0$ if X is normal. In practice the test which is used assumes $\xi = 0$; that is, F^* is not corrected by an estimate of ξ. This is one reason why the usual F-test for ratio in variances does not possess robustness of validity; that is, the significance level is not asymptotically distribution-free. Unlike the t-test, the F-test for variances is not even asymptotically distribution-free under H_0.

In order to obtain the **efficacy** of the F-test, consider the sequence of contiguous alternatives (2.10.29). Assume without loss of generality that the locations of X and Y are the same. Under this sequence of alternatives we have $Y_j = e^{\Delta_n}U_j$, where U_j is a random variable with cdf $F(x)$ while Y_j has cdf $F(e^{\Delta_n}x)$. We also get $\widehat{\sigma}_2^2 = \exp\{2\Delta_n\}\widehat{\sigma}_U^2$, where $\widehat{\sigma}_U^2$ denotes the sample variance of $U_1, \ldots, U_{n_2}$. Let $\gamma_F(\Delta)$ denote the power function of the F-test. The asymptotic power lemma for the F test is as follows:

Theorem 2.10.2 *Assuming that X has a finite fourth moment, with $\xi = (\mu_{X,4}/\sigma_1^4) - 3$,*

$$\lim_{n \to \infty} \gamma_F(\Delta_n) = P(Z \geq z_\alpha - c_F \delta),$$

where Z has a standard normal distribution and efficacy

$$c_F = 2\sqrt{\lambda_1 \lambda_2}/\sqrt{\xi + 2}. \qquad (2.10.40)$$

Proof. The conclusion follows directly upon observing

$$\sqrt{n} \log F^* = \sqrt{n}(\log \hat{\sigma}_2^2 - \log \hat{\sigma}_1^2)$$
$$= \sqrt{n}(\log \hat{\sigma}_U^2 + 2(\delta/\sqrt{n}) - \log \hat{\sigma}_1^2)$$
$$= 2\delta + \sqrt{\frac{n}{n_2}}\sqrt{n_2}(\log \hat{\sigma}_U^2 - \log \sigma^2) - \sqrt{\frac{n}{n_1}}\sqrt{n_1}(\log \hat{\sigma}_1^2 - \log \sigma^2)$$

and that the last quantity converges in distribution to a $N(2\delta, (\xi + 2)/(\lambda_1 \lambda_2))$ variate.

Assuming locations are the same, by the asymptotic power lemma, the asymptotic efficiency of a linear rank statistic relative to the F-test is the ratio of the squares of their efficacies, i.e. $e(S, F) = c_S^2/c_F^2$. Assuming that X is normal, it is shown in Exercise 2.13.38, that $e(\text{Mood}, F) = 15/(2\pi^2) \doteq 0.76$, $e(\text{Ansari-Bradley}, F) = 6/\pi^2 \doteq 0.61$, and $e(\text{Klotz}, F) = 1$. The low efficiency of the Ansari-Bradley relative to the F-statistic at the normal is not surprising since it is optimal for the density $f(x) = 2^{-1}(1 + |x|)^{-2}$, which has very heavy tails; see Exercise 2.13.39.

In practice usually the **location parameters are not known**. As with the folded rank statistics, a simple strategy is to align the observations first and then employ a linear rank test. Let $X_i^* = X_i - \widehat{\theta}_X$ and $Y_j^* = Y_j - \widehat{\theta}_Y$ denote the aligned observations, where $\widehat{\theta}_X$ and $\widehat{\theta}_Y$ are the sample medians of the X and Y samples, respectively. We then consider the linear rank statistic

$$S^* = \sum_{j=1}^{n_2} a(R(Y_j^*)), \qquad (2.10.41)$$

where the rankings are over the aligned observations. The statistic S^* is no longer distribution-free for finite samples, but, as shown by Puri (1968) it is asymptotically distribution free and has the same efficiency properties as S, provided it is additionally assumed that f is symmetric; see also Fligner and Hettmansperger (1979). As a final remark, though, this class of aligned rank tests did not perform nearly as well as the folded rank statistics, (2.10.20), in the large Monte Carlo study of Conover *et al.* (1981).

2.11 Behrens–Fisher Problem

Consider the general model in Section 2.1, where $X_1, \ldots, X_{n_1}$ is a random sample on the random variable X which has distribution function $F(x)$ and density function $f(x)$, and $Y_1, \ldots, Y_{n_2}$ is a second random sample, independent of the first, on the random variable Y which has common distribution function $G(x)$ and density $g(x)$. Let θ_X and θ_Y denote the medians of X and Y, respectively, and let $\Delta = \theta_Y - \theta_X$. In Section 2.4 we showed that the MWW test was consistent for the stochastically ordered alternative. In the location model, where the distributions of X and Y differ by at most a shift in location, the hypothesis $F = G$ is equivalent to the null hypothesis that $\Delta = 0$. In this section we drop the location model assumption, that is, we will assume that X and Y have distribution functions F and G respectively, but we still consider the null hypothesis that $\Delta = 0$. In order to avoid confusion with Section 2.4, we explicitly state the hypotheses of this section as

$$H_0 : \Delta = 0 \text{ versus } H_A : \Delta > 0,$$

$$\text{where } \Delta = \theta_Y - \theta_X, \text{ and } \mathcal{L}(X) = F, \text{ and } \mathcal{L}(Y) = G. \quad (2.11.1)$$

As in the previous sections, we have selected a specific alternative for the discussion.

The above hypothesis is our most general hypothesis of this section and the modified Mathisen test defined below is consistent for it. We will also consider the case where the forms of F and G are the same; that is, $G(x) = F(x/\eta)$, for some parameter η. Note in this case that $\mathcal{L}(Y) = \mathcal{L}(\eta X)$; hence, $\eta = T(Y)/T(X)$, where $T(X)$ is any scale functional ($T(X) > 0$ and $T(aX) = aT(X)$ for $a \geq 0$). If $T(X) = \sigma_X$, the standard deviation of X, then this is a Behrens–Fisher problem with F unknown. If we further assume that the distributions of X and Y are symmetric then the modified MWW, defined below, can be used to test that $\Delta = 0$. The most restrictive case is when both F and G are assumed to be normal distribution functions. This is, of course, the classical Behrens–Fisher problem and the classical solution to it is the Welch-type t-test, discussed below. For motivation we first show the behavior of the usual MWW statistic. We then consider general rank procedures and finally specialize to analogs of the L_1 and MWW analyses.

2.11.1 Behavior of the Usual MWW Test

In order to motivate the problem, consider the null behavior of the usual MWW test under (2.11.1) with the further restriction that the distributions of X and Y are symmetric. Under H_0, since we are examining null behavior there is no loss of generality if we assume that $\theta_X = \theta_Y = 0$. The asymptotic form of the MWW test rejects H_0 in favor of H_A if

$$S_R^+ = \sum_{i=1}^{n_1} \sum_{j=1}^{n_2} I(Y_j - X_i > 0) \geq \frac{n_1 n_2}{2} + z_\alpha \sqrt{\frac{n_1 n_2 (n+1)}{12}}.$$

This test would have asymptotic level α if $F = G$. As Exercise 2.13.40 shows, we still have $E_{H_0}(S_R^+) = n_1 n_2 / 2$ when the densities of X and Y are symmetric.

From Theorem 2.4.5, part (a), the variance of the MWW statistic under H_0 satisfies the limit,

$$\frac{\text{Var}_{H_0}(S_R^+)}{n_1 n_2 (n+1)} \to \lambda_1 \text{Var}(F(Y)) + \lambda_2 \text{Var}(G(X)).$$

Recall that we obtained the asymptotic distribution of S_R^+ (Theorem 2.4.10) under general conditions which cover the current assumptions; hence, the true significance level of the MWW test has the following limiting behavior:

$$\alpha_{S_R^+} = P_{H_0}\left[S_R^+ \geq \frac{n_1 n_2}{2} + z_\alpha \sqrt{\frac{n_1 n_2 (n+1)}{12}}\right]$$

$$= P_{H_0}\left[\frac{S_R^+ - \frac{n_1 n_2}{2}}{\sqrt{\text{Var}_{H_0}(S_R^+)}} \geq z_\alpha \sqrt{\frac{n_1 n_2 (n+1)}{12 \text{Var}_{H_0}(S_R^+)}}\right]$$

$$\to 1 - \Phi\left[z_\alpha (12)^{-1/2}(\lambda_1 \text{Var}(F(Y)) + \lambda_2 \text{Var}(G(X)))^{-1/2}\right]. \quad (2.11.2)$$

Under the assumptions that the sample sizes are the same and that $\mathcal{L}(X)$ and the $\mathcal{L}(Y)$ have the same form, we can simplify expression (2.11.2) further. We express the result in the following theorem.

Theorem 2.11.1 *Suppose that the null hypothesis in (2.11.1) is true. Assume that the distributions of Y and X are symmetric, $n_1 = n_2$, and $G(x) = F(x/\eta)$, where η is an unknown parameter. Then the maximum observed significance level is $1 - \Phi(0.816 z_\alpha)$ which is approached as $\eta \to 0$ or $\eta \to \infty$.*

Proof. Under these assumptions, $\text{Var}(F(Y)) = \int F^2(\eta t) dF(t) - \frac{1}{4}$ and $\text{Var}(G(X)) = \int F^2(x/\eta) dF(x) - \frac{1}{4}$. Differentiating (2.11.2) with respect to η, we get

$$\phi\left[z_\alpha(12)^{-1/2}(\tfrac{1}{2}\text{Var}(F(Y)) + \tfrac{1}{2}\text{Var}(G(X)))^{-1/2}\right]z_\alpha(12)^{-1/2}$$

$$\times \left\{\int F(\eta t) t f(\eta t) f(t) dt + \int F(t/\eta) f(t/\eta)(-t/\eta^2) f(t) dt\right\}^{-3/2}. \quad (2.11.3)$$

Making the substitution $u = \eta t$ in the first integral, the quantity in braces reduces to $\eta^{-2} \int (F(u) - F(u/\eta)) u f(u) f(u/\eta) du$. Note that the other factors in (2.11.3) are strictly positive. Thus to determine the graphical behavior of (2.11.2) with respect to η, we need only consider the factor in braces. First note that it has a critical point at $\eta = 1$. Next consider the case $\eta > 1$. In this case $F(u) - F(u/\eta)$ is negative on the interval $(-\infty, 0)$ and positive on the interval $(0, \infty)$; hence, the factor in braces is positive for $\eta > 1$. Using a similar argument, this factor is negative for $0 < \eta < 1$. Therefore the limit of the function $\alpha_{S_R^+}(\eta)$ is decreasing on the interval $(0, 1)$, has a minimum at $\eta = 1$, and is increasing on the interval $(1, \infty)$.

Thus the minimum level of significance occurs at $\eta = 1$ (the location model), where it is α. By the graphical behavior of the function, maximum levels would occur at the extremes of 0 and ∞. But it follows that

$$\text{Var}(F(Y)) = \int F^2(\eta t) dF(t) - \tfrac{1}{4} \rightarrow \begin{cases} 0 & \text{if } \eta \to 0 \\ \tfrac{1}{4} & \text{if } \eta \to \infty \end{cases}$$

and

$$\text{Var}(G(X)) = \int F^2(x/\eta) dF(x) - \tfrac{1}{4} \rightarrow \begin{cases} \tfrac{1}{4} & \text{if } \eta \to 0 \\ 0 & \text{if } \eta \to \infty. \end{cases}$$

From these two results and (2.11.2), the true significance level of the MWW test satisfies

$$\alpha_{S_R^+} \rightarrow \begin{cases} 1 - \Phi(z_\alpha (3/2)^{-1/2}) & \text{if } \eta \to 0 \\ 1 - \Phi(z_\alpha (3/2)^{-1/2}) & \text{if } \eta \to \infty. \end{cases}$$

Hence,

$$\alpha_{S_R^+} \rightarrow 1 - \Phi(z_\alpha (3/2)^{-1/2}) = 1 - \Phi(0.816 z_\alpha),$$

whether $\eta \to 0$ or ∞. Thus the maximum observed significance level is $1 - \Phi(0.816 z_\alpha)$ which is approached as $\eta \to 0$ or $\eta \to \infty$.

For example, if $\alpha = 0.05$ then $0.816 z_\alpha = 1.34$ and $\alpha_{S_R^+} \to 1 - \Phi(1.34) = 0.09$. Thus in the equal sample size case when F and G differ only in scale parameter and are symmetric, the nominal 5% level of the MWW test will not be worse than 0.09. In order to guarantee that $\alpha \leqslant 0.05$ choose z_α so that $1 - \Phi(0.816 z_\alpha) = 0.05$. This leads to $z_\alpha = 2.02$, which is the critical value for $\alpha = 0.02$. Hence, another way of saying this is: by performing a 2% MWW test we are guaranteed that the true (asymptotic) level is at most 5%.

2.11.2 General Rank Tests

Assuming the most general hypothesis, (2.11.1), we will follow the development of Fligner and Policello (1981) to construct general tests. Suppose T represents a rank test statistic, used in the case $F = G$, and that the test rejects $H_0 : \Delta = 0$ in favor of $H_A : \Delta > 0$ for large values of T. Suppose further that $n^{1/2}(T - \mu_{F,G})/\sigma_{F,G}$ converges in distribution to a standard normal. Let μ_0 denote the null mean of T and assume that it is independent of F. Next suppose that $\hat{\sigma}$ is a consistent estimate of $\sigma_{F,G}$ which is a function only of the ranks of the combined sample. This will ensure distribution-freeness under H_0; otherwise, the test statistic will only be asymptotically distribution-free. The modified test statistic is

$$\hat{T} = \frac{n^{1/2}(T - \mu_0)}{\hat{\sigma}}. \qquad (2.11.4)$$

Such a test can be used for the general hypothesis (2.11.1). Fligner and Policello (1981) applied this approach to Mood's statistic; see also

Hettmansperger and Malin (1975). In the next section, we consider Mathisen's test.

2.11.3 Modified Mathisen Test

We next present a modified version of Mathisen's test for the most general hypothesis (2.11.1). Let $\theta_X = \text{med}_i X_i$ and define the sign process

$$S_2(\theta) = \sum_{j=1}^{n_2} \text{sgn}(Y_j - \theta). \tag{2.11.5}$$

Recall from expression (2.6.8), that Mathisen's test statistic (centered version) is given by $S_2(\widehat{\theta}_X)$. This will be our test statistic. The modification lies in its asymptotic distribution which is given in the next theorem.

Theorem 2.11.2 *Assume the null hypothesis in expression (2.11.1) is true. Then under assumption (D.1), (2.4.7), $(1/\sqrt{n_2})S_2(\widehat{\theta}_X)$ is asymptotically normal with mean 0 and asymptotic variance $1 + K_{12}^2$, where K_{12}^2 is defined by*

$$K_{12}^2 = \frac{\lambda_2}{\lambda_1} \frac{g^2(\theta_Y)}{f^2(\theta_X)}. \tag{2.11.6}$$

Proof. Assume without loss of generality that $\theta_X = \theta_Y = 0$. From the asymptotic linearity results discussed in Example 1.5.2 of Chapter 1, we have that

$$\frac{1}{\sqrt{n_2}} S_2(\theta_n) \doteq \frac{1}{\sqrt{n_2}} S_2(0) - 2g(0)\sqrt{n_2}\theta_n,$$

for $\sqrt{n}|\theta_n| \leq c$, $c > 0$. Since $\sqrt{n_2}\widehat{\theta}_X$ is bounded in probability, upon substitution in this expression we get

$$\frac{1}{\sqrt{n_2}} S_2(\widehat{\theta}_X) \doteq \frac{1}{\sqrt{n_2}} S_2(0) - 2g(0)\sqrt{n_2}\widehat{\theta}_X. \tag{2.11.7}$$

In Example 1.5.2, we also have the approximation

$$\widehat{\theta}_X \doteq \frac{1}{n_1 2f(0)} S_1(0), \tag{2.11.8}$$

where $S_1(0) = \sum_{i=1}^{n_1} \text{sgn}(X_i)$. Combining (2.11.7) and (2.11.8), we get

$$\frac{1}{\sqrt{n_2}} S_2(\widehat{\theta}_X) \doteq \frac{1}{\sqrt{n_2}} S_2(0) - \frac{g(0)}{f(0)} \sqrt{\frac{n_2}{n_1}} \frac{1}{\sqrt{n_1}} S_1(0). \tag{2.11.9}$$

The result follows because of independent samples and because $S_i(0)/\sqrt{n_i} \xrightarrow{\mathcal{D}} N(0, 1)$, for $i = 1, 2$.

In order to use this test we need an estimate of K_{12}. As in Chapter 1, selected order statistics from the sample $X_1, \ldots, X_{n_1}$ will provide a confidence interval for the median of X. Hence, given a level α, the interval (L_1, U_1), where $L_1 = X_{(k+1)}$, $U_1 = X_{(n_1-k)}$, and $k = n_1/2 - z_{\alpha/2}(\sqrt{n_1}/2)$, is an approximate $(1-\alpha)100\%$ confidence interval for the median of X. Let D_X denote the length of this confidence interval. By Theorem 1.5.10,

$$\frac{\sqrt{n_1}D_X}{2z_{\alpha/2}} \xrightarrow{P} 2f(0). \tag{2.11.10}$$

In the same way, let D_Y denote the length of the corresponding $(1-\alpha)100\%$ confidence interval for the median of Y. Define

$$\widehat{K}_{12} = \frac{D_Y}{D_X}. \tag{2.11.11}$$

From (2.11.10) and the corresponding result for D_Y, the estimate $\widehat{K}_{12}$ is a consistent estimate of K_{12}, under both H_0 and H_A.

Thus the modified Mathisen test for the general hypotheses (2.11.1) is to reject H_0 at approximately level α if

$$Z_M = \frac{S_2(\widehat{\theta}_X)}{\sqrt{n_2(1 + \widehat{K}_{12}^2)}} \geq z_\alpha. \tag{2.11.12}$$

To derive the efficacy of this statistic we will use the development of Section 1.5.2. The average to consider is $n^{-1}S_2(\widehat{\theta}_X)$. Let Δ denote the shift in medians and without loss of generality let $\theta_X = 0$. Then the mean function we need is

$$\lim_{n \to \infty} E_\Delta(n^{-1}S_2(\widehat{\theta}_X)) = \mu(\Delta).$$

Note that we can re-express the expansion (2.11.9) as

$$\frac{1}{n}S_2(\widehat{\theta}_X) = \frac{n_2}{n}\frac{1}{n_2}S_2(\widehat{\theta}_X)$$

$$\doteq \frac{n_2}{n}\left\{\frac{1}{n_2}S_2(0) - \frac{g(0)}{f(0)}\sqrt{\frac{n_2}{n_1}}\sqrt{\frac{n_1}{n_2}}\frac{1}{n_1}S_1(0)\right\}$$

$$\xrightarrow{P_\Delta} \lambda_2\left\{E_\Delta[\text{sgn}(Y)] - \frac{g(0)}{f(0)}E[\text{sgn}(X)]\right\}$$

$$= \lambda_2 E_\Delta[\text{sgn}(Y)] = \mu(\Delta), \tag{2.11.13}$$

where the penultimate equality holds since $\theta_X = 0$. Using $E_\Delta(\text{sgn}(Y)) = 1 - 2G(-\Delta)$, we obtain the derivative

$$\mu'(0) = 2\lambda_2 g(0). \tag{2.11.14}$$

By Theorem 2.11.2 we have the asymptotic null variance of the test statistic $S_2(\widehat{\theta}_X)/\sqrt{n}$. From the above discussion, then, the statistic $S_2(\widehat{\theta}_X)$ is Pitman regular with efficacy

$$c_{MM} = \frac{2\lambda_2 g(0)}{\sqrt{\lambda_2(1 + K_{12}^2)}} = \frac{\sqrt{\lambda_1 \lambda_2} 2g(0)}{\sqrt{\lambda_1 + \lambda_2(g^2(0)/f^2(0))}}. \tag{2.11.15}$$

Using Theorem 1.5.4, consistency of the modified Mathisen test for the hypotheses (2.11.1) is obtained provided $\mu(\Delta) > \mu(0)$. But this follows immediately from the inequality $G(-\Delta) > G(0)$.

2.11.4 Modified MWW Test

Recall by Theorem 2.4.10 that the mean of the MWW test statistic S_R^+ is $n_1 n_2 P(Y > X) = 1 - \int G(x) f(x) dx$. For general F and G, though, this mean may not be $\frac{1}{2}$ under H_0. Since this section is concerned with methods for testing the specific hypothesis that $\Delta = 0$, we add the further restriction that the **distributions** of X and Y **are symmetric**. Recall from Section 2.11.1 that, under this assumption and $\Delta = 0$, $E(S_R^+) = n_1 n_2 / 2$; see Exercise 2.13.40.

Using the general development of rank tests (Section 2.11.2), our modified rank test is given by: reject $H_0 : \Delta = 0$ in favor of $H_A : \Delta > 0$ if $Z > z_\alpha$, where

$$Z = \frac{S_R^+ - (n_1 n_2)/2}{\sqrt{\widehat{\mathrm{Var}}(S_R^+)}}, \qquad (2.11.16)$$

where $\widehat{\mathrm{Var}}(S_R^+)$ is a consistent estimate of $\mathrm{Var}(S_R^+)$, under H_0. From the asymptotic distribution theory obtained for S_R^+ under general conditions (Theorem 2.4.10), it follows that this test has approximate level α. By Theorem 2.4.5, we can express the variance as

$$\mathrm{Var}(S_R^+) = n_1 n_2 \left(\int G dF - \left(\int G dF \right)^2 \right) + n_1 n_2 (n_1 - 1) \left(\int F^2 dG - \left(\int F dG \right)^2 \right)$$

$$+ n_1 n_2 (n_2 - 1) \left(\int (1-G)^2 dF - \left(\int (1-G) dF \right)^2 \right). \qquad (2.11.17)$$

Following the suggestion of Fligner and Policello (1981), we estimate $\mathrm{Var}(S_R^+)$ by replacing F and G by the empirical cdfs F_{n_1} and G_{n_2} respectively. As Exercise 2.13.41 demonstrates, this estimate is consistent and, further, it is a function of the ranks of the combined sample. Thus the test is distribution-free when $F(x) = G(x)$ and is asymptotically distribution-free when F and G have symmetric densities.

The efficacy for the modified MWW follows using an argument similar to that for the MWW in Section 2.4. As there, the function $S_R^+(\Delta)$ is a decreasing function of Δ. Its mean function is given by

$$E_\Delta(S_R^+) = E_0(S_R^+(-\Delta)) = n_1 n_2 \int (1 - G(x - \Delta)) f(x) dx.$$

The average to consider here is $\overline{S}_R = (n_1 n_2)^{-1} S_R^+$. Letting $\mu(\Delta)$ denote the mean of $\overline{S}_R$ under Δ, we have $\mu'(0) = \int g(x) f(x) dx > 0$. The variance we need is $\sigma^2(0) = \lim_{n \to \infty} n \mathrm{Var}_0(\overline{S}_R)$, which, using the above result on variance, simplifies to

$$\sigma^2(0) = \lambda_2^{-1} \left(\int F^2 dG - \left(\int F dG \right)^2 \right) + \lambda_1^{-1} \left(\int (1-G)^2 dF - \left(\int (1-G) dF \right)^2 \right).$$

134   Two-Sample Problems

The process $S_R^+(\Delta)$ is Pitman regular and, in particular, its efficacy is given by

$$c_{MMWW} = \frac{\sqrt{\lambda_1 \lambda_2} \int g(x)f(x)}{\sqrt{\lambda_1 \left(\int F^2 dG - \left(\int F dG \right)^2 \right) + \lambda_2 \left(\int (1-G)^2 dF - \left(\int (1-G) dF \right)^2 \right)}}.$$

(2.11.18)

As with the modified Mathisen test, we show consistency of the modified MWW test by using Theorem 1.5.4. Again we need only show that $\mu(0) < \mu(\Delta)$. But this follows immediately provided the supports of F and G overlap in a neighborhood of 0. Note that this shows that the modified MWW is consistent for the hypotheses (2.11.1) under the further restriction that the densities of X and Y are symmetric.

2.11.5   Efficiencies and Discussion

Before obtaining the asymptotic relative efficiencies of the above procedures, we shall briefly discuss traditional methods. Suppose we restrict F and G to have symmetric densities of the same form with finite variance; that is, $F(x) = F_0((x - \theta_X)/\sigma_X)$ and $G(x) = F_0((x - \theta_Y)/\sigma_Y)$, where F_0 is some distribution function with symmetric density f_0, and σ_X and σ_Y are the standard deviations of X and Y, respectively.

Under these assumptions, it follows that $\sqrt{n}(\overline{Y} - \overline{X} - \Delta)$ converges in distribution to $N(0, (\sigma_X^2/\lambda_1) + (\sigma_Y^2/\lambda_2))$; see Exercise 2.13.42. The test is to reject $H_0 : \Delta = 0$ in favor of $H_A : \Delta > 0$ if $t_W > z_\alpha$, where

$$t_W = \frac{\overline{Y} - \overline{X}}{\sqrt{\dfrac{s_X^2}{n_1} + \dfrac{s_Y^2}{n_2}}},$$

in which s_X^2 and s_Y^2 are the sample variances of X_i and Y_j, respectively. Under these assumptions, it follows that these sample variances are consistent estimates of σ_X^2 and σ_Y^2, respectively; hence, the test has approximate level α. If F_0 is also normal then, under H_0, t_W has an approximate t-distribution with a degrees of freedom correction proposed by Welch (1937). This test is frequently used in practice and we shall subsequently call it the **Welch t-test**.

In contrast, the pooled t-test can behave poorly in this situation, since we have

$$t_p = \frac{\overline{Y} - \overline{X}}{\sqrt{\dfrac{(n_1 - 1)s_X^2 + (n_2 - 1)s_Y^2}{n_1 + n_2 - 2} \left(\dfrac{1}{n_1} + \dfrac{1}{n_2} \right)}}$$

$$\doteq \frac{\overline{Y} - \overline{X}}{\sqrt{\dfrac{s_X^2}{n_2} + \dfrac{s_Y^2}{n_1}}};$$

that is, the sample variances are divided by the wrong sample sizes. Hence, unless the sample sizes are fairly close the pooled t is not asymptotically distribution-free. Exercise 2.13.43 obtains the true asymptotic level of t_p.

In order to get the efficacy of the Welch t, consider the statistic $\overline{Y} - \overline{X}$. The mean function at Δ is $\mu(\Delta) = \Delta$; hence, $\mu'(0) = 1$. It follows from the asymptotic distribution discussed above that

$$\sqrt{n}\left[\frac{\sqrt{\lambda_1\lambda_2}(\overline{Y} - \overline{X})}{\sqrt{(\sigma_X^2/\lambda_1) + (\sigma_Y^2/\lambda_2)}}\right] \xrightarrow{\mathcal{D}} N(0, 1);$$

hence, $\sigma(0) = \sqrt{(\sigma_X^2/\lambda_1) + (\sigma_Y^2/\lambda_2)}/\sqrt{\lambda_1\lambda_2}$. Thus the efficacy of t_W is given by

$$c_{tw} = \frac{\mu'(0)}{\sigma(0)} = \frac{\sqrt{\lambda_1\lambda_2}}{\sqrt{(\sigma_X^2/\lambda_1) + (\sigma_Y^2/\lambda_2)}}. \tag{2.11.19}$$

We obtain the AREs of the above procedures for the case where $G(x) = F(x/\eta)$ and $F(x)$ has density $f(x)$ symmetric about 0 with variance 1. Thus η is the ratio of standard deviations σ_Y/σ_X. For this case the efficacies (2.11.15), (2.11.18), and (2.11.19) reduce to

$$c_{MM} = \frac{2\sqrt{\lambda_1\lambda_2}f(0)}{\sqrt{\lambda_2 + \lambda_1\eta^2}}$$

$$c_{MMWW} = \frac{\sqrt{\lambda_1\lambda_2}\int gf}{\sqrt{\lambda_1\left[\int F^2 dG - \left(\int F dG\right)^2\right] + \lambda_2\left[\int (1 - G)^2 dF - \left(\int (1 - G)dF\right)^2\right]}}$$

$$c_{tw} = \frac{\sqrt{\lambda_1\lambda_2}}{\sqrt{\lambda_2 + \lambda_1\eta^2}}.$$

Thus the ARE between the modified Mathisen's procedure and the Welch procedure is the ratio $c_{MM}^2/c_{tw}^2 = 4\sigma_X^2 f^2(0) = 4f_0^2(0)$. This is the same ARE as in the location problem. In particular, the ARE does not depend on $\eta = \sigma_Y/\sigma_X$. Thus the modified Mathisen test in comparison to t_W would have poor efficiency at the normal distribution, 0.63, but in general it would be much more efficient than t_W for heavy-tailed distributions. Similar to the modified Mathisen test, the Mocd test can also be modified for these problems; see Exercise 2.13.44. Its efficacy is the same as that of the Mathisen's test.

Asymptotic relative efficiencies involving the modified Wilcoxon do depend on the ratio of scale parameters η. Fligner and Rust (1982) show that if the variances of X and Y are quite different then the modified Mathisen test may be as efficient as the modified MWW irrespective of the shape of the underlying distribution.

Fligner and Policello (1981) conducted a simulation study of the pooled t, Welch's t, MWW and modified MWW over situations where F and G differ in scale only. The unmodified tests did not maintain their level. Welch's t performed well when F and G were normal whereas the modified MWW performed well over all situations, including unequal sample sizes and normal

136 Two-Sample Problems

and contaminated normal distributions. In their simulation study, Fligner and Rust (1982) found that the modified Mood test maintains its level over the situations that were considered by Fligner and Policello (1981).

As a final note, Welch's t requires distributions with the same shape and the modified MWW requires symmetric densities. The modified Mathisen test and the modified Mood test, though, are consistent tests for the general problem stated in expression (2.11.1).

2.12 Paired Designs

Consider the situation where we have two treatments of interest, say, A and B, which can be applied to subjects from a population of interest. Suppose we are interested in a particular response after these treatments have been applied. Let X denote the response of a subject after treatment A has been applied and let Y be the corresponding measurement for a subject after treatment B has been applied. The natural null hypothesis, H_0, is that there is no difference in treatment effects. A one-sided alternative would be that the response of a subject under treatment B is in general larger than that of a subject under treatment A. Reversing the roles of A and B would yield the other one-sided alternative, while the union of the these two alternatives would result in the two-sided alternative. Again for definiteness we choose as our alternative, H_A, the first one-sided alternative.

The completely randomized design and the paired design are two experimental designs which are often employed in this situation. In the completely randomized design, n subjects are selected at random from the population of interest and n_1 of them are randomly assigned to treatment A while the remaining $n_2 = n - n_1$ are assigned to treatment B. At the end of the treatment period, we then have two samples, one on X while the other is on Y. The two-sample procedures discussed in the previous sections can be used to analyze the data. Proper randomization along with carefully controlled experimental conditions give credence to the assumptions that the samples are random and are independent of one another. The design that produced the data of Example 2.3.1 was a completely randomized design.

While the completely randomized design is often used in practice, the underlying variability may impair the power of any procedure, robust or classical, to detect alternative hypotheses. The design discussed next usually results in a more powerful analysis but it does require a pairing device; i.e. a block of length two.

Suppose we have a pairing device. Some examples include identical twins for a study on human subjects, litter mates for a study on animal subjects, or the same exterior wall of a house for a study on the durability of exterior house paints. In the paired design, n pairs of subjects are randomly selected from the population of interest. Within each pair, one member is randomly assigned to treatment A while the other receives treatment B. Again let X and Y denote the responses of subjects after treatments A and B respectively have been applied. This experimental design results in a sample of pairs $(X_1, Y_1), \ldots, (X_n, Y_n)$. The sample differences $D_1 = X_1 - Y_1, \ldots D_n = X_n - Y_n$, however, become the *single* sample of interest. Note that the random pairing in this design induces under the null hypothesis a symmetrical distribution for the differences.

Theorem 2.12.1 *In a randomized paired design, under the null hypothesis of no treatment effect, the differences D_i are symmetrically distributed about 0.*

Proof. Let $F(x, y)$ denote the joint distribution of (X, Y). Under the null hypothesis of no treatment effect and randomized pairing, it follows that X and Y are exchangable random variables; that is, $P(X \leqslant x, Y \leqslant y) = P(X \leqslant y, Y \leqslant x)$. Hence, for a difference $D = Y - X$ we have

$$P[D \leqslant t] = P[Y - X \leqslant t] = P[X - Y \leqslant t] = P[-D \leqslant t].$$

Thus D and $-D$ have the same distribution; hence D is symmetrically distributed about 0.

Let θ be a location functional for the distribution of D_i. We shall further assume that D_i is symmetrically distributed under alternative models also. Then we can express the above hypotheses by $H_0 : \theta = 0$ versus $H_A : \theta > 0$.

Note that the one-sample analyses based on signs and signed ranks discussed in Chapter 1 are appropriate for the randomly paired design. The appropriate sign test statistic is $S = \sum \text{sgn}(D_i)$, while the signed-rank statistic is $T = \sum \text{sgn}(D_i) R(|D_i|)$.

From Chapter 1 we shall summarize the analysis based on the signed-rank statistic. A level α test would reject H_0 in favor of H_A if $T \geqslant c_\alpha$, where c_α is determined from the null distribution of the Wilcoxon signed-rank test or from the asymptotic approximation to the distribution. The test is consistent for $\theta > 0$ and it has the efficiency results discussed in Chapter 1. In particular, for normal errors the efficiency of T with respect to the usual paired t-test is 0.955. The associated point estimate of θ is the Hodges–Lehmann estimate given by $\hat{\theta} = \text{med}_{i \leqslant j}\{(D_i + D_j)/2\}$. A distribution-free confidence interval for θ is constructed based on the Walsh averages $\{(D_i + D_j)/2\}$, $i \leqslant j$, as discussed in Chapter 1. Instead of using Wilcoxon scores, general signed-rank scores as discussed in Chapter 1, can also be used.

A similar summary holds for the analysis based on the sign statistic. In fact for the sign scores we need not assume that $D_1, \ldots, D_n$ are identically distributed; that is, there can be a block effect. This is discussed further in Chapter 4.

We should mention that if the pairing is not done randomly then D_i may or may not be symmetrically distributed. If the symmetry assumption is realistic, then both sign and signed-rank analyses can be used. If, however, it is not realistic then the sign analysis would still be valid but caution would be necessary in interpreting the results of the signed-rank analysis.

Example 2.12.1 *Darwin Data.*

The data in Table 2.12.1 are some measurements recorded by Charles Darwin in 1878. They consist of 15 pairs of heights in inches of cross-fertilized plants and self-fertilized plants (*Zea mays*), each pair grown in the same pot.

138 Two-Sample Problems

Table 2.12.1: Plant growth

Pot	1	2	3	4	5	6	7	8
Cross-	23.500	12.000	21.000	22.000	19.125	21.500	22.125	20.375
Self-	17.375	20.375	20.000	20.000	18.375	18.625	18.625	15.250

Pot	9	10	11	12	13	14	15
Cross-	18.250	21.625	23.250	21.000	22.125	23.000	12.000
Self-	16.500	18.000	16.250	18.000	12.750	15.500	18.000

Let D_i denote the difference between the heights of the cross-fertilized and self-fertilized plants of the ith pot and let θ denote the median of the distribution of D_i. Suppose we are interested in testing for an effect; that is, the hypotheses are $H_0 : \theta = 0$ versus $H_A : \theta \neq 0$. The boxplot of the differences is displayed in Figure 2.12.1(a), while Figure 2.12.1(b) gives the normal $q-q$ plot of the differences. As the plots indicate, the differences for pot 2 and, perhaps, pot 15 are possible outliers. The value of the signed-rank Wilcoxon statistic for these data is $T = 36$ with the approximate p-value of 0.044. The corresponding estimate of θ is 3.13 inches and the 95% confidence interval, using Minitab, is (0.50, 5.21).

There are 13 positive differences; hence, the sign test has an approximate p-value of 0.0074. The corresponding estimate of θ is 3 inches and the 95% interpolated confidence interval of Section 1.2 is (1.280, 5.751). The paired t-test statistic has the value of 2.15 with p-value 0.050. The difference in sample means is 2.62 inches and the corresponding 95% confidence interval is (0, 5.23).

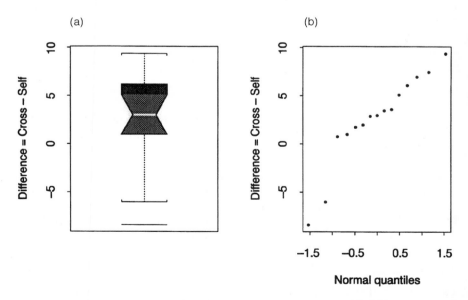

Figure 2.12.1: (a) Boxplot with 95% notched confidence interval of Darwin data; (b) normal $q-q$ plot of Darwin data

Note that the outliers impaired the *t*-test and to a lesser degree the Wilcoxon signed-rank test; see Exercise 2.13.45 for further analyses.

2.12.1 Behavior under Alternatives

In this subsection we will compare sample size determination for the paired design with sample size determination for the completely randomized design. For the paired design, let $\gamma^+(\theta)$ denote the power function of Wilcoxon signed-rank test statistic for the alternative θ. Then the asymptotic power lemma (Theorem 1.5.9) with $c = \tau^{-1} = \sqrt{12} \int f^2(t)\,dt$, for the signed-rank Wilcoxon test from Chapter 1 states that at significance level α and under the sequence of contiguous alternatives, $\theta_n = \theta/\sqrt{n}$,

$$\lim_{n \to \infty} \gamma^+(\theta_n) = P_{\theta_n}\left[Z \geq z_\alpha - \frac{\theta}{\tau}\right].$$

We will only consider the case where the random vector (Y, X) is jointly normal with variance-covariance matrix

$$\mathbf{V} = \sigma^2 \begin{bmatrix} 1 & \rho \\ \rho & 1 \end{bmatrix}.$$

Then $\tau = \sqrt{\pi/3}\sigma\sqrt{2(1-\rho)}$.

Now suppose we select the sample size n^* so that the Wilcoxon signed-rank test has power $\gamma^+(\theta_0)$ to detect the one-sided alternative $\theta_0 > 0$ for a level α test. Then, writing $\theta_0 = \sqrt{n^*}\theta_0/\sqrt{n^*}$, we have by the asymptotic power lemma and (1.5.25) that

$$\gamma^+(\theta_0) \doteq 1 - \Phi(z_\alpha - \sqrt{n^*}\theta_0/\tau)$$

and

$$n^* \doteq \frac{(z_\alpha - z_{\gamma^+(\theta_0)})^2}{\theta_0^2}\tau^2.$$

Substituting the value of τ into this final equation, we have that the necessary sample size for the paired design to have the desired local power is

$$n^* \doteq \frac{(z_\alpha - z_{\gamma^+(\theta_0)})^2}{\theta_0^2}(\pi/3)\sigma^2 2(1-\rho). \tag{2.12.1}$$

Next consider a two-sample design with equal sample sizes $n_i = n^*$. Assume that X and Y are iid normal with variance σ^2. Then $\tau^2 = (\pi/3)\sigma^2$. Hence by (2.4.25), the necessary sample size for the completely randomized design to achieve power $\gamma^+(\theta_0)$ for the one-sided alternative $\theta_0 > 0$ for a level α test is given by

$$n = \left(\frac{z_\alpha - z_{\gamma^+(\theta_0)}}{\theta_0}\right)^2 2(\pi/3)\sigma^2. \tag{2.12.2}$$

Based on expressions (2.12.1) and (2.12.2), the sample size needed for the paired design is $(1-\rho)$ times the sample size needed for the completely randomized design. If the pairing device is such that X and Y are strongly positively correlated then it pays to use the paired design. The paired design is a disaster, of course, if the variables are negatively correlated.

2.13 Exercises

2.13.1 (a) Derive the L_2 estimates of intercept and shift based on the L_2 norm on model (2.2.4).

(b) Next apply the pseudo-norm, (2.2.16), to (2.2.4) and derive the estimating function. Show that the natural test statistic is the pooled t-statistic.

2.13.2 Show that (2.2.17) is a pseudo-norm. Show also that it can be written in terms of ranks; see the formula following (2.2.17).

2.13.3 In the proof of Theorem 2.4.2, verify that $\mathcal{L}(Y_j - X_i) = \mathcal{L}(X_i - Y_j)$.

2.13.4 Prove Theorem 2.4.3.

2.13.5 Prove that if a continuous random variable Z has cdf $H(z)$, then the random variable $H(Z)$ has a uniform distribution on $(0, 1)$.

2.13.6 In Theorem 2.4.4, show that $E(F(Y)) = \int F(y) dG(y) = \int (1 - G(x)) dF(x) = E(1 - G(X))$.

2.13.7 Prove that if Z_n converges in distribution to Z and if $\text{Var}(Z_n - W_n)$ and $EZ_n - EW_n$ converge to 0, then W_n also converges in distribution to Z.

2.13.8 Verify (2.4.10).

2.13.9 Explain what happens to the MWW statistic when one support is shifted completely to the right of the other support. What does this imply about the consistency of the MWW in this case?

2.13.10 Show that the L_2 estimating function is Pitman regular and derive the efficacy of the pooled t-test. Also, establish the asymptotic power lemma (Theorem 2.4.17), for the L_2 case. Finally, establish the asymptotic distribution of $\sqrt{n}(\overline{Y} - \overline{X})$.

2.13.11 Prove that the Hodges–Lehmann estimate of shift, (2.2.18), is translation- and scale-equivariant. (See the discussion in Section 2.4.4.)

2.13.12 Prove Theorem 2.4.19.

2.13.13 In Example 2.4.1, form the residuals $Z_i - \widehat{\Delta} c_i$, $i = 1, \ldots, n$. Then, similarly to Section 1.5.5, use these residuals to estimate τ based on (1.3.30).

2.13.14 Simulate independent random samples from $N(20, 5^2)$ and $N(22, 5^2)$ distributions of sizes 10 and 15 respectively. Let Δ denote the shift in the locations of the distributions.

(a) Obtain comparison boxplots for your samples.

(b) Use the Wilcoxon procedure to test $H_0 : \Delta = 0$ versus $H_A : \Delta \neq 0$ at level 0.05.

(c) Use the Wilcoxon procedure to estimate Δ and obtain a 95% confidence interval for it.

(d) Obtain the true value of τ. Use your confidence interval in the last item to obtain an estimate of τ. Obtain a symmetric 95% confidence interval for Δ based on your estimate.

(e) Form a pooled estimate of τ based on the Wilcoxon signed-rank process for each sample. Obtain a symmetric 95% confidence interval for Δ based on your estimate. Compare it with the estimate from the last item and the true value.

2.13.15 Write Minitab macros to bootstrap the distribution of $\widehat{\Delta}$. Obtain the bootstrap distribution for 500 bootstraps for the data in Exercise 2.13.14. What is your bootstrap estimate of τ? Compare with the true value and the other estimates.

2.13.16 Verify the scalar multiple condition for the pseudo-norm in the proof of Theorem 2.5.1.

2.13.17 Verify (2.5.9) and (2.5.10).

2.13.18 Consider the process $S_\varphi(\Delta)$, (2.5.11):
(a) Show that $S_\varphi(\Delta)$ is a decreasing step function, with steps occurring at $Y_j - X_i$.
(b) Verify expressions (2.5.14), (2.5.15), and (2.5.16).

2.13.19 Consider the the optimal score function (2.5.22):
(a) Show it is location-invariant and scale-equivariant. Hence, show if $g(x) = \frac{1}{\sigma} f((x-\mu)/\sigma)$, then $\varphi_g = \sigma^{-1}\varphi_f$.
(b) Use (2.5.22) to show that the MWW is asymptotically efficient when the underlying distribution is logistic. ($F(x) = (1 + \exp(-x))^{-1}$, $-\infty < x < \infty$.)
(c) Show that (2.6.1) is optimal for a Laplace or double exponential distribution. ($f(x) = \frac{1}{2}\exp(-|x|)$, $-\infty < x < \infty$.)
(d) Show that the optimal score function for the extreme value distribution, ($f(x) = \exp\{x - e^x\}$, $-\infty < x < \infty$), is given by (2.8.8).
(e) Show that the optimal score function for the normal distribution is given by (2.5.33). Show that it is standardized.
(f) Show that (2.5.34) is the optimal score function for an underlying distribution that has a left logistic tail and a right exponential tail.

2.13.20 Show that when the underlying density f is symmetric then $\varphi_f(-u) = -\varphi_f(1-u)$.

2.13.21 Show that expression (2.6.6) is true and that the $n = 2r$ differences,
$$Y_{(1)} - X_{(r)} < Y_{(2)} - X_{(r-1)} < \cdots < Y_{(n_2)} - X_{(r-n_2+1)},$$
can be ordered only knowing the order statistics from the individual samples.

2.13.22 Develop the asymptotic linearity formula for Mood's estimating function given in (2.6.3). Then give an alternative proof of Theorem 2.6.1 based on this result.

2.13.23 Verify the moment formulas (2.6.9) and (2.6.10).

142   Two-Sample Problems

2.13.24 Show that any estimator based on the pseudo-norm (2.5.2) is equivariant. Hence, if we multiply the combined sample observations by a constant, then the estimator is multiplied by that same constant.

2.13.25 Suppose X is a continuous random variable representing the time until failure of some process. The hazard function for a continuous random variable X with cdf F is defined to be the instantaneous rate of failure at $X = t$, conditional on survival to time t. It is formally given by:

$$h_X(t) = \lim_{\Delta t \to 0^+} \frac{P(t \leq X < t + \Delta t | X \geq t)}{\Delta t}.$$

(a) Show that

$$h_X(t) = \frac{f(t)}{1 - F(t)}.$$

(b) Suppose that Y has cdf given by (2.8.1). Show that the hazard function is given by $h_Y(t) = \alpha h_X(t)$.

2.13.26 Verify (2.8.4).

2.13.27 Apply the delta method of finding the asymptotic distribution of a function to (2.8.3) to find the asymptotic distribution of $\hat{\alpha}$. Then verify (2.8.5). Explain how this can be used to find an approximate $(1 - \alpha)100\%$ confidence interval for α.

2.13.28 Verify (2.8.14).

2.13.29 Show that the asymptotic efficiency of the Mann–Whitney–Wilcoxon test relative to the Savage test for the log exponential model is 3/4.

2.13.30 Verify (2.10.13).

2.13.31 Show that if $|X|$ has an $F(2, 2)$ distribution then $\log|X|$ has a logistic distribution.

2.13.32 (a) Verify (2.10.14).

(b) Apply (2.10.14) to the normal distribution.

(c) Apply (2.10.14) to the Laplace or double exponential distribution.

2.13.33 We consider the Siegel and Tukey (1960) test for the equality of variances when the underlying centers are equal but possibly unknown. The test statistic is the sum of ranks of the Y sample in the combined sample (MWW statistic). However, the ranks are assigned in a different way: In the ordered combined sample assign rank 1 to the smallest value, rank 2 to the largest value, rank 3 to the second largest value, rank 4 to the second smallest value, and so on, alternatively assigning ranks to end-values. To test $H_0 : \text{Var} X = \text{Var} Y$ vs $H_A : \text{Var} X > \text{Var} Y$, reject H_0 when the sum of ranks of the Y sample is large. Find the mean, variance and the limiting distribution of the test statistic. Show how to find an approximate size α test.

2.13.34 Develop a sample size formula for the scale problem similar to the sample size formula in the location problem, (2.4.25).

2.13.35 Verify (2.10.35).

2.13.36 Compute the efficacy of Mood's scale test, Ansari–Bradley scale test, and Klotz's scale test discussed in Section 2.10.3.

2.13.37 Verify the asymptotic properties given in (2.10.36), (2.10.37) and (2.10.38).

2.13.38 Compute the efficiency of Mood's scale test and the Ansari–Bradley scale test relative to the classical F-test for equality of variances.

2.13.39 Show that the Ansari–Bradley scale test is optimal for $f(x) = \frac{1}{2}(1 + |x|)^{-2}$, $-\infty < x < \infty$.

2.13.40 Show that when F and G have densities symmetric at 0 (or any common point), the expected value of $S_{R^+} = n_1 n_2/2$, where S_{R^+} is MWW.

2.13.41 Show that the estimate of (2.11.17) based on the empirical cdfs is consistent and that it is a function only of the combined sample ranks.

2.13.42 Under the general model in Section 2.11.5, derive the limiting distribution of $\sqrt{n}(\overline{Y} - \Delta - \overline{X})$.

2.13.43 Find the true asymptotic level of the pooled t-test under the null hypothesis in (2.11.1).

2.13.44 Develop a modified Mood test similar to the modified Mathisen test discussed in Section 2.11.5.

2.13.45 Construct and discuss a normal quantile plot of the differences from Table 2.12.1. Carry out the Boos test for asymmetry (1.9.2). Why do these results suggest that the L_1 analysis may be the best analysis in this example?

2.13.46 Consider the data on professional baseball players given in Exercise 1.13.25. Let Δ denote the shift parameter of the difference between the height of a pitcher and the height of a hitter.

(a) Obtain comparison dotplots between the heights of the pitchers and hitters. Does a shift model seem appropriate?

(b) Use the MWW test statistic to test the hypotheses $H_0 : \Delta = 0$ versus $H_A : \Delta > 0$. Compute the p-value.

(c) Determine a point estimate for Δ and a 95% confidence interval for Δ based on the MWW procedure.

(d) Obtain an estimate of the standard deviation of $\widehat{\Delta}$. Use it to obtain an approximate 95% confidence interval for Δ.

2.13.47 Repeat Exercise 2.13.46 when Δ is the shift parameter for the difference in pitchers' and hitters' weights.

2.13.48 Repeat Exercise 2.13.46 when Δ is the shift parameter for the difference in left-handed ($A=1$) and right-handed ($A=0$) pitchers' ERAs and the hypotheses are $H_0 : \Delta = 0$ versus $H_A : \Delta \neq 0$.

3
Linear Models

3.1 Introduction

In this chapter we discuss the theory for a rank-based analysis of a general linear model. Applications of this analysis to experimental design models will be discussed in Chapter 4. The rank-based analysis is complete, consisting of estimation, testing, and diagnostic tools for checking the adequacy of fit of the model, outlier detection, and detection of influential cases. As in the earlier chapters, we present the analysis in terms of its geometry; see McKean and Schrader (1980).

The analysis could be based on either rank scores or signed-rank scores. We have chosen to use the general rank scores of Chapter 2. This allows the error distribution to be either asymmetric or symmetric. An analysis based on signed-rank scores would parallel the one based on rank scores except that the theory would require a symmetric error distribution; see Hettmansperger and McKean (1983) for discussion. Although the results are established for general score functions, we illustrate the methods with Wilcoxon and sign scores throughout. We will commonly use the subscripts R and S for results based on Wilcoxon and sign scores, respectively.

3.2 Geometry of Estimation and Tests

For $i = 1, \ldots, n$, let Y_i denote the ith observation and let $\mathbf{x}_i$ denote a $p \times 1$ vector of explanatory variables. Consider the linear model

$$Y_i = \mathbf{x}_i'\boldsymbol{\beta} + e_i^*, \qquad (3.2.1)$$

where $\boldsymbol{\beta}$ is a $p \times 1$ vector of unknown parameters. In this chapter, the components of $\boldsymbol{\beta}$ are the parameters of interest. We are interested in estimating $\boldsymbol{\beta}$ and testing linear hypotheses concerning it. However, it will be convenient also to have a location parameter. Accordingly, let $\alpha = T(e_i^*)$ be a location functional.

One that we will frequently use is the median. Let $e_i = e_i^* - \alpha$; then $T(e_i) = 0$ and the model can be written as

$$Y_i = \alpha + \mathbf{x}_i'\boldsymbol{\beta} + e_i. \tag{3.2.2}$$

The parameter α is called an intercept parameter. An argument similar to the one concerning the shift parameter Δ of Chapter 2 shows that $\boldsymbol{\beta}$ does not depend on the location functional used.

Let $\mathbf{Y} = (Y_1, \ldots, Y_n)'$ denote the $n \times 1$ vector of observations and let $\mathbf{X}$ denote the $n \times p$ matrix whose ith row is $\mathbf{x}_i'$. We can then express the model as

$$\mathbf{Y} = \mathbf{1}\alpha + \mathbf{X}\boldsymbol{\beta} + \mathbf{e}, \tag{3.2.3}$$

where $\mathbf{1}$ is an $n \times 1$ vector of ones, and $\mathbf{e}' = (e_1, \ldots, e_n)$. Since the model includes an intercept parameter, α, there is no loss in generality in assuming that $\mathbf{X}$ is centered; i.e. the columns of $\mathbf{X}$ sum to 0. Further, in this chapter, we will assume that $\mathbf{X}$ has full column rank p. Let Ω_F denote the column space spanned by the columns of $\mathbf{X}$. Note that we can then write the model as

$$\mathbf{Y} = \mathbf{1}\alpha + \boldsymbol{\eta} + \mathbf{e}, \quad \text{where } \boldsymbol{\eta} \in \Omega_F. \tag{3.2.4}$$

This model is often called the **coordinate-free model**.

Besides estimation of the regression coefficients, we are interested in tests of general linear hypotheses of the form

$$H_0 : \mathbf{M}\boldsymbol{\beta} = \mathbf{0} \text{ versus } H_A : \mathbf{M}\boldsymbol{\beta} \neq \mathbf{0}, \tag{3.2.5}$$

where $\mathbf{M}$ is a $q \times p$ matrix of full row rank. In this section, we discuss the geometry of estimation and testing with rank-based procedures for the linear model.

3.2.1 Estimation

With respect to model (3.2.4), we will estimate $\boldsymbol{\eta}$ by minimizing the distance between $\mathbf{Y}$ and the subspace Ω_F. In this chapter we will define distance in terms of the norms or pseudo-norms presented in Chapter 2. Consider, first, the general R pseudo-norm discussed in Chapter 2 which is given by expression (2.5.2) and which we write for convenience as

$$\|\mathbf{v}\|_\varphi = \sum_{i=1}^n a(R(v_i))v_i, \tag{3.2.6}$$

where $a(1) \leq a(2) \leq \cdots \leq a(n)$ is a set of scores generated as $a(i) = \varphi(i/(n+1))$ for some nondecreasing score function $\varphi(u)$ defined on the interval $(0, 1)$ and standardized such that $\int \varphi(u)du = 0$ and $\int \varphi^2(u)du = 1$. This was shown to be a pseudo-norm in Chapter 2. Recall that the Wilcoxon pseudo-norm is generated by the linear score function $\varphi(u) = \sqrt{12}(u - \frac{1}{2})$. We will also discuss the sign pseudo-norm which is generated by $\varphi(u) = \text{sgn}(u - \frac{1}{2})$ and show that it is equivalent to using the L_1 norm. In Section 3.10 we will also discuss a class of score functions appropriate for survival-type analyses.

For the general R pseudo-norm given above by (3.2.6), an R-estimate of $\boldsymbol{\eta}$ is a vector $\widehat{\mathbf{Y}}_\varphi$ such that

$$D_\varphi(\mathbf{Y}, \Omega_F) = \|\mathbf{Y} - \widehat{\mathbf{Y}}_\varphi\|_\varphi = \min_{\boldsymbol{\eta} \in \Omega_F} \|\mathbf{Y} - \boldsymbol{\eta}\|_\varphi. \qquad (3.2.7)$$

These quantities are represented geometrically in Figure 3.2.1.

Once $\boldsymbol{\eta}$ has been estimated, $\boldsymbol{\beta}$ can be estimated by solving the equation $\mathbf{X}\boldsymbol{\beta} = \widehat{\mathbf{Y}}_\varphi$; that is, the **R-estimate** of $\boldsymbol{\beta}$ is $\widehat{\boldsymbol{\beta}}_\varphi = (\mathbf{X}'\mathbf{X})^{-1}\mathbf{X}'\widehat{\mathbf{Y}}_\varphi$. As discussed later in Section 3.7, the intercept α can be estimated by a location estimate based on the residuals $\widehat{\mathbf{e}} = \mathbf{Y} - \widehat{\mathbf{Y}}_\varphi$. One that we will frequently use is the median of the residuals, which we denote as $\widehat{\alpha}_S = \text{med}\{Y_i - \mathbf{x}_i'\widehat{\boldsymbol{\beta}}_\varphi\}$. Theorem 3.5.11 shows, under regularity conditions, that

$$\begin{pmatrix} \widehat{\alpha}_S \\ \widehat{\boldsymbol{\beta}}_\varphi \end{pmatrix} \text{ has an approximate } N_{p+1}\left(\begin{pmatrix} \alpha \\ \boldsymbol{\beta} \end{pmatrix}, \begin{bmatrix} n^{-1}\tau_S^2 & \mathbf{0}' \\ \mathbf{0} & \tau_\varphi^2(\mathbf{X}'\mathbf{X})^{-1} \end{bmatrix}\right) \text{ distribution,}$$

(3.2.8)

where τ_φ and τ_S are the scale parameters defined in (3.4.4) and (3.4.6), respectively. From this result, an asymptotic confidence interval for the linear function $\mathbf{h}'\boldsymbol{\beta}$ is given by

$$\mathbf{h}'\widehat{\boldsymbol{\beta}}_\varphi \pm t_{(\alpha/2, n-p-1)}\widehat{\tau}_\varphi\sqrt{\mathbf{h}'(\mathbf{X}'\mathbf{X})^{-1}\mathbf{h}}, \qquad (3.2.9)$$

where the estimate $\widehat{\tau}_\varphi$ is discussed in Section 3.7.1. The use of t critical values instead of z critical values is documented in the small-sample studies cited in Section 3.7. Note the close analogy between this confidence interval and those based on least-squares estimates. The only difference is that $\widehat{\sigma}$ has been replaced by $\widehat{\tau}_\varphi$.

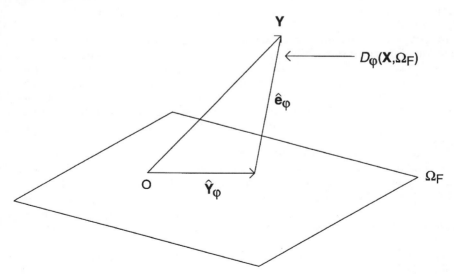

Figure 3.2.1: The R-estimate of $\boldsymbol{\eta}$ is a vector $\widehat{\mathbf{Y}}_\varphi$ which minimizes the normed differences, (3.2.6), between $\mathbf{Y}$ and Ω_F. The distance between $\mathbf{Y}$ and the space Ω_F is $D_\varphi(\mathbf{Y}, \Omega_F)$.

We will make use of the coordinate-free model, especially in Chapter 4; however, in this chapter we are primarily concerned with the properties of the estimator $\widehat{\boldsymbol{\beta}}_\varphi$ and it will be more convenient to use the coordinate model (3.2.3). Define the **dispersion function** by

$$D_\varphi(\boldsymbol{\beta}) = \|\mathbf{Y} - \mathbf{X}\boldsymbol{\beta}\|_\varphi. \qquad (3.2.10)$$

Then $D_\varphi(\widehat{\boldsymbol{\beta}}_\varphi) = D_\varphi(\mathbf{Y}, \Omega_F) = \|\mathbf{Y} - \widehat{\mathbf{Y}}_\varphi\|_\varphi$ is the R-distance between $\mathbf{Y}$ and the subspace Ω_F. It is also the residual dispersion.

Because D_φ is expressed in terms of a norm it is a continuous and convex function of $\boldsymbol{\beta}$; see Exercise 1.13.2. Exercise 3.12.2 shows that the ranks of the residuals can only change at the boundaries of the regions defined by the $\binom{n}{2}$ equations $y_i - \mathbf{x}_i'\boldsymbol{\beta} = y_j - \mathbf{x}_j'\boldsymbol{\beta}$. Note that in the simple linear regression case, these equations define the sample slopes $(Y_j - Y_i)/(x_j - x_i)$. Hence, in the interior of these regions the ranks are constant. Therefore, $D_\varphi(\boldsymbol{\beta})$ is a piecewise linear, continuous, convex function of $\boldsymbol{\beta}$ with gradient (defined almost everywhere) given by

$$\nabla D_\varphi(\boldsymbol{\beta}) = -\mathbf{S}_\varphi(\mathbf{Y} - \mathbf{X}\boldsymbol{\beta}), \qquad (3.2.11)$$

where

$$\mathbf{S}_\varphi(\mathbf{Y} - \mathbf{X}\boldsymbol{\beta}) = \mathbf{X}'\mathbf{a}(R(\mathbf{Y} - \mathbf{X}\boldsymbol{\beta})) \qquad (3.2.12)$$

and $\mathbf{a}(R(\mathbf{Y} - \mathbf{X}\boldsymbol{\beta}))' = (a(R(Y_1 - \mathbf{x}_1'\boldsymbol{\beta})), \ldots, a(R(Y_n - \mathbf{x}_n'\boldsymbol{\beta})))$. Thus $\widehat{\boldsymbol{\beta}}_\varphi$ solves the equations

$$\mathbf{S}_\varphi(\mathbf{Y} - \mathbf{X}\boldsymbol{\beta}) = \mathbf{X}'\mathbf{a}(R(\mathbf{Y} - \mathbf{X}\boldsymbol{\beta})) \doteq \mathbf{0}, \qquad (3.2.13)$$

which are called the **R normal equations**. A quadratic form in $\mathbf{S}_\varphi(\mathbf{Y} - \mathbf{X}\boldsymbol{\beta}_0)$ serves as the gradient R test statistic for testing $H_0 : \boldsymbol{\beta} = \boldsymbol{\beta}_0$ versus $H_A : \boldsymbol{\beta} \neq \boldsymbol{\beta}_0$.

In terms of the simple regression problem $S_\varphi(\beta)$ is a decreasing step function of β, which steps down at each sample slope. There may be an interval of solutions of $S_\varphi(\beta) = 0$, or $S_\varphi(\beta)$ may step across the horizontal axis. Let $\widehat{\beta}_\varphi$ denote any point in the interval in the former case and the crossing point in the latter case. The gradient test statistic is $S_\varphi(\beta_0) = \sum x_i a(R(y_i - x_i\beta_0))$. If the xs are distinct and equally spaced then for Wilcoxon scores this test statistic is equivalent to the test for correlation based on Spearman's r_S; see Exercise 3.12.4.

For the asymptotic distribution theory of estimation and testing, we note that the estimate is location- and scale-equivariant. Let $\widehat{\boldsymbol{\beta}}_\varphi(\mathbf{Y})$ denote the R-estimate $\boldsymbol{\beta}$ for the linear model (3.2.3). Then, as shown in Exercise 3.12.6, $\widehat{\boldsymbol{\beta}}_\varphi(\mathbf{Y} + \mathbf{X}\boldsymbol{\delta}) = \widehat{\boldsymbol{\beta}}_\varphi(\mathbf{Y}) + \boldsymbol{\delta}$ and $\widehat{\boldsymbol{\beta}}_\varphi(k\mathbf{Y}) = k\widehat{\boldsymbol{\beta}}_\varphi(\mathbf{Y})$. In particular these results imply, without loss of generality, that the theory developed in the following sections can be accomplished under the assumption that the true $\boldsymbol{\beta}$ is $\mathbf{0}$.

As a final note, we outline the least squares estimates. The LS estimate of $\boldsymbol{\eta}$ in model (3.2.4) is given by

$$\widehat{\mathbf{Y}}_{LS} = \text{Argmin } \|\mathbf{Y} - \boldsymbol{\eta}\|_{LS}^2,$$

Geometry of Estimation and Tests   149

where $\|\cdot\|_{LS}$ denotes the LS pseudo-norm given by (2.2.16) of Chapter 2. The value of η which minimizes this pseudo-norm is

$$\widehat{\eta}_{LS} = \mathbf{HY}, \tag{3.2.14}$$

where $\mathbf{H}$ is the projection matrix onto the space Ω_F i.e.; $\mathbf{H} = \mathbf{X}(\mathbf{X}'\mathbf{X})^{-1}\mathbf{X}'$. Denote the sum of squared residuals by $SSE = \min_{\eta \in \Omega_F} \|\mathbf{Y} - \eta\|_{LS}^2 = \|(\mathbf{I} - \mathbf{H})\mathbf{Y}\|_{LS}^2$. In order to have similar notation we shall denote this minimum by $D_{LS}^2(\mathbf{Y}, \Omega_F)$. Also, it is easy to show that the LS estimate of β is $\widehat{\beta}_{LS} = (\mathbf{X}'\mathbf{X})^{-1}\mathbf{X}'\mathbf{Y}$.

3.2.2 The Geometry of Testing

We next discuss the geometry behind rank-based tests of the general linear hypotheses given by (3.2.5). As above, consider the model (3.2.4),

$$\mathbf{Y} = \mathbf{1}\alpha + \eta + \mathbf{e}, \quad \text{where } \eta \in \Omega_F, \tag{3.2.15}$$

and Ω_F is the column space of the full model design matrix $\mathbf{X}$. Let $\widehat{\mathbf{Y}}_{\varphi,\Omega_F}$ denote the R-fitted value in the full model. Note that $D_\varphi(\mathbf{Y}, \Omega_F)$ is the amount of residual dispersion not accounted for in fitting model (3.2.4). These are shown geometrically in Figure 3.2.2.

Next let ω denote the subspace of Ω_F subject to H_0. In symbols $\omega = \{\eta \in \Omega_F : \eta = \mathbf{X}\beta$, for some β such that $\mathbf{M}\beta = \mathbf{0}\}$. In Exercise 3.12.7 the reader is asked to show that ω is a subspace of Ω_F of dimension $p - q$. Let $\widehat{\mathbf{Y}}_{\varphi,\omega}$ denote the R-estimate of η when the reduced model is fit and let $D_\varphi(\mathbf{Y}, \omega) = \|\mathbf{Y} - \widehat{\mathbf{Y}}_{\varphi,\omega}\|_R$ denote the distance between Y and the subspace ω. These are illustrated by Figure 3.2.2. The nonnegative quantity

$$RD_\varphi = D_\varphi(\mathbf{Y}, \omega) - D_\varphi(\mathbf{Y}, \Omega_F), \tag{3.2.16}$$

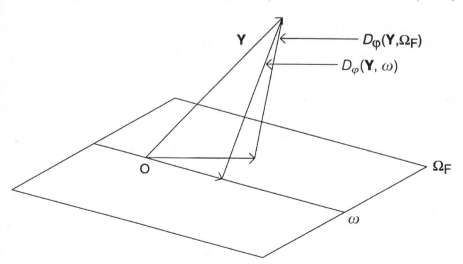

Figure 3.2.2: The reduction in dispersion RD_φ is the difference in normed distances between **Y** and the subspaces Ω_F and ω.

150   Linear Models

denotes the **reduction in residual dispersion** when we pass from the reduced model to the full model. Large values of RD_φ indicate H_A while small values support H_0.

This drop in residual dispersion, RD_φ, is analogous to the drop in residual sums of squares for the LS analysis. In fact to obtain this reduction in sums of squares, we need only replace the R-norm with the square of the Euclidean norm in the above development. Thus the drop in sums of squared errors is

$$SS = D_{LS}^2(\mathbf{Y}, \omega) - D_{LS}^2(\mathbf{Y}, \Omega_F),$$

where $D_{LS}^2(\mathbf{Y}, \Omega_F)$ is defined above. Hence the reduction in sums of squared residuals can be written as

$$SS = \|(\mathbf{I} - \mathbf{H}_\omega)\mathbf{Y}\|_{LS}^2 - \|(\mathbf{I} - \mathbf{H}_{\Omega_F})\mathbf{Y}\|_{LS}^2.$$

The traditional least squares F-test is given by

$$F_{LS} = \frac{SS/q}{\widehat{\sigma}^2}, \qquad (3.2.17)$$

where $\widehat{\sigma}^2 = D_{LS}^2(\mathbf{Y}, \Omega_F)/(n - p)$. Other than replacing one norm with another, Figures 3.2.1 and 3.2.2 remain the same for the two analyses, LS and R.

In order to be useful as a test statistic, similar to least squares, the reduction in dispersion $RD\varphi$ must be standardized. The asymptotic distribution theory that follows suggests the standardization

$$F_\varphi = \frac{RD\varphi/q}{\widehat{\tau}_\varphi/2}, \qquad (3.2.18)$$

where $\widehat{\tau}_\varphi$ is the estimate of τ_φ discussed in Section 3.7. Small-sample studies cited in Section 3.7 indicate that F_φ should be compared with F critical values with q and $n - (p + 1)$ degrees of freedom analogous to the LS classical F-test statistic. Similar to the LS F-test, the test based on F_φ can be summarized in the analysis of variance (ANOVA) table, Table 3.2.1. Note that the reduction in dispersion replaces the reduction in sums of squares in the classical table. These robust ANOVA tables were first discussed by Schrader and McKean (1977).

Tests That All Regression Coefficients are 0

As discussed more fully in Section 3.6, there are three R-test statistics for the hypotheses (3.2.5). These are the R analogs of the classical tests: the likelihood ratio test, the scores test, and the Wald test. We shall introduce them here for the special null hypothesis that all the regression parameters are 0, i.e.

$$H_0 : \boldsymbol{\beta} = \mathbf{0} \text{ versus } H_0 : \boldsymbol{\beta} \neq \mathbf{0}. \qquad (3.2.19)$$

Table 3.2.1: Robust ANOVA Table for $H_0 : \mathbf{M}\boldsymbol{\beta} = \mathbf{0}$

Source in Dispersion	Reduction in Dispersion	df	Mean Reduction in Dispersion	F_φ
Regression	$RD_\varphi = \left(D_\varphi(\mathbf{Y}, \omega) - D_\varphi(\mathbf{Y}, \Omega_F)\right)$	q	RD/q	F_φ
Error		$n - p - 1$	$\widehat{\tau}_\varphi/2$	

Table 3.2.2: Robust ANOVA Table for $H_0 : \boldsymbol{\beta} = \mathbf{0}$

Source in Dispersion	Reduction in Dispersion	df	Mean Reduction in Dispersion	
Regression	$RD = \big(D_\varphi(\mathbf{0}) - D_\varphi(\mathbf{Y}, \Omega_F)\big)$	p	RD/p	F_φ
Error		$n-p-1$	$\widehat{\tau}_\varphi/2$	

Their asymptotic theory and small-sample properties are discussed in more detail in later sections.

In this case, the reduced model dispersion is just the dispersion of the response vector $\mathbf{Y}$, i.e. $D_\varphi(\mathbf{0})$. Hence, the **R-test based on the reduction in dispersion** is

$$F_\varphi = \frac{(D_\varphi(\mathbf{0}) - D_\varphi(\mathbf{Y}, \Omega_F))/p}{\widehat{\tau}_\varphi/2}. \qquad (3.2.20)$$

As discussed above, F_φ should be compared with $F(\alpha, p, n-p-1)$ critical values. Similar to the general hypothesis, the test based on F_φ can be expressed as in the robust ANOVA table given in Table 3.2.2. This is the robust analog of the traditional ANOVA table that is printed out for a regression analysis by most least-squares regression packages.

The **R-scores test** is the test based on the gradient. Theorem 3.5.2, below, gives the asymptotic distribution of the gradient $\mathbf{S}_\varphi(\mathbf{0})$ under the null hypothesis. This leads to the asymptotic level α test rejecting H_0 if

$$\mathbf{S}'_\varphi(\mathbf{0})(\mathbf{X}'\mathbf{X})^{-1}\mathbf{S}_\varphi(\mathbf{0}) \geqslant \chi^2(\alpha, p). \qquad (3.2.21)$$

where $\chi^2(\alpha, p)$ denotes the upper level α critical of the χ^2 distribution with p degrees of freedom. Note that this test avoids the estimation of τ_φ.

The **R-Wald test** is a quadratic form in the full model estimates. Based on the asymptotic distribution of the full model estimate $\widehat{\boldsymbol{\beta}}_\varphi$ given in Corollary 3.5.6, an asymptotic level α test rejects H_0 if

$$\frac{\widehat{\boldsymbol{\beta}}'_\varphi(\mathbf{X}'\mathbf{X})\widehat{\boldsymbol{\beta}}_\varphi/p}{\widehat{\tau}^2_\varphi} \geqslant F(\alpha, p, n-p-1). \qquad (3.2.22)$$

3.3 Examples

We offer several examples to illustrate the rank-based estimates and test procedures discussed in the previous section. For all the examples, we use Wilcoxon scores, $\varphi(u) = \sqrt{12}(u - \frac{1}{2})$, for the rank-based estimates of the regression coefficients. We estimate the intercept by the median of the residuals and we estimate the scale parameter τ_φ as discussed in Section 3.7. We begin with a simple regression data set and proceed to multiple regression problems.

Example 3.3.1 *Telephone Data*

The response for this data set is the number of telephone calls (tens of millions) made in Belgium for the years 1950 through 1973. Time, the years,

Table 3.3.1: Data for Example 3.3.1. The number of calls is in tens of millions and the years are 1950–1973

Year	50	51	52	53	54	55	56	57	58	59	60	61
No. of Calls	0.44	0.47	0.47	0.59	0.66	0.73	0.81	0.88	1.06	1.20	1.35	1.49
Year	62	63	64	65	66	67	68	69	70	71	72	73
No. of Calls	1.61	2.12	11.90	12.40	14.20	15.90	18.20	21.20	4.30	2.40	2.70	2.90

serves as our only predictor variable. The data are discussed in Rousseeuw and Leroy (1987) and, for convenience, are displayed in Table 3.3.1.

The Wilcoxon estimates of the intercept and slope are −7.13 and 0.145, respectively, while the LS estimates are −26 and 0.504. The reason for this disparity in fits is easily seen in Figure 3.3.1(a), which is a scatterplot of the data overlaid with the LS and Wilcoxon fits. Note that the years 1964 through 1969 had a profound effect on the LS fit while the Wilcoxon fit was much less sensitive to these years. As discussed in Rousseeuw and Leroy, the recording system for the years 1964 through 1969 differed from that for the other years. The studentized residual plots of the fits are given by Figure 3.3.1(b,c); see (3.9.31) of Section 3.9. As with internal LS studentized residuals, values of the internal R studentized residuals which exceed 2 in absolute value are potential outliers. Note that the internal Wilcoxon studentized residuals clearly show that the years 1964–1969 are outliers while the internal LS studentized residuals only detect 1969. The Wilcoxon studentized residuals also mildly detect the year 1970. Based on the scatterplot, this point does not follow the trend of the early (before 1964) years either. The scatterplot and Wilcoxon residual plot indicate that there may be a quadratic trend over the years before the outliers occur. The last few years, though, do not seem to follow this trend. Hence, a linear model for those data is questionable. On the basis of these plots, we will not discuss any formal inference for this data set.

Figure 3.3.1(d) depicts the Wilcoxon dispersion function over the interval (−0.2, 0.6). Note that Wilcoxon estimate $\hat{\beta}_R = 0.145$ is the minimizing value. Next consider the hypotheses $H_0 : \beta = 0$ versus $H_A : \beta \neq 0$. The basis for the test statistic F_φ can be read from this plot. The reduction in dispersion is given by $RD = D(0) - D(0.145)$. Also the gradient test of these hypotheses would be the negative of the slope of the dispersion function at 0, i.e. $-D'(0)$.

Example 3.3.2 *Baseball Salaries*

As a large data set, we consider data on the salaries of professional baseball pitchers for the 1987 baseball season. This data set was taken from the data set on baseball salaries which was used in the 1988 ASA Graphics Section Poster Session. It can be obtained at the Web site http://lib.stat.cmu.edu/datasets. Our analysis concerns a data subset of 176 pitchers, which can be obtained from the authors upon request. Our response variable is the 1987 beginning salary (in log dollars) of these pitchers. As predictors, we took the career summary statistics through the end of the 1986 season. The names of

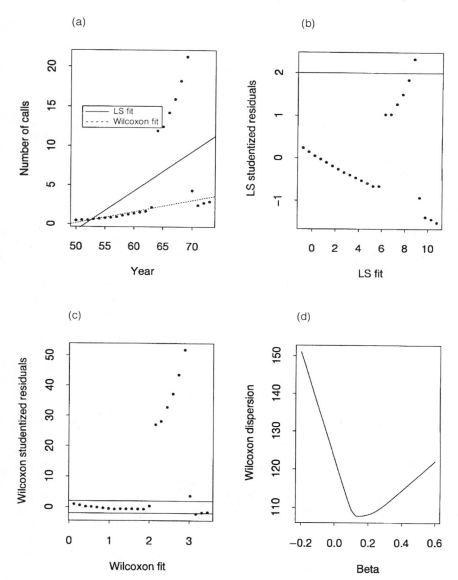

Figure 3.3.1: (a) Scatterplot of the telephone data, overlaid with the LS and Wilcoxon fits; (b) internal LS studentized residual plot; (c) internal Wilcoxon studentized residual plot; (d) Wilcoxon dispersion function

these variables are listed in Table 3.3.2. The scatterplots of the log of salary versus each of the predictors are shown in Figure 3.3.2(a)–(g). Certainly the strongest predictor on the basis of these plots is log years, although linearity in this plot is questionable.

Table 3.3.2: Predictors for baseball salaries of pitchers and their estimated (Wilcoxon fit) coefficients

Predictor	Estimate	Stand. error	t-ratio
Log years in professional baseball	0.839	0.044	19.15
Average wins per year	0.045	0.028	1.63
Average losses per year	−0.024	0.026	−0.921
Earned run average	−0.146	0.070	−2.11
Average games per year	−0.006	0.004	1.60
Average innings per year	0.004	0.003	1.62
Average saves per year	0.012	0.011	1.07
Intercept	4.220	0.324	
Scale (τ)	0.388		

The internal Wilcoxon studentized residuals, (3.9.31), versus fitted values are displayed in the Figure 3.3.2(h). Based on Figure 3.3.2(a,h), the pattern in the residual plot follows from the fact that log years is not a linear predictor. Better-fitting models are pursued in Exercise 3.12.1. Note that there are several large outliers. The three identified outliers, circled points in Figure 3.3.2(h), are interesting. These correspond to the pitchers Steve Carlton, Phil Niekro, and Rick Sutcliff. These were very good pitchers, but in 1987 they were at the end of their careers (21, 23, and 21 years of pitching, respectively); hence, they missed the rapid rise in baseball salaries. A diagnostic analysis (see Section 3.9 and Exercise 3.12.1), indicates a few mildly influential points, also. For illustration, though, we will consider the model that we fit. Table 3.3.2 also displays the estimated coefficients and their standard errors. The outliers impaired the LS fit somewhat. The LS estimate of σ is 0.515 in comparison to the estimate of τ which is 0.388.

Table 3.3.3 displays the robust ANOVA table for testing that all the coefficients, except the intercept, are 0. Based on the large value of F_φ, (3.2.20), the predictors are helpful in explaining the response. In particular, based on Table 3.3.2, the predictors years in professional baseball, earned run average, average innings per year, and average number of saves per year seem more important than the variables wins, losses, and games. These last three variables form a similar group of variables; hence, as an illustration of the rank-based statistic F_φ, the hypothesis that the coefficients for these three predictors are 0 was tested. The reduction in dispersion for this hypothesis is $RD = 1.24$, which leads to $F_\varphi = 2.12$, significant at the 10% level. This confirms the above observations on the regression coefficients.

Table 3.3.3: Wilcoxon ANOVA table for $H_0 : \boldsymbol{\beta} = \boldsymbol{0}$

Source in dispersion	Reduction in dispersion	df	Mean reduction in dispersion	F_φ
Regression	78.287	7	11.18	57.65
Error		168	0.194	

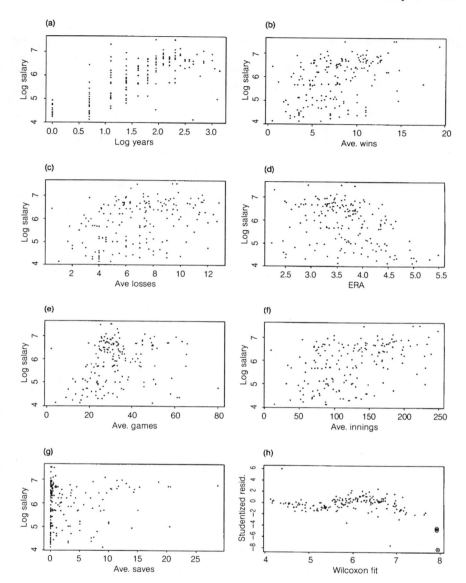

Figure 3.3.2: (a)–(g) Plots of log salary versus each of the predictors for the baseball data of Example 3.3.2; (h) internal Wilcoxon studentized residual plot

Example 3.3.3 *Potency Data*

This example is part of an $n = 34$ multivariate data set discussed in Chapter 6; see Table 6.6.2 for the data. The experiment concerned the potency of drug compounds which were manufactured under different levels of four factors.

156 *Linear Models*

Here we shall consider only one of the response variables, POT2, which is the potency of a drug compound at the end of 2 weeks. The factors are: SAI, the amount of intragranular steric acid, which was set at the three levels, $-1, 0$ and 1; SAE, the amount of extragranular steric acid, which was set at the three levels, $-1, 0$ and 1; ADS, the amount of cross carmellose sodium, which was set at the three levels, $-1, 0$ and 1; and TYPE of steric acid, which was set at two levels, -1 and 1. The initial potency of the compound, POT0, served as a covariate.

In Example 3.9.2 of Section 3.9 a residual analysis of this data set is performed. This analysis indicates that the model which includes the covariate, the linear terms of the factors, the simple two-way interaction terms of the factors, and the quadratic terms of the three factors SAE, SAI and ADS is adequate. Let x_j, $j = 1, \ldots, 4$, denote the level of the factors SAI, SAE, ADS, and TYPE, respectively, and let c_i denote the value of the covariate. Then the model is expressed as

$$y_i = \alpha + \beta_1 x_{1,i} + \beta_2 x_{2,i} + \beta_3 x_{3,i} + \beta_4 x_{4,i} + \beta_5 x_{1,i} x_{2,i} + \beta_6 x_{1,i} x_{3,i}$$
$$+ \beta_7 x_{1,i} x_{4,i} + \beta_8 x_{2,i} x_{3,i} + \beta_9 x_{2,i} x_{4,i} + \beta_{10} x_{3,i} x_{4,i}$$
$$+ \beta_{11} x_{1,i}^2 + \beta_{12} x_{2,i}^2 + \beta_{13} x_{3,i}^2 + \beta_{14} c_i + e_i. \qquad (3.3.1)$$

The Wilcoxon and LS estimates of the regression coefficients and their standard errors are given in Table 3.3.4. The Wilcoxon estimates are more precise. As the diagnostic analysis of Example 3.9.2 shows, this is due to the outliers in this data set.

Note that the Wilcoxon estimate of the parameter β_{13}, the quadratic term of the factor ADS, is significant. Again referring to the residual analysis given

Table 3.3.4: Wilcoxon and LS estimates for the potency data

Terms	Parameter	Wilcoxon estimates Est.	SE	LS estimates Est.	SE
Intercept	α	7.184	2.96	5.998	4.50
	β_1	0.072	0.05	0.000	0.08
Linear	β_2	0.023	0.05	−0.018	0.07
	β_3	0.166	0.05	0.135	0.07
	β_4	0.020	0.04	−0.011	0.05
	β_5	0.042	0.05	0.086	0.08
	β_6	−0.040	0.05	0.035	0.08
Two−way	β_7	0.040	0.05	0.102	0.07
Inter.	β_8	−0.085	0.06	−0.030	0.09
	β_9	0.024	0.05	0.070	0.07
	β_{10}	−0.049	0.05	−0.011	0.07
	β_{11}	−0.002	0.10	0.117	0.15
Quad.	β_{12}	−0.222	0.09	−0.240	0.13
	β_{13}	0.022	0.09	−0.007	0.14
Covariate	β_{14}	0.092	0.31	0.217	0.47
Scale	τ or σ	0.204		0.310	

Table 3.3.5: Wilcoxon ANOVA table for $H_0 : \beta_{12} = \beta_{13} = \beta_{14} = 0$

Source of dispersion	Reduction in dispersion	df	Mean reduction in dispersion	F_φ
Quadratic Terms	0.977	3	0.326	3.20
Error		19	0.102	

in Example 3.9.2, there is some graphical evidence to retain the three quadratic coefficients in the model. In order to confirm this evidence statistically, we will test the hypotheses

$$H_0 : \beta_{12} = \beta_{13} = \beta_{14} = 0 \text{ versus } H_A : \beta_i \neq 0 \text{ for some } i = 12, 13, 14.$$

The Wilcoxon test is summarized in Table 3.3.5 and it is based on the test statistic (3.2.18). The value of the test statistic is significant at the 0.05 level. The LS F-test statistic, though, has the value 1.19. As with its estimates of the regression coefficients, the LS F-test statistic has been impaired by the outliers.

3.4 Assumptions for Asymptotic Theory

For the asymptotic theory developed in this chapter certain assumptions on the distribution of the errors, the design matrix, and the scores are needed. The required assumptions for each section may differ, but for easy reference we have placed them in this section.

The major assumption on the error density function f for much of the rank-based analyses, is:

(E.1) f is absolutely continuous, $0 < I(f) < \infty$.   (3.4.1)

where $I(f)$ denotes the Fisher information (2.4.16). Since f is absolutely continuous, we can write

$$f(s) - f(t) = \int_t^s f'(x)dx$$

for some function f'. An application of the Cauchy–Schwarz inequality yields

$$|f(s) - f(t)| \leq I(f)^{1/2}\sqrt{|F(s) - F(t)|}; \qquad (3.4.2)$$

see Exercise 1.13.15. It follows from (3.4.2) that assumption (E.1) implies that f is uniformly bounded and is uniformly continuous.

An assumption that will be used for analyses based on the L_1 norm is:

(E.2) $f(\theta_e) > 0$,   (3.4.3)

where θ_e denotes the median of the error distribution, i.e. $\theta_e = F^{-1}(1/2)$.

For easy reference, we list again the scale parameter τ_φ (2.5.23):

$$\tau_\varphi^{-1} = \int \varphi(u)\varphi_f(u)du, \qquad (3.4.4)$$

where

$$\varphi_f(u) = -\frac{f'(F^{-1}(u))}{f(F^{-1}(u))}. \tag{3.4.5}$$

Under (E.1) the scale parameter τ_φ is well defined. Another scale parameter that will be needed is τ_S, defined as

$$\tau_S = (2f(\theta_e))^{-1}; \tag{3.4.6}$$

see (1.5.21). Note that it is well defined under assumption (E.2).

As above, let $\mathbf{H} = \mathbf{X}(\mathbf{X}'\mathbf{X})^{-1}\mathbf{X}'$ denote the projection matrix onto Ω, the column space of $\mathbf{X}$. Our asymptotic theory assumes that the design matrix $\mathbf{X}$ is imbedded in a sequence of design matrices which satisfy the next two properties. We should subscript quantities such as $\mathbf{X}$ and the projection matrix with n to show this, but as a matter of convenience we have not done so. We will subscript the leverage values h_{iin} which are the diagonal entries of the projection matrix $\mathbf{H}$. We will often impose the next two conditions on the design matrix:

$$(D.2) \quad \lim_{n \to \infty} \max_{1 \leq i \leq n} h_{iin} = 0 \tag{3.4.7}$$

$$(D.3) \quad \lim_{n \to \infty} n^{-1}\mathbf{X}'\mathbf{X} = \Sigma, \tag{3.4.8}$$

where Σ is a $p \times p$ positive definite matrix. The first condition has become known as Huber's condition. Huber (1981) showed that (D.2) is a necessary and sufficient design condition for the least-squares estimates to have an asymptotic normal distribution provided the errors, e_i, are iid with finite variance. Condition (D.3) reduces to assumption (D.1), (2.4.7) of Chapter 2, for the two-sample problem.

Another design condition is Noether's condition, which is given by

$$(N.1) \quad \max_{1 \leq i \leq n} \frac{x_{ik}^2}{\sum_{j=1}^{n} x_{jk}^2} \to 0 \text{ for all } k = 1, \ldots, p. \tag{3.4.9}$$

Although this condition will be convenient, the next lemma shows it is implied by Huber's condition.

Lemma 3.4.1 *(D.2) implies (N.1)*.

Proof. By the generalized Cauchy–Schwarz inequality (see Graybill, 1976, p. 224), for all $i = 1, \ldots, n$ we have the following equalities:

$$\sup_{\|\delta\|=1} \frac{\delta'\mathbf{x}_i\mathbf{x}_i'\delta}{\delta'\mathbf{X}'\mathbf{X}\delta} = \mathbf{x}_i'(\mathbf{X}'\mathbf{X})^{-1}\mathbf{x}_i = h_{iin}.$$

Next, for $k = 1, \ldots, p$ take δ to be δ_k, the $p \times 1$ vector of zeros except for 1 in the kth component. Then the above equalities imply that

$$\frac{x_{ik}^2}{\sum_{j=1}^{n} x_{jk}^2} \leqslant h_{iin}, \quad i=1,\ldots,n, \; k=1,\ldots,p.$$

Hence

$$\max_{1 \leqslant k \leqslant p} \max_{1 \leqslant i \leqslant n} \frac{x_{ik}^2}{\sum_{j=1}^{n} x_{jk}^2} \leqslant \max_{1 \leqslant i \leqslant n} h_{iin}.$$

Therefore Huber's condition implies Noether's condition.

As in Chapter 2, we will often assume that the score generating function $\varphi(u)$ satisfies assumption (2.5.5). We will, in addition, assume that it is bounded. For reference, we will assume that $\varphi(u)$ is a function defined on $(0, 1)$ such that

(S.1) $\begin{cases} \varphi(u) \text{ is a nondecreasing, square-integrable, and bounded function} \\ \int_0^1 \varphi(u)\, du = 0 \text{ and } \int_0^1 \varphi^2(u)\, du = 1. \end{cases}$

(3.4.10)

Occasionally we will need further assumptions on the score function. In Section 3.7, we will need to assume that

(S.2) φ is differentiable.   (3.4.11)

When estimating the intercept parameter based on signed-rank scores, we need to assume that the score function is odd about $\frac{1}{2}$, i.e.

(S.3) $\varphi(1-u) = -\varphi(u);$ (3.4.12)

see also (2.5.5).

3.5 Theory of Rank-Based Estimates

Consider the linear model given by (3.2.3). To avoid confusion, we will denote the true vector of parameters by $(\alpha_0, \boldsymbol{\beta}_0)'$; that is, the true model is $\mathbf{Y} = \mathbf{1}\alpha_0 + \mathbf{X}\boldsymbol{\beta}_0 + \mathbf{e}$. In this section we will derive the asymptotic theory for the R analysis, estimation and testing, under the assumptions (E.1), (D.2), (D.3), and (S.1). We will occasionally supress the subscripts φ and R from the notation. For example, we will denote the R-estimate simply by $\widehat{\boldsymbol{\beta}}$.

3.5.1 R-Estimators of the Regression Coefficients

A key result for both estimation and testing concerns the gradient $\mathbf{S}(\mathbf{Y} - \mathbf{X}\boldsymbol{\beta})$, given by (3.2.12). We first derive its mean and covariance matrix and then obtain its asymptotic distribution.

Theorem 3.5.1 *Under model (3.2.3),*

$$E[\mathbf{S}(\mathbf{Y} - \mathbf{X}\boldsymbol{\beta}_0)] = \mathbf{0}$$
$$\text{Var}[\mathbf{S}(\mathbf{Y} - \mathbf{X}\boldsymbol{\beta}_0)] = \sigma_a^2 \mathbf{X}'\mathbf{X},$$

where $\sigma_a^2 = (n-1)^{-1} \sum_{i=1}^n a^2(i) \doteq 1$.

Proof. Note that $\mathbf{S}(\mathbf{Y} - \mathbf{X}\boldsymbol{\beta}_0) = \mathbf{X}'\mathbf{a}(R(\mathbf{e}))$. Under model (3.2.3), $e_1, \ldots, e_n$ are iid; hence, the ith component $\mathbf{a}(R(\mathbf{e}))$ has mean

$$E[a(R(e_i))] = \sum_{j=1}^n a(j) n^{-1} = 0,$$

from which the result for the expectation follows.

For the result on the variance-covariance matrix, note that $\text{Var}[\mathbf{S}(\mathbf{Y} - \mathbf{X}\boldsymbol{\beta}_0)] = \mathbf{X}'\text{Var}[\mathbf{a}(R(\mathbf{e}))]\mathbf{X}$. The diagonal entries for the covariance matrix on the right-hand side are:

$$\text{Var}[a(R(e_i))] = E[a^2(R(e_i))] = \sum_{j=1}^n a(j)^2 n^{-1} = \frac{n-1}{n} \sigma_a^2.$$

The off-diagonal entries are the covariances given by

$$\text{Cov}(a(R(e_i)), a(R(e_l))) = E[a(R(e_i)) a(R(e_l))]$$
$$= \sum_{\substack{j=1 \\ j \neq k}}^n \sum_{k=1}^n a(j) a(k) (n(n-1))^{-1}$$
$$= -(n(n-1))^{-1} \sum_{j=1}^n a^2(j)$$
$$= -\sigma_a^2/n, \qquad (3.5.1)$$

where the third step in the derivation follows from $0 = \left(\sum_{j=1}^n a(j) \right)^2$. The result, (3.5.1), is obtained directly from these variances and covariances.

Under (D.3), we have that

$$\text{Var}[n^{-1/2} \mathbf{S}(\mathbf{Y} - \mathbf{X}\boldsymbol{\beta}_0)] \to \Sigma. \qquad (3.5.2)$$

This anticpates our next result.

Theorem 3.5.2 *Under model (3.2.3), (E.1), (D.2), (D.3), and (S.1) in Section 3.4,*

$$n^{-1/2} \mathbf{S}(\mathbf{Y} - \mathbf{X}\boldsymbol{\beta}_0) \xrightarrow{\mathcal{D}} N_p(\mathbf{0}, \Sigma). \qquad (3.5.3)$$

Proof. Let $\mathbf{S}(\mathbf{0}) = \mathbf{S}(\mathbf{Y} - \mathbf{X}\boldsymbol{\beta}_0)$ and let $\mathbf{T}(\mathbf{0}) = \mathbf{X}'\varphi(F(\mathbf{Y} - \mathbf{X}\boldsymbol{\beta}_0))$. Under the above assumptions, the discussion around Theorem A.3.1 of the Appendix shows that $(\mathbf{T}(\mathbf{0}) - \mathbf{S}(\mathbf{0}))/\sqrt{n}$ converges to $\mathbf{0}$ in probability. Hence we need only show that $\mathbf{T}(\mathbf{0})/\sqrt{n}$ converges to the intended distribution. Letting

$W^* = n^{-1/2}\mathbf{t}'\mathbf{T}(\mathbf{e})$ where $\mathbf{t} \neq \mathbf{0}$ is an arbitrary $p \times 1$ vector, it suffices to show that W^* converges in distribution to a $N(0, \mathbf{t}'\Sigma\mathbf{t})$ distribution. Note that we can write W^* as

$$W^* = n^{-1/2} \sum_{k=1}^{n} \mathbf{t}'\mathbf{x}_k \varphi(F(e_k)). \qquad (3.5.4)$$

Since F is the distribution function of e_k, it follows from $\int \varphi = 0$ that $E[W^*] = 0$, and from $\int \varphi^2 = 1$ and (D.3) that

$$\text{Var}[W^*] = n^{-1} \sum_{k=1}^{n} (\mathbf{t}'\mathbf{x}_k)^2 = \mathbf{t}'n^{-1}\mathbf{X}'\mathbf{X}\mathbf{t} \to \mathbf{t}'\Sigma\mathbf{t} > 0. \qquad (3.5.5)$$

Since W^* is a sum of independent random variables which are not identically distributed we establish the limit distribution by the Lindeberg–Feller central limit theorem; see Theorem A.1.1 of the Appendix. In the notation of this theorem, let $B_n^2 = \text{Var}[W^*]$. By (3.5.5), B_n^2 converges to a positive real number. We need to show

$$\lim B_n^{-2} \sum_{k=1}^{n} E\left[\frac{1}{n}(\mathbf{x}_k'\mathbf{t})^2 \varphi^2(F(e_k))I\left(\left|\frac{1}{\sqrt{n}}(\mathbf{x}_k'\mathbf{t})\varphi(F(e_k))\right| > \varepsilon B_n\right)\right] = 0. \qquad (3.5.6)$$

The key is the factor $n^{-1/2}(\mathbf{x}_k'\mathbf{t})$ in the indicator function. By the Cauchy–Schwarz inequality and (D.2) we have the string of inequalities

$$n^{-1/2}|(\mathbf{x}_k'\mathbf{t})| \leq n^{-1/2}\|\mathbf{x}_k\|\|\mathbf{t}\|$$

$$= \left[n^{-1}\sum_{j=1}^{p} x_{kj}^2\right]^{1/2} \|\mathbf{t}\|$$

$$\leq \left[p \max_j n^{-1} x_{kj}^2\right]^{1/2} \|\mathbf{t}\|. \qquad (3.5.7)$$

By assumptions (D.2) and (D.3), it follows that the quantity in brackets in equation (3.5.7), and hence $n^{-1/2}|(\mathbf{x}_k'\mathbf{t})|$, converges to zero as $n \to \infty$. Call the term on the right-hand side of equation (3.5.7) M_n. Note that it does not depend on k and $M_n \to 0$. From this string of inequalities, the limit on the left-hand side of (3.5.6) is less than or equal to

$$\lim B_n^{-2} \lim E\left[\varphi^2(F(e_1))I\left(|\varphi(F(e_1))| > \frac{\varepsilon B_n}{M_n}\right)\right] \lim n^{-1}\sum_{k=1}^{n}(\mathbf{x}_k'\mathbf{t})^2.$$

The first and third limits are positive reals. For the second limit, note that the random variable inside the expectation is bounded; hence, by the Lebesgue dominated convergence theorem we can interchange the limit and expectation. Since $\varepsilon B_n/M_n \to \infty$ the expectation goes to 0 and our desired result is obtained.

Similar to Chapter 2, Exercise 3.12.9 obtains the proof of the above theorem for the special case of the Wilcoxon scores by first getting the projection of the statistic W.

Note from this theorem we have the gradient test that all the regression coefficients are 0; that is, $H_0: \boldsymbol{\beta} = \mathbf{0}$ versus $H_A: \boldsymbol{\beta} \neq \mathbf{0}$. Consider the test statistic

$$T = \sigma_a^{-2} \mathbf{S}(\mathbf{Y})'(\mathbf{X}'\mathbf{X})^{-1}\mathbf{S}(\mathbf{Y}). \tag{3.5.8}$$

From Theorem 3.5.1 an approximate level α test for H_0 versus H_A is:

Reject H_0 in favor of H_A if $T \geq \chi^2(\alpha, p)$, (3.5.9)

where $\chi^2(\alpha, p)$ denotes the upper level α critical value with p degrees of freedom.

Theorem A.3.8 of the Appendix gives the following linearity result for the process $\mathbf{S}(\boldsymbol{\beta}_n)$:

$$\frac{1}{\sqrt{n}} \mathbf{S}(\boldsymbol{\beta}_n) = \frac{1}{\sqrt{n}} \mathbf{S}(\boldsymbol{\beta}_0) - \tau_\varphi^{-1} \Sigma \sqrt{n}(\boldsymbol{\beta}_n - \boldsymbol{\beta}_0) + o_p(1), \tag{3.5.10}$$

for $\sqrt{n}(\boldsymbol{\beta}_n - \boldsymbol{\beta}_0) = O(1)$, where the scale parameter τ_φ is given by (3.4.4). Recall that we made use of this result in Section 2.5 when we showed that the two-sample location process under general scores functions is Pitman regular. If we integrate the right-hand side of this result we obtain a locally smooth approximation of the dispersion function $D(\boldsymbol{\beta}_n)$ which is given by the following quadratic function:

$$Q(\mathbf{Y} - \mathbf{X}\boldsymbol{\beta}) = (2\tau_\varphi)^{-1}(\boldsymbol{\beta} - \boldsymbol{\beta}_0)'\mathbf{X}'\mathbf{X}(\boldsymbol{\beta} - \boldsymbol{\beta}_0)$$
$$- (\boldsymbol{\beta} - \boldsymbol{\beta}_0)'\mathbf{S}(\mathbf{Y} - \mathbf{X}\boldsymbol{\beta}_0) + D(\mathbf{Y} - \mathbf{X}\boldsymbol{\beta}_0). \tag{3.5.11}$$

Note that Q depends on τ_φ and $\boldsymbol{\beta}_0$ so it cannot be used to estimate $\boldsymbol{\beta}$. As we will show, the function Q is quite useful for establishing asymptotic properties of the R-estimates and test statistics. As discussed in Section 3.7.3, it also leads to a Gauss–Newton type algorithm for obtaining R-estimates.

The following theorem shows that Q provides a local approximation to D. This is an asymptotic quadraticity result which was proved by Jaeckel (1972). It in turn is based on an asymptotic linearity result derived by Jurečková (1971) and given in (3.5.10). It is proved in the Appendix; see Theorem A.3.8.

Theorem 3.5.3 *Under model (3.2.3) and assumptions (E.1), (D.1), (D.2), and (S.1) of Section 3.4, for any $\varepsilon > 0$ and $c > 0$,*

$$P\left[\max_{\|\boldsymbol{\beta} - \boldsymbol{\beta}_0\| < c/\sqrt{n}} |D(\mathbf{Y} - \mathbf{X}\boldsymbol{\beta}) - Q(\mathbf{Y} - \mathbf{X}\boldsymbol{\beta})| \geq \varepsilon\right] \to 0, \tag{3.5.12}$$

as $n \to \infty$.

We will use this result to obtain the asymptotic distribution of the R-estimate. Without loss of generality, assume that the true $\boldsymbol{\beta}_0 = \mathbf{0}$. Then we can write $Q(\mathbf{Y} - \mathbf{X}\boldsymbol{\beta}) = (2\tau_\varphi)^{-1}\boldsymbol{\beta}'\mathbf{X}'\mathbf{X}\boldsymbol{\beta} - \boldsymbol{\beta}'\mathbf{S}(\mathbf{Y}) + D(\mathbf{Y})$. Because Q is a quadratic function it follows from differentiation that it is minimized by

$$\widetilde{\boldsymbol{\beta}} = \tau_\varphi(\mathbf{X}'\mathbf{X})^{-1}\mathbf{S}(\mathbf{Y}). \tag{3.5.13}$$

Hence, $\widetilde{\boldsymbol{\beta}}$ is a linear function of $\mathbf{S}(\mathbf{Y})$. Thus we immediately have from Theorem 3.5.2:

Theorem 3.5.4 *Under model (3.2.3) and assumptions (E.1), (D.1), (D.2) and (S.1) of Section 3.4,*

$$\sqrt{n}(\widetilde{\boldsymbol{\beta}} - \boldsymbol{\beta}_0) \xrightarrow{D} N_p(0, \tau_\varphi^2 \Sigma^{-1}). \tag{3.5.14}$$

Since Q is a local approximation to D, it would seem that their minimizing values are close also. As the next result shows, this indeed is the case. The proof first appeared in Jaeckel (1972) and is sketched in the Appendix; see Theorem A.3.9.

Theorem 3.5.5 *Under model (3.2.3) and assumptions (E.1), (D.1), (D.2) and (S.1),*

$$\sqrt{n}(\widehat{\boldsymbol{\beta}} - \widetilde{\boldsymbol{\beta}}) \xrightarrow{P} 0.$$

Combining this result with Theorem 3.5.4, we get the next corollary which gives the asymptotic distribution of the R-estimate.

Corollary 3.5.6 *Under model (3.2.3) and assumptions (E.1), (D.1), (D.2) and (S.1),*

$$\sqrt{n}(\widehat{\boldsymbol{\beta}}_\varphi - \boldsymbol{\beta}_0) \xrightarrow{D} N_p(0, \tau_\varphi^2 \Sigma^{-1}). \tag{3.5.15}$$

Under the further restriction that the errors have finite variance σ^2, Exercise 3.12.10 shows that the least-squares estimate $\widehat{\boldsymbol{\beta}}_{LS}$ of $\boldsymbol{\beta}$ satisfies $\sqrt{n}(\widehat{\boldsymbol{\beta}}_{LS} - \boldsymbol{\beta}) \xrightarrow{D} N_p(0, \sigma^2 \Sigma^{-1})$. Hence, as in the location problems of Chapters 1 and 2, the asymptotic efficiency of R-estimates relative to least-squares is the ratio σ^2/τ_φ^2, where τ_φ is the scale parameter (3.4.4). Thus the R-estimates of regression coefficients have the same high efficiency relative to LS estimates as do the rank-based estimates in the location problem. In particular, the efficiency of the Wilcoxon estimates relative to the LS estimates for the normal distribution is 0.955. For longer-tailed error distributions this relative efficency is much higher; see the efficiency discussion for contaminated normal distributions in Example 1.7.1.

From the above corollary, R-estimates are asymptotically unbiased. It follows from the invariance properties, if we additionally asume that the errors have a symmetric distribution, that R-estimates are unbiased for all sample sizes; see Exercise 3.12.11 for details.

The random vector $\widetilde{\boldsymbol{\beta}}$ in (3.5.13) is an asymptotic representation of the R-estimate $\widehat{\boldsymbol{\beta}}$. The following representation will be useful later:

Corollary 3.5.7 *Under model (3.2.3) and assumptions (E.1), (D.1), (D.2) and (S.1),*

$$n^{1/2}(\widehat{\boldsymbol{\beta}}_\varphi - \boldsymbol{\beta}_0) = \tau_\varphi(n^{-1}\mathbf{X}'\mathbf{X})^{-1} n^{-1/2}\mathbf{X}'\varphi(F(\mathbf{Y} - \mathbf{X}\boldsymbol{\beta}_0)) + \mathbf{o}_p(1), \tag{3.5.16}$$

where the notation $\varphi(F(\mathbf{Y}))$ means the $n \times 1$ vector whose ith component is $\varphi(F(Y_i))$.

164   Linear Models

Proof. This follows immediately from (A.3.9), (A.3.10), the proof of Theorem 3.5.2, and equation (3.5.12).

Based on this last corollary, we have that the **influence function** of the R-estimate is given by

$$\Omega(\mathbf{x}_0, y_0; \widehat{\boldsymbol{\beta}}_\varphi) = \tau_\varphi \Sigma^{-1} \varphi(F(y_0)) \mathbf{x}_0. \tag{3.5.17}$$

A more rigorous derivation of this result, based on Fréchet derivatives, is given in the Appendix; see Section A.5.2. Note that the influence function is bounded in the Y-space but it is unbounded in the x-space. Hence an outlier in the x-space can seriously impair an R-estimate. Although, as noted above, the R-estimates are highly efficient relative to the LS estimates, it follows from its influence function that the breakdown of the R-estimate is 0. In Chapter 5 we present the GR- and the HBR-estimates whose influence functions are bounded in both spaces and which have, respectively, positive breakdown and 50% breakdown; however, both are less efficient than the R-estimate.

3.5.2   R-Estimates of the Intercept

As discussed in Section 3.2, the intercept parameter requires the specification of a location functional, $T(e_i)$. In this section we shall take $T(e_i) = \text{med}(e_i)$. Since we assume, without loss of generality, that $T(e_i) = 0$, $\alpha = T(Y_i - \mathbf{x}_i'\boldsymbol{\beta})$. This leads immediately to estimating α by the median of the R residuals. Note that this is analogous to LS, since the LS estimate of the intercept is the arithmetic average of the LS residuals. Further, this estimate is associated with the sign-test statistic and the L_1 norm. More generally we could also consider estimates associated with signed-rank test statistics. For example, if we consider the signed-rank Wilcoxon scores of Chapter 1 then the corresponding estimate is the median of the Walsh averages of the residuals. The theory of such estimates based on signed-rank tests, however, requires symmetrically distributed errors. Thus, while we briefly discuss these later, we now concentrate on the median of the residuals which does not require this symmetry assumption. We will make use of assumption (E.2), (3.4.3), i.e. $f(0) > 0$.

The process we consider is the sign process based on residuals given by

$$S_1(\mathbf{Y} - \alpha\mathbf{1} - \mathbf{X}\widehat{\boldsymbol{\beta}}_\varphi) = \sum_{i=1}^{n} \text{sgn}(Y_i - \alpha - \mathbf{x}_i'\widehat{\boldsymbol{\beta}}_\varphi). \tag{3.5.18}$$

As with the sign process in Chapter 1, this process is a nondecreasing step function of α which steps down at the residuals. The solution to the equation

$$S_1(\mathbf{Y} - \alpha\mathbf{1} - \mathbf{X}\widehat{\boldsymbol{\beta}}_\varphi) \doteq 0 \tag{3.5.19}$$

is the median of the residuals which we shall denote by $\widehat{\alpha}_S = \text{med}\{Y_i - \mathbf{x}_i'\widehat{\boldsymbol{\beta}}_\varphi\}$. Our goal is to obtain the asymptotic joint distribution of the estimate $\widehat{\mathbf{b}}_\varphi = (\widehat{\alpha}_S, \widehat{\boldsymbol{\beta}}_\varphi')'$.

Similar to the R-estimate of $\boldsymbol{\beta}$ the estimate of the intercept is location- and scale-equivariant; hence, without loss of generality, we will assume that the true intercept and regression parameters are 0. We begin with a lemma.

Theory of Rank-Based Estimates 165

Lemma 3.5.8 *Assume conditions (E.1), (E.2), (S.1), (D.1) and (D.2) of Section 3.4. For any $\varepsilon > 0$ and for any $a \in \mathcal{R}$,*

$$\lim_{n \to \infty} P[|S_1(\mathbf{Y} - an^{-1/2}\mathbf{1} - \mathbf{X}\widehat{\boldsymbol{\beta}}_\varphi) - S_1(\mathbf{Y} - an^{-1/2}\mathbf{1})| \geq \varepsilon\sqrt{n}] = 0.$$

The proof of this lemma was first given by Jurečková (1971) for general signed-rank scores and it is briefly sketched in the Appendix for the sign scores; see Lemma A.3.12. This lemma leads to the asymptotic linearity result for the process (3.5.18).

We need the following linearity result:

Theorem 3.5.9 *Assume conditions (E.1), (E.2), (S.1), (D.1) and (D.2) of Section 3.4. For any $\varepsilon > 0$ and $c > 0$,*

$$\lim_{n \to \infty} P\left[\sup_{|a| \leq c} |n^{-1/2}S_1(\mathbf{Y} - an^{-1/2}\mathbf{1} - \mathbf{X}\widehat{\boldsymbol{\beta}}_\varphi) - n^{-1/2}S_1(\mathbf{Y} - \mathbf{X}\widehat{\boldsymbol{\beta}}_\varphi) + a\tau_S^{-1}| \geq \varepsilon\right] = 0,$$

where τ_S is the scale parameter defined in expression (3.4.6).

Proof. For any fixed a write

$$|n^{-1/2}S_1(\mathbf{Y} - an^{-1/2}\mathbf{1} - \mathbf{X}\widehat{\boldsymbol{\beta}}_\varphi) - n^{-1/2}S_1(\mathbf{Y} - \mathbf{X}\widehat{\boldsymbol{\beta}}_\varphi) + a\tau_S^{-1}|$$

$$\leq |n^{-1/2}S_1(\mathbf{Y} - an^{-1/2}\mathbf{1} - \mathbf{X}\widehat{\boldsymbol{\beta}}_\varphi) - n^{-1/2}S_1(\mathbf{Y} - an^{-1/2}\mathbf{1})|$$

$$+ |n^{-1/2}S_1(\mathbf{Y} - an^{-1/2}\mathbf{1}) - n^{-1/2}S_1(\mathbf{Y}) + a\tau_S^{-1}|$$

$$+ |n^{-1/2}S_1(\mathbf{Y}) - n^{-1/2}S_1(\mathbf{Y} - \mathbf{X}\widehat{\boldsymbol{\beta}}_\varphi)|.$$

We can apply Lemma 3.5.8 to the first and third terms on the right-hand side of the above inequality. For the middle term we can use the asymptotic linearity result in Chapter 1 for the sign process, (1.5.22). This yields the result for any a and the sup will follow from the monotonicity of the process, similar to the proof of Theorem 1.5.6 of Chapter 1.

Letting $a = 0$ in Lemma 3.5.8, we have that the difference $n^{-1/2}S_1(\mathbf{Y} - \mathbf{X}\widehat{\boldsymbol{\beta}}_\varphi) - n^{-1/2}S_1(\mathbf{Y})$ goes to zero in probability. Thus the asymptotic distribution of $n^{-1/2}S_1(\mathbf{Y} - \mathbf{X}\widehat{\boldsymbol{\beta}}_\varphi)$ is the same as that of $n^{-1/2}S_1(\mathbf{Y})$, namely, $N(0, 1)$. We have two applications of these results. The first is found in the next lemma.

Lemma 3.5.10 *Assume conditions (E.1), (E.2), (D.1), (D.2), and (S.1) of Section 3.4. The random variable, $n^{1/2}\widehat{\alpha}_S$ is bounded in probability.*

Proof. Let $\varepsilon > 0$ be given. Since $n^{-1/2}S_1(\mathbf{Y} - \mathbf{X}\widehat{\boldsymbol{\beta}}_\varphi)$ is asymptotically $N(0, 1)$ there exists a $c < 0$ such that

$$P[n^{-1/2}S_1(\mathbf{Y} - \mathbf{X}\widehat{\boldsymbol{\beta}}_\varphi) < c] < \frac{\varepsilon}{2}. \qquad (3.5.20)$$

166   Linear Models

Take $c^* = \tau_S^{-1}(c - \varepsilon)$. By the process's monotonicity and the definition of $\widehat{\alpha}$, we have the implication $n^{1/2}\widehat{\alpha}_S < c^* \Rightarrow n^{-1/2}S_1(\mathbf{Y} - c^*n^{-1/2}\mathbf{1} - \mathbf{X}\widehat{\boldsymbol{\beta}}_\varphi) \leq 0$. Adding in and subtracting out the above linearity result leads to

$$P[n^{1/2}\widehat{\alpha}_S < c^*] \leq P[n^{-1/2}S_1(\mathbf{Y} - n^{-1/2}c^*\mathbf{1} - \mathbf{X}\widehat{\boldsymbol{\beta}}_\varphi) \leq 0]$$

$$\leq P[|n^{-1/2}S_1(\mathbf{Y} - c^*n^{-1/2}\mathbf{1} - \mathbf{X}\widehat{\boldsymbol{\beta}}_\varphi) - (n^{-1/2}S_1(\mathbf{Y} - \mathbf{X}\widehat{\boldsymbol{\beta}}_\varphi) - c^*\tau_S^{-1})| \geq \varepsilon]$$

$$+ P[n^{-1/2}S_1(\mathbf{Y} - \mathbf{X}\widehat{\boldsymbol{\beta}}_\varphi) - c^*\tau_S^{-1} < \varepsilon]. \tag{3.5.21}$$

The first term on the right-hand side can be made less than $\varepsilon/2$ for sufficiently large n, whereas the second term is (3.5.20). From this it follows that $n^{1/2}\widehat{\alpha}_S$ is bounded below in probability. To finish the proof, a similar argument shows that $n^{1/2}\widehat{\alpha}_S$ is bounded above in probability.

As a second application, we can write the linearity result of Theorem 3.5.9 as

$$n^{-1/2}S_1(\mathbf{Y} - an^{-1/2}\mathbf{1} - \mathbf{X}\widehat{\boldsymbol{\beta}}_\varphi) = n^{-1/2}S_1(\mathbf{Y}) - a\tau_S^{-1} + o_p(1) \tag{3.5.22}$$

uniformly for all $|a| \leq c$ and for $c > 0$.

Because $\widehat{\alpha}_S$ is a solution to equation (3.5.19) and $n^{1/2}\widehat{\alpha}_S$ is bounded in probability, the second linearity result, (3.5.22), yields, after some simplification, the following asymptotic representation of our result for the estimate of the intercept for the true intercept α_0:

$$n^{1/2}(\widehat{\alpha}_S - \alpha_0) = \tau_S n^{-1/2} \sum_{i=1}^{n} \text{sgn}(Y_i - \alpha_0) + o_p(1), \tag{3.5.23}$$

where τ_S is given in (3.4.6). From this we have that $n^{1/2}(\widehat{\alpha}_S - \alpha_0) \xrightarrow{\mathcal{D}} N(0, \tau_S^2)$. Our interest, though, is in the joint distribution of $\widehat{\alpha}_S$ and $\widehat{\boldsymbol{\beta}}_\varphi$.

By Corollary 3.5.7 the corresponding asymptotic representation of $\widehat{\boldsymbol{\beta}}_\varphi$ for the true vector of regression coefficients $\boldsymbol{\beta}_0$ is

$$n^{1/2}(\widehat{\boldsymbol{\beta}}_\varphi - \boldsymbol{\beta}_0) = \tau_\varphi(n^{-1}\mathbf{X}'\mathbf{X})^{-1}n^{-1/2}\mathbf{X}'\varphi(F(\mathbf{Y})) + o_p(1), \tag{3.5.24}$$

where τ_φ is given by (3.4.4). The joint asymptotic distribution is given in the following theorem.

Theorem 3.5.11 *Under (D.1), (D.2), (S.1), (E.1) and (E.2) in Section 3.4,*

$$\widehat{\mathbf{b}}_\varphi = \begin{pmatrix} \widehat{\alpha}_S \\ \widehat{\boldsymbol{\beta}}_\varphi \end{pmatrix} \text{ has an approximate }$$

$$N_{p+1}\left(\begin{pmatrix} \alpha_0 \\ \boldsymbol{\beta}_0 \end{pmatrix}, \begin{bmatrix} n^{-1}\tau_S^2 & \mathbf{0}' \\ \mathbf{0} & \tau_\varphi^2(\mathbf{X}'\mathbf{X})^{-1} \end{bmatrix} \right) \text{ distribution.}$$

Proof. As above, assume without loss of generality that the true parameters are 0. It is easier to work with the random vector $\mathbf{T}_n = (\tau_s^{-1}\sqrt{n}\widehat{\alpha}_S, \sqrt{n}(\tau_\varphi^{-1}(n^{-1}\mathbf{X}'\mathbf{X})\widehat{\boldsymbol{\beta}}_\varphi)')'$. Let $\mathbf{t} = (t_1, \mathbf{t}_2')'$ be an arbitrary, nonzero, vector in $\mathcal{R}^{p+1}$. We need only show that $Z_n = \mathbf{t}'\mathbf{T}_n$ has an asymptotically univariate normal distribution. Based on the above asymptotic representations of $\widehat{\alpha}_S$, (3.5.23), and $\widehat{\boldsymbol{\beta}}_\varphi$, (3.5.24), we have

$$Z_n = n^{-1/2} \sum_{k=1}^{n} (t_1 \operatorname{sgn}(Y_k) + (\mathbf{t}_2' \mathbf{x}_k)\varphi(F(Y_k))) + o_p(1). \tag{3.5.25}$$

Denote the sum on the right side of (3.5.25) as Z_n^*. We need only show that Z_n^* converges in distribution to a univariate normal distribution. Denote the kth summand as Z_{nk}^*. We shall use the Lindeberg–Feller central limit theorem. Our application of this theorem is similar to its use in the proof of Theorem 3.5.2. First, since the score function φ is standardized ($\int \varphi = 0$), note that $E(Z_n^*) = 0$. Let $B_n^2 = \operatorname{Var}(Z_n^*)$. Because the individual summands are independent, the Y_k are identically distributed, φ is standardized ($\int \varphi^2 = 1$), and the design is centered, B_N^2 simplifies to

$$B_n^2 = n^{-1} \left(\sum_{k=1}^{n} t_1^2 + \sum_{k=1}^{n} (\mathbf{t}_2'\mathbf{x}_k)^2 + 2t_1 \operatorname{cov}(\operatorname{sgn}(Y_1), \varphi(F(Y_1))) \mathbf{t}_2' \sum_{k=1}^{n} \mathbf{x}_k \right)$$

$$= t_1^2 + \mathbf{t}_2'(n^{-1}\mathbf{X}'\mathbf{X})\mathbf{t}_2 + 0.$$

Hence, by (D.2),

$$\lim_{n \to \infty} B_n^2 = t_1^2 + \mathbf{t}_2' \Sigma \mathbf{t}_2, \tag{3.5.26}$$

which is a positive number. To satisfy the Lindeberg–Feller condition, we need to show that for any $\varepsilon > 0$,

$$\lim_{n \to \infty} B_n^{-2} \sum_{k=1}^{n} E[Z_{nk}^{*2} I(|Z_{nk}^*| > \varepsilon B_n)] = 0. \tag{3.5.27}$$

Since B_n^2 converges to a positive constant we need only show that the sum converges to 0. By the triangle inequality we can show that the indicator function satisfies

$$I(n^{-1/2}|t_1| + n^{-1/2}|\mathbf{t}_2'\mathbf{x}_k||\varphi(F(Y_k))| > \varepsilon B_n) \geq I(|Z_{nk}^*| > \varepsilon B_n). \tag{3.5.28}$$

Following the discussion after expression (3.5.7), we have that $n^{-1/2}|(\mathbf{x}_k'\mathbf{t})| \leq M_n$ where M_n, is independent of k and, furthermore, $M_n \to 0$. Hence, we have

$$I\left(|\varphi(F(Y_k))| > \frac{\varepsilon B_n - n^{-1/2}t_1}{M_n}\right) \geq I(n^{-1/2}|t_1| + n^{-1/2}|\mathbf{t}_2'\mathbf{x}_k||\varphi(F(Y_k))| > \varepsilon B_n).$$

$$\tag{3.5.29}$$

Thus the sum in expression (3.5.27) is less than or equal to

$$\sum_{k=1}^{n} E\left[Z_{nk}^{*2} I\left(|\varphi(F(Y_k))| > \frac{\varepsilon B_n - n^{-1/2} t_1}{M_n}\right)\right] = t_1 E\left[I\left(|\varphi(F(Y_1))| > \frac{\varepsilon B_n - n^{-1/2} t_1}{M_n}\right)\right]$$

$$+ (2/n) E\left[\text{sgn}(Y_1) \varphi(F(Y_1)) I\left(|\varphi(F(Y_1))| > \frac{\varepsilon B_n n^{-1/2} t_1}{M_n}\right)\right] \mathbf{t}_2' \sum_{k=1}^{n} \mathbf{x}_k$$

$$+ E\left[\varphi^2(F(Y_1)) I\left(|\varphi(F(Y_1))| > \frac{\varepsilon B_n n^{-1/2} t_1}{M_n}\right)\right] (1/n) \sum_{k=1}^{n} (\mathbf{t}_2' \mathbf{x}_k)^2.$$

Because the design is centered the middle term on the right-hand side is 0. As remarked above, the term $(1/n) \sum_{k=1}^{n} (\mathbf{t}_2' \mathbf{x}_k)^2 = (1/n) \mathbf{t}_2' \mathbf{X}' \mathbf{X} \mathbf{t}_2$ converges to a positive constant. In the expression $(\varepsilon B_n - n^{-1/2} t_1)/M_n$, the numerator converges to a positive constant as the denominator converges to 0; hence, the expression goes to ∞. Therefore since φ is bounded, the indicator function converges to 0. Again using the boundedness of φ, we can interchange limit and expectation by the Lebesgue dominated convergence theorem. Thus condition (3.5.27) is true and, hence, Z_n^* converges in distribution to a univariate normal distribution. Therefore, $\mathbf{T}_n$ converges to a multivariate normal distribution. Note that, by (3.5.26), it follows that the asymptotic covariance of $\widehat{\mathbf{b}}_\varphi$ is the result displayed in the theorem.

In the above development, we considered the centered design. In practice, though, we are often concerned with an uncentered design. Let α^* denote the intercept for the uncentered model. Then $\alpha^* = \alpha - \overline{\mathbf{x}}' \boldsymbol{\beta}$, where $\overline{\mathbf{x}}$ denotes the vector of column averages of the uncentered design matrix. An estimate of α^* based on R-estimates is given by $\widehat{\alpha}_S^* = \widehat{\alpha}_S - \overline{\mathbf{x}}' \widehat{\boldsymbol{\beta}}_\varphi$. Based on the previous theorem, it follows (Exercise 3.12.14) that

$$\begin{pmatrix} \widehat{\alpha}_S^* \\ \widehat{\boldsymbol{\beta}}_\varphi \end{pmatrix} \text{ is approximately } N_{p+1}\left(\begin{pmatrix} \alpha_0 \\ \boldsymbol{\beta}_0 \end{pmatrix}, \begin{bmatrix} \kappa_n & -\tau_\varphi^2 \overline{\mathbf{x}}' (\mathbf{X}'\mathbf{X})^{-1} \\ -\tau_\varphi^2 (\mathbf{X}'\mathbf{X})^{-1} \overline{\mathbf{x}} & \tau_\varphi^2 (\mathbf{X}'\mathbf{X})^{-1} \end{bmatrix}\right),$$

(3.5.30)

where $\kappa_n = n^{-1} \tau_S^2 + \tau_\varphi^2 \overline{\mathbf{x}}' (\mathbf{X}'\mathbf{X})^{-1} \overline{\mathbf{x}}$ and τ_S and and τ_φ are given respectively by (3.4.6) and (3.4.4).

Intercept Estimate Based on Signed-Rank Scores

Suppose we additionally assume that the errors have a symmetric distribution; i.e. $f(-x) = f(x)$. In this case, all location functionals are the same. Let $\varphi_f(u) = -f'(F^{-1}(u))/f(F^{-1}(u))$ denote the optimal scores for the density $f(x)$. Then, as Exercise 3.12.12 shows, $\varphi_f(1 - u) = -\varphi_f(u)$; that is, the scores are odd about $\frac{1}{2}$. Hence, in this subsection we will additionally assume that the scores satisfy property (S.3), (3.4.12).

For scores satisfying (S.3), the corresponding signed-rank scores are generated as $a^+(i) = \varphi^+(i/(n+1))$, where $\varphi^+(u) = \varphi((u+1)/2)$; see the discussion in Section 2.5.3. For example if Wilcoxon scores are used, $\varphi(u) = \sqrt{12}(u - \frac{1}{2})$, then the signed-rank score function is $\varphi^+(u) = \sqrt{3}u$. Recall from Chapter 1 that these signed-rank scores can be used to define a norm and a subsequent R analysis. Here we only want to apply the associated one-sample signed-rank procedure to the residuals in order to obtain an estimate of the intercept. So consider the process

$$T^+(\widehat{\mathbf{e}}_R - \alpha \mathbf{1}) = \sum_{i=1}^{n} \text{sgn}(\widehat{e}_{Ri} - \alpha \mathbf{1}) a^+(R|\widehat{e}_{Ri} - \alpha|), \tag{3.5.31}$$

where $\widehat{e}_{Ri} = y_i - \mathbf{x}_i'\widehat{\boldsymbol{\beta}}_\varphi$; see (1.8.2). Note that this is the process discussed in Section 1.8, except now the iid observations are replaced by residuals. The process is still a nonincreasing function of α which steps down at the Walsh averages of the residuals; see Exercise 1.13.23. The estimate of the intercept is a value $\widehat{\alpha}_\varphi^+$ which solves the equation $T^+(\widehat{\mathbf{e}}_R - \alpha) \doteq 0$. If Wilcoxon scores are used then the estimate is the median of the Walsh averages, (1.3.25), while if sign scores are used the estimate is the median of the residuals.

Let $\widehat{\mathbf{b}}_\varphi^+ = (\widehat{\alpha}_\varphi^+, \widehat{\boldsymbol{\beta}}_\varphi')'$. We next briefly sketch the development of the asymptotic distribution of $\widehat{\mathbf{b}}_\varphi^+$. Assume, without loss of generality, that the true parameter vector $(\alpha_0, \boldsymbol{\beta}_0')'$ is $\mathbf{0}$. Suppose instead of the residuals we had the true errors in (3.5.31). Theorem A.2.11 of the Appendix then yields an asymptotic linearity result for the process. McKean and Hettmansperger (1976) show that this result also holds for the residuals; that is,

$$\frac{1}{\sqrt{n}} S^+(\widehat{\mathbf{e}}_R - \alpha \mathbf{1}) = S^+(\mathbf{e}) - \alpha \tau_\varphi^{-1} + o_p(1), \tag{3.5.32}$$

for all $|\alpha| \leq c$, where $c > 0$. Using arguments similar to those in McKean and Hettmansperger (1976), we can show that $\sqrt{n}\widehat{\alpha}_\varphi^+$ is bounded in probability; hence, by (3.5.32), we have that

$$\sqrt{n}\widehat{\alpha}_\varphi^+ = \tau_\varphi \frac{1}{\sqrt{n}} S^+(\mathbf{e}) + o_p(1). \tag{3.5.33}$$

But by (A.2.43) and (A.2.45) of the Appendix, we have the second representation given by

$$\sqrt{n}\widehat{\alpha}_\varphi^+ = \tau_\varphi \frac{1}{\sqrt{n}} \sum_{i=1}^{n} \varphi^+(F^+|e_i|)\text{sgn}(e_i) + o_p(1)$$

$$= \tau_\varphi \frac{1}{\sqrt{n}} \sum_{i=1}^{n} \varphi^+(2F(e_i) - 1) + o_p(1), \tag{3.5.34}$$

where F^+ is the distribution function of the absolute errors $|e_i|$. Due to symmetry, $F^+(t) = 2F(t) - 1$. Then using the relationship between the rank and the signed-rank scores, $\varphi^+(u) = \varphi((u+1)/2)$, we finally obtain

$$\sqrt{n}\widehat{\alpha}_\varphi^+ = \tau_\varphi \frac{1}{\sqrt{n}} \sum_{i=1}^{n} \varphi(F(Y_i)) + o_p(1). \tag{3.5.35}$$

Therefore, using expression (3.5.17), we have

$$\sqrt{n} \begin{bmatrix} \widehat{\alpha}_\varphi^+ \\ \widehat{\boldsymbol{\beta}}_\varphi \end{bmatrix} = \frac{\tau_\varphi}{\sqrt{n}} \begin{bmatrix} \mathbf{1}'\varphi(F(\mathbf{Y})) \\ \mathbf{X}'\varphi(F(\mathbf{Y})) \end{bmatrix} + o_p(1). \tag{3.5.36}$$

170   Linear Models

This and an application of the Lindeberg central limit theorem, similar to the proof of Theorem 3.5.11, leads to the following theorem,

Theorem 3.5.12 *Under assumptions (D.1), (D.2), (E.1), (E.2), (S.1) and (S.3) of Section 3.4,*

$$\begin{bmatrix} \widehat{\alpha}_\varphi^+ \\ \widehat{\boldsymbol{\beta}}_\varphi \end{bmatrix} \text{ has an approximate } N_{p+1}\left(\begin{pmatrix} \alpha_0 \\ \boldsymbol{\beta}_0 \end{pmatrix}, \tau_\varphi^2(\mathbf{X}_1'\mathbf{X}_1)^{-1}\right) \text{ distribution,} \quad (3.5.37)$$

where $\mathbf{X}_1 = [\mathbf{1}\ \mathbf{X}]$.

3.6 Theory of Rank-Based Tests

Consider the general linear hypotheses discussed in Section 3.2,

$$H_0 : \mathbf{M}\boldsymbol{\beta} = \mathbf{0} \text{ versus } H_A : \mathbf{M}\boldsymbol{\beta} \neq \mathbf{0}, \quad (3.6.1)$$

where $\mathbf{M}$ is a $q \times p$ matrix of full row rank. The geometry of R testing (Section 3.2.2), indicated the statistic based on the reduction of dispersion between the reduced and full models, $F_\varphi = (RD/q)/(\widehat{\tau}_\varphi/2)$, as a test statistic; see (3.2.18). In this section we develop the asymptotic theory for this test statistic under null and alternative hypotheses. This theory will be sufficient for two other rank-based tests which we will discuss later. See Table 3.2.2 and the discussion relating to that table for the special case when $\mathbf{M} = \mathbf{I}$.

3.6.1 Null Theory of Rank-Based Tests

We proceed with two lemmas about the dispersion function $D(\boldsymbol{\beta})$ and its quadratic approximation $Q(\boldsymbol{\beta})$ given by expression (3.5.11).

Lemma 3.6.1 *Let $\widehat{\boldsymbol{\beta}}$ denote the R-estimate of $\boldsymbol{\beta}$ in the full model (3.2.3), then under (E.1), (S.1), (D.1) and (D.2) of Section 3.4,*

$$D(\widehat{\boldsymbol{\beta}}) - Q(\widehat{\boldsymbol{\beta}}) \xrightarrow{P} 0. \quad (3.6.2)$$

Proof: Assume without loss of generality that the true $\boldsymbol{\beta}$ is $\mathbf{0}$. Let $\varepsilon > 0$ be given. Choose c_0 such that $P[\sqrt{n}\|\widehat{\boldsymbol{\beta}}\| > c_0] < \varepsilon/2$, for n sufficiently large. Using asymptotic quadraticity (Theorem A.3.8), we have for n sufficiently large

$$P\left[|D(\widehat{\boldsymbol{\beta}}) - Q(\widehat{\boldsymbol{\beta}})| < \varepsilon\right] \geq P\left[\left\{\max_{\|\boldsymbol{\beta}\| \leq c_0/\sqrt{n}} |D(\boldsymbol{\beta}) - Q(\boldsymbol{\beta})| < \varepsilon\right\} \cap \left\{\sqrt{n}\|\widehat{\boldsymbol{\beta}}\| < c_0\right\}\right]$$

$$> 1 - \varepsilon. \quad (3.6.3)$$

From this we obtain the result.

This result shows that D and Q are close for the R-estimate of $\boldsymbol{\beta}$. Our next result shows that $Q(\widehat{\boldsymbol{\beta}})$ is close to the minimum of Q.

Lemma 3.6.2 *Let $\tilde{\boldsymbol{\beta}}$ denote the minimizing value of the quadratic function Q; then under (E.1), (S.1), (D.1) and (D.2) of Section 3.4,*

$$Q(\tilde{\boldsymbol{\beta}}) - Q(\widehat{\boldsymbol{\beta}}) \xrightarrow{P} 0. \tag{3.6.4}$$

Proof. By simple algebra we have

$$Q(\tilde{\boldsymbol{\beta}}) - Q(\widehat{\boldsymbol{\beta}}) = (2\tau_\varphi)^{-1}(\tilde{\boldsymbol{\beta}} - \widehat{\boldsymbol{\beta}})'\mathbf{X}'\mathbf{X}(\tilde{\boldsymbol{\beta}} + \widehat{\boldsymbol{\beta}}) - (\tilde{\boldsymbol{\beta}} - \widehat{\boldsymbol{\beta}})'\mathbf{S}(\mathbf{Y})$$

$$= (2\tau_\varphi)^{-1}\sqrt{n}(\tilde{\boldsymbol{\beta}} - \widehat{\boldsymbol{\beta}})'\left[n^{-1}\mathbf{X}'\mathbf{X}\sqrt{n}((\tilde{\boldsymbol{\beta}} + \widehat{\boldsymbol{\beta}})) - n^{-1/2}\mathbf{S}(\mathbf{Y}))\right].$$

It is shown in Exercise 3.12.15 that the term in brackets in this equation is bounded in probability. Since the left-hand factor converges to zero in probability by Theorem 3.5.5 the desired result follows.

It is easier to work with the equivalent formulation of the linear hypotheses given by the following lemma.

Lemma 3.6.3 *An equivalent formulation of the model and the hypotheses is*

$$\mathbf{Y} = \mathbf{1}\alpha + \mathbf{X}_1^*\boldsymbol{\beta}_1^* + \mathbf{X}_2^*\boldsymbol{\beta}_2^* + \mathbf{e}, \tag{3.6.5}$$

with the hypotheses $H_0 : \boldsymbol{\beta}_2^* = \mathbf{0}$ versus $H_A : \boldsymbol{\beta}_2^* \neq \mathbf{0}$, where $\mathbf{X}_i^*$ and $\boldsymbol{\beta}_i^*$, $i = 1, 2$, are defined in (3.6.7).

Proof. Consider the QR decomposition of $\mathbf{M}$ given by

$$\mathbf{M}' = [\mathbf{Q}_2 \ \mathbf{Q}_1] = \begin{bmatrix} \mathbf{R} \\ \mathbf{O} \end{bmatrix} = \mathbf{Q}_2\mathbf{R}, \tag{3.6.6}$$

where the columns of $\mathbf{Q}_1$ form an orthonormal basis for the kernel of the matrix $\mathbf{M}$, the columns of $\mathbf{Q}_2$ form an orthonormal basis for the column space of $\mathbf{M}'$, $\mathbf{O}$ is a $(p-q) \times q$ matrix of 0s, and $\mathbf{R}$ is a $q \times q$ upper triangular, nonsingular matrix. Define

$$\mathbf{X}_i^* = \mathbf{X}\mathbf{Q}_i \text{ and } \boldsymbol{\beta}_i^* = \mathbf{Q}_i'\boldsymbol{\beta} \text{ for } i = 1, 2. \tag{3.6.7}$$

It follows that

$$\mathbf{Y} = \mathbf{1}\alpha + \mathbf{X}\boldsymbol{\beta} + \mathbf{e}$$
$$= \mathbf{1}\alpha + \mathbf{X}_1^*\boldsymbol{\beta}_1^* + \mathbf{X}_2^*\boldsymbol{\beta}_2^* + \mathbf{e}.$$

Further, $\mathbf{M}\boldsymbol{\beta} = \mathbf{0}$ if and only if $\boldsymbol{\beta}_2^* = \mathbf{0}$, which yields the desired result.

Without loss of generality, by the previous lemma, for the remainder of the section, we will consider a model of the form

$$\mathbf{Y} = \mathbf{1}\alpha + \mathbf{X}_1\boldsymbol{\beta}_1 + \mathbf{X}_2\boldsymbol{\beta}_2 + \mathbf{e}, \tag{3.6.8}$$

with the hypotheses

$$H_0 : \boldsymbol{\beta}_2 = \mathbf{0} \text{ versus } H_A : \boldsymbol{\beta}_2 \neq \mathbf{0}. \tag{3.6.9}$$

With these lemmas we are now ready to obtain the asymptotic distribution of F_φ. Let $\boldsymbol{\beta}_r = (\boldsymbol{\beta}_1', \mathbf{0}')'$ denote the reduced model vector of parameters, let $\widehat{\boldsymbol{\beta}}_{r,1}$ denote the reduced model R-estimate of $\boldsymbol{\beta}_1$, and let $\widehat{\boldsymbol{\beta}}_r = (\widehat{\boldsymbol{\beta}}_{r,1}', \mathbf{0}')'$. We shall use similar notation with the minimizing value of the approximating quadratic Q. With this notation the drop in dispersion becomes $RD_\varphi = D(\widehat{\boldsymbol{\beta}}_r) - D(\widehat{\boldsymbol{\beta}})$. McKean and Hettmansperger (1976) proved the following:

Theorem 3.6.4 *Suppose the assumptions (E.1), (D.1), (D.2), and (S.1) of Section 3.4 hold. Then under H_0,*

$$\frac{RD_\varphi}{\tau_\varphi/2} \xrightarrow{\mathcal{D}} \chi^2(q),$$

where RD_φ is formally defined in expression (3.2.16).

Proof. Assume that the true vector of parameters is $\boldsymbol{0}$ and suppress the subscript φ on RD. Write RD as the sum of five differences:

$$\begin{aligned} RD &= D(\widehat{\boldsymbol{\beta}}_r) - D(\widehat{\boldsymbol{\beta}}) \\ &= \left(D(\widehat{\boldsymbol{\beta}}_r) - Q(\widehat{\boldsymbol{\beta}}_r)\right) + \left(Q(\widehat{\boldsymbol{\beta}}_r) - Q(\widetilde{\boldsymbol{\beta}}_r)\right) + \left(Q(\widetilde{\boldsymbol{\beta}}_r) - Q(\widetilde{\boldsymbol{\beta}})\right) \\ &\quad + \left(Q(\widetilde{\boldsymbol{\beta}}) - Q(\widehat{\boldsymbol{\beta}})\right) + \left(Q(\widehat{\boldsymbol{\beta}}) - D(\widehat{\boldsymbol{\beta}})\right). \end{aligned}$$

By Lemma 3.6.1 the first and fifth differences go to zero in probability and by Lemma 3.6.2 the second and fourth differences go to zero in probability. Hence we need only show that the third difference converges in distribution to the intended distribution. As in Lemma 3.6.2, algebra leads to

$$Q(\widetilde{\boldsymbol{\beta}}) = -2^{-1}\tau_\varphi \mathbf{S}(\mathbf{Y})'(\mathbf{X}'\mathbf{X})^{-1}\mathbf{S}(\mathbf{Y}) + D(\mathbf{Y}),$$

while

$$Q(\widetilde{\boldsymbol{\beta}}_r) = -2^{-1}\tau_\varphi \mathbf{S}(\mathbf{Y})' \begin{bmatrix} (\mathbf{X}_1'\mathbf{X}_1)^{-1} & \mathbf{0} \\ \mathbf{0} & \mathbf{0} \end{bmatrix} \mathbf{S}(\mathbf{Y}) + D(\mathbf{Y}).$$

Combining these last two results, we obtain

$$Q(\widetilde{\boldsymbol{\beta}}_r) - Q(\widetilde{\boldsymbol{\beta}}) = 2^{-1}\tau_\varphi \mathbf{S}(\mathbf{Y})'\left((\mathbf{X}'\mathbf{X})^{-1} - \begin{bmatrix} (\mathbf{X}_1'\mathbf{X}_1)^{-1} & \mathbf{0} \\ \mathbf{0} & \mathbf{0} \end{bmatrix}\right)\mathbf{S}(\mathbf{Y}).$$

Using a well-known matrix identity, (see Searle, 1971, p. 27),

$$(\mathbf{X}'\mathbf{X})^{-1} = \begin{bmatrix} (\mathbf{X}_1'\mathbf{X}_1)^{-1} & \mathbf{0} \\ \mathbf{0} & \mathbf{0} \end{bmatrix} + \begin{bmatrix} -\mathbf{A}_1^{-1}\mathbf{B} \\ \mathbf{I} \end{bmatrix} \mathbf{W}[-\mathbf{B}'\mathbf{A}_1^{-1} \ \mathbf{I}],$$

where

$$\mathbf{X}'\mathbf{X} = \begin{bmatrix} \mathbf{A}_1 & \mathbf{B} \\ \mathbf{B}' & \mathbf{A}_2 \end{bmatrix}$$

$$\mathbf{W} = \left(\mathbf{A}_2 - \mathbf{B}'\mathbf{A}_1^{-1}\mathbf{B}\right)^{-1}. \tag{3.6.10}$$

Hence after some simplification we have

$$\frac{RD}{\tau_\varphi/2} = S(Y)' \left[\begin{bmatrix} -A_1^{-1}B \\ I \end{bmatrix} W [-B'A_1^{-1} \; I] \right] S(Y) + o_p(1)$$

$$= ([-B'A_1^{-1} \; I]S(Y))' W ([-B'A_1^{-1} \; I]S(Y)) + o_p(1)$$

$$= ([-B'A_1^{-1} \; I]n^{-1/2}S(Y))' nW ([-B'A_1^{-1} \; I]n^{-1/2}S(Y)) + o_p(1). \quad (3.6.11)$$

Using $n^{-1}X'X \to \Sigma$ and the asymptotic distribution of $n^{-1/2}S(Y)$ from Theorem 3.5.2, it follows that the right-hand side of (3.6.11) converges in distribution to a χ^2 random variable with q degrees of freedom, which completes the proof of the theorem.

A consistent estimate of τ_φ is discussed in Section 3.7. We shall denote this estimate by $\hat{\tau}_\varphi$. The test statistic we shall subsequently use is given by

$$F_\varphi = \frac{RD_\varphi/q}{\hat{\tau}_\varphi/2}. \quad (3.6.12)$$

Although the test statistic qF_φ has an asymptotic χ^2 distribution, small-sample studies (see below) have indicated that it is best to compare the test statistic with F critical values having q and $n - p$ degrees of freedom; that is, the test at nominal level α is

Reject $H_0 : M\beta = 0$ in favor of $H_A : M\beta \neq 0$ if $F_\varphi \geq F(\alpha, q, n - p - 1)$.

$$(3.6.13)$$

McKean and Sheather (1991) review numerous small-sample studies concerning the validity of the rank-based analysis based on the test statistic F_φ. These small-sample studies demonstrate that the empirical α level of F_φ over a variety of designs, sample sizes, and error distributions is close to the nominal value.

In classical inference there are three tests of general hypotheses: the likelihood ratio test (reduction in sums of squares test), Wald's test and Rao's scores (gradient) test. A good discussion of these tests can be found in Rao (1973). When the hypotheses are the general linear hypotheses, (3.6.1), the errors have a normal distribution, and the least-squares procedure is used, the three test statistics are algebraically equivalent. Actually the equivalence holds without normality, although in this case the reduction in sums of squares statistic is not the likelihood ratio test; see the discussion in Hettmansperger and McKean (1983).

There are also three rank-based tests for the general linear hypotheses. The **reduction in dispersion** test statistic F_φ is the analog of the likelihood ratio test, i.e. the reduction in sums of squares test. Since **Wald's test statistic** is a quadratic form in full model estimates, its rank analog is given by

$$F_{\varphi,Q} = \frac{(M\hat{\beta})' [M(X'X)^{-1}M']^{-1} (M\hat{\beta})/q}{\hat{\tau}_\varphi^2}. \quad (3.6.14)$$

Provided $\widehat{\tau}_\varphi$ is a consistent estimate of τ_φ, it follows from the asymptotic distribution of $\widehat{\boldsymbol{\beta}}_R$ (Corollary 3.5.6) that, under H_0, $qF_{\varphi,Q}$ has an asymptotic χ^2 distribution. Hence the test statistics F_φ and $F_{\varphi,Q}$ have the same null asymptotic distributions. Actually, as Exercise 3.12.16 shows, the difference of the test statistics converges to zero in probability under H_0. Unlike the classical methods, though, they are not algebraically equivalent; see Hettmansperger and McKean (1983).

The **rank gradient scores** test is easiest to define in terms of the reparameterized model (3.6.20); that is, the null hypothesis is $H_0 : \boldsymbol{\beta}_2 = \mathbf{0}$. Rewrite the random vector defined in (3.6.11) of Theorem 3.6.4 using as the true parameter under H_0, $\boldsymbol{\beta}_0 = (\boldsymbol{\beta}_{01}, \mathbf{0}')'$, i.e.

$$([-\mathbf{B}'\mathbf{A}_1^{-1}\ \mathbf{I}]n^{-1/2}\mathbf{S}(\mathbf{Y}-\mathbf{X}\boldsymbol{\beta}_0))'n\mathbf{W}([-\mathbf{B}'\mathbf{A}_1^{-1}\ \mathbf{I}]n^{-1/2}\mathbf{S}(\mathbf{Y}-\mathbf{X}\boldsymbol{\beta}_0)). \quad (3.6.15)$$

From the proof of Theorem 3.6.4 this quadratic form has an asymptotic χ^2 distribution with q degrees of freedom. Since it does depend on $\boldsymbol{\beta}_0$, it cannot be used as a test statistic. Suppose we substitute the reduced model R-estimate of $\boldsymbol{\beta}_1$; i.e. the first $p-q$ components of $\widehat{\boldsymbol{\beta}}_r$, defined immediately after expression (3.6.9). We shall call it $\widehat{\boldsymbol{\beta}}_{01}$. Now since this is the reduced model R-estimate, we have

$$\mathbf{S}(\mathbf{Y}-\mathbf{X}\widehat{\boldsymbol{\beta}}_r) \doteq \begin{pmatrix} \mathbf{0} \\ \mathbf{S}_2(\mathbf{Y}-\mathbf{X}_1\widehat{\boldsymbol{\beta}}_{r,1}) \end{pmatrix}, \quad (3.6.16)$$

where the subscript 2 on $\mathbf{S}$ denotes the last $p-q$ components of $\mathbf{S}$. This yields

$$A_\varphi = \mathbf{S}_2(\mathbf{Y}-\mathbf{X}_1\widehat{\boldsymbol{\beta}}_{r,1})'\left\{\mathbf{X}_2'\mathbf{X}_2 - \mathbf{X}_2'\mathbf{X}_1(\mathbf{X}_1'\mathbf{X}_1)^{-1}\mathbf{X}_1'\mathbf{X}_2\right\}^{-1}\mathbf{S}_2(\mathbf{Y}-\mathbf{X}_1\widehat{\boldsymbol{\beta}}_{r,1}) \quad (3.6.17)$$

as a test statistic. This is often called the **aligned rank test**, since the observations are aligned by the reduced model estimate. Exercise 3.12.17 shows that, under H_0, A_φ has an asymptotic χ^2 distribution. As the proof shows, the difference between qF_φ and A_φ converges to zero in probability under H_0. Aligned rank tests were introduced by Hodges and Lehmann (1962) and are developed in the linear model by Puri and Sen (1985).

Suppose in (3.6.16) we use a reduced model estimate $\widehat{\boldsymbol{\beta}}_{r,1}^*$ which is not the R-estimate; for example, it may be the LS estimate. Then we have

$$\mathbf{S}(\mathbf{Y}-\mathbf{X}\widehat{\boldsymbol{\beta}}_r^*) \doteq \begin{pmatrix} \mathbf{S}_1(\mathbf{Y}-\mathbf{X}_1\widehat{\boldsymbol{\beta}}_{r,1}^*) \\ \mathbf{S}_2(\mathbf{Y}-\mathbf{X}_1\widehat{\boldsymbol{\beta}}_{r,1}^*) \end{pmatrix}. \quad (3.6.18)$$

The reduced model estimate must satisfy $\sqrt{n}(\widehat{\boldsymbol{\beta}}_r^* - \boldsymbol{\beta}_0) = O_p(1)$, under H_0. Then the statistic in (3.6.17) is

$$A_\varphi^* = \mathbf{S}_2^{*'}\left\{\mathbf{X}_2'\mathbf{X}_2 - \mathbf{X}_2'\mathbf{X}_1(\mathbf{X}_1'\mathbf{X}_1)^{-1}\mathbf{X}_1'\mathbf{X}_2\right\}^{-1}\mathbf{S}_2^*, \quad (3.6.19)$$

where, from (3.6.11),

$$\mathbf{S}_2^* = \mathbf{S}_2(\mathbf{Y}-\mathbf{X}_1\widehat{\boldsymbol{\beta}}_{r,1}^*) - \mathbf{X}_2'\mathbf{X}_1(\mathbf{X}_1'\mathbf{X}_1)^{-1}\mathbf{S}_1(\mathbf{Y}-\mathbf{X}_1\widehat{\boldsymbol{\beta}}_{r,1}^*). \quad (3.6.20)$$

Note that when the R-estimate is used, the second term in $\mathbf{S}_2^*$ vanishes and we have (3.6.17); see Adichi (1978) and Chiang and Puri (1984).

Hettmansperger and McKean (1983) give a general discussion of these three tests. Note that both $F_{\varphi,Q}$ and F_φ require estimation of full model estimates and the scale parameter τ_φ, while A_φ does not. However, when using a linear model, one is usually interested in more than hypothesis testing. Of primary interest is checking the quality of the fit, i.e. whether the model fits the data. This requires estimation of the full model parameters and an estimate of τ_φ. Diagnostics for fits based on R-estimates are discussed in Section 3.9;. One is also usually interested in estimating contrasts and their standard errors. For R-estimates this requires an estimate of τ_φ. Moreover, as discussed in Hettmansperger and McKean (1983), the small-sample properties of the aligned rank test can be poor on certain designs.

The **influence function of the test statistic** F_φ is derived in Appendix A.5.2. As discussed there, it is easier to work with the $\sqrt{qF_\varphi}$. The result is given by

$$\Omega(\mathbf{x}_0, y_0; \sqrt{qF_\varphi}) = |\varphi[F(y_0 - \mathbf{x}_0'\boldsymbol{\beta}_r)]| \left\{ \mathbf{x}_0' \left((\mathbf{X}'\mathbf{X})^{-1} - \begin{bmatrix} (\mathbf{X}_1'\mathbf{X}_1)^{-1} & 0 \\ 0 & 0 \end{bmatrix} \right) \mathbf{x}_0 \right\}^{1/2}.$$

(3.6.21)

As shown in the Appendix, the null distribution of F_φ can be read from this result. Note that similar to the R-estimates, the influence function of F_φ is bounded in the Y-space but not in the x-space; see (3.5.17).

3.6.2 Theory of Rank-Based Tests under Alternatives

In the last section, we developed the null asymptotic theory of the rank-based tests based on a general score function. In this section we obtain some properties of these tests under alternative models. We show first that the test based on the reduction of dispersion, RD_φ, (3.2.16), is consistent under any alternative to the general linear hypothesis. We then show that the efficiency of these tests is the same as the efficiency results obtained in Chapter 2.

Consistency

We want to show that the test statistic F_φ is consistent for the general linear hypothesis, (3.2.5). Without loss of generality, we will again reparameterize the model as in (3.6.20) and consider as our hypothesis $H_0: \boldsymbol{\beta}_2 = \mathbf{0}$ versus $H_A: \boldsymbol{\beta}_2 \neq \mathbf{0}$. Let $\boldsymbol{\beta}_0 = (\boldsymbol{\beta}_{01}', \boldsymbol{\beta}_{02}')'$ be the true parameter. We will assume that the alternative is true; hence, $\boldsymbol{\beta}_{02} \neq \mathbf{0}$. Let α be a given level of significance. Let $T(\tau_\varphi) = RD_\varphi/(\tau_\varphi/2)$, where $RD_\varphi = D(\widehat{\boldsymbol{\beta}}_r) - D(\widehat{\boldsymbol{\beta}})$. Because we estimate τ_φ under the full model by a consistent estimate, to show consistency of F_φ it suffices to show

$$P_{\boldsymbol{\beta}_0}[T(\tau_\varphi) \geq \chi^2_{\alpha,q}] \to 1, \qquad (3.6.22)$$

as $n \to \infty$.

As in the proof under the null hypothesis, it is convenient to work with the approximating quadratic function $Q(\mathbf{Y} - \mathbf{X}\boldsymbol{\beta})$, (3.5.11). As above, let $\widetilde{\boldsymbol{\beta}}$ and $\widehat{\boldsymbol{\beta}}$ denote the minimizing values of Q and D respectively under the full model. The present argument simplifies if, for the full model, we replace $\widehat{\boldsymbol{\beta}}$ by $\widetilde{\boldsymbol{\beta}}$ in $T(\tau_\varphi)$. We can do this because we can write

$$D(\mathbf{Y} - \mathbf{X}\widetilde{\boldsymbol{\beta}}) - D(\mathbf{Y} - \mathbf{X}\widehat{\boldsymbol{\beta}}) = \left(D(\mathbf{Y} - \mathbf{X}\widetilde{\boldsymbol{\beta}}) - Q(\mathbf{Y} - \mathbf{X}\widetilde{\boldsymbol{\beta}})\right)$$
$$+ \left(Q(\mathbf{Y} - \mathbf{X}\widetilde{\boldsymbol{\beta}}) - Q(\mathbf{Y} - \mathbf{X}\widehat{\boldsymbol{\beta}})\right)$$
$$+ \left(Q(\mathbf{Y} - \mathbf{X}\widehat{\boldsymbol{\beta}}) - D(\mathbf{Y} - \mathbf{X}\widehat{\boldsymbol{\beta}})\right).$$

Applying asymptotic quadraticity (Theorem A.3.8), the first and third differences go to 0 in probability while the second difference goes to 0 in probability by Lemma 3.6.2; hence the left-hand side goes to 0 in probability under the alternative model. Thus we need only show that

$$P_{\boldsymbol{\beta}_0}[(2/\tau_\varphi)(D(\widehat{\boldsymbol{\beta}}_r) - D(\widetilde{\boldsymbol{\beta}})) \geq \chi^2_{\alpha,q}] \to 1, \qquad (3.6.23)$$

where, as above, $\widehat{\boldsymbol{\beta}}_r$ denotes the reduced model R-estimate. We state the result next. The proof can be found in the Appendix; see Theorem A.3.11.

Theorem 3.6.5 *Suppose conditions (E.1), (D.1), (D.2), and (S.1) of Section 3.4 hold. The test statistic F_φ is consistent for the hypotheses (3.2.5).*

Efficiency Results

The above result establishes that the rank-based test statistic F_φ is consistent for the general linear hypothesis, (3.2.5). We next derive the efficiency results of the test. Our first step is to obtain the asymptotic power of F_φ along a sequence of alternatives. This generalizes the asymptotic power lemmas discussed in Chapters 1 and 2. From this the efficiency results will follow. As with the consistency discussion it is more convenient to work with model (3.6.20).

The sequence of alternative models to the hypothesis $H_0: \boldsymbol{\beta}_2 = \mathbf{0}$ is

$$\mathbf{Y} = \mathbf{1}\alpha + \mathbf{X}_1\boldsymbol{\beta}_1 + \mathbf{X}_2(\boldsymbol{\theta}/\sqrt{n}) + \mathbf{e}, \qquad (3.6.24)$$

where $\boldsymbol{\theta}$ is a nonzero vector. Because R-estimates are invariant to location shifts, we can assume without loss of generality that $\boldsymbol{\beta}_1 = \mathbf{0}$. Let $\boldsymbol{\beta}_n = (\mathbf{0}', \boldsymbol{\theta}'/\sqrt{n})'$ and let H_n denote the hypothesis that (3.6.24) is the true model. The concept of contiguity will prove helpful with the asymptotic theory of the statistic F_φ under this sequence of models. A discussion of contiguity is given in the Appendix; see Section A.2.2.

Theorem 3.6.6 *Under the sequence of models (3.6.24) and the assumptions (E.1), (D.1), (D.2), and (S.1) of Section 3.4,*

$$P_{\boldsymbol{\beta}_n}(T(\widehat{\tau_\varphi}) \leq t) \to P(\chi^2_q(\eta_\varphi) \leq t), \qquad (3.6.25)$$

where $\chi_q^2(\eta_\varphi)$ has a noncentral χ^2 distribution with q degrees of freedom and noncentrality parameter

$$\eta_\varphi = \tau_\varphi^{-2}\theta'\mathbf{W}_0^{-1}\theta, \tag{3.6.26}$$

where $\mathbf{W}_0 = \lim_{n \to \infty} n\mathbf{W}$ and $\mathbf{W}$ is as defined in (3.6.10).

Proof. As in the proof of Theorem 3.6.4, we can write the drop in dispersion as the sum of the same five differences. Since the first two and last two differences go to zero in probability under the null model, it follows from the discussion on contiguity, (Section A.2.2) that these differences go to zero in probability under model (3.6.24). Hence we need only be concerned about the third difference. Since $\boldsymbol{\beta}_1 = \mathbf{0}$, the third difference reduces to the same quantity as in Theorem 3.6.4; i.e. we obtain

$$\frac{RD_\varphi}{\tau_\varphi/2} = ([-\mathbf{B}'\mathbf{A}_1^{-1}\ \mathbf{I}]\mathbf{S}(\mathbf{Y}))'\mathbf{W}([-\mathbf{B}'\mathbf{A}_1^{-1}\ \mathbf{I}]\mathbf{S}(\mathbf{Y})) + o_p(1).$$

The asymptotic linearity result derived in the Appendix, (Theorem A.3.8) is

$$\sup_{\sqrt{n}\|\boldsymbol{\beta}\| \leq c} \|n^{-1/2}\mathbf{S}(\mathbf{Y} - \mathbf{X}\boldsymbol{\beta}) - \left(n^{-1/2}\mathbf{S}(\mathbf{Y}) - \tau_\varphi^{-1}\boldsymbol{\Sigma}\sqrt{n}\boldsymbol{\beta}\right)\| = o_p(1),$$

for all $c > 0$. Since $\sqrt{n}\|\boldsymbol{\beta}_n\| = \|\boldsymbol{\theta}\|$, we can take $c = \|\boldsymbol{\theta}\|$ and get

$$\|n^{-1/2}\mathbf{S}(\mathbf{Y} - \mathbf{X}\boldsymbol{\beta}_n) - \left(n^{-1/2}\mathbf{S}(\mathbf{Y}) - \tau_\varphi^{-1}\boldsymbol{\Sigma}(\mathbf{0}', \boldsymbol{\theta}')'\right)\| = o_p(1). \tag{3.6.27}$$

The above probability statements hold under the null model and, hence, by contiguity also under the sequence of models (3.6.24). Under (3.6.24), however,

$$n^{-1/2}\mathbf{S}(\mathbf{Y} - \mathbf{X}\boldsymbol{\beta}_n) \xrightarrow{\mathcal{D}} N_p(\mathbf{0}, \boldsymbol{\Sigma}).$$

Hence, under (3.6.24)

$$n^{-1/2}\mathbf{S}(\mathbf{Y}) \xrightarrow{\mathcal{D}} N_p(\tau_\varphi^{-1}\boldsymbol{\Sigma}(\mathbf{0}', \boldsymbol{\theta}')', \boldsymbol{\Sigma}). \tag{3.6.28}$$

Then under the sequence of models (3.6.24),

$$[-\mathbf{B}'\mathbf{A}_1^{-1}\ \mathbf{I}]n^{-1/2}\mathbf{S}(\mathbf{Y}) \xrightarrow{\mathcal{D}} N_q(\tau_\varphi^{-1}\mathbf{W}_0, \mathbf{W}_0).$$

From this last result, the conclusion readily follows.

Several interesting remarks follow from this theorem. First, since $\mathbf{W}_0$ is positive definite, under alternatives the noncentrality parameter $\eta > 0$. Thus the asymptotic distribution of $T(\tau_\varphi)$ under the sequence of models (3.6.24) has mean $q + \eta$. Furthermore, the asymptotic power of a level α test based on $T(\tau_\varphi)$ is $P[\chi_q^2(\eta) \geq \chi_{\alpha,q}^2]$.

Second, note that that we can write the noncentrality parameter as

$$\eta = (\tau_\varphi^2 n)^{-1}[\theta'\mathbf{A}_2\theta - (\mathbf{B}\theta)'\mathbf{A}_1^{-1}\mathbf{B}\theta].$$

Both matrices $\mathbf{A}_2$ and $\mathbf{A}_1^{-1}$ are positive definite; hence, the noncentrality parameter is maximized when θ is in the kernel of $\mathbf{B}$. One way of assuring this

178   Linear Models

for a design is to take $\mathbf{B} = \mathbf{0}$. Because $\mathbf{B} = \mathbf{X}_1'\mathbf{X}_2$, this condition holds for orthogonal designs. Therefore orthogonal designs are generally more efficient than nonorthogonal designs.

We next obtain the asymptotic relative efficiency of the test statistic F_φ with respect to the least squares classical F-test, F_{LS}, defined by (3.2.17) in Section 3.2.2. The theory for F_{LS} under local alternatives is outlined in Exercise 3.12.18, where it is shown that, under the additional assumption that the random errors e_i have finite variance σ^2, the null asymptotic distribution of qF_{LS} is a central χ_q^2 distribution. Thus both F_φ and F_{LS} have the same asymptotic null distribution. As outlined in Exercise 3.12.18, under the sequence of models (3.6.24) qF_{LS} has an asymptotic noncentral $\chi^2_{q,\eta_{LS}}$ with noncentrality parameter

$$\eta_{LS} = (\sigma^2)^{-1}\boldsymbol{\theta}'\mathbf{W}_0^{-1}\boldsymbol{\theta}. \qquad (3.6.29)$$

Based on Theorem 3.6.6, the asymptotic relative efficiency of F_φ and F_{LS} is the ratio of their noncentrality parameters, i.e.

$$e(F_\varphi, F_{LS}) = \frac{\eta_\varphi}{\eta_{LS}} = \frac{\sigma^2}{\tau_\varphi^2}.$$

Thus the efficiency results for the rank-based estimates and tests discussed in this section are the same as the efficiency results presented in Chapters 1 and 2. An asymptotically efficient analysis can be obtained if the selected rank-score function is $\varphi_f(u) = -f_0'(F_0^{-1}(u))/f_0(F_0^{-1}(u))$, where f_0 is the form of the density of the error distribution. If the errors have a logistic distribution then the Wilcoxon scores will result in an asymptotically efficient analysis.

Usually we have no knowledge of the distribution of the errors. In that case, we would recommend using Wilcoxon scores. With them, the loss in relative efficiency with respect to the classical analysis for the normal distribution is only 5%, while the gain in efficiency over the classical analysis for long-tailed error distributions can be substantial, as discussed in Chapters 1 and 2.

Many of the studies reviewed in the article by McKean and Sheather (1991) included power comparisons of the rank-based analyses with the least-squares F-test, F_{LS}. The empirical power of F_{LS} for normal error distributions was slightly better than the empirical power of F_φ, under Wilcoxon scores. Under error distributions with heavier tails than the normal distribution, the empirical power of F_φ was generally larger, often much larger, than the empirical power of F_{LS}. These studies provide empirical evidence that the good asymptotic efficiency properties of the rank-based analysis hold in the small-sample setting.

As discussed above, the noncentrality parameters of the test statistics F_φ and F_{LS} differ in only the scale parameters. Hence, in practice, planning designs based on the noncentrality parameter of F_φ can proceed similar to the planning of a design using the noncentrality parameter of F_{LS}; see, for example, the discussion in Chapter 4 of Graybill (1976).

3.6.3 Further Remarks on the Dispersion Function

Let $\widehat{\mathbf{e}}$ denote the rank-based residuals when the linear model (3.2.4), is fit using the scores based on the function φ. Suppose the same assumptions hold as above; i.e. (E.1), (D.1), and (D.2) in Section 3.4. In this section, we explore further properties of the residual dispersion $D(\widehat{\mathbf{e}})$; see also Sections 3.9.2 and 3.11.

The functional corresponding to the dispersion function evaluated at the errors e_i is determined as follows: letting F_n denote the empirical distribution function of the iid errors $e_1, \ldots, e_n$, we have

$$\frac{1}{n} D(\mathbf{e}) = \sum_{i=1}^n a(R(e_i)) e_i \frac{1}{n}$$

$$= \sum_{i=1}^n \varphi\left(\frac{n}{n+1} F_n(e_i)\right) e_i \frac{1}{n}$$

$$= \int \varphi\left(\frac{n}{n+1} F_n(x)\right) x \, dF_n(x)$$

$$\xrightarrow{P} \int \varphi(F(x)) x \, dF(x) = \overline{D}_e. \tag{3.6.30}$$

As Exercise 3.12.19 shows, $\overline{D}_e$ is a scale parameter; see also the examples below.

Let $D(\widehat{\mathbf{e}})$ denote the residual dispersion $D(\widehat{\boldsymbol{\beta}}) = D(\mathbf{Y}, \Omega)$. We next show that $n^{-1} D(\widehat{\mathbf{e}})$ also converges in probability to $\overline{D}_e$, a result which will prove useful in Sections 3.9.2 and 3.11. Assume without loss of generality that the true $\boldsymbol{\beta}$ is $\mathbf{0}$. We can write

$$D(\widehat{\mathbf{e}}) = (D(\widehat{\mathbf{e}}) - Q(\widehat{\boldsymbol{\beta}})) + (Q(\widehat{\boldsymbol{\beta}}) - Q(\widetilde{\boldsymbol{\beta}})) + Q(\widetilde{\boldsymbol{\beta}}).$$

By Lemmas 3.6.1 and 3.6.2 the two differences on the right-hand side converge to 0 in probability. After some algebra, we obtain

$$Q(\widetilde{\boldsymbol{\beta}}) = -\frac{\tau_\varphi}{2} \left\{ \frac{1}{\sqrt{n}} \mathbf{S}(\mathbf{e})' \left(\frac{1}{n} \mathbf{X}'\mathbf{X}\right)^{-1} \frac{1}{\sqrt{n}} \mathbf{S}(\mathbf{e}) \right\} + D(\mathbf{e}).$$

By Theorem 3.5.2 the term in braces on the right-hand side converges in distribution to a χ^2 random variable with p degrees of freedom. This implies that $(D(\mathbf{e}) - D(\widehat{\mathbf{e}}))/(\tau_\varphi/2)$ also converges in distribution to a χ^2 random variable with p degrees of freedom. Although this is a stronger result than we need, it does imply that $n^{-1}(D(\mathbf{e}) - D(\widehat{\mathbf{e}}))$ converges to 0 in probability. Hence, $n^{-1} D(\widehat{\mathbf{e}})$ converges in probability to $\overline{D}_e$.

The natural analog to the least-squares F-test statistic is

$$F_\varphi^* = \frac{RD/q}{\widehat{\sigma}_D/2}, \tag{3.6.31}$$

180   Linear Models

where $\widehat{\sigma}_D = D(\widehat{\mathbf{e}})/(n-p-1)$, rather than F_φ. But we have

$$qF_\varphi^* = \frac{\widehat{\tau}_\varphi/2}{n^{-1}D(\widehat{\mathbf{e}})/2} qF_\varphi \xrightarrow{D} \kappa_F \chi^2(q), \qquad (3.6.32)$$

where κ_F is defined by

$$\frac{\widehat{\tau}_\varphi}{n^{-1}D(\widehat{\mathbf{e}})} \xrightarrow{P} \kappa_F. \qquad (3.6.33)$$

Hence, to have a limiting χ^2 distribution for qF_φ^* we need to have $\kappa_F = 1$. Below we give several examples where this occurs. In the first example, the form of the error distribution is known while in the second example the errors are normally distributed; however, these cases rarely occur in practice.

There is a more acute problem with using F_φ^*. As discussed above, (3.6.21), the influence function of the statistic F_φ^* is bounded in the Y-space; hence, it is robust. As shown in Section A.5.2 of the Appendix, though, the influence function of $\overline{D}(\widehat{\mathbf{e}})$ is unbounded in the Y-space. Thus similar to F_{LS}, the statistic F_φ^* is not robust. The denominator $\overline{D}(\widehat{\mathbf{e}})$ of F_φ^*, however, is a measure of scale while the denominator $\widehat{\sigma}^2$ of F_{LS} is a measure of variance. Hence, while not robust, F_φ^* is still somewhat less sensitive to outliers than F_{LS}; see Hettmansperger and McKean (1978).

Example 3.6.1 *Form of Error Density Known.*

Assume that the errors have density $f(x) = \sigma^{-1} f_0(x/\sigma)$, where f_0 is known. Our choice of scores would then be the optimal scores given by

$$\varphi_0(u) = -\frac{1}{\sqrt{I(f_0)}} \frac{f_0'(F_0^{-1}(u))}{f_0(F_0^{-1}(u))}, \qquad (3.6.34)$$

where $I(f_0)$ denotes the Fisher information corresponding to f_0. These scores yield an asymptotically efficient rank-based analysis. Exercise 3.12.20 shows that with these scores

$$\tau_\varphi = \overline{D}_e. \qquad (3.6.35)$$

Thus $\kappa_F = 1$ for this example and $qF_{\varphi_0}^*$ has a limiting $\chi^2(q)$ distribution under H_0.

Example 3.6.2 *Errors are Normally Distributed.*

In this case the form of the error density is $f_0(x) = (\sqrt{2\pi})^{-1} \exp\{-x^2/2\}$, the standard normal density. This is of course a subcase of Example 3.6.1. The optimal scores in this case are the normal scores $\varphi_0(u) = \Phi^{-1}(u)$, where Φ denotes the standard normal distribution function. Using these scores, the statistic $qF_{\varphi_0}^*$ has a limiting $\chi^2(q)$ distribution under H_0. Note here that the score function $\varphi_0(u) = \Phi^{-1}(u)$ is unbounded; hence the above theory must be modified to obtain this result. Under further regularity conditions on the design matrix, Jurečková (1969) obtained asymptotic linearity for the unbounded score function case; see also Koul (1992, p. 51). Using these results, the limiting distribution of $qF_{\varphi_0}^*$ can be obtained. The R-estimates based on these scores, however, have an unbounded influence function; see Section 1.8.1. We next consider this analysis for Wilcoxon and sign scores.

If Wilcoxon scores are employed then Exercise 3.12.21 shows that

$$\tau_\varphi = \sigma\sqrt{\frac{\pi}{3}} \tag{3.6.36}$$

$$\overline{D}_e = \sigma\sqrt{\frac{3}{\pi}}. \tag{3.6.37}$$

Thus, in this case, a consistent estimate of $\tau_\varphi/2$ is $n^{-1}D(\widehat{e})(\pi/6)$.

For sign scores a similar computation yields

$$\tau_S = \sigma\sqrt{\frac{\pi}{2}} \tag{3.6.38}$$

$$\overline{D}_e = \sigma\sqrt{\frac{2}{\pi}}. \tag{3.6.39}$$

Hence $n^{-1}D(\widehat{e})(\pi/4)$ is a consistent estimate of $\tau_S/2$.

Note that both examples are overly restrictive and again in all cases the resulting rank-based test of the general linear hypothesis H_0 has an unbounded influence function, even in the case when the errors have a normal density and the analysis is based on Wilcoxon or sign scores. In general, then, we recommend using a bounded score function φ and the corresponding test statistic F_φ, (3.2.18) which is highly efficent and whose influence function, (3.6.21), is bounded in the Y-space.

3.7 Implementation of the R Analysis

Up to this point, we have presented the geometry and asymptotic theory of the R-analysis. In order to implement the analysis we need to discuss the estimation of the scale parameter τ_φ. We also discuss algorithms for obtaining the rank-based analysis.

3.7.1 Estimates of the Scale Parameter τ_φ

The estimators of τ_φ that we dicuss are based on the R residuals formed after estimating $\boldsymbol{\beta}$. In particular, the estimators do not depend on the estimate of the intercept parameter α. Suppose, then, we have fit model (3.2.3) based on a score function φ which satisfies (S.1), (3.4.10), i.e. φ is bounded, and is standardized so that $\int \varphi = 0$ and $\int \varphi^2 = 1$. Let $\widehat{\boldsymbol{\beta}}_\varphi$ denote the R-estimate of $\boldsymbol{\beta}$ and let $\widehat{\mathbf{e}}_R = \mathbf{Y} - \mathbf{X}\widehat{\boldsymbol{\beta}}_\varphi$ denote the residuals based on the R-fit.

There have been several estimates of τ_φ proposed. McKean and Hettmansperger (1976) proposed a Lehmann-type estimator based on the standardized length of a confidence interval for the intercept parameter α. This estimator is a function of residuals and is consistent provided the density of the errors is symmetric. It is similar to the estimators of τ_φ discussed in Chapter 1. For Wilcoxon scores, Aubuchon and Hettmansperger (1984, 1989) obtained a density-type estimator for τ_φ and showed it was consistent for symmetric and asymmetric error distributions. Both of these estimators are available as options in the command RREGR in Minitab. In this section we briefly sketch the development of an estimator of τ_φ for bounded score functions proposed by Koul, Sievers and McKean (1987). It is a density-type estimator based on residuals which is consistent for symmetric and asymmetric error distributions which satisfy (E.1), (3.7.1). It further satisfies a uniform consistency property as stated in Theorem 3.7.1. A bootstrap percentile-t procedure based on this estimator did quite well in terms of empirical validity and efficiency in the Monte Carlo study performed by George et al. (1995).

Let the score function φ satisfy (S.1), (S.2), and (S.3) of Section 3.4. Since it is bounded, consider its standardization given by

$$\varphi^*(u) = \frac{\varphi(u) - \varphi(0)}{\varphi(1) - \varphi(0)}. \qquad (3.7.1)$$

Since φ^* is a linear function of φ the inference properties under either score function are the same. The score function φ^* will be useful since it is also a distribution function on $(0, 1)$. Recall that $\tau_\varphi = 1/\gamma$, where

$$\gamma = \int_0^1 \varphi(u)\varphi_f(u)du \text{ and } \varphi_f(u) = -\frac{f'(F^{-1}(u))}{f(F^{-1}(u))}.$$

Note that $\gamma^* = \int \varphi^*(u)\varphi_f(u)du = (\varphi(1) - \varphi(0))^{-1}\gamma$. For the present it will be more convenient to work with γ^*.

If we make the change of variable $u = F(x)$ in γ^*, we can rewrite it as

$$\gamma^* = -\int_{-\infty}^{\infty} \varphi^*(F(x))f'(x)dx$$

$$= \int_{-\infty}^{\infty} \varphi^{*\prime}(F(x))f^2(x)dx$$

$$= \int_{-\infty}^{\infty} f(x)d\varphi^*(F(x)),$$

where the second equality is obtained upon integration by parts using $dv = f'(x)dx$ and $u = \varphi^*(F(x))$.

From the above assumptions on φ^*, $\varphi^*(F(x))$ is a distribution function. Suppose Z_1 and Z_2 are independent random variables with distributions functions $F(x)$ and $\varphi^*(F(x))$, respectively. Let $H(y)$ denote the distribution function of $|Z_1 - Z_2|$. It then follows that

$$H(y) = \begin{cases} P[|Z_1 - Z_2| \leq y] = \int_{-\infty}^{\infty} [F(z_2 + y) - F(z_2 - y)]d\varphi^*(F(z_2)) & y > 0 \\ 0 & y \leq 0. \end{cases}$$
$$(3.7.2)$$

Let $h(y)$ denote the density of $H(y)$. Upon differentiating under the integral sign in (3.7.2) it easily follows that

$$h(0) = 2\gamma^*. \qquad (3.7.3)$$

So to estimate γ we need to estimate $h(0)$.

Using the transformation $t = F(z_2)$, rewrite (3.7.2) as

$$H(y) = \int_0^1 [F(F^{-1}(t) + y) - F(F^{-1}(t) - y)] d\varphi^*(t). \tag{3.7.4}$$

Next let $\widehat{F}_n$ denote the empirical distribution function of the R-residuals and let $\widehat{F}_n^{-1}(t) = \inf\{x : \widehat{F}_n(x) \geq t\}$ denote the usual inverse of $\widehat{F}_n$. Let $\widehat{H}_n$ denote the estimate of H which is obtained by replacing F by $\widehat{F}_n$. Some simplification follows by noting that for $t \in ((j-1)/n, j/n]$, $\widehat{F}_n^{-1}(t) = \widehat{e}_{(j)}$. This leads to the following form of $\widehat{H}_n$:

$$\widehat{H}_n(y) = \int_0^1 \left[\widehat{F}_n(\widehat{F}_n^{-1}(t) + y) - \widehat{F}_n(\widehat{F}_n^{-1}(t) - y)\right] d\varphi^*(t)$$

$$= \sum_{j=1}^n \int_{(\frac{j-1}{n}, \frac{j}{n}]} \left[\widehat{F}_n(\widehat{F}_n^{-1}(t) + y) - \widehat{F}_n(\widehat{F}_n^{-1}(t) - y)\right] d\varphi^*(t)$$

$$= \sum_{j=1}^n \left[\widehat{F}_n(\widehat{e}_{(j)} + y) - \widehat{F}_n(\widehat{e}_{(j)} - y)\right]\left(\varphi^*\left(\frac{j}{n}\right) - \varphi^*\left(\frac{j-1}{n}\right)\right)$$

$$= \frac{1}{n} \sum_{i=1}^n \sum_{j=1}^n \left(\varphi^*\left(\frac{j}{n}\right) - \varphi^*\left(\frac{j-1}{n}\right)\right) I(|\widehat{e}_{(i)} - \widehat{e}_{(j)}| \leq y). \tag{3.7.5}$$

An estimate of $h(0)$ and hence γ^*, (3.7.3), is an estimate of the form $\widehat{H}_n(t_n)/(2t_n)$ where t_n is chosen close to 0. Since $\widehat{H}_n$ is a distribution function, let $\widehat{t}_{n,\delta}$ denote the δth quantile of $\widehat{H}_n$, i.e. $\widehat{t}_{n,\delta} = \widehat{H}_n^{-1}(\delta)$. Then take $t_n = t_{n,\delta}/\sqrt{n}$. Our estimate of γ is given by

$$\widehat{\gamma}_{n,\delta} = \frac{(\varphi(1) - \varphi(0))\widehat{H}_n(t_{n,\delta}/\sqrt{n})}{2t_{n,\delta}/\sqrt{n}}. \tag{3.7.6}$$

Its consistency is given by the following theorem:

Theorem 3.7.1 *Under (E.1),(D.1), (S.1), and (S.2) of Section 3.4, and for any $0 < \delta < 1$,*

$$\sup_{\varphi \in C} |\widehat{\gamma}_{n,\delta} - \gamma| \xrightarrow{P} 0,$$

where C denotes the class of all bounded, right continuous, nondecreasing score functions defined on the interval $(0, 1)$.

The proof can be found in Koul et al. (1987). It follows immediately that $\widehat{\tau}_\varphi = 1/\widehat{\gamma}_{n,\delta}$ is a consistent estimate of τ_φ. Note that the uniformity condition on the scores in the theorem is more than we need here. This result, though, proves useful in adaptive procedures which estimate the score function; see McKean and Sievers (1989).

Since the scores are differentiable, an approximation of $\widehat{H}_n$ is obtained by an application of the mean value theorem to (3.7.5) which results in

$$\widehat{H}_n^*(y) = \frac{1}{c_n n} \sum_{i=1}^n \sum_{j=1}^n \varphi^{*\prime}\left(\frac{j}{n+1}\right) I(|\widehat{e}_{(i)} - \widehat{e}_{(j)}| \leq y), \qquad (3.7.7)$$

where $c_n = \sum_{j=1}^n \varphi^{*\prime}(j/(n+1))$ is such that $\widehat{H}_n^*$ is a distribution function.

The expression (3.7.5) for $\widehat{H}_n$ contains a density estimate of f based on a rectangular kernel. Hence, in choosing δ we are really choosing a bandwidth for a density estimator. As most kernel-type density estimates are sensitive to the bandwidth, so is γ^* sensitive to δ. Several small-sample studies have been done on this estimate of τ_φ; see McKean and Sheather (1991) for a summary. In these studies the quality of an estimator of τ_φ is based on how well it standardizes test statistics such as F_φ in terms of how close the empirical α levels of the test statistic are to nominal α levels. In the same way, scale estimators used in confidence intervals were judged by how close empirical confidence levels were to nominal confidence levels. The major concern is thus the validity of the inference procedure. For moderate sample sizes where the ratio of n/p exceeds 5, the value of $\delta = 0.80$ yielded valid estimates. For ratios less than 5, larger values of δ, around 0.90, gave valid estimates. In all cases it was found that the following simple degrees of freedom correction benefited the analysis:

$$\widehat{\tau}_\varphi = \sqrt{\frac{n}{n-p-1}}\, \widehat{\gamma}^{-1}. \qquad (3.7.8)$$

Note that this is similar to the least-squares correction on the maximum likelihood estimate (under normality) of the variance.

3.7.2 Algorithms for Computing the R Analysis

As we saw in Section 3.2, the dispersion function $D(\boldsymbol{\beta})$ is a continuous convex function of $\boldsymbol{\beta}$. Gradient-type algorithms, such as steepest descent, can be use to minimize $D(\boldsymbol{\beta})$ but they are often agonizingly slow. The algorithm which we describe next is a Newton-type algorithm based on the asymptotic quadraticity of $D(\boldsymbol{\beta})$. It is generally much faster than gradient-type algorithms and is currently used in the RREGR command in Minitab and in the program rglm (Kapenga, McKean and Vidmar, 1988). A finite algorithm to minimize $D(\boldsymbol{\beta})$ is discussed by Osborne (1985).

The Newton-type algorithm needs an initial estimate which we denote as $\widehat{\boldsymbol{\beta}}^{(0)}$. Let $\widehat{\mathbf{e}}^{(0)} = \mathbf{Y} - \mathbf{X}\widehat{\boldsymbol{\beta}}^{(0)}$ denote the initial residuals and let $\widehat{\tau}_\varphi^{(0)}$ denote the initial estimate of τ_φ based on these residuals. By (3.5.11) the quadratic approximating to D, based on $\widehat{\boldsymbol{\beta}}^{(0)}$, is given by

$$Q(\boldsymbol{\beta}) = (2\widehat{\tau}_\varphi^{(0)})^{-1}\left(\boldsymbol{\beta} - \widehat{\boldsymbol{\beta}}^{(0)}\right)'\mathbf{X}'\mathbf{X}\left(\boldsymbol{\beta} - \widehat{\boldsymbol{\beta}}^{(0)}\right)$$

$$- \left(\boldsymbol{\beta} - \widehat{\boldsymbol{\beta}}^{(0)}\right)'\mathbf{S}\left(\mathbf{Y} - \mathbf{X}\widehat{\boldsymbol{\beta}}^{(0)}\right) + D\left(\mathbf{Y} - \mathbf{X}\widehat{\boldsymbol{\beta}}^{(0)}\right).$$

By (3.5.13), the value of $\boldsymbol{\beta}$ which minimizes $Q(\boldsymbol{\beta})$ is given by

$$\widehat{\boldsymbol{\beta}}^{(1)} = \widehat{\boldsymbol{\beta}}^{(0)} + \widehat{\tau}_\varphi^{(0)}(\mathbf{X}'\mathbf{X})^{-1}\mathbf{S}(\mathbf{Y} - \mathbf{X}\widehat{\boldsymbol{\beta}}^{(0)}). \qquad (3.7.9)$$

This is the first Newton step. In the same way that the first step was defined in terms of the initial estimate, so can a second step be defined in terms of the first step. We shall call these iterated estimates or k-step estimates. In practice, though, we would want to know if $D(\widehat{\boldsymbol{\beta}}^{(1)})$ is less than $D(\widehat{\boldsymbol{\beta}}^{(0)})$ before proceeding. A more formal algorithm is presented below.

These **k-step estimates** satisfy some interesting properties themselves which we briefly discuss; details can be found in McKean and Hettmansperger (1978). Provided the initial estimate is such that $\sqrt{n}(\widehat{\boldsymbol{\beta}}^{(0)} - \boldsymbol{\beta})$ is bounded in probability, then for any $k \geq 1$ we have

$$\sqrt{n}(\widehat{\boldsymbol{\beta}}^{(k)} - \widehat{\boldsymbol{\beta}}_\varphi) \xrightarrow{P} 0,$$

where $\widehat{\boldsymbol{\beta}}_\varphi$ denotes a minimizing value of D. Hence the k-step estimates have the same asymptotic distribution as $\widehat{\boldsymbol{\beta}}_\varphi$. Furthermore $\widehat{\tau}_\varphi^{(k)}$ is a consistent estimate of τ_φ, if it is any of the scale estimates discussed in Section 3.7.1 based on k-step residuals. Let $F_\varphi^{(k)}$ denote the R-test of a general linear hypothesis based on reduced and full model k-step estimates. Then it can be shown that $F_\varphi^{(k)}$ satisfies the same asymptotic properties as the test statistic F_φ under the null hypothesis and contiguous alternatives. Also it is consistent for any alternative H_A.

Formal Algorithm

In order to outline the algorithm used by rglm, first consider the QR-decomposition of $\mathbf{X}$ which is given by

$$\mathbf{Q}'\mathbf{X} = \mathbf{R}, \tag{3.7.10}$$

where $\mathbf{Q}$ is an $n \times n$ orthogonal matrix and $\mathbf{R}$ is an $n \times p$ upper triangular matrix of rank p. As discussed in Stewart (1973), $\mathbf{Q}$ can be expressed as a product of p Householder transformations. Writing $\mathbf{Q} = [\mathbf{Q}_1 \ \mathbf{Q}_2]$ where $\mathbf{Q}_1$ is $n \times p$, it is easy to show that the columns of $\mathbf{Q}_1$ form an orthonormal basis for the column space of $\mathbf{X}$. In particular, the projection matrix onto the column space of $\mathbf{X}$ is given by $\mathbf{H} = \mathbf{Q}_1 \mathbf{Q}_1'$. The software package LINPACK (Dongarra et al., 1979) is a collection of subroutines which efficiently computes QR decompositions and it further has routines which obtain projections of vectors. Note that we can write the kth Newton step in terms of residuals as

$$\widehat{\mathbf{e}}^{(k)} = \widehat{\mathbf{e}}^{(k-1)} - \widehat{\tau}_\varphi \mathbf{H} \mathbf{a}(R(\widehat{\mathbf{e}}^{(k-1)})), \tag{3.7.11}$$

where $\mathbf{a}(R(\widehat{\mathbf{e}}^{(k-1)}))$ denotes the vector whose ith component is $a(R(\widehat{e}_i^{(k-1)}))$. Let $D^{(k)}$ denote the dispersion function evaluated at $\widehat{\mathbf{e}}^{(k)}$. The Newton step is a step from $\widehat{\mathbf{e}}^{(k-1)}$ along the direction $\widehat{\tau}_\varphi \mathbf{H} \mathbf{a}(R(\widehat{\mathbf{e}}^{(k-1)}))$. If $D^{(k)} < D^{(k-1)}$ the step has been successful; otherwise, a linear search can be made along the direction to find a value which minimizes D. This would then become the kth step residual. Such a search can be performed using methods such as false position as discussed below in Section 3.7.3. Stopping rules can be based on the relative drop in dispersion, i.e. stop when

$$\frac{D^{(k)} - D^{(k-1)}}{D^{(k-1)}} < \varepsilon_D, \tag{3.7.12}$$

186 *Linear Models*

where ε_D is a specified tolerance. A similar stopping rule can be based on the relative size of the step. Upon stopping at step k, obtain the fitted value $\widehat{\mathbf{Y}} = \mathbf{Y} - \widehat{\mathbf{e}}^{(k)}$ and then the estimate of $\boldsymbol{\beta}$ by solving $\mathbf{X}\boldsymbol{\beta} = \widehat{\mathbf{Y}}$.

A formal algorithm is as follows. Let ε_D and ε_s be the given stopping tolerances.

1. Set $k = 1$. Obtain initial residuals $\widehat{\mathbf{e}}^{(k-1)}$ and, based upon these, obtain an initial estimate $\widehat{\tau}_\varphi^{(0)}$ of τ_φ.
2. Obtain $\widehat{\mathbf{e}}^{(k)}$ as in (3.7.11). If the step is successful proceed to the next step; otherwise search along the Newton direction for a value which minimizes D, then go to the next step. An algorithm for this search is discussed in Section 3.7.3.
3. If the relative drop in dispersion or length of step is within its respective tolerance ε_D or ε_s then stop; otherwise set $\widehat{\mathbf{e}}^{(k-1)} = \widehat{\mathbf{e}}^{(k)}$ and go to step 2.
4. Obtain the estimate of $\boldsymbol{\beta}$ and the final estimate of τ_φ.

The QR decomposition can readily be used to form a reduced model design matrix for testing the general linear hypotheses (3.2.5), $\mathbf{M}\boldsymbol{\beta} = \mathbf{0}$, where $\mathbf{M}$ is a specified $q \times p$ matrix. Recall that we called the column space of $\mathbf{X}$, Ω_F, and the space Ω_F constrained by $\mathbf{M}\boldsymbol{\beta} = \mathbf{0}$ the reduced model space, ω. The key result lies in the following theorem.

Theorem 3.7.2 *Denote the row space of* $\mathbf{M}$ *by* $R(\mathbf{M}')$. *Let* $\mathbf{Q}_\mathbf{M}$ *be a* $p \times (p - q)$ *matrix whose columns consist of an orthonormal basis for the space* $(R(\mathbf{M}'))^\perp$. *If* $\mathbf{U} = \mathbf{X}\mathbf{Q}_\mathbf{M}$, *then* $\mathcal{R}(\mathbf{U}) = \omega$.

Proof. If $\mathbf{u} \in \omega$ then $\mathbf{u} = \mathbf{X}\mathbf{b}$ for some $\mathbf{b}$ where $\mathbf{M}\mathbf{b} = \mathbf{0}$. Hence $\mathbf{b} \in (R(\mathbf{M}'))^\perp$; i.e. $\mathbf{b} = \mathbf{Q}_\mathbf{M}\mathbf{c}$ for some $\mathbf{c}$. Conversely, if $\mathbf{u} \in R(\mathbf{U})$ then for some $\mathbf{c} \in R^{p-q}$, $\mathbf{u} = \mathbf{X}(\mathbf{Q}_\mathbf{M}\mathbf{c})$. Hence $\mathbf{u} \in R(\mathbf{X})$ and $\mathbf{M}(\mathbf{Q}_\mathbf{M}\mathbf{c}) = (\mathbf{M}\mathbf{Q}_\mathbf{M})\mathbf{c} = \mathbf{0}$.

Thus using the LINPACK subroutines mentioned above, it is easy to write an algorithm which obtains the reduced model design matrix $\mathbf{U}$ defined above in the theorem. The package rglm uses such an algorithm to test linear hypotheses; see Kapenga, McKean, and Vidmar (1988).

3.7.3 An Algorithm for a Linear Search

The computation for many of the quantities needed in a rank-based analysis involves simple linear searches. Examples include the estimate of the location parameter for a signed-rank procedure, the estimate of the shift in location in the two-sample location problem, the estimate of τ_φ discussed in Section 3.7.1, and the search along the Newton direction for a minimizing value in step 2 of the algorithm for the R-fit in a regression problem discussed in Section 3.7.2. The following is a generic setup for these problems: solve the equation

$$S(b) = K, \tag{3.7.13}$$

where $S(b)$ is a decreasing step function and K is a specified constant. Without loss of generality we will take $K = 0$ for the remainder of the discussion. By the monotonicity, a solution always exists, although it may be an interval of solutions. In almost all cases, $S(b)$ is asymptotically linear; so the search problem becomes relatively more efficient as the sample size increases.

There are certainly many search algorithms that can be used for solving (3.7.13). One that we have successfully employed is the Illinois version of *regula falsi*; see Dowell and Jarratt (1971). McKean and Ryan (1977) employed this routine to obtain the estimate and confidence interval for the two-sample Wilcoxon location problem. We will write the generic asymptotic linearity result as

$$S(b) \doteq S(b^{(0)}) - \zeta(b - b^{(0)}). \qquad (3.7.14)$$

The parameter ζ is often of the form $\delta^{-1}C$, where C is some constant. Since δ is a scale parameter, initial estimates of it include such estimates as the median absolute deviation (MAD), given in (3.9.27), or the sample standard deviation. We have usually found MAD to be preferable. An outline of an algorithm for the search is as follows.

1. *Bracket Step.* Beginning with an initial estimate $b^{(0)}$, step along the b-axis to $b^{(1)}$ where the interval $(b^{(0)}, b^{(1)})$, or vice versa, brackets the solution. Asymptotic linearity can be used here to make these steps; for instance, if $\zeta^{(0)}$ is an estimate of ζ based on $b^{(0)}$ then the first step is

$$b^{(1)} = b^{(0)} + S(b^{(0)})/\zeta^{(0)}.$$

2. *Regula Falsi.* Assume the interval $(b^{(0)}, b^{(1)})$ brackets the solution and that $b^{(1)}$ is the more recent value of $b^{(0)}, b^{(1)}$. If $|b^{(1)} - b^{(0)}| < \varepsilon$ then stop; otherwise, the next step is where the secant line determined by $b^{(0)}, b^{(1)}$ intersects the b-axis, i.e.

$$b^{(2)} = b^{(0)} - \frac{b^{(1)} - b^{(0)}}{S(b^{(1)}) - S(b^{(0)})} S(b^{(0)}). \qquad (3.7.15)$$

   (a) If $(b^{(0)}, b^{(2)})$ brackets the solution then replace $b^{(1)}$ by $b^{(2)}$ and begin step 2 again but use $S(b^{(0)})/2$ in place of $S(b^{(0)})$ in determination of the secant line, (this is the Illinois modification).
   (b) If $(b^{(2)}, b^{(1)})$ brackets the solution then replace $b^{(0)}$ by $b^{(2)}$ and begin step 2 again.

The above algorithm is easy to implement. Such searches are used in the package rglm; see Kapenga, McKean, and Vidmar (1988).

3.8 L_1 Analysis

This section is devoted to L_1 procedures. These are widely used; see, for example, Bloomfield and Steiger (1983). We first show that they are equivalent to R-estimates based on the sign-score function under model (3.2.4). Hence the asymptotic theory for L_1 estimation and subsequent analysis is contained in Section 3.5. The asymptotic theory for L_1 estimation can also be found in Bassett and Koenker (1978) and Rao (1988) from an L_1 point of view.

Consider the sign scores; i.e. the scores generated by $\varphi(u) = \text{sgn}(u - \frac{1}{2})$. In this section we shall denote the associated pseudo-norm by

$$\|\mathbf{v}\|_S = \sum_{i=1}^{n} \text{sgn}(R(v_i) - (n+1)/2)v_i, \quad \mathbf{v} \in \mathcal{R}^n;$$

see also Section 2.6.1. This score function is optimal if the errors follow a double exponential (Laplace) distribution; see Exercise 2.13.19 of Chapter 2. We shall summarize the analysis based on the sign scores, but first we show that indeed the R-estimates based on sign scores are also L_1 estimates, provided that the intercept is estimated by the median of residuals.

Consider the intercept model, (3.2.4), as given in Section 3.2 and let Ω denote the column space of $\mathbf{X}$ and Ω_1 denote the column space of the augmented matrix $\mathbf{X}_1 = [\mathbf{1} \ \mathbf{X}]$.

First consider the R-estimate of $\mathbf{\eta} \in \Omega$ based on the L_1 pseudo-norm. This is a vector $\widehat{\mathbf{Y}}_S \in \Omega$ such that

$$\widehat{\mathbf{Y}}_S = \text{Argmin}_{\mathbf{\eta} \in \Omega} \|\mathbf{Y} - \mathbf{\eta}\|_S.$$

Next consider the L_1 estimate for the space Ω_1; i.e. the L_1 estimate of $\alpha \mathbf{1} + \mathbf{\eta}$. This is a vector $\widehat{\mathbf{Y}}_{L_1} \in \Omega_1$ such that

$$\widehat{\mathbf{Y}}_{L_1} = \text{Argmin}_{\mathbf{\theta} \in \Omega_1} \|\mathbf{Y} - \mathbf{\theta}\|_{L_1},$$

where $\|\mathbf{v}\|_{L_1} = \sum |v_i|$ is the L_1-norm.

Theorem 3.8.1 *R-estimates based on sign scores are equivalent to L_1 estimates; that is,*

$$\widehat{\mathbf{Y}}_{L_1} = \widehat{\mathbf{Y}}_S + \text{med}\{\mathbf{Y} - \widehat{\mathbf{Y}}_S\}\mathbf{1}. \tag{3.8.1}$$

Proof. Any vector $\mathbf{v} \in \Omega_1$ can be written uniquely as $\mathbf{v} = a\mathbf{1} + \mathbf{v}_c$, where a is a scalar and $\mathbf{v}_c \in \Omega$. Since the sample median minimizes the L_1 distance between a vector and the space spanned by $\mathbf{1}$, we have

$$\|\mathbf{Y} - \mathbf{v}\|_{L_1} = \|\mathbf{Y} - a\mathbf{1} - \mathbf{v}_c\|_{L_1} \geq \|\mathbf{Y} - \text{med}\{\mathbf{Y} - \mathbf{v}_c\}\mathbf{1} - \mathbf{v}_c\|_{L_1}.$$

But it is easy to show that $\text{sgn}(Y_i - \text{med}\{\mathbf{Y} - \mathbf{v}_c\} - v_{ci}) = \text{sgn}(R(Y_i - v_{ci}) - (n+1)/2)$ for $i = 1, \ldots, n$. Putting these two results together, along with the fact that the sign scores sum to 0, we have

$$\|\mathbf{Y} - \mathbf{v}\|_{L_1} = \|\mathbf{Y} - a\mathbf{1} - \mathbf{v}_c\|_{L_1} \geq \|\mathbf{Y} - \text{med}\{\mathbf{Y} - \mathbf{v}_c\}\mathbf{1} - \mathbf{v}_c\|_{L_1} = \|\mathbf{Y} - \mathbf{v}_c\|_S, \tag{3.8.2}$$

for any vector $\mathbf{v} \in \Omega_1$. Once more using the sign argument above, we can show that

$$\|\mathbf{Y} - \text{med}\{\mathbf{Y} - \widehat{\mathbf{Y}}_S\}\mathbf{1} - \widehat{\mathbf{Y}}_S\|_{L_1} = \|\mathbf{Y} - \widehat{\mathbf{Y}}_S\|_S. \tag{3.8.3}$$

Putting (3.8.2) and (3.8.3) together establishes the result.

Let $\widehat{\mathbf{b}}'_S = (\widehat{\alpha}_S, \widehat{\mathbf{\beta}}'_S)$ denote the R-estimate of the vector of regression coefficients $\mathbf{b} = (\beta_0, \mathbf{\beta}')'$. It follows that these R-estimates are the maximum likelihood estimates if the errors e_i are double exponentially distributed; see Exercise 3.12.13.

From the discussions in Section 3.5, $\widehat{\mathbf{b}}_S$ has an approximate $N(\mathbf{b}, \tau_S^2(\mathbf{X}_1'\mathbf{X}_1)^{-1})$ distribution, where $\tau_S = (2f(0))^{-1}$. From this the efficiency properties of the L_1 procedures discussed in the first two chapters carry over to the L_1 linear model procedures. In particular, the efficiency of the latter relative to LS for the normal distribution is 0.63, and it can be much more efficient than LS for heavier-tailed error distributions.

As Exercise 3.12.22 shows, the drop in dispersion test based on sign scores, F_S, is, except for the scale parameter, the likelihood ratio test of the general linear hypothesis (3.2.5), provided the errors have a double exponential distribution. For other error distributions, the same comments about efficiency of the L_1 estimates can be made about the test F_S.

In terms of implementation, Schrader and McKean (1987) found it more difficult to standardize the L_1 statistics than other R procedures, such as the Wilcoxon. Their most successful standardization of F_S was based on the following bootstrap procedure:

1. Compute the full model L_1 estimates $\widehat{\boldsymbol{\beta}}_S$ and $\widehat{\alpha}_S$, the full model residuals $\widehat{e}_1, \ldots, \widehat{e}_n$, and the test statistic F_S.
2. Select $\widetilde{e}_1, \ldots, \widetilde{e}_{\tilde{n}}$, the $\tilde{n} = n - (p+1)$ nonzero residuals.
3. Draw a bootstrap random sample $e_1^*, \ldots, e_{\tilde{n}}^*$ with replacement from $\widetilde{e}_1, \ldots, \widetilde{e}_{\tilde{n}}$. Calculate $\widehat{\boldsymbol{\beta}}_S^*$ and F_S^*, the L_1 estimate and test statistic, from the model $y_i^* = \widehat{\alpha}_S + \mathbf{x}_i'\widehat{\boldsymbol{\beta}}_S + e_i^*$.
4. Independently repeat step 3 a large number (B) of times. The bootstrap p value, $p^* = \#\{F_S^* \geq F_S\}/B$.
5. Reject H_0 at level α if $p^* \leq \alpha$.

Notice that by using full model residuals, the algorithm estimates the null distribution of F_S. The algorithm depends on the number B of bootstrap samples taken. We suggest at least 2000.

3.9 Diagnostics

One of the most important parts in the analysis of a linear model is the examination of the resulting fit. Tools for doing this include residual plots and diagnostic techniques. Since the early 1980s, these tools have been developed for fits based on least squares; see, for example, Cook and Weisberg (1982) and Belsley, Kuh, and Welsch (1980). Least-squares residual plots can be used to detect such things as curvature not accounted for by the fitted model; see Cook and Weisberg (1994) for a recent discussion. Further diagnostic techniques can be used to detect outliers and to measure the influence of individual cases on the least-squares fit.

In this section we explore the properties of the residuals from the rank-based fits, showing how they can be used to determine model misspecification. We present diagnostic techniques for rank-based residuals that detect outlying

190   Linear Models

and influential cases. Together these tools offer the user a residual analysis for the rank-based method for the fit of a linear model similar to the residual analysis based on least-squares estimates.

In this section we consider the same linear model, (3.2.3), as in Section 3.2. For a given score function φ, let $\widehat{\boldsymbol{\beta}}_\varphi$ and $\widehat{\mathbf{e}}_R$ denote the R-estimate of $\boldsymbol{\beta}$ and residuals from the R fit of the model based on these scores. Much of the discussion is taken from the articles by McKean, Sheather, and Hettmansperger (1990; 1991; 1993). See also Dixon and McKean (1996) for a robust rank-based approach to modeling heteroscedasticity.

3.9.1   Properties of R Residuals and Model Misspecification

As we discussed above, a primary use of least-squares residuals is in detection of model misspecification. In order to show that the R residuals can also be used to detect model misspecification, consider the sequence of models

$$\mathbf{Y} = \mathbf{1}\alpha + \mathbf{X}\boldsymbol{\beta} + \mathbf{Z}\boldsymbol{\gamma} + \mathbf{e}, \qquad (3.9.1)$$

where $\mathbf{Z}$ is an $n \times q$ centered matrix of constants and $\boldsymbol{\gamma} = \boldsymbol{\theta}/\sqrt{n}$, for $\boldsymbol{\theta} \neq \mathbf{0}$. Note that this sequence of models is contiguous to model (3.2.3). Suppose we fit model (3.2.3), i.e. $\mathbf{Y} = \mathbf{1}\alpha + \mathbf{X}\boldsymbol{\beta} + \mathbf{e}$, when model (3.9.1) is the true model. Hence the model has been misspecified. As a first step in examining the residuals in this situation, we consider the limiting distribution of the corresponding R-estimate.

Theorem 3.9.1 *Assume model (3.9.1) is the true model. Let $\widehat{\boldsymbol{\beta}}_\varphi$ be the R-estimate for the model (3.2.3). Suppose that conditions (E.1) and (S.1) of Section 3.4 are true and that conditions (D.1) and (D.2) are true for the augmented matrix [$\mathbf{X}\ \mathbf{Z}$]. Then*

$$\widehat{\boldsymbol{\beta}}_\varphi \text{ has an approximate } N_p\!\left(\boldsymbol{\beta} + (\mathbf{X}'\mathbf{X})^{-1}\mathbf{X}'\mathbf{Z}\boldsymbol{\theta}/\sqrt{n},\, \tau_\varphi^2(\mathbf{X}'\mathbf{X})^{-1}\right) \text{ distribution.}$$

(3.9.2)

Proof: Without loss of generality assume that $\boldsymbol{\beta} = \mathbf{0}$. Note that the situation here is the same as the situation in Theorem 3.6.6, except now the null hypothesis corresponds to $\boldsymbol{\gamma} = \mathbf{0}$ and $\widehat{\boldsymbol{\beta}}_\varphi$ is the reduced model estimate. Thus we seek the asymptotic distribution of the reduced model estimate. As in Section 3.5.1, it is easier to consider the corresponding pseudo-estimate $\widetilde{\boldsymbol{\beta}}$ which is the reduced model estimate which minimzes the quadratic $Q(\mathbf{Y} - \mathbf{X}\boldsymbol{\beta})$ of (3.5.11). Under the null hypothesis, $\boldsymbol{\gamma} = \mathbf{0}$, $\sqrt{n}(\widehat{\boldsymbol{\beta}}_\varphi - \widetilde{\boldsymbol{\beta}}) \xrightarrow{P} 0$; hence, by contiguity, $\sqrt{n}(\widehat{\boldsymbol{\beta}}_\varphi - \widetilde{\boldsymbol{\beta}}) \xrightarrow{P} 0$ under the sequence of models (3.9.1). Thus $\widehat{\boldsymbol{\beta}}_\varphi$ and $\widetilde{\boldsymbol{\beta}}$ have the same distributions under (3.9.1); hence, it suffices to find the distribution of $\widetilde{\boldsymbol{\beta}}$. But by (3.5.13),

$$\widetilde{\boldsymbol{\beta}} = \tau_\varphi (\mathbf{X}'\mathbf{X})^{-1}\mathbf{S}(\mathbf{Y}), \qquad (3.9.3)$$

where $\mathbf{S}(\mathbf{Y})$ is the first p components of the vector $T(\mathbf{Y}) = [\mathbf{X}\ \mathbf{Z}]'\mathbf{a}(R(\mathbf{Y}))$. By (3.6.28) of Theorem 3.6.6,

$$n^{-1/2}T(\mathbf{Y}) \xrightarrow{D} N_{p+q}(\tau_\varphi^{-1}\Sigma^*(\mathbf{0}', \boldsymbol{\theta}')', \Sigma^*), \qquad (3.9.4)$$

where Σ^* is the limit

$$\lim_{n \to \infty} \frac{1}{n} \begin{bmatrix} \mathbf{X'X} & \mathbf{X'Z} \\ \mathbf{Z'X} & \mathbf{Z'Z} \end{bmatrix} = \Sigma^*.$$

Because $\widetilde{\boldsymbol{\beta}}$ is defined by (3.9.3), the result is just an algebraic computation applied to (3.9.4).

With a few more steps we can write a first-order expression for $\widehat{\boldsymbol{\beta}}_\varphi$, which is given in the following corollary:

Corollary 3.9.2 *Under the assumptions of the Theorem 3.9.1,*

$$\widehat{\boldsymbol{\beta}}_\varphi = \boldsymbol{\beta} + \tau_\varphi(\mathbf{X'X})^{-1}\mathbf{X'}\varphi(F(\mathbf{e})) + (\mathbf{X'X})^{-1}\mathbf{X'Z}\boldsymbol{\theta}/\sqrt{n} + \mathbf{o}_p(n^{-1/2}). \qquad (3.9.5)$$

Proof: Without loss of generality assume that the regression coefficients are 0. By (A.3.10) and expression (3.6.27) of Theorem 3.6.6, we can write

$$\frac{1}{\sqrt{n}} T(\mathbf{Y}) = \frac{1}{\sqrt{n}} \begin{bmatrix} \mathbf{X'}\varphi(F(\mathbf{e})) \\ \mathbf{Z'}\varphi(F(\mathbf{e})) \end{bmatrix} + \tau_\varphi^{-1} \frac{1}{n} \begin{bmatrix} \mathbf{X'Z}\boldsymbol{\theta} \\ \mathbf{Z'Z}\boldsymbol{\theta} \end{bmatrix} + \mathbf{o}_p(1);$$

hence, the first p components of $(1/\sqrt{n})T(\mathbf{Y})$ satisfy

$$\frac{1}{\sqrt{n}} S(\mathbf{Y}) = \frac{1}{\sqrt{n}} \mathbf{X'}\varphi(F(\mathbf{e})) + \tau_\varphi^{-1} \frac{1}{n} \mathbf{X'Z}\boldsymbol{\theta} + \mathbf{o}_p(1).$$

By (3.9.3) and the fact that $\sqrt{n}(\widehat{\boldsymbol{\beta}} - \widetilde{\boldsymbol{\beta}}) \xrightarrow{P} 0$, the result follows.

From this corollary we obtain the following first-order expressions of the R residuals and R fitted values:

$$\widehat{\mathbf{Y}}_R \doteq \alpha\mathbf{1} + \mathbf{X}\boldsymbol{\beta} + \tau_\varphi\mathbf{H}\varphi(F(\mathbf{e})) + \mathbf{HZ}\boldsymbol{\gamma} \qquad (3.9.6)$$

$$\widehat{\mathbf{e}}_R \doteq \mathbf{e} - \tau_\varphi\mathbf{H}\varphi(F(\mathbf{e})) + (\mathbf{I} - \mathbf{H})\mathbf{Z}\boldsymbol{\gamma}, \qquad (3.9.7)$$

where $\mathbf{H} = \mathbf{X}(\mathbf{X'X})^{-1}\mathbf{X'}$. In Exercise 3.12.23 the reader is asked to show that the least-squares fitted values and residuals satisfy

$$\widehat{\mathbf{Y}}_{LS} = \alpha\mathbf{1} + \mathbf{X}\boldsymbol{\beta} + \mathbf{He} + \mathbf{HZ}\boldsymbol{\gamma} \qquad (3.9.8)$$

$$\widehat{\mathbf{e}}_{LS} = \mathbf{e} - \mathbf{He} + (\mathbf{I} - \mathbf{H})\mathbf{Z}\boldsymbol{\gamma}. \qquad (3.9.9)$$

In terms of model mispecification the coefficients of interest are the regression coefficients. Hence, at this time we need not consider the effect of the estimation of the intercept. This avoids the problem of which estimate of the intercept to use. In practice, though, for both R and LS fits, the intercept will also be fitted and its effect will be removed from the residuals. We will also include the effect of estimation of the intercept in our discussion of the standardization of residuals and fitted values in Sections 3.9.2 and 3.9.3, respectively.

Suppose that the linear model (3.2.3) is correct. Based on its first-order expression when $\boldsymbol{\gamma} = \mathbf{0}$, $\widehat{\mathbf{e}}_R$ is a function of the random errors similar to $\widehat{\mathbf{e}}_{LS}$; hence, it follows that a plot of $\widehat{\mathbf{e}}_R$ versus $\widehat{\mathbf{Y}}_R$ should generally be a random scatter, similar to the least-squares residual plot.

In the case of model misspecification, note that the R residuals and least-squares residuals have the same asymptotic bias, namely $(\mathbf{I} - \mathbf{H})\mathbf{Z}\boldsymbol{\gamma}$. Hence R residual plots, similar to those of least squares, are useful in identifying model misspecification.

For least-squares residual plots, since least-squares residuals and the fitted values are uncorrelated, any pattern in this plot is due to model misspecification and not the fitting procedure used. The converse, however, is not true. As the example on the potency of drug compounds below illustrates, the least-squares residual plot can exhibit a random scatter for a poorly fitted model. This orthogonality in the LS residual plot does, however, make it easier to pick out patterns in the plot. Of course the R residuals are not orthogonal to the R-fitted values, but they are usually close to orthogonality; see Naranjo et al. (1994). We introduce the following parameter v to measure the extent of departure from orthogonality.

Denote general fitted values and residuals by $\widehat{\mathbf{Y}}$ and $\widehat{\mathbf{e}}$, respectively. The **expected departure from orthogonality** is the parameter v defined by

$$v = E\left[\widehat{\mathbf{e}}'\widehat{\mathbf{Y}}\right]. \tag{3.9.10}$$

For least squares, v_{LS} is of course 0. For R fits, we have the following first-order expression for it:

Theorem 3.9.3 *Under the assumptions of Theorem 3.9.1 and either model (3.2.3) or model (3.9.1),*

$$v_R \doteq p\tau_\varphi(E[\varphi(F(e_1))e_1] - \tau_\varphi). \tag{3.9.11}$$

Proof. Suppose model (3.9.1) holds. Using the above first-order expressions, we have

$$v_R \doteq E[(\mathbf{e} + \alpha\mathbf{1} - \tau_\varphi \mathbf{H}\boldsymbol{\varphi}(F(\mathbf{e})) + (\mathbf{I} - \mathbf{H})\mathbf{Z}\boldsymbol{\gamma})'(\mathbf{X}\boldsymbol{\beta} + \tau_\varphi \mathbf{H}\boldsymbol{\varphi}(F(\mathbf{e})) + \mathbf{H}\mathbf{Z}\boldsymbol{\gamma})].$$

Using $E[\boldsymbol{\varphi}(F(\mathbf{e}))] = 0$, $E[\mathbf{e}] = E(e_1)\mathbf{1}$, and the fact that $\mathbf{X}$ is centered, this expression simplifies to

$$v_R \doteq \tau_\varphi E[\text{tr}(\mathbf{H}\boldsymbol{\varphi}(F(\mathbf{e}))\mathbf{e}')] - \tau_\varphi^2 E[\text{tr}(\mathbf{H}\boldsymbol{\varphi}(F(\mathbf{e}))\boldsymbol{\varphi}(F(\mathbf{e}))')].$$

Since the components of $\mathbf{e}$ are independent, the result follows. The result is invariant to either of the models.

Although in general, $v_R \neq 0$ for R-estimates, if, as the next corollary shows, optimal scores (see Examples 3.6.1 and 3.6.2) are used the expected departure from orthogonality is 0.

Corollary 3.9.4 *Under the hypothesis of Theorem 3.9.3, if optimal R scores are used then $v_R = 0$.*

Proof. Let $\varphi(u) = -cf'(F^{-1}(u))/f(F^{-1}(u))$, where c is chosen so that $\int \varphi^2(u)du = 1$. Then

$$\tau_\varphi = \left[\int \varphi(u)\left(-\frac{f'(F^{-1}(u))}{f(F^{-1}(u))}\right)du\right]^{-1} = c.$$

Some simplification and an integration by parts shows

$$\int \varphi(F(e)) e\, dF(e) = -c \int f'(e)\, de = c.$$

Naranjo et al. (1994) conducted a simulation study to investigate the above properties of rank-based and LS residuals over several small-sample situations of null (the true model was fitted) models and misspecified models. Error distributions included the normal distribution and a contaminated normal distribution. Wilcoxon scores were used. The first part of the study concerned the amount of association between residuals and fitted values where the association was measured by several correlation coefficients, including Pearson's r and Kendall's τ. Because of orthogonality between the LS residuals and fitted values, Pearson's r is always 0 for LS. On the other measures of association, however, the results for the Wilcoxon analysis and LS were about the same. In general, there was little association. The second part investigated measures of randomness in a residual plot, including a runs test and a quadrant count test (the quadrants were determined by the medians of the residuals and fitted values). The results were similar for the LS and Wilcoxon fits. Both showed validity over the null models and exhibited similar power over the misspecified models. In a power study over a quadratic misspecified model, the Wilcoxon analysis exhibited more power for long-tailed error distributions. In summary, the simulation study provided empirical evidence that residual analyses based on Wilcoxon fits are similar to LS-based residual analyses.

There are other useful residual plots. Two that we will briefly discuss are $q-q$ **plots** and **added variable plots**. As with standard residual plots, the internal R studentized residuals (see Section 3.9.2) can be used in place of the residuals. Since the R-estimates of $\boldsymbol{\beta}$ are consistent, the distribution of the residuals should resemble the distribution of the errors. This leads to consideration of another useful residual plot, the $q-q$ plot. In this plot, the quantiles of the target distribution form the horizontal coordinates while the sample quantiles (ordered residuals) form the vertical coordinates. Linearity of this plot indicates the appropriateness of the target distribution as the true model distribution; see Exercise 3.12.24. McKean and Sievers (1989) discuss how to use these plots adaptively to select appropriate rank scores. In the next example, we use them to examine how well the R fit fits the bulk of the data and to highlight outliers.

For the added variable plot, let $\widehat{\mathbf{e}}_R$ denote the residuals from the R fit of the model $\mathbf{Y} = \alpha \mathbf{1} + \mathbf{X}\boldsymbol{\beta} + \mathbf{e}$. In this case, $\mathbf{Z}$ is a known vector and we wish to decide whether or not to add it to the regression model. For the added variable plot, we regress $\mathbf{Z}$ on $\mathbf{X}$. We will denote the residuals from this fit as $\widehat{\mathbf{e}}(\mathbf{Z} \mid \mathbf{X}) = (\mathbf{I} - \mathbf{H})\mathbf{Z}$. The added variable plot consists of the scatterplot of the residuals $\widehat{\mathbf{e}}_R$ versus $\widehat{\mathbf{e}}(\mathbf{Z} \mid \mathbf{X})$. Under model misspecification $\boldsymbol{\gamma} \neq \mathbf{0}$ from expression (3.9.7), the residuals $\widehat{\mathbf{e}}_R$ are also a function of $(\mathbf{I} - \mathbf{H})\mathbf{Z}$. Hence, the plot can be quite powerful in determining the potential of $\mathbf{Z}$ as a predictor.

Example 3.9.1 *Cloud Data*

The data for this example can be found in Table 3.9.1. They are taken from Draper and Smith (1966, p. 162). The dependent variable is the cloud point of

Table 3.9.1: Cloud data

%I-8	0	1	2	3	4	5	6	7	8	0
Cloud point	22.1	24.5	26.0	26.8	28.2	28.9	30.0	30.4	31.4	21.9
%I-8	2	4	6	8	10	0	3	6	9	
Cloud point	26.1	28.5	30.3	31.5	33.1	22.8	27.3	29.8	31.8	

Table 3.9.2: Wilcoxon and LS estimates of the regression coefficients for the cloud data. Standard errors are in parentheses.

Method	Intercept	Linear	Quadratic	Cubic	Scale
Wilcoxon	22.35 (0.18)	2.24 (0.17)	−0.23 (0.04)	0.01 (0.003)	$\widehat{\tau}_\varphi = 0.307$
Least Squares	22.31 (0.15)	2.22 (0.15)	−0.22 (0.04)	0.01 (0.003)	$\widehat{\sigma} = 0.281$

a liquid, a measure of degree of crystallization in a stock. The independent variable is the percentage of I-8 in the base stock. The subsequent R fits for this data set were all based on Wilcoxon scores with the intercept estimate $\widehat{\alpha}_S$, the median of the residuals.

Figure 3.9.1(a) displays the residual plot (R residuals versus R fitted values) of the R fit of the simple linear model. The curvature in the plot indicates that this model is a poor choice and that a higher-degree polynomial model would be more appropriate. Figure 3.9.1(b) displays the residual plot from the R fit of a quadratic model. Some curvature is still present in the plot. A cubic polynomial was fitted next. Its R residual plot, found in Figure 3.9.1(c), is much more of a random scatter than the first two plots. On the basis of residual plots the cubic polynomial is an adequate model. Least-squares residual plots would also lead to a third-degree polynomial.

In the R residual plot of the cubic model, several points appear to be outlying from the bulk of the data. These points are also apparent in Figure 3.9.1(d) which displays the $q-q$ plot of the R residuals. Based on these plots, the R regression appears to have fit the bulk of the data well. The $q-q$ plot suggests that the underlying error distribution has slightly heavier tails than the normal distribution. A scale would be helpful in interpreting these residual plots as discussed the next section. Table 3.9.2 displays the estimated coefficients along with their standard errors. The Wilcoxon and least-squares fits are practically the same.

Example 3.9.2 *Potency Data, Example 3.3.3 continued*

Recall from Section 3.3 that the data were the result of an experiment concerning the potency of drug compounds manufactured under different levels of four factors and one covariate. Here we want to discuss a residual analysis of the rank-based fits of the two models that were fit in Example 3.3.3.

First consider model (3.3.1) without the quadratic terms, i.e. without the parameters β_{11}, β_{12} and β_{13}. The residuals used are the internal R studentized residuals defined below; see (3.9.31). They provide a convenient scale for

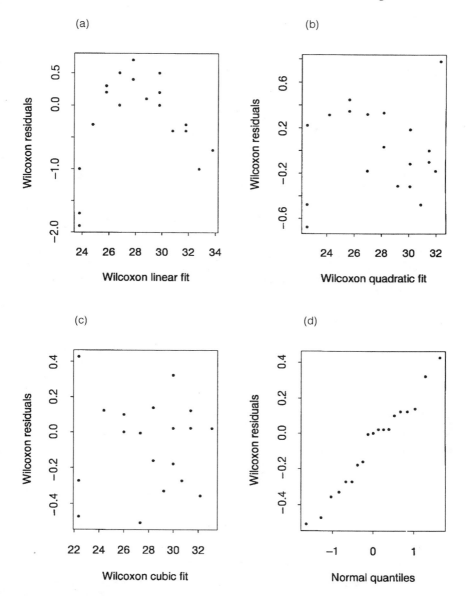

Figure 3.9.1: (a)–(c) Residual plots of the Wilcoxon fits of the linear, quadratic and cubic models, respectively, for the cloud data. (d) The $q-q$ plot based on the Wilcoxon fit of the cubic model

detecting outliers. The curvature in the Wilcoxon residual plot of this model, Figure 3.9.2(a), is quite apparent, indicating the need for quadratic terms in the model; whereas the LS residual plot, Figure 3.9.2(c), does not exhibit this quadratic effect. As the R residual plot indicates, there are outliers in the data

196   *Linear Models*

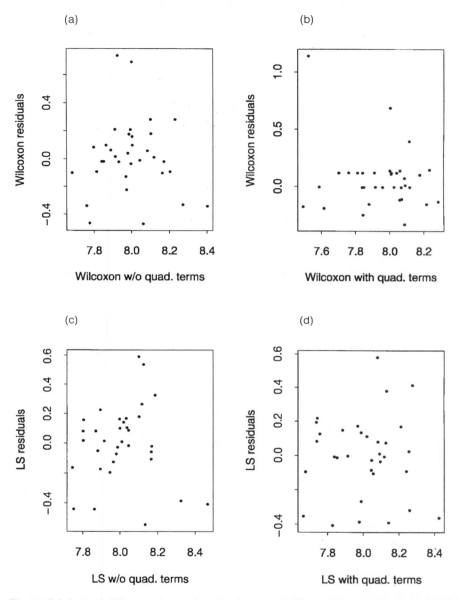

Figure 3.9.2: (a,b) Wilcoxon internal studentized residuals plots for models respectively without and with the three quadratic terms β_{11}, β_{12} and β_{13}. (c,d) The analogous plots for the LS fit.

and these had an effect on the LS fit. In Figure 3.9.2(b,d) are displayed the residual plots, when the squared terms of the factors are added to model, i.e. model (3.3.1) was fit. This R residual plot no longer exhibits the quadratic effect indicating a better-fitting model. Also by examining the R plots for both models, it is seen that the outlyingness of some of the outliers indicated in the plot for the first model was accounted for by the larger model.

3.9.2 Standardization of R Residuals

In this subsection we want to obtain an expression for the variance of the R residuals under model (3.2.3). We will assume here that σ^2, the variance of the errors, is finite. As we show below, similar to the least-squares residual, the variance of an R residual depends both on its location in the x-space and the underlying variation of the errors. The internal studentized least-squares residuals (residuals divided by their estimated standard errors) have proved useful in diagnostic procedures since they correct for both the model and the underlying variance. The internal R studentized residuals defined below, (3.9.31), are similarly studentized R residuals.

A diagnostic use of a studentized residual is in detecting outlying observations. The R method provides a robust fit to the bulk of the data. Thus any case with a large studentized residual can be considered an outlier from this model. Even though a robust fit is resistant to outliers, it is still useful to detect such points. Indeed in practice these are often the points of most interest. The value of an internally studentized residual is in its simplicity. It tells how many estimated standard errors a residual is away from the center of the data.

The standardization depends on which estimate of the intercept is selected. We shall obtain the result for $\widehat{\alpha}_S$ the median of $\widehat{e}_{Ri}$ and only state the results for the intercept based on symmetric errors. Thus the residuals we seek to standardize are given by

$$\widehat{\mathbf{e}}_R = \mathbf{Y} - \widehat{\alpha}_S \mathbf{1} - \mathbf{X}\widehat{\boldsymbol{\beta}}_\varphi. \tag{3.9.12}$$

We will obtain a first-order approximation of $\mathrm{Cov}(\widehat{\mathbf{e}}_R)$. Since the residuals are invariant to the regression coefficients, we can assume without loss of generality that the true parameters are zero. Recall that h_{ci} is the ith diagonal element of $\mathbf{H} = \mathbf{X}(\mathbf{X}'\mathbf{X})^{-1}\mathbf{X}'$ and $h_i = n^{-1} + h_{ci}$.

Theorem 3.9.5 *Under the conditions (E.1), (E.2), (D.1), (D.2) and (S.1) of Section 3.4, if the intercept estimate is $\widehat{\alpha}_s$ then a first-order representation of the variance of $\widehat{e}_{R,i}$ is*

$$\mathrm{Var}(\widehat{e}_{R,i}) \doteq \sigma^2(1 - K_1 n^{-1} - K_2 h_{ci}), \tag{3.9.13}$$

where K_1 and K_2 are defined in expressions (3.9.18) and (3.9.19), respectively. In the case of a symmetric error distribution when the estimate of the intercept is given by $\widehat{\alpha}_\varphi^+$, discussed in Section 3.5.2, and (S.3) also holds,

$$\mathrm{Var}(\widehat{e}_{R,i}) \doteq \sigma^2(1 - K_2 h_i). \tag{3.9.14}$$

Proof. Using the first-order expression for $\widehat{\boldsymbol{\beta}}_\varphi$ given in (3.5.24) and the asymptotic representation of $\widehat{\alpha}_S$ given by (3.5.23), we have

$$\widehat{\mathbf{e}}_R \doteq \mathbf{e} - \tau_S \overline{\mathrm{sgn}(\mathbf{e})}\mathbf{1} - \mathbf{H}\tau_\varphi \boldsymbol{\varphi}(F(\mathbf{e})), \tag{3.9.15}$$

where $\overline{\text{sgn}(\mathbf{e})} = \sum \text{sgn}(e_i)/n$ and τ_S and τ_φ are defined in expressions (3.4.6) and (3.4.4), respectively. Because the median of e_i is 0 and $\int \varphi(u)\,du = 0$, we have

$$E[\widehat{\mathbf{e}}_R] \doteq E(e_1)\mathbf{1}.$$

Hence,

$$\text{Cov}(\widehat{\mathbf{e}}_R) \doteq E[(\mathbf{e} - \tau_S \overline{\text{sgn}(\mathbf{e})}\mathbf{1} - \mathbf{H}\tau_\varphi \varphi(F(\mathbf{e})) - E(e_1)\mathbf{1}) \times$$
$$(\mathbf{e} - \tau_S \overline{\text{sgn}(\mathbf{e})}\mathbf{1} - \mathbf{H}\tau_\varphi \varphi(F(\mathbf{e})) - E(e_1)\mathbf{1})']. \qquad (3.9.16)$$

Let $\mathbf{J} = \mathbf{1}\mathbf{1}'/n$ denote the projection onto the space spanned by $\mathbf{1}$. Since our design matrix is $[\mathbf{1}\ \mathbf{X}]$, the leverage of the ith case is $h_i = n^{-1} + h_{ci}$, where h_{ci} is the ith diagonal entry of the projection matrix $\mathbf{H}$. By expanding the above expression and using the independence of the components of $\mathbf{e}$, we obtain, after some simplification (see Exercise 3.12.25):

$$\text{Cov}(\widehat{\mathbf{e}}_R) \doteq \sigma^2\{\mathbf{I} - K_1\mathbf{J} - K_2\mathbf{H}\}, \qquad (3.9.17)$$

where

$$K_1 = \left(\frac{\tau_S}{\sigma}\right)^2 \left(2\frac{\delta_S}{\tau_S} - 1\right), \qquad (3.9.18)$$

$$K_2 = \left(\frac{\tau_\varphi}{\sigma}\right)^2 \left(2\frac{\delta}{\tau_\varphi} - 1\right), \qquad (3.9.19)$$

$$\delta_S = E[e_i \text{sgn}(e_i)], \qquad (3.9.20)$$

$$\delta = E[e_i \varphi(F(e_i))], \qquad (3.9.21)$$

$$\sigma^2 = \text{Var}(e_i) = E((e_i - E(e_i))^2). \qquad (3.9.22)$$

This yields the first result, (3.9.13). Next consider the case of a symmetric error distribution. If the estimate of the intercept is given by $\widehat{\alpha}_\varphi^+$, discussed in Section 3.5.2, the result simplifies to (3.9.14).

From Cook and Weisberg (1982, p. 11) in the least-squares case, $\text{Var}(\widehat{e}_{LS,i}) = \sigma^2(1 - h_i)$ so that K_1 and K_2 are correction factors due to using the rank-score function.

Based on the results in the theorem, an estimate of the variance-covariance matrix of $\widehat{\mathbf{e}}_R$ is

$$\widetilde{\mathbf{S}} = \widehat{\sigma}^2\{\mathbf{I} - \widehat{K}_1\mathbf{J} - \widehat{K}_2\mathbf{H}_c\}, \qquad (3.9.23)$$

where

$$\widehat{K}_1 = \frac{\widehat{\tau}_S^2}{\widehat{\sigma}^2}\left(\frac{2\widehat{\delta}_S}{\widehat{\tau}_S} - 1\right), \qquad (3.9.24)$$

$$\widehat{K}_2 = \frac{\widehat{\tau}_\varphi^2}{\widehat{\sigma}^2}\left(\frac{2\widehat{\delta}}{\widehat{\tau}_\varphi} - 1\right), \qquad (3.9.25)$$

$$\widehat{\delta}_S = \frac{1}{n-p}\sum |\widehat{e}_{R,i}|, \qquad (3.9.26)$$

and

$$\hat{\delta} = \frac{1}{n-p} D(\hat{\boldsymbol{\beta}}_\varphi).$$

The estimators $\hat{\tau}_S$ and $\hat{\tau}_\varphi$ are discussed in Section 3.7.1.

To complete the estimate of the $\text{Cov}(\widehat{\mathbf{e}}_R)$ we need to estimate σ. A robust estimate of it is given by the MAD,

$$\hat{\sigma} = 1.483 \text{med}_i\{|\hat{e}_{Ri} - \text{med}_j \hat{e}_{Rj}|\}, \tag{3.9.27}$$

which is a consistent estimate of σ if the errors have a normal distribution. For the examples discussed here, we used this estimate in (3.9.23)–(3.9.25).

It follows from (3.9.23) that an estimate of $\text{Var}(\widehat{e}_{R,i})$ is

$$\widetilde{s}_{R,i}^2 = \hat{\sigma}^2 \left(1 - \widehat{K}_1 \frac{1}{n} - \widehat{K}_2 h_{c,i}\right), \tag{3.9.28}$$

where $h_{ci} = \mathbf{x}_i(\mathbf{X}'\mathbf{X})^{-1}\mathbf{x}_i$.

Let $\hat{\sigma}_{LS}^2$ denote the usual least-squares estimate of the variance. Least-squares residuals are standardized by $\widetilde{s}_{LS,i}$, where

$$\widetilde{s}_{LS,i}^2 = \hat{\sigma}_{LS}^2(1 - h_i); \tag{3.9.29}$$

see Cook and Weisberg (1982, p. 11) and recall that $h_i = n^{-1} + \mathbf{x}_i'(\mathbf{X}'\mathbf{X})^{-1}\mathbf{x}_i$.

If the error distribution is symmetric (3.9.28) reduces to

$$\widetilde{s}_{R,i}^2 = \hat{\sigma}^2(1 - \widehat{K}_2 h_i). \tag{3.9.30}$$

Internal R Studentized Residual

We define the **internal R studentized residuals** as

$$r_{R,i} = \frac{\widehat{e}_{R,i}}{\widetilde{s}_{R,i}} \text{ for } i = 1, \ldots, n, \tag{3.9.31}$$

where $\widetilde{s}_{R,i}$ is the square root of either (3.9.28) or (3.9.30) depending on whether one assumes an asymmetric or symmetric error distribution, respectively.

It is interesting to compare expression (3.9.30) with the estimate of the variance of the least-squares residual $\hat{\sigma}_{LS}^2(1 - h_i)$. The correction factor $\widehat{K}_2$ depends on the score function $\varphi(\cdot)$ and the underlying symmetric error distribution. If, for example, the error distribution is normal and if we use normal scores, then $\widehat{K}_2$ converges in probability to 1; see Exercise 3.12.26. In general, however, we will not wish to specify the error distribution and then $\widehat{K}_2$ provides a natural adjustment.

A simple benchmark is useful in declaring whether or not a case is an outlier. We are certainly not advocating eliminating such cases but recommend flagging them as potential outliers and targeting them for further study. As we discussed in the previous section, the distribution of the R residuals should resemble the true distribution of the errors. Hence a simple rule for all cases is not apparent. In general, unless the residuals appear to be from a highly skewed distribution, a simple rule is to declare a case to be a

potential outlier if its residual exceeds two standard errors in absolute value, i.e. $|r_{R,i}| > 2$.

The matrix $\widetilde{S}$, (3.9.23), is an estimate of a first-order approximation of $\text{Cov}(\widehat{e}_R)$. It is not necessarily positive semidefinite and we have not constrained it to be so. In practice this has not proved troublesome since only occasionally have we encountered negative estimates of the variance of the residuals. For instance, the R fit for the cloud data resulted in one case with a negative variance. Presently, we replace (3.9.28) by $\widehat{\sigma}\sqrt{1-h_i}$, where $\widehat{\sigma}$ is the MAD estimate (3.9.27), in these situations.

We have already illustrated the internal R studentized residuals for the potency of Example 3.9.2. We use them next on the cloud data.

Example 3.9.3 *Cloud Data, Example 3.9.1, continued*

Returning to cloud data example, Figure 3.9.3(a) displays a residual plot of the internal Wilcoxon studentized residuals versus the fitted values. It is similar to Figure 3.9.1(c) but it has a meaningful scale on the vertical axis. The residuals for three of the cases (4, 10, and 16) are over two standard errors from the center of the data. These should be flagged as potential outliers. Figure 3.9.3(b) displays the normal $q-q$ plot of the internal Wilcoxon studentized residuals. The underlying error structure appears to have heavier tails than the normal distribution.

As with their least-squares counterparts, we think the chief benefit of the internal R studentized residuals is their usefulness in diagnostic plots and flagging potential outliers.

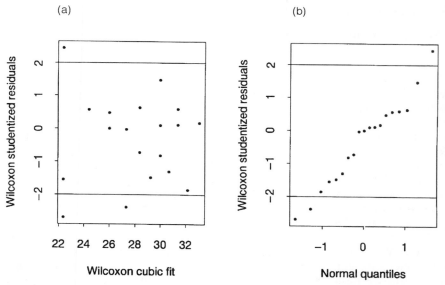

Figure 3.9.3: (a) Internal Wilcoxon studentized residual plot and (b) corresponding normal $q-q$ plot for the cloud data.

External R Studentized Residual

Another statistic that is useful for flagging outliers is a robust version of the external t-statistic. The LS version of this diagnostic is discussed in detail in Cook and Weisberg (1982). A robust version is discussed in McKean, Sheather and Hettmansperger (1991). We briefly describe the latter approach.

Suppose we want to examine the ith case to see if its an outlier. Consider the **mean shift model** given by

$$\mathbf{Y} = \mathbf{X}_1 \mathbf{b} + \theta_i \mathbf{d}_i + \mathbf{e}, \tag{3.9.32}$$

where $\mathbf{X}_1$ is the augmented matrix $[\mathbf{1}\,\mathbf{X}]$ and $\mathbf{d}_i$ is an $n \times 1$ vector of zeros except for its ith component which is a 1. A formal hypothesis that the the ith case is an outlier is given by

$$H_0: \theta_i = 0 \text{ versus } H_A: \theta_i \neq 0. \tag{3.9.33}$$

One way of testing these hypotheses is to use the test procedures described in Section 3.6. This requires fitting model (3.9.32) for each value of i. A second approach is described next.

Note that we can rewrite model (3.9.32) equivalently as

$$\mathbf{Y} = \mathbf{X}_1 \mathbf{b}^* + \theta_i \mathbf{d}_i^* + \mathbf{e}, \tag{3.9.34}$$

where $\mathbf{d}_i^* = (\mathbf{I} - \mathbf{H}_1)\mathbf{d}_i$, $\mathbf{H}_1$ is the projection matrix onto the column space of $\mathbf{X}_1$ and $\mathbf{b}^* = \mathbf{b} + \mathbf{H}_1 \mathbf{d}_i \theta_i$; see Exercise 3.12.27. Because of the orthogonality between $\mathbf{X}$ and $\mathbf{d}_i^*$, the least-squares estimate of θ_i can be obtained by a simple linear regression of $\mathbf{Y}$ on $\mathbf{d}_i^*$ or equivalently of $\widehat{\mathbf{e}}_{LS}$ on $\mathbf{d}_i^*$. For the rank-based estimate, the asymptotic distribution theory of the regression estimates suggests a similar approach. Accordingly, let $\widehat{\theta}_{R,i}$ denote the R-estimate when $\widehat{\mathbf{e}}_R$ is regressed on $\mathbf{d}_i^*$. This is a simple regression and the estimate can be obtained by a linear search algorithm; see Section 3.7.2. As Exercise 3.12.29 shows, this estimate is the inversion of an aligned rank statistic to test the hypotheses (3.9.33). Next let $\widehat{\tau}_{\varphi,i}$ denote the estimate of τ_φ produced from this regression. We define the **external R studentized residual** to be the statistic

$$t_R(i) = \frac{\widehat{\theta}_{R,i}}{\widehat{\tau}_{\varphi,i}/\sqrt{1-h_{1,i}}}, \tag{3.9.35}$$

where $h_{1,i}$ is the ith diagonal entry of $\mathbf{H}_1$. Note that we have standardized $\widehat{\theta}_{R,i}$ by its asymptotic standard error.

A final remark on these external t-statistics is in order. In the mean shift model, (3.9.32), the leverage value of the ith case is 1. Hence, the design assumption (D.2), (3.4.7), is not true. This invalidates both the LS and rank-based asymptotic theory for the external t-statistics. In light of this, we do not propose the statistic $t_R(i)$ as a test statistic for the hypotheses (3.9.33) but as a diagnostic for flagging potential outliers. As a benchmark, we suggest the value 2.

3.9.3 Measures of Influential Cases

Since R-estimates have bounded influence in the y-space but not in the x-space, the R fit may be affected by outlying points in the x-space. We next introduce a statistic which measures the influence of the ith case on the robust fit. We work with the usual model (3.2.3). First, we need the first-order representation of $\widehat{\mathbf{Y}}_R$. Similar to the proof of Theorem 3.9.5 which obtained the first-order representation of the residuals, (3.9.15), we have

$$\widehat{\mathbf{Y}}_R \doteq \alpha \mathbf{1} + \mathbf{X}\boldsymbol{\beta} + \tau_S \overline{\mathrm{sgn}(\mathbf{e})} \mathbf{1} + \mathbf{H}\tau_\varphi \varphi(F(\mathbf{e})); \quad (3.9.36)$$

see Exercise 3.12.28.

Let $\widehat{Y}_R(i)$ denote the R predicted value of Y_i when the ith case is deleted from the model. We shall call this model the **delete i model**. Then the change in the robust fit due to the ith case is

$$RDFFIT_i = \widehat{Y}_{R,i} - \hat{Y}_R(i). \quad (3.9.37)$$

$RDFFIT_i$ is our measure of the influence of case i. Computation of this statistic is discussed later. Clearly, in order to be useful, $RDFFIT_i$ must be assessed relative to some scale.

$RDFFIT$ is a change in the fitted value; hence, a natural scale for assessing $RDFFIT$ is a fitted value scale. Using as our estimate of the intercept $\widehat{\alpha}_S$, it follows from (3.9.36) with $\boldsymbol{\gamma} = \mathbf{0}$ that

$$\mathrm{Var}(\widehat{Y}_{R,i}) \doteq n^{-1}\tau_S^2 + h_{c,i}\tau_\varphi^2. \quad (3.9.38)$$

Hence, based on a fitted scale assessment, we standardize $RDFFIT$ by an estimate of the square root of this quantity.

For least-squares diagnostics there is some discussion on whether to use the original model or the model with the ith point deleted for the estimation of scale. Cook and Weisberg (1982) advocate the original model. In this case the scale estimate is the same for all n cases. This allows casewise comparisons involving the diagnostic. Belsley, Kuh, and Welsch (1980), however, advocate scale estimation based on the delete i model. Note that both standardizations correct for the model and the underlying variation of the errors.

Let $\widehat{\tau}_S(i)$ and $\widehat{\tau}_\varphi(i)$ denote the estimates of τ_S and τ_φ for the delete i model as discussed above. Then our diagnostic in which $RDFFIT_i$ is assessed relative to a fitted value scale with estimates of scale based on the delete i model is given by

$$RDFFITS_i = \frac{RDFFIT_i}{(n^{-1}\widehat{\tau}_S^2(i) + h_{c,i}\widehat{\tau}_\varphi^2(i))^{\frac{1}{2}}}. \quad (3.9.39)$$

This is an R analog of the least squares diagnostic $DFFITS_i$ proposed by Belsley *et al.* (1980). For standardization based on the original model, replace $\widehat{\tau}_S(i)$ and $\widehat{\tau}_\varphi(i)$ by $\widehat{\tau}_S$ and $\widehat{\tau}_\varphi$, respectively. We shall define

$$RDCOOK_i = \frac{RDFFIT_i}{(n^{-1}\widehat{\tau}_S^2 + h_{c,i}\widehat{\tau}_\varphi^2)^{\frac{1}{2}}}. \quad (3.9.40)$$

If $\widehat{\alpha}_R^+$ is used as the estimate of the intercept then, provided the errors have a symmetric distribution, the R diagnostics are obtained by replacing $\text{Var}(\widehat{Y}_{R,i})$ with $\text{Var}(\widehat{Y}_{R,i}) = h_i \widehat{\tau}_\varphi^2$; see Exercise 3.12.30 for details. This results in the diagnostics

$$RDFFITS_{\text{symm},i} = \frac{RDFFIT_i}{\sqrt{h_i}\,\widehat{\tau}_\varphi(i)}, \qquad (3.9.41)$$

and

$$RDCOOK_{\text{symm},i} = \frac{RDFFIT_i}{\sqrt{h_i}\,\widehat{\tau}_\varphi}, \qquad (3.9.42)$$

eliminating the need to estimate τ_S.

There is also disagreement on what benchmarks to use for flagging points of potential influence. As Belsley et al. (1980) discuss in some detail, the LS *DFFITS* is inversely influenced by sample size. They advocate a size-adjusted benchmark of $2\sqrt{p/n}$ for the LS *DFFITS*. Cook and Weisberg (1982) suggest a more conservative value which results in $\sqrt{p}$. We shall use both benchmarks in the examples. We realize these diagnostics only flag potentially influential points that require investigation. Similar to the two references cited above, we would never recommend indiscriminately deleting observations solely because their diagnostic values exceed the benchmark. Rather these are potential points of influence which should be investigated.

The diagnostics described above are formed with the leverage values based on the projection matrix. These leverage values are nonrobust (see Rousseeuw and van Zomeren, 1990). For data sets with clusters of outliers in factor space robust leverage values can be formulated in terms of high-breakdown estimates of the center and scatter matrix in factor space. One such choice would be the minimum volume ellipsoid (MVE), proposed by Rousseeuw and van Zomeren (1990). Other estimates could be based on the robust singular value decomposition discussed by Ammann (1993). See also Simpson, Ruppert, and Carroll (1992). We recommend computing $\widehat{Y}_R(i)$ with a one- or two-step R-estimate based on the residuals from the original model; see Section 3.7.2. Each step involves a single ordering of the residuals which are nearly in order (in fact on the first step they are in order) and a single projection onto the range of X (easily obtained by using the routines in LINPACK as discussed in Section 3.7.2).

The diagnostic $RDFITTS_i$ measures the change in the fitted values when the ith case is deleted. Similarly, we can also measure changes in the estimates of the regression coefficients. For the LS analysis, this is the diagnostic *DBETAS* proposed by Belsley, Kuh and Welsch (1980). The corresponding diagnostics for the rank-based analysis are:

$$RDBETAS_{ij} = \frac{\widehat{\beta}_{\varphi,j} - \widehat{\beta}_{\varphi,j}(i)}{\widehat{\tau}_\varphi(i)\sqrt{(X'X)_{jj}}}, \qquad (3.9.43)$$

where $\widehat{\boldsymbol{\beta}}_\varphi(i)$ denotes the R-estimate of $\boldsymbol{\beta}$ in the delete i model. A similar statistic can be constructed for the intercept parameter. Furthermore, a *DCOOK* version can also be constructed as above. These diagnostics are often

204 *Linear Models*

used when $|RDFFITS_i|$ is large. In such cases, it may be of interest to know which components of the regression coefficients are more influential than other components. The benchmark suggested by Belsley, Kuh and Welsch (1980) is $2/\sqrt{n}$.

Example 3.9.4 *Free Fatty Acid (FFA) Data.*

The data for this example can be found in Morrison (1983, p. 64) and for convenience in Table 3.9.3. The response is the level of free fatty acid of prepubescent boys while the independent variables are age, weight, and skin fold

Table 3.9.3: Free fatty acid (FFA) data

Case	Age (months)	Weight (lb)	Skin fold thickness	Free fatty Acid
1	105	67	0.96	0.759
2	107	70	0.52	0.274
3	100	54	0.62	0.685
4	103	60	0.76	0.526
5	97	61	1.00	0.859
6	101	62	0.74	0.652
7	99	71	0.76	0.349
8	101	48	0.62	1.120
9	107	59	0.56	1.059
10	100	51	0.44	1.035
11	100	80	0.74	0.531
12	101	57	0.58	1.333
13	104	58	1.10	0.674
14	99	58	0.72	0.686
15	101	54	0.72	0.789
16	110	66	0.54	0.641
17	109	59	0.68	0.641
18	109	64	0.44	0.355
19	110	76	0.52	0.256
20	111	50	0.60	0.627
21	112	64	0.70	0.444
22	117	73	0.96	1.016
23	109	68	0.82	0.582
24	112	67	0.52	0.325
25	111	81	1.14	0.368
26	115	74	0.82	0.818
27	115	63	0.56	0.384
28	125	74	0.72	0.509
29	131	70	0.58	0.634
30	121	63	0.90	0.526
31	123	67	0.66	0.337
32	125	82	0.94	0.307
33	122	62	0.62	0.748
34	124	67	0.74	0.401
35	122	60	0.60	0.451
36	129	98	1.86	0.344
37	128	76	0.82	0.545
38	127	63	0.26	0.781
39	140	79	0.74	0.501
40	141	60	0.62	0.524
41	139	81	0.78	0.318

thickness. The sample size is 41. Figure 3.9.4(a) depicts the residual plot based on the least-squares internal t residuals. From this plot there appear to be several outliers. Certainly cases 12, 22, 26, and 9 are outlying and perhaps cases 8, 10, and 38. In fact, the first four of these cases probably control the least-squares fit, obscuring cases 8, 10, and 38.

As our first R fit of this data, we used the Wilcoxon scores with the intercept estimated by the median of the residuals, $\hat{\alpha}_S$. Note that all seven cases stand out in the Wilcoxon residual plot based on the internal R studentized residuals, (3.9.31); see Figure 3.9.4(b). This is further confirmed by the fits displayed in Table 3.9.4, where the LS fit with these seven cases deleted is very similar to the Wilcoxon fit using all the cases. The $q-q$ plot of the internal R studentized residuals, Figure 3.9.4(c), also highlights these outlying cases. Similar to the residual plot, the $q-q$ plot suggests that the underlying error distribution is positively skewed with a light left tail. The estimates of the regression coefficients and their standard errors are displayed in Table 3.9.4. Due to the skewness in the data, it is not surprising that the LS and R-estimates of the intercept are different since the former estimates the mean of the residuals while the later estimates the median of the residuals.

Table 3.9.5 displays the values of the R and LS diagnostics for the cases of interest. For the seven cases cited above, the internal Wilcoxon studentized residuals, (3.9.31), definitely flag three of the cases and for two of the others we have scores exceeding 1.70; see Figure 3.9.4(b). As *RDFFITS*, (3.9.39), indicates none of these seven cases seems to have an effect on the Wilcoxon fit (the liberal benchmark is 0.62), whereas the 12th case appears to have an effect on the least-squares fit. *RDFFITS* exceeded the benchmark only for case 2 for which it had the value -0.64. Case 36 with $h_{36} = 0.53$ has high leverage but it did not have an adverse effect on either the Wilcoxon fit or the LS fit. This is true also of cases 11 and 40 which were the only other cases whose leverage values exceeded the benchmark of $2p/n$.

As we noted above, both the residual and the $q-q$ plots indicate that the distribution of the residuals is positively skewed. This suggests a transformation as discussed below, or perhaps a prudent choice of a score function which would be more appropriate for skewed error distributions than the Wilcoxon scores. The score function $\varphi_{0.5}(u)$, (2.5.34), is more suited to positively skewed errors. Figure 3.9.4(d) displays the internal R studentized residuals based on the R fit using this **bent score function**. From this plot and the tabulated

Table 3.9.4: Estimates of β (first cell entry) and $\hat{\sigma}_\beta$ (second cell entry) for free fatty acid data

	Original Data								log y	
Par.	LS		Wilcoxon		LS (w/o 7 pts.)		R-Bent Score		LS		Wilcoxon
β_0	1.70	0.33	1.49	0.27	1.24	0.21	1.37	0.21	1.12	0.52	0.99	0.54
β_1	-0.002	0.003	-0.001	0.003	-0.001	0.002	-0.001	0.002	-0.001	0.005	0.000	0.005
β_2	-0.015	0.005	-0.015	0.004	-0.013	0.003	-0.015	0.003	-0.029	0.008	-0.031	0.008
β_3	0.205	0.167	0.274	0.137	0.285	0.103	0.355	0.104	0.444	0.263	0.555	0.271
Scale	0.215		0.178		0.126		0.134		0.341		0.350	

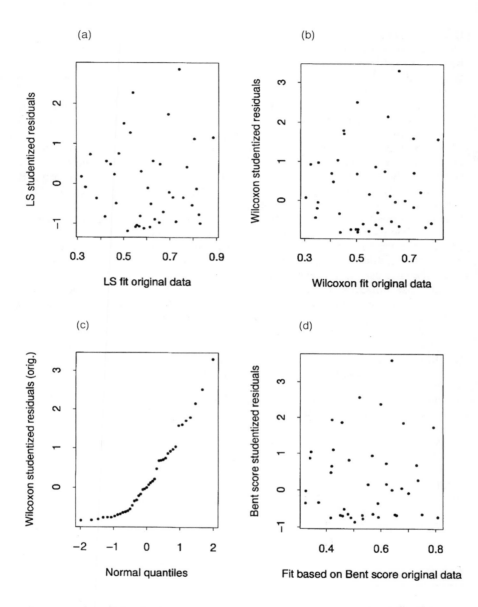

Figure 3.9.4: (a) Internal LS studentized residual plot on the original free fatty acid data; (b) internal Wilcoxon studentized residual plot on the original free fatty acid data; (c) internal Wilcoxon studentized normal $q-q$ plot on the original free fatty acid data; (d) internal R studentized residual plot on the original free fatty acid data based on the score function $\varphi_{0.5}(u)$

Table 3.9.5: Regression diagnostics for cases of interest for the fatty acid data

Case	h_i	LS Int. t	LS DFFIT	Wilcoxon Int. t	Wilcoxon DFFIT	Bent Score Int. t	Bent Score DFFIT
8	0.12	1.16	0.43	1.57	0.44	1.73	0.31
9	0.04	1.74	0.38	2.14	0.13	2.37	0.26
10	0.09	1.12	0.36	1.59	0.53	1.84	0.30
12	0.06	2.84	0.79	3.30	0.33	3.59	0.30
22	0.05	2.26	0.53	2.51	−0.06	2.55	0.11
26	0.04	1.51	0.32	1.79	0.20	1.86	0.10
38	0.15	1.27	0.54	1.70	0.53	1.93	0.19
2	0.10	−1.19	−0.40	−0.17	−0.64	−0.75	−0.48
7	0.11	−1.07	−0.37	−0.75	−0.44	−0.74	−0.64
11	0.22	0.56	0.30	0.97	0.31	1.03	0.07
40	0.25	−0.51	−0.29	−0.31	−0.21	−0.35	0.06
36	0.53	0.18	0.19	−0.04	−0.27	−0.66	−0.34

diagnostics, the outliers stand out more from this fit than the previous two fits. The *RDFFITS* values for this fit are even smaller than those of the Wilcoxon fit, which is expected since this score function protects on the right. While case 7 has a little influence on the bent score fit, no other cases have *RDFFITS* exceeding the benchmark.

Table 3.9.4 displays the estimates of the betas for the three fits along with their standard errors. At the 0.05 level, coefficients 2 and 3 are significant for the robust fits while only coefficient 2 is significant for the LS fit. The robust fits appear to be an improvement over LS. Of the two robust fits, the bent score fit appears to be more precise than the Wilcoxon fit.

A practical transformation on the response variable suggested by the Box–Cox transformation is the log. Figure 3.9.5(a) shows the internal R-studentized residuals plot based on the Wilcoxon fit of the log-transformed response. Note that five of the cases still stand out in the plot. The residuals from the transformed response still appear to be skewed as is evident in the $q-q$ plot, Figure 3.9.5(b). From Table 3.9.4, the Wilcoxon fit seems slightly more precise in terms of standard errors.

3.10 Survival Analysis

In this section we discuss scores which are appropriate for lifetime distributions when the log of lifetime follows a linear model. These are called **accelerated failure time** models; see Kalbfleisch and Prentice (1980). Let T denote the lifetime of a subject and let $\mathbf{x}$ be a $p \times 1$ vector of covariates associated with T. Let $h(t; \mathbf{x})$ denote the hazard function of T at time t; see Section 2.8. Suppose T follows a log-linear model; that is, $Y = \log T$ follows the linear model

$$Y = \alpha + \mathbf{x}'\boldsymbol{\beta} + e, \qquad (3.10.1)$$

where e is a random error with density f. Exponentiating both sides, we get $T = \exp\{\alpha + \mathbf{x}'\boldsymbol{\beta}\}T_0$, where $T_0 = \exp\{e\}$. Let $h_0(t)$ denote the hazard function of

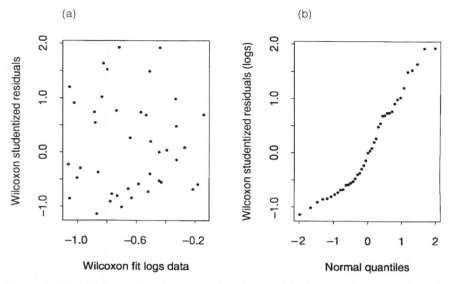

Figure 3.9.5: (a) internal R studentized residuals plot of the log-transfomed free fatty acid data; (b) corresponding normal $q-q$ plot

T_0. This is called the baseline hazard function. Then the hazard function of T is given by

$$h(t; \mathbf{x}) = h_0(t \exp\{-(\alpha + \mathbf{x}'\boldsymbol{\beta})\}) \exp\{-(\alpha + \mathbf{x}'\boldsymbol{\beta})\}. \tag{3.10.2}$$

Thus the covariate $\mathbf{x}$ accelerates or decelerates the failure time of T; hence the name accelerated failure time for these models.

An important subclass of the accelerated failure time models are those where T_0 follows a Weibull distribution, i.e.

$$f_{T_0}(t) = \lambda\gamma(\lambda t)^{\gamma-1} \exp\{-(\lambda t)^\gamma\}, \quad t > 0, \tag{3.10.3}$$

where λ and γ are unknown parameters. In this case it follows that the hazard function of T is proportional to the baseline hazard function with the covariate acting as the factor of proportionality; i.e.

$$h(t; \mathbf{x}) = h_0(t) \exp\{-(\alpha + \mathbf{x}'\boldsymbol{\beta})\}. \tag{3.10.4}$$

Hence these models are called **proportional hazards models**. Kalbfleisch and Prentice (1980) show that the only proportional hazards models which are also accelerated failure time models are those for which T_0 has the Weibull density. We can write the random error $e = \log T_0$ as $e = \xi + \gamma^{-1} W_0$, where $\xi = -\log \gamma$ and W_0 has the extreme value distribution discussed in Section 2.8. Thus the optimal rank scores for these log-linear models are generated by the function

$$\varphi_{f_\varepsilon}(u) = -1 - \log(1 - u); \tag{3.10.5}$$

see (2.8.8).

Next we consider suitable score functions for the general failure time models, (3.10.1). As noted in Kalbfleisch and Prentice (1980), many of the error distributions currently used for these models are contained in the log-F class. In this class, $e = \log T$ is distributed down to an unknown scale parameter, as the log of an F random variable with $2m_1$ and $2m_2$ degrees of freedom. In this case we shall say that e has a $GF(2m_1, 2m_2)$ distribution. The distribution of T is Weibull if $(m_1, m_2) \to (1, \infty)$, lognormal if $(m_1, m_2) \to (\infty, \infty)$, and generalized gamma if $(m_1, m_2) \to (\infty, 1)$; see Kalbfleisch and Prentice (1980). If $(m_1, m_2) = (1, 1)$ then the e has a logistic distribution. In general this class contains a variety of shapes. The distributions are symmetric for $m_1 = m_2$, positively skewed for $m_1 > m_2$, and negatively skewed for $m_1 < m_2$. While Kalbfleisch and Prentice discuss this class for $m_1, m_2 \geq 1$, we will extend the class to $m_1, m_2 > 0$ in order to include heavier-tailed error distributions.

For random errors with distribution $GF(2m_1, 2m_2)$, the optimal rank score function is given by

$$\varphi_{m_1,m_2}(u) = (m_1 m_2 (\exp\{F^{-1}(u)\} - 1))/(m_2 + m_1 \exp\{F^{-1}(u)\}), \qquad (3.10.6)$$

where F is the cdf of the $GF(2m_1, 2m_2)$ distribution; see Exercise 3.12.31. We shall label these scores as $GF(2m_1, 2m_2)$ scores. It follows that the scores are strictly increasing and bounded below by $-m_1$ and above by m_2. Hence, an R analysis based on these scores will have bounded influence in the Y-space.

This class of scores can be conveniently divided into the four subclasses C_1 through C_4 which are represented by the four quadrants with center (1, 1) as depicted in Figure 3.10.1. The point (1, 1) in this figure corresponds to the linear rank, Wilcoxon scores. These scores are optimal for the logistic distribution, $GF(2, 2)$, and form a 'natural' center point for the scores. One score function from each class with the density for which it is optimal is plotted in Figure 3.10.2. These plots are generally representative. The score functions in C_2 change from concave to convex as u increases and, hence, are suitable for light-tailed error structure, while those in C_4 pass from convex to concave and are suitable for heavy-tailed error structure. The score functions in C_3 are always convex and are suitable for negatively skewed error structure with heavy left tails and moderate right tails, while those in C_1 are suitable for positively skewed errors with heavy right tails and moderate left tails.

Figure 3.10.2 shows how a score function corresponds to its density. If the density has a heavy right tail then the score function will tend to be flat on the right; hence, the resulting estimate will be less sensitive to outliers on the right. On the other hand, if the density has a light right tail then the scores will tend to rise on the right in order to accentuate points on the right. The plots in Figure 3.10.2 suggest approximating these scores by scores consisting of two or three line segments such as the bent score function, (2.5.34).

Generally the $GF(2m_1, 2m_2)$ scores cannot be obtained in closed form due to F^{-1}, but programs such as Minitab and S-Plus can easily produce them. There are two interesting subclasses for which closed forms are possible. These are the subclasses $GF(2, 2m_2)$ and $GF(2m_1, 2)$. As Exercise 3.12.32

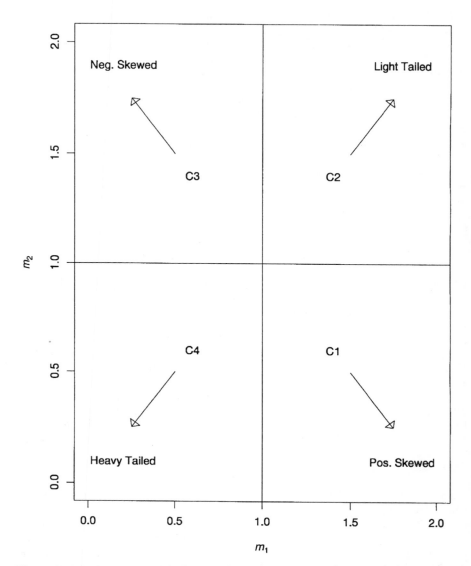

Figure 3.10.1: Schematic of the four classes, C1–C4, of the $GF(2m_1, 2m_2)$ scores

shows, the random variables for these classes are the logs of variates having Pareto distributions. For the subclass $GF(2, 2m_2)$ the score generating function is

$$\varphi_{m_2}(u) = \left(\frac{m_2 + 2}{m_2}\right)^{1/2} \left(m_2 - (m_2 + 1)(1 - u)^{1/m_2}\right). \qquad (3.10.7)$$

These are the powers of rank scores discussed by Mielke (1972) in the context of two-sample problems.

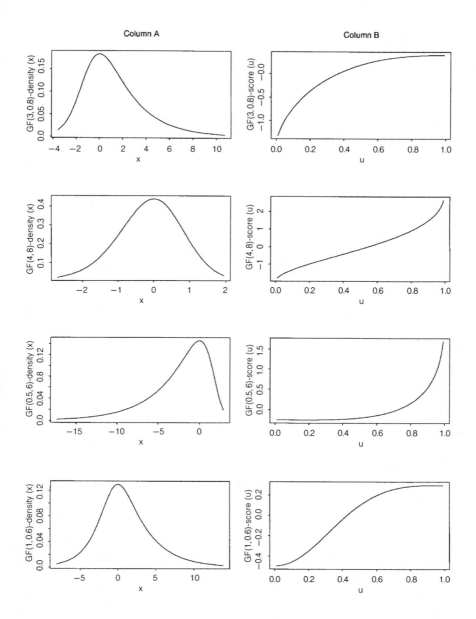

Figure 3.10.2: Column A contains plots of the densities: the Class C1 distribution $GF(3, 0.8)$; the Class C2 distribution $GF(4, 8)$; the Class C3 distribution $GF(0.5, 6)$; and the Class C4 distribution $GF(1, 0.6)$. Column B contains the corresponding optimal score functions.

212   Linear Models

It is interesting to note that the asymptotic efficiency of the Wilcoxon relative to the optimal rank-score function at the $GF(2m_1, 2m_2)$ distribution is given by

$$ARE = \frac{12\ \Gamma^4(m_1 + m_2)\Gamma^2(2m_1)\Gamma^2(2m_2)(m_1 + m_2 + 1)}{\Gamma^4(m_1)\Gamma^4(m_2)\Gamma^2(2m_1 + 2m_2)m_1 m_2}; \qquad (3.10.8)$$

see Exercise 3.12.31. This efficiency can be arbitrarily small. For instance, in the subclass $GF(2, 2m_2)$ the efficiency reduces to

$$ARE = \frac{3m_2(m_2 + 2)}{(2m_2 + 1)^2}, \qquad (3.10.9)$$

which approaches 0 as $m_2 \to 0$ and $\frac{3}{4}$ as $m_2 \to \infty$. Thus in the presence of severely skewed errors, the Wilcoxon scores can have arbitrarily low efficiency compared to a fully efficient R-estimate based on the optimal scores.

For a given problem, the choice of scores presents a problem. McKean and Sievers (1989) discuss several methods for score selection, one of which is illustrated in the next example. This method is adaptive in nature, with the adaption depending on residuals from an initial fit. In practice, this can lead to overfitting. Its use, however, can lead to insight and may prove beneficial for fitting future data sets of the same type; see McKean et al. (1989) for such an application. Using XLISP-STAT (Tierney, 1990), Wang (1996) presents a graphical interface for methods of score selection.

Example 3.10.1 *Insulating Fluid Data*

We consider a problem discussed by Nelson (1982, p. 227) and Lawless (1982, p. 185). The data consist of breakdown times T of an electrical insulating fluid subject to seven different levels of voltage stress v. Figure 3.10.3(a) displays a scatterplot of $Y = \log T$ versus $\log v$.

As a full model we consider a one-way layout, as discussed in Chapter 4, with the response variable $Y = \log T$ and with the seven voltage levels as treatments. The comparison boxplots, Figure 3.10.3(b), are an appropriate display for this model. The one method for score selection that we briefly touch on here is based on $q-q$ plots; see McKean and Sievers (1989). Using Wilcoxon scores we obtained an initial fit of the one-way layout model as discussed in Chapter 4. Figure 3.10.3(c) displays the $q-q$ plot of the ordered residuals versus the logistic quantiles based on this fit. Although the left tail of the logistic distribution appears adequate, the right-hand side of the plot indicates that distributions with lighter right tails might be more appropriate. This is confirmed by the near-linearity of the $GF(2, 10)$ quantiles versus the Wilcoxon residuals. After trying several R fits using $GF(2m_1, 2m_2)$ scores with $m_1, m_2 \geq 1$, we decided that the $q-q$ plot of the $GF(2, 10)$ fit, Figure 3.10.3(d), appeared to be most linear and we used it to conduct the following R analysis.

For the fit of the full model using the scores $GF(2, 10)$, the minimum value of the dispersion function, D, is 103.298 and the estimate of τ_φ is 1.38. Note that this minimum value of D is the analog of the 'pure' sum of squared errors in a least-squares analysis; hence, we will use the notation $DPE = 103.298$ for **pure error dispersion**. We first test the goodness of fit of a simple linear model.

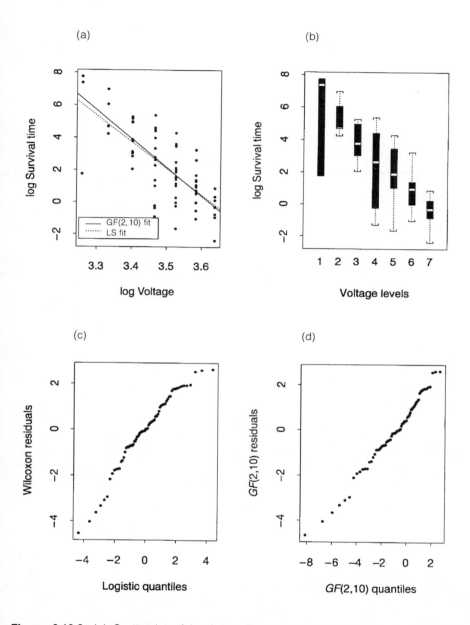

Figure 3.10.3: (a) Scatterplot of insulating fluid data, Example 3.10.1, overlaid with GF(2, 10) and LS fits; (b) comparison boxplots of log breakdown times over levels of voltage stress; (c) $q-q$ plot of Wilcoxon fit versus logistic population quantiles of full model (one-way layout); (d) $q-q$ plot of GF(2, 10) fit versus GF(2, 10) population quantiles of full model (one-way layout)

214 Linear Models

Table 3.10.1: Sensitivity analysis for insulating data.

Estimate	Original (0.05) $\hat{\alpha}$	$\hat{\beta}$	7.75 $\hat{\alpha}$	$\hat{\beta}$	Value of Y_5 10.05 $\hat{\alpha}$	$\hat{\beta}$	16.05 $\hat{\alpha}$	$\hat{\beta}$	30.05 $\hat{\alpha}$	$\hat{\beta}$
LS	59.4	−16.4	60.8	−16.8	62.7	−17.3	67.6	−18.7	79.1	−21.9
Wilcoxon	62.7	−17.2	63.1	−17.4	63.0	−17.4	63.1	−17.4	63.1	−17.4
GF(2,10)	64.0	−17.7	65.5	−18.1	67.0	−18.5	67.1	−18.5	67.1	−18.5
REXT	64.1	−17.7	65.5	−18.1	68.3	−18.9	68.3	−18.9	68.3	−18.9
EXT	64.8	−17.7	68.4	−18.7	79.3	−21.8	114.6	−31.8	191.7	−53.5

The reduced model in this case is a simple linear model. The alternative hypothesis is that the model is not linear but, other than this, it is not specified; hence, the full model is the one-way layout. Thus the hypotheses are

$$H_0 : Y = \alpha + \beta \log v + e \text{ versus } H_A : \text{the model is not linear.} \quad (3.10.10)$$

To test H_0, we fit the reduced model $Y = \alpha + \beta \log v + e$. The dispersion at the reduced model is 104.399. Since, as noted above, the dispersion at the full model is 103.298, the **lack of fit** is the reduction in dispersion $RDLOF = 104.399 - 103.298 = 1.101$. Therefore the value of the robust test statistic is $F_\varphi = 0.319$. There is no evidence on the basis of this test to contest a linear model.

The $GF(2, 10)$ fit of the simple linear model is $\hat{Y} = 64 - 17.67 \log v$, which is graphed in Figure 3.10.3(a). Under this linear model, the estimate of the scale parameter τ_φ is 1.57. From this we compute a 95% confidence interval for the slope parameter β to be -17.67 ± 3.67; hence, it appears that the slope parameter differs significantly from 0. In Lawless (1982) there was interest in computing a confidence interval for $E(Y|x = \log 20)$. The robust estimate of this conditional mean is $\hat{Y} = 11.07$ and a confidence interval is 11.07 ± 1.9. Similar to the other robust confidence intervals, this interval is the same as in the least-squares analysis, except that $\hat{\tau}_\varphi$ replaces $\hat{\sigma}$. A fuller discussion of the R analysis of this data set can be found in McKean and Sievers (1989).

Example 3.10.2 *Sensitivity Analysis for Insulating Fluid Data.*

As noted by Lawless (1982), engineers may suggest a Weibull distribution for breakdown times in this problem. As discussed earlier, this means the errors have an extreme value distribution. This distribution is essentially the limit of a $GF(2, 2m)$ distribution as $m \to \infty$. For completeness we obtained, using the International Mathematical and Statistical Libraries (1987) subroutine UMIAH, estimates based on an extreme value likelihood function. These estimates are labeled EXT. R-estimates based on the the optimum R score function (2.8.8) for the extreme value distribution are labeled as $REXT$. The influence functions for EXT and $REXT$ estimates are unbounded in Y-space and, hence, neither estimate is robust; see (3.5.17).

In order to illustrate this lack of robustness, we conducted a small sensitivity analysis. We replaced the fifth point, which had the value 6.05 (log units), in the data with an outlying observation. Table 3.10.1 summarizes the results for several different choices of the outlier. Note that even for the first case when the changed point is 7.75, which is the maximum of the original data, there is a substantial change in the EXT-estimates. The EXT fit is a disaster when the point is changed to 10.05, whereas the R estimates exhibit robustness. This is even more so for succeeding cases. Although the $REXT$ estimates have an unbounded influence function, they behaved well in this sensitivity analysis.

3.11 Correlation Model

In this section, we are concerned with the **correlation model** defined by

$$Y = \alpha + \mathbf{x}'\boldsymbol{\beta} + e, \quad (3.11.1)$$

where $\mathbf{x}$ is a p-dimensional random vector with distribution function M and density function m, e is a random variable with distribution function F and density f, and $\mathbf{x}$ and e are independent. Let H and h denote the joint distribution function and joint density function of Y and $\mathbf{x}$. It follows that

$$h(\mathbf{x}, y) = f(y - \alpha - \mathbf{x}'\boldsymbol{\beta})m(\mathbf{x}) . \quad (3.11.2)$$

Denote the marginal distribution and density functions of Y by G and g.

The hypotheses of interest are:

H_0 : Y and $\mathbf{x}$ are independent versus H_A : Y and $\mathbf{x}$ are dependent. (3.11.3)

By (3.11.2) this is equivalent to the hypotheses $H_0 : \boldsymbol{\beta} = \mathbf{0}$ versus $H_A : \boldsymbol{\beta} \neq \mathbf{0}$. For this section, we will use the additional assumptions:

$$(\text{E.2}) \quad \text{Var}(e) = \sigma_e^2 < \infty \quad (3.11.4)$$

$$(\text{M.1}) \quad E[\mathbf{x}\mathbf{x}'] = \Sigma, \; \Sigma > 0. \quad (3.11.5)$$

Without loss of generality assume that $E[\mathbf{x}] = \mathbf{0}$ and $E(e) = 0$.

Let $(\mathbf{x}_1, Y_1), \ldots, (\mathbf{x}_n, Y_n)$ be a random sample from the above model. Define the $n \times p$ matrix $\mathbf{X}_1$ to be the matrix whose ith row is the vector $\mathbf{x}_i$ and let $\mathbf{X}$ be the corresponding centered matrix, i.e. $\mathbf{X} = (\mathbf{I} - n^{-1}\mathbf{1}\mathbf{1}')\mathbf{X}_1$. Thus, the notation here agrees with that found in the previous sections.

We intend to describe briefly the rank-based analysis for this model. As we will show using conditional arguments, the asymptotic inference we developed for the fixed $\mathbf{x}$ case will hold for the stochastic case also. We then want to explore measures of association between $\mathbf{x}$ and Y. These will be analogs of the classical **coefficient of multiple determination** (CMD), $\overline{R}^2$. As with $\overline{R}^2$, these robust CMDs will be 0 when $\mathbf{x}$ and Y are independent and positive when they are dependent. Besides defining these measures, we will obtain consistent estimates of them. First we show that, conditionally, the assumptions of Section 3.4 hold. Much of the discussion in this section is taken from Witt, Naranjo, and McKean (1995).

3.11.1 Huber's Condition for the Correlation Model

The key assumption on the design matrix for the nonstochastic $\mathbf{x}$ linear model was Huber's condition, (D.2), (3.4.7). As we next show, it holds almost

216   Linear Models

surely (a.s.) for the correlation model. This will allow us to easily obtain inference methods for the correlation model as discussed below.

First define the **modulus of a matrix A** to be

$$m(\mathbf{A}) = \max_{i,j} |a_{ij}|. \tag{3.11.6}$$

As Exercise 3.12.33 shows, the following three facts follow from this definition: $m(\mathbf{AB}) \leq p\, m(\mathbf{A}) m(\mathbf{B})$, where p is the common dimension of $\mathbf{A}$ and $\mathbf{B}$; $m(\mathbf{AA'}) \geq m(\mathbf{A})^2$; and $m(\mathbf{A}) = \max a_{ii}$ if $\mathbf{A}$ is positive semidefinite. We next need a preliminary lemma found in Arnold (1980).

Lemma 3.11.1 *Let $\{a_n\}$ be a sequence of nonnegative real numbers. If $n^{-1} \sum_{i=1}^{n} a_i \to a_0$ then $n^{-1} \sup_{1 \leq i \leq n} a_i \to 0$.*

Proof. We have

$$\frac{a_n}{n} = \frac{1}{n} \sum_{i=1}^{n} a_i - \frac{n-1}{n} \frac{1}{n-1} \sum_{i=1}^{n-1} a_i \to 0. \tag{3.11.7}$$

Now suppose that $n^{-1} \sup_{1 \leq i \leq n} a_n \not\to 0$. Then for some $\varepsilon > 0$ and for all integers N there exists an n_N such that $n_N \geq N$ and $n_N^{-1} \sup_{1 \leq i \leq n_N} a_i \geq \varepsilon$. Thus we can find a subsequence of integers $\{n_j\}$ such that $n_j \to \infty$ and $n_j^{-1} \sup_{1 \leq i \leq n_j} a_i \geq \varepsilon$. Let $a_{i_{n_j}} = \sup_{1 \leq i \leq n_j} a_i$. Then

$$\varepsilon \leq \frac{a_{i_{n_j}}}{n_j} \leq \frac{a_{i_{n_j}}}{i_{n_j}}. \tag{3.11.8}$$

Also, since $n_j \to \infty$ and $\varepsilon > 0$, $i_{n_j} \to \infty$; hence, expression (3.11.8) leads to a contradiction of expression (3.11.7).

The following theorem is due to Arnold (1980).

Theorem 3.11.2 *Under (3.11.5),*

$$\lim_{n \to \infty} \max \operatorname{diag}\{\mathbf{X}(\mathbf{X'X})^{-1}\mathbf{X'}\} = 0, \text{ a.s.}; \tag{3.11.9}$$

Proof. Using the facts cited above on the modulus of a matrix, we have

$$m\left(\mathbf{X}(\mathbf{X'X})^{-1}\mathbf{X'}\right) \leq p^2 n^{-1} m(\mathbf{XX'}) m\left(\left(\frac{1}{n}\mathbf{X'X}\right)^{-1}\right). \tag{3.11.10}$$

Using the assumptions on the correlation model, the law of large numbers yields $(n^{-1}\mathbf{X'X}) \to \Sigma$ a.s. Hence we need only show that $n^{-1} m(\mathbf{XX'}) \to 0$ a.s. Let U_i denote the ith diagonal element of $\mathbf{XX'}$. We then have,

$$\frac{1}{n} \sum_{i=1}^{n} U_i = \frac{1}{n} \operatorname{tr} \mathbf{X'X} \overset{\text{a.s.}}{\to} \operatorname{tr} \Sigma.$$

By Lemma 3.11.1 we have $n^{-1} \sup_{i \leq n} U_i \overset{\text{a.s.}}{\to} 0$. Since $\mathbf{XX'}$ is positive semidefinite, the desired conclusion is obtained from the facts which followed expression (3.11.6).

Thus, given **X**, we have the same assumptions on the design matrix as we did in the previous sections. By conditioning on **X**, the theory derived in Section 3.5 also holds for the correlation model. Such a conditional argument is demonstrated in Theorem 3.11.3 below. For later discussion we summarize the rank-based inference for the correlation model. Given a specified score function φ, let $\widehat{\boldsymbol{\beta}}_\varphi$ denote the R-estimate of $\boldsymbol{\beta}$ defined in Section 3.2. Under the correlation model (3.11.1) and the assumptions (3.11.4), (S.1), (3.4.10), and (3.11.5) $\sqrt{n}(\widehat{\boldsymbol{\beta}}_\varphi - \boldsymbol{\beta}) \xrightarrow{\mathcal{D}} N_p(\mathbf{0}, \tau_\varphi^2 \Sigma^{-1})$. Also the estimates of τ_φ discussed in Section 3.7.1 will be consistent estimates of τ_φ under the correlation model. Let $\widehat{\tau}_\varphi$ denote such an estimate. In terms of testing, consider the R test statistic, $F_\varphi = (RD/p)/(\widehat{\tau}_\varphi/2)$, of the above hypothesis H_0 of independence. Employing the usual conditional argument, it follows that $pF_\varphi \xrightarrow{\mathcal{D}} \chi^2(p, \delta_R)$, a.e. M under $H_n: \boldsymbol{\beta} = \boldsymbol{\theta}/\sqrt{n}$ where the noncentrality parameter δ_R is given by $\delta = \boldsymbol{\theta}'\Sigma\boldsymbol{\theta}/\tau_\varphi^2$.

Likewise for the LS estimate $\widehat{\boldsymbol{\beta}}_{LS}$ of $\boldsymbol{\beta}$. Using the conditional argument, (see Arnold (1980) for details), $\sqrt{n}(\widehat{\boldsymbol{\beta}}_{LS} - \boldsymbol{\beta}) \xrightarrow{\mathcal{D}} N_p(\mathbf{0}, \sigma^2 \Sigma^{-1})$ and under H_n, $pF_{LS} \xrightarrow{\mathcal{D}} \chi^2(p, \delta_{LS})$ with noncentrality parameter $\delta_{LS} = \boldsymbol{\theta}'\Sigma\boldsymbol{\theta}/\sigma^2$. Thus the ARE of the R test F_φ to the least squares test F_{LS} is the ratio of noncentrality parameters, σ^2/τ_φ^2. This is the usual ARE of rank tests to tests based on least squares in simple location models. Hence the test statistic F_φ has efficiency robustness. The theory of rank-based tests in Section 3.6 applies to the correlation model.

We return to measures of association and their estimates. For motivation, we consider the least-squares measure first.

3.11.2 Traditional Measure of Association and its Estimate

The traditional population coefficient of multiple determination is defined by

$$\overline{R}^2 = \frac{\boldsymbol{\beta}'\Sigma\boldsymbol{\beta}}{\sigma_e^2 + \boldsymbol{\beta}'\Sigma\boldsymbol{\beta}}, \qquad (3.11.11)$$

see Arnold (1981). Note that $\overline{R}^2$ is a measure of association between Y and **x**. It lies between 0 and 1, and it is 0 if and only if Y and **x** are independent (because Y and **x** are independent if and only if $\boldsymbol{\beta} = \mathbf{0}$).

In order to obtain a consistent estimate of $\overline{R}^2$, treat $\mathbf{x}_i$ as nonstochastic and fit by least squares the model $Y_i = \alpha + \mathbf{x}_i'\boldsymbol{\beta} + e_i$, which will be called the full model. The residual amount of variation is $SSE = \sum_{i=1}^n (Y_i - \widehat{\alpha}_{LS} - \mathbf{x}_i'\widehat{\boldsymbol{\beta}}_{LS})^2$, where $\widehat{\boldsymbol{\beta}}_{LS}$ and $\widehat{\alpha}_{LS}$ are the least-squares estimates. Next fit the reduced model defined as the full model subject to $H_0: \boldsymbol{\beta} = \mathbf{0}$. The total amount of variation is $SST = \sum_{i=1}^n (Y_i - \overline{Y})^2$. The reduction in variation in fitting the full model over the reduced model is $SSR = SST - SSE$. An estimate of $\overline{R}^2$ is the proportion of explained variation given by

$$R^2 = \frac{SSR}{SST}. \qquad (3.11.12)$$

The least-squares test statistic for H_0 versus H_A is $F_{LS} = (SSR/p)/\hat{\sigma}_{LS}^2$ where $\hat{\sigma}_{LS}^2 = SSE/(n-p-1)$. Recall that R^2 can be expressed as

$$R^2 = \frac{SSR}{SSR + (n-p-1)\hat{\sigma}_{LS}^2} = \frac{\frac{p}{n-p-1}F_{LS}}{1 + \frac{p}{n-p-1}F_{LS}} . \quad (3.11.13)$$

Now consider the general correlation model. As shown in Arnold (1980), under (3.11.4) and (3.11.5), R^2 is a consistent estimate of $\overline{R}^2$. Under the multivariate normal model R^2 is the maximum likelihood estimate of $\overline{R}^2$.

3.11.3 Robust Measure of Association and its Estimate

The rank-based analog to the reduction in residual variation is the reduction in residual dispersion, which is given by $RD = D(0) - D(\hat{\boldsymbol{\beta}}_R)$. Hence, the proportion of dispersion explained by fitting $\boldsymbol{\beta}$ is

$$R_1 = RD/D(0) . \quad (3.11.14)$$

This is a natural CMD for any robust estimate and, as we shall show below, the population CMD for which R_1 is a consistent estimate does satisfy interesting properties. As expression (A.5.11) of the Appendix shows, however, the influence function of the denominator is not bounded in the Y-space. Hence, the statistic R_1 is not robust.

In order to obtain a CMD which is robust, consider the test statistic of H_0, $F_\varphi = (RD/p)/(\hat{\tau}_\varphi/2)$, (3.6.12). As we indicated above, the test statistic F_φ has efficiency robustness. Furthermore, as shown in the Appendix, the influence function of F_φ is bounded in the Y-space. Hence, the test statistic is robust.

Consider the relationship between the classical F-test and R^2 given by (3.11.13). In the same way but using the robust test F_φ, we can define a second **R-coefficient of multiple determination**

$$R_2 = \frac{\frac{p}{n-p-1}F_R}{1 + \frac{p}{n-p-1}F_R}$$

$$= \frac{RD}{RD + (n-p-1)(\hat{\tau}_\varphi/2)} . \quad (3.11.15)$$

It follows from the above discussion on the R test statistic that the influence function of R_2 has bounded influence in the Y-space.

The parameters that respectively correspond to the statistics $D(0)$ and $D(\hat{\boldsymbol{\beta}}_R)$ are $\overline{D}_y = \int \varphi(G(y))y dG(y)$ and $\overline{D}_e = \int \varphi(F(e))e dF(e)$; see the discussion in Section 3.6.3. The population CMDs associated with R_1 and R_2 are:

$$\overline{R}_1 = \overline{RD}/\overline{D}_y \quad (3.11.16)$$

$$\overline{R}_2 = \overline{RD}/(\overline{RD} + (\tau_\varphi/2)), \quad (3.11.17)$$

where $\overline{RD} = \overline{D}_y - \overline{D}_e$. The properties of these parameters are discussed in the next subsection. The consistency of R_1 and R_2 is given in the following theorem:

Theorem 3.11.3 *Under the correlation model (3.11.1) and the assumptions (E.1), (2.4.16), (S.1), (3.4.10), (S.2), (3.4.11), and (3.11.5),*

$$R_i \xrightarrow{P} \overline{R}_i \text{ a.e. } M, \ i = 1, 2.$$

Proof. Note that we can write

$$D(0) = \sum_{i=1}^{n} \varphi\left(\frac{n}{n+1}\widehat{F}_n(Y_i)\right) Y_i \frac{1}{n}$$

$$= \int \varphi\left(\frac{n}{n+1}\widehat{F}_n(t)\right) t d\widehat{F}_n(t),$$

where $\widehat{F}_n$ denotes the empirical distribution function of the random sample $Y_1, \ldots, Y_n$. As $n \to \infty$ the integral converges to $\overline{D}_y$.

Next consider the reduction in dispersion. By Theorem 3.11.2, with probability 1, we can restrict the sample space to a space on which Huber's design condition (D.1) holds and on which $n^{-1}\mathbf{X}'\mathbf{X} \to \Sigma$. Then conditionally given $\mathbf{X}$, we have the assumptions found in Section 3.4 for the nonstochastic model. Hence, from the discussion found in Section 3.6.3, $(1/n)D(\widehat{\boldsymbol{\beta}}_R) \xrightarrow{P} \overline{D}_e$. Hence it is true unconditionally, a.e. M. The consistency of $\widehat{\tau}_\varphi$ was discussed above. The result then follows.

Example 3.11.1 *Measures of Association for Wilcoxon Scores.*

For the Wilcoxon score function, $\varphi_W(u) = \sqrt{12}(u - \frac{1}{2})$, as Exercise 3.12.34 shows, $\overline{D}_y = \int \varphi(G(y))y \, dy = \sqrt{3/4}E|Y_1 - Y_2|$ where Y_1, Y_2 are iid with distribution function G. Likewise, $\overline{D}_e = \sqrt{3/4}E|e_1 - e_2|$ where e_1, e_2 are iid with distribution function F. Finally $\tau_\varphi = (\sqrt{12} \int f^2)^{-1}$. Hence, for Wilcoxon scores these CMDs simplify to

$$\overline{R}_{W1} = \frac{E|Y_1 - Y_2| - E|e_1 - e_2|}{E|Y_1 - Y_2|} \quad (3.11.18)$$

$$\overline{R}_{W2} = \frac{E|Y_1 - Y_2| - E|e_1 - e_2|}{E|Y_1 - Y_2| - E|e_1 - e_2| + 1/(6\int f^2)}. \quad (3.11.19)$$

As discussed above, in general, $\overline{R}_{W1}$ is not robust but $\overline{R}_{W2}$ is.

Example 3.11.2 *Measures of Association for Sign Scores*

For the sign-score function, Exercise 3.12.34 shows that $\overline{D}_y = \int \varphi(G(y))y \, dy = E|Y - \text{med}Y|$ where $\text{med}Y$ denotes the median of Y. Likewise $\overline{D}_e = E|e - \text{med}e|$. Hence for sign scores, the coefficients of multiple determination are

$$\overline{R}_{S1} = \frac{E|Y - \operatorname{med} Y| - E|e - \operatorname{med} e|}{E|Y - \operatorname{med} Y|} \qquad (3.11.20)$$

$$\overline{R}_{S2} = \frac{E|Y - \operatorname{med} Y| - E|e - \operatorname{med} e|}{E|Y - \operatorname{med} Y| - E|e - \operatorname{med} e| + (4f(\operatorname{med} e))^{-1}}. \qquad (3.11.21)$$

These were obtained by McKean and Sievers (1987) from a L_1 point of view.

3.11.4 Properties of R-Coefficients of Multiple Determination

In this subsection we explore further properties of the population CMDs proposed in the previous subsection. To show that $\overline{R}_1$ and $\overline{R}_2$, (3.11.16) and (3.11.17), are indeed measures of association we have the following two theorems. The proof of the first theorem is quite similar to corresponding proofs of properties of the dispersion function for the nonstochastic model.

Theorem 3.11.4 *Suppose f and g satisfy the condition (E.1), (3.4.1), and their first moments are finite; then $\overline{D}_y > 0$ and $\overline{D}_e > 0$, where $\overline{D}_y = \int \varphi(G(y)) y \, dy$.*

Proof. It suffices to exhibit the proof for $\overline{D}_y$ since the proof for $\overline{D}_e$ is the same. The function φ is increasing and $\int \varphi = 0$; hence, φ must take on both negative and positive values. Thus the set $A = \{y : \varphi(G(y)) < 0\}$ is not empty and is bounded above. Let $y_0 = \sup A$. Then

$$\overline{D}_y = \int_{-\infty}^{y_0} \varphi(G(y))(y - y_0) dG(y) + \int_{y_0}^{\infty} \varphi(G(y))(y - y_0) dG(y). \qquad (3.11.22)$$

Since both integrands are nonnegative, it follows that $\overline{D}_y \geq 0$. If $\overline{D}_y = 0$ then it follows from (E.1) that $\varphi(G(y)) = 0$ for all $y \neq y_0$, which contradicts the facts that φ takes on both positive and negative values and that G is absolutely continuous.

The next theorem is taken from Witt (1989).

Theorem 3.11.5 *Suppose f and g satisfy the conditions (E.1) and (E.2) in Section 3.4 and that φ satisfies assumption (S.2), (3.4.11). Then $\overline{RD}$ is a strictly convex function of $\boldsymbol{\beta}$ and has a minimum value of 0 at $\boldsymbol{\beta} = \mathbf{0}$.*

Proof. We will show that the gradient of $\overline{RD}$ is zero at $\boldsymbol{\beta} = \mathbf{0}$ and that its second matrix derivative is positive definite. Note first that the distribution function, G, and density, g, of Y can be expressed as $G(y) = \int F(y - \boldsymbol{\beta}'\mathbf{x}) dM(\mathbf{x})$ and $g(y) = \int f(y - \boldsymbol{\beta}'\mathbf{x}) dM(\mathbf{x})$. We have

$$\frac{\partial \overline{RD}}{\partial \boldsymbol{\beta}} = -\int\int\int \varphi'[G(y)] y f(y - \boldsymbol{\beta}'\mathbf{x}) f(y - \boldsymbol{\beta}'\mathbf{u}) \mathbf{u} \, dM(\mathbf{x}) dM(\mathbf{u}) dy$$

$$- \int\int \varphi[G(y)] y f'(y - \boldsymbol{\beta}'\mathbf{x}) \mathbf{x} \, dM(\mathbf{x}) dy. \qquad (3.11.23)$$

Since $E[\mathbf{x}] = \mathbf{0}$, both terms on the right-hand side of the above expression are 0 at $\boldsymbol{\beta} = \mathbf{0}$. Before obtaining the second derivative, we rewrite the first term of (3.11.23) as

$$-\int\left[\int\int\varphi'[G(y)]yf(y-\boldsymbol{\beta}'\mathbf{x})f(y-\boldsymbol{\beta}'\mathbf{u})dy dM(\mathbf{x})\right]\mathbf{u}dM(\mathbf{u}) =$$

$$-\int\left[\int\varphi'[G(y)]g(y)yf(y-\boldsymbol{\beta}'\mathbf{u})dy\right]\mathbf{u}dM(\mathbf{u}).$$

Next integrate by parts the expression in brackets with respect to y using $dv = \varphi'[G(y)]g(y)dy$ and $t = yf(y - \boldsymbol{\beta}'\mathbf{u})$. Since φ is bounded and f has a finite second moment this leads to

$$\frac{\partial \overline{RD}}{\partial \boldsymbol{\beta}} = \int\int\varphi[G(y)]f(y-\boldsymbol{\beta}'\mathbf{u})dy dM(\mathbf{u}) + \int\int\varphi[G(y)]yf'(y-\boldsymbol{\beta}'\mathbf{u})\mathbf{u}dy dM(\mathbf{u})$$

$$-\int\int\varphi[G(y)]yf'(y-\boldsymbol{\beta}'\mathbf{x})\mathbf{x}dy dM(\mathbf{x})$$

$$= \int\int\varphi[G(y)]f(y-\boldsymbol{\beta}'\mathbf{u})\mathbf{u}dy dM(\mathbf{u}).$$

Hence the second derivative of $\overline{RD}$ is

$$\frac{\partial^2 \overline{RD}}{\partial\boldsymbol{\beta}\partial\boldsymbol{\beta}'} = -\int\int\varphi[G(y)]f'(y-\boldsymbol{\beta}'\mathbf{x})\mathbf{x}\mathbf{x}'dy dM(\mathbf{x})$$

$$-\int\int\int\varphi'[G(y)]f(y-\boldsymbol{\beta}'\mathbf{x})f(y-\boldsymbol{\beta}'\mathbf{u})\mathbf{x}\mathbf{u}'dy dM(\mathbf{x})dM(\mathbf{u}). \quad (3.11.24)$$

Now integrate the first term on the right-hand side of (3.11.24) by parts with respect to y by using $dt = f'(y - \boldsymbol{\beta}'\mathbf{x})dy$ and $v = \varphi[(G(y)]$. This leads to

$$\frac{\partial^2 \overline{RD}}{\partial\boldsymbol{\beta}\partial\boldsymbol{\beta}'} = -\int\int\int\varphi'[G(y)]f(y-\boldsymbol{\beta}'\mathbf{x})f(y-\boldsymbol{\beta}'\mathbf{u})\mathbf{x}(\mathbf{u}-\mathbf{x})'dy dM(\mathbf{x})dM(\mathbf{u}).$$

$$(3.11.25)$$

We have, however, the following identity

$$\int\int\int\varphi'[G(y)]f(y-\boldsymbol{\beta}'\mathbf{x})f(y-\boldsymbol{\beta}'\mathbf{u})(\mathbf{u}-\mathbf{x})(\mathbf{u}-\mathbf{x})'dy dM(\mathbf{x})dM(\mathbf{u}) =$$

$$\int\int\int\varphi'[G(y)]f(y-\boldsymbol{\beta}'\mathbf{x})f(y-\boldsymbol{\beta}'\mathbf{u})\mathbf{u}(\mathbf{u}-\mathbf{x})'dy dM(\mathbf{x})dM(\mathbf{u})$$

$$-\int\int\int\varphi'[G(y)]f(y-\boldsymbol{\beta}'\mathbf{x})f(y-\boldsymbol{\beta}'\mathbf{u})\mathbf{x}(\mathbf{u}-\mathbf{x})'dy dM(\mathbf{x})dM(\mathbf{u}).$$

Since the two integrals on the right-hand side of the last expression are negatives of each other, this combined with expression (3.11.24) leads to

$$2\frac{\partial^2 \overline{RD}}{\partial\boldsymbol{\beta}\partial\boldsymbol{\beta}'} = \int\int\int\varphi'[G(y)]f(y-\boldsymbol{\beta}'\mathbf{x})f(y-\boldsymbol{\beta}'\mathbf{u})(\mathbf{u}-\mathbf{x})(\mathbf{u}-\mathbf{x})'dy dM(\mathbf{x})dM(\mathbf{u}).$$

Since the functions f and M are continuous and the score function is increasing, it follows that the right-hand side of this last expression is a positive definite matrix.

It follows from these theorems that the $\overline{R}_i$s satisfy properties of association similar to $\overline{R}^2$. We have $0 \leq \overline{R}_i \leq 1$. By Theorem 3.11.5, $\overline{R}_i = 0$ if and only if $\boldsymbol{\beta} = \mathbf{0}$ if and only if Y and $\mathbf{x}$ are independent.

Example 3.11.3 *Multivariate Normal Model*

Further understanding of $\overline{R}_i$ can be gleaned from their direct relationship with R^2 for the multivariate normal model.

Theorem 3.11.6 *Suppose Model (3.11.1) holds. Assume further that $(\mathbf{x}, Y)$ follows a multivariate normal distribution with the variance-covariance matrix*

$$\Sigma_{(\mathbf{x},Y)} = \begin{bmatrix} \Sigma & \boldsymbol{\beta}'\Sigma \\ \Sigma\boldsymbol{\beta} & \sigma_e^2 + \boldsymbol{\beta}'\Sigma\boldsymbol{\beta} \end{bmatrix}. \tag{3.11.26}$$

Then, from (3.11.16) and (3.11.17),

$$\overline{R}_1 = 1 - \sqrt{1 - \overline{R}^2} \tag{3.11.27}$$

$$\overline{R}_2 = \frac{1 - \sqrt{1 - \overline{R}^2}}{1 - \sqrt{1 - \overline{R}^2[1 - 1/(2T^2)]}}, \tag{3.11.28}$$

where $T = \int \varphi[\Phi(t)] t \, d\Phi(t)$, Φ is the standard normal distribution function, and $\overline{R}^2$ is the traditional CMD given by (3.11.11).

Proof. Note that $\sigma_y^2 = \sigma_e^2 + \boldsymbol{\beta}'\Sigma\boldsymbol{\beta}$ and $E(Y) = \alpha + \boldsymbol{\beta}'E[\mathbf{x}]$. Further, the distribution function of Y is $G(y) = \Phi((y - \alpha - \boldsymbol{\beta}'E(\mathbf{x}))/\sigma_y)$, where Φ is the standard normal distribution function. Then

$$\overline{D}_y = \int_{-\infty}^{\infty} \varphi[\Phi(y/\sigma_y)] y \, d\Phi(y/\sigma_y) \tag{3.11.29}$$

$$= \sigma_y T. \tag{3.11.30}$$

Similarly, $\overline{D}_e = \sigma_e T$. Hence,

$$\overline{RD} = (\sigma_y - \sigma_e) T. \tag{3.11.31}$$

By the definition of $\overline{R}^2$, we have $\overline{R}^2 = 1 - \sigma_e^2/\sigma_y^2$. This leads to the relationship

$$1 - \sqrt{1 - \overline{R}^2} = \frac{\sigma_y - \sigma_e}{\sigma_y}. \tag{3.11.32}$$

The result (3.11.27) follows from the expressions (3.11.31) and (3.11.32).

For the result (3.11.28), by the assumptions on the distribution of $(\mathbf{x}, Y)$, the distribution of e is $N(0, \sigma_e^2)$, i.e. $f(x) = (2\pi\sigma_e^2)^{-1/2} \exp\{-x^2/(2\sigma_e^2)\}$ and

$F(x) = \Phi(x/\sigma_e)$. It follows that $f'(x)/f(x) = -\sigma_e^{-2}x$, which leads to

$$-\frac{f'(F^{-1}(u))}{f'(F(u))} = \frac{1}{\sigma_e}\Phi^{-1}(u).$$

Hence,

$$\tau_\varphi^{-1} = \int_0^1 \varphi(u)\left\{\frac{1}{\sigma_e}\Phi^{-1}(u)\right\} du$$

$$= \frac{1}{\sigma_e}\int_0^1 \varphi(u)\Phi^{-1}(u)\, du.$$

Upon making the substitution $u = \Phi(t)$, we obtain the relationship $T = \sigma_e/\tau_\varphi$. Using this, the result (3.11.31), and the definition of $\overline{R}_2$, (3.11.11), we get

$$\overline{R}_2 = \frac{\dfrac{\sigma_y - \sigma_e}{\sigma_y}}{\dfrac{\sigma_y - \sigma_e}{\sigma_y} + \dfrac{\sigma_e}{\sigma_y}\dfrac{1}{2T^2}}.$$

The result for $\overline{R}_2$ follows from this and (3.11.32).

Note that T is free of all parameters. It can be shown directly that the $\overline{R}_i$s are one-to-one increasing functions of $\overline{R}^2$; see Exercise 3.12.35. Hence, for the multivariate normal model the parameters $\overline{R}^2$, $\overline{R}_1$, and $\overline{R}_2$ are equivalent.

Although the CMDs are equivalent for the normal model, they measure dependence between **x** and Y on different scales. We can use the above relationships derived in Theorem 3.11.6 to have these coefficients measure the same quantity for the normal model by simply solving for $\overline{R}^2$ in terms of R_1 and R_2 in (3.11.27) and (3.11.28) respectively. These parameters will be useful later so we will call them $\overline{R}_1^*$ and $\overline{R}_2^*$, respectively. Hence, solving as indicated, we obtain

$$\overline{R}_1^{*2} = 1 - (1 - \overline{R}_1)^2 \tag{3.11.33}$$

$$\overline{R}_2^{*2} = 1 - \left[\frac{1 - \overline{R}_2}{1 - \overline{R}_2(1 - (1/(2T^2)))}\right]^2. \tag{3.11.34}$$

Again, for the multivariate normal model we have $\overline{R}^2 = \overline{R}_1^{*2} = \overline{R}_2^{*2}$.

For Wilcoxon scores and sign scores the reader is ask to show in Exercise 3.12.36 that $1/(2T^2) = \pi/6$ and $1/(2T^2) = \pi/4$, respectively.

Example 3.11.4 *A Contaminated Normal Model.*

As an illustration of these population CMDs, we evaluate them for the situation where the random error e has a contaminated normal distribution with proportion of contamination ε and the ratio of contaminated variance to uncontaminated σ_c^2, the random variable x has a univariate normal $N(0, 1)$ distribution, and the parameter $\beta = 1$. So $\boldsymbol{\beta}'\boldsymbol{\Sigma}\boldsymbol{\beta} = 1$. Without loss of generality, we take $\alpha = 0$ in (3.11.1). Hence, Y and x are dependent. We consider the CMDs based on the Wilcoxon score function only.

The density of $Y = x + e$ is given by

$$g(y) = \frac{1-\varepsilon}{\sqrt{2}} \phi\left(\frac{y}{\sqrt{2}}\right) + \frac{\varepsilon}{\sqrt{1+\sigma_c^2}} \phi\left(\frac{y}{\sqrt{1+\sigma_c^2}}\right).$$

This leads to the expressions

$$\overline{D}_y = \frac{\sqrt{12}}{\sqrt{2\pi}} \left\{ 2^{-1/2}(1-\varepsilon)^2\sqrt{2} + 2^{-1/2}\varepsilon^2\sqrt{1+\sigma_c^2} + \varepsilon(1-\varepsilon)\sqrt{3+\sigma_c^2} \right\}$$

$$\overline{D}_e = \frac{\sqrt{12}}{\sqrt{2\pi}} \left\{ 2^{-1/2}(1-\varepsilon)^2 + 2^{-1/2}\varepsilon^2\sigma_c + \varepsilon(1-\varepsilon)\sqrt{1+\sigma_c^2} \right\}$$

$$\tau_\varphi = \left[\frac{\sqrt{12}}{\sqrt{2\pi}} \left\{ \frac{(1-\varepsilon)^2}{\sqrt{2}} + \frac{\varepsilon^2}{\sigma_c\sqrt{2}} + \frac{2\varepsilon(1-\varepsilon)}{\sqrt{\sigma_c^2+1}} \right\} \right]^{-1};$$

see Exercise 3.12.37. Based on these quantities the CMDs $\overline{R}^2$, $\overline{R}_1$, and $\overline{R}_2$ can be readily formulated.

Table 3.11.1 displays these parameters for several values of ε and for $\sigma_c^2 = 9$ and 100. For ease of interpretation we rescaled the robust CMDs as discussed above. Thus for the normal ($\varepsilon = 0$) we have $\overline{R}_1^{*2} = \overline{R}_2^{*2} = \overline{R}^2$ with the common value of 0.5 in these situations. Certainly as either ε or σ_c changes, the amount of dependence between Y and x changes; hence all the coefficients change somewhat. However, R^2 decays as the percentage of contamination increases, and the decay is rapid in the case $\sigma_c^2 = 100$. This is true also, to a lesser degree, for $\overline{R}_1^*$, which is predictable since its denominator has unbounded influence in the Y-space. The coefficient $\overline{R}_2^*$ shows stability with the increase in contamination. For instance, when $\sigma_c^2 = 100$, R^2 decays 0.44 units while $\overline{R}_2^*$ decays only 0.14 units. See Witt et al. (1995) for more discussion on this example.

Ghosh and Sen (1971) proposed the mixed rank test statistic to test the hypothesis of independence (3.11.3). It is essentially the gradient test of the hypothesis $H_0: \beta = 0$. As we showed in Section 3.6, this test statistic is asymptotically equivalent to F_φ. Ghosh and Sen (1971) also proposed a pure rank statistic in which both variables are ranked and scored.

Table 3.11.1: Coefficients of multiple determination under contaminated errors (e)

CMD	$e \sim CN(\varepsilon, \sigma_c^2 = 9)$						$e \sim CN(\varepsilon, \sigma_c^2 = 100)$					
	ε						ε					
	0.00	0.01	0.02	0.05	0.10	0.15	0.00	0.01	0.02	0.05	0.10	0.15
$\overline{R}^2$	0.50	0.48	0.46	0.42	0.36	0.31	0.50	0.33	0.25	0.14	0.08	0.06
$\overline{R}_1^*$	0.50	0.50	0.48	0.45	0.41	0.38	0.50	0.47	0.42	0.34	0.26	0.19
$\overline{R}_2^*$	0.50	0.50	0.49	0.47	0.44	0.42	0.50	0.49	0.47	0.45	0.40	0.36

3.11.5 Coefficients of Determination for Regression

We have mainly been concerned with CMDs as measures of dependence between the random variables Y and $\mathbf{x}$. In the regression setting, however, R^2 is one of the most widely used statistics, not in the sense of estimating dependence but in the sense of comparing models. As the proportion of variance accounted for, R^2 is intuitively appealing. Likewise R_1, the proportion of dispersion accounted for in the fit, is an intuitive statistic. But neither of these statistics is robust. The statistic R_2, however, is robust and is directly linked (by a one-to-one function) to the robust test statistic F_φ. Furthermore, it lies between 0 and 1, having the values 1 for a perfect fit and 0 for a complete lack of fit. These properties make R_2 an attractive coefficient of determination for regression, as the following example illustrates.

Example 3.11.5 *Hald Data*

This data set consists of 13 observations and four predictors. It can be found in Hald (1952) but is also discussed in Draper and Smith (1966), where it serves to illustrate a method of predictor subset selection based on R^2. The data are given in Table 3.11.2. The response is the heat evolved in calories per gram of cement. The predictors are the percentage weight of ingredients used in the cement and are given by:

$$x_1 = \text{amount of tricalcium aluminate}$$
$$x_2 = \text{amount of tricalcium silicate}$$
$$x_3 = \text{amount of tetracalcium alumino ferrite}$$
$$x_4 = \text{amount of dicalcium silicate.}$$

To illustrate the use of the coefficients of determination R_1 and R_2, suppose we are interested in the best two-variable predictor model based on coefficients of determination. Table 3.11.3 gives the results for two data sets. The first is

Tabel 3.11.2: Hald data used in Example 3.11.5

x_1	x_2	x_3	x_4	Response
7	26	6	60	78.5
1	29	15	52	74.3
11	56	8	20	104.3
11	31	8	47	87.6
7	52	6	33	95.9
11	55	9	22	109.2
3	71	17	6	102.7
1	31	22	44	72.5
2	54	18	22	93.1
21	47	4	26	115.9
1	40	23	34	83.8
11	66	9	12	113.3
10	68	8	12	109.4

226   Linear Models

Table 3.11.3: Coefficients of multiple determination on Hald data

Subset of Predictors	Original Data			Changed Data		
	R^2	R_1	R_2	R^2	R_1	R_2
$\{x_1, x_2\}$	0.98	0.86	0.92	0.57	0.55	0.92
$\{x_1, x_3\}$	0.55	0.33	0.52	0.47	0.24	0.41
$\{x_1, x_4\}$	0.97	0.84	0.90	0.52	0.51	0.88
$\{x_2, x_3\}$	0.85	0.63	0.76	0.66	0.46	0.72
$\{x_2, x_4\}$	0.68	0.46	0.62	0.34	0.27	0.57
$\{x_3, x_4\}$	0.94	0.76	0.89	0.67	0.52	0.83

the original Hald data, while in the second we changed the 11th response observation from 83.8 to 8.8.

Note that on the original data all three coefficients choose the subset $\{x_1, x_2\}$. For the changed data, though, the outlier severely affects the LS coefficient R^2 and the nonrobust coefficient R_1, but the robust coefficient R_2 was much less sensitive to the outlier. It chooses the same subset $\{x_1, x_2\}$ as it did with the original data; however, the LS coefficient selects the subset $\{x_3, x_4\}$, two different predictors than its selection for the original data. The nonrobust coefficient R_1 still chooses $\{x_1, x_2\}$, although at a relativity much smaller value.

This example illustrates that the coefficient R_2 can be used in the **selection of predictors** in a regression problem. This selection could be formalized like the MAXR procedure in SAS. In a similar vein, the **stepwise model building** criteria based on LS estimation (Draper and Smith, 1966) could easily be robustified by using R-estimates in place of LS estimates and the robust test statistic F_φ in place of F_{LS}.

3.12 Exercises

3.12.1. *For the baseball data in Example 3.3.2, explore other transformations of the predictor Years in order to obtain a better-fitting model than the one discussed in the example.*

3.12.2 *Consider the linear model (3.2.2).*
(a) Show that the ranks of the residuals can only change values at the $\binom{n}{2}$ equations $y_i - \mathbf{x}_i'\boldsymbol{\beta} = y_j - \mathbf{x}_j'\boldsymbol{\beta}$.
(b) Determine the drop in dispersion as $\boldsymbol{\beta}$ moves across one of these defining planes.
(c) For the telephone data, Example 3.3.1, obtain the plot shown in Figure 3.3.1(d), i.e. a plot of the dispersion function $D(\beta)$ for a set of values β in the interval $(-0.2, 0.6)$. Locate the estimate of slope on the plot.
(d) Plot the gradient function $S(\beta)$ for the same set of values β in the interval $(-0.2, 0.6)$. Locate the estimate of slope on the plot.

3.12.3 *In Section 2.2 the two-sample location problem was modeled as a regression problem; see (2.2.2). Consider fitting this model using Wilcoxon scores.*
(a) Show that the gradient test statistic (3.5.8) simplifies to the square of the standardized MWW test statistic (2.2.21).

(b) Show that the regression estimate of the slope parameter is the Hodges–Lehmann estimator given by expression (2.2.18).

(c) Verify (a) and (b) by fitting the data in the two-sample problem of Exercise 2.13.47 as a regression model.

3.12.4 For the simple linear regression problem, if the values of the independent variable x are distinct and equally spaced show that the Wilcoxon test statistic is equivalent to the test for correlation based on Spearman's r_s, where

$$r_s = \frac{\sum\left(R(x_i) - \frac{n+1}{2}\right)\left(R(y_i) - \frac{n+1}{2}\right)}{\sqrt{\sum\left(R(x_i) - \frac{n+1}{2}\right)^2}\sqrt{\sum\left(R(y_i) - \frac{n+1}{2}\right)^2}}.$$

Note that the denominator of r_s is a constant. Obtain its value.

3.12.5 For the simple linear regression model consider the process

$$T(\beta) = \sum_{i=1}^{n}\sum_{j=1}^{n} \text{sgn}(x_i - x_j)\text{sgn}((Y_i - x_i\beta) - (Y_j - x_j\beta)).$$

(a) Show, under the null hypothesis $H_0 : \beta = 0$, that $E(T(0)) = 0$ and that $\text{Var}(T(0)) = 2(n-1)n(2n+5)/9$.

(b) Determine the estimate of β based on inverting the test statistic $T(0)$, i.e. the value of β which solves

$$T(\beta) \doteq 0.$$

(c) Show that when the two-sample problem is written as a regression model, (2.2.2), this estimate of β is the Hodges–Lehmann estimate (2.2.18).

Note: **Kendall's τ** is a measure of association between x_i and Y_i given by $\tau = T(0)/(n(n-1))$; see Chapter 4 of Hettmansperger (1984a) for further discussion.

3.12.6 Show that the R-estimate $\widehat{\boldsymbol{\beta}}_\varphi$ is an equivariant estimator; that is, $\widehat{\boldsymbol{\beta}}_\varphi(\mathbf{Y} + \mathbf{X}\boldsymbol{\delta}) = \widehat{\boldsymbol{\beta}}_\varphi(\mathbf{Y}) + \boldsymbol{\delta}$ and $\widehat{\boldsymbol{\beta}}_\varphi(k\mathbf{Y}) = k\widehat{\boldsymbol{\beta}}_\varphi(\mathbf{Y})$.

3.12.7 Consider model (3.2.1) and the hypotheses (3.2.5). Let Ω_F denote the column space of the full model design matrix $\mathbf{X}$ and let ω denote the subspace of Ω_F subject to H_0. Show that ω is a subspace of Ω_F and determine its dimension. Hint: One way of establishing the dimension is to show that $\mathbf{C} = \mathbf{X}(\mathbf{X}'\mathbf{X})^{-1}\mathbf{M}'$ is a basis matrix for $\Omega_F \cap \omega^c$.

3.12.8 Show that assumptions (3.4.9) and (3.4.8) imply assumption (3.4.7).

3.12.9 For the special case of Wilcoxon scores, obtain the proof of Theorem 3.5.2 by first getting the projection of the statistic $\mathbf{S}(\mathbf{0})$.

3.12.10 Assume that the errors e_i in model (3.2.2) have finite variance σ^2. Let $\widehat{\boldsymbol{\beta}}_{LS}$ denote the least-squares estimate of $\boldsymbol{\beta}$. Show that $\sqrt{n}(\widehat{\boldsymbol{\beta}}_{LS} - \boldsymbol{\beta}) \xrightarrow{\mathcal{D}} N_p(\mathbf{0}, \sigma^2\Sigma^{-1})$. Hint: First show that the LS estimate is location- and scale-equivariant. Then without loss of generality we can assume that the true $\boldsymbol{\beta}$ is $\mathbf{0}$.

3.12.11 Under the additional assumption that the errors have a symmetric distribution, show that R-estimates are unbiased for all sample sizes.

3.12.12 Let $\varphi_f(u) = -f'(F^{-1}(u))/f(F^{-1}(u))$ denote the optimal scores for the density $f(x)$ and suppose that f is symmetric. Show that $\varphi_f(1-u) = -\varphi_f(u)$; that is, the optimal scores are odd about $1/2$.

3.12.13 Suppose the errors e_i are double exponentially distributed. Show that the L_1 estimate, i.e. the R estimate based on sign scores, is the maximum likelihood estimate.

3.12.14 Using Theorem 3.5.11, show that

$$\begin{pmatrix} \widehat{\alpha}_S^* \\ \widehat{\boldsymbol{\beta}}_\varphi \end{pmatrix} \text{ is approximately } N_{p+1}\left(\begin{pmatrix} \alpha_0 \\ \boldsymbol{\beta}_0 \end{pmatrix}, \begin{bmatrix} \kappa_n & -\tau_\varphi^2 \overline{\mathbf{x}}'(\mathbf{X}'\mathbf{X})^{-1} \\ -\tau_\varphi^2 (\mathbf{X}'\mathbf{X})^{-1} \overline{\mathbf{x}} & \tau_\varphi^2 (\mathbf{X}'\mathbf{X})^{-1} \end{bmatrix} \right), \tag{3.12.1}$$

where $\kappa_n = n^{-1}\tau_S^2 + \tau_\varphi^2 \overline{\mathbf{x}}'(\mathbf{X}'\mathbf{X})^{-1}\overline{\mathbf{x}}$ and τ_S and and τ_φ are given respectively by (3.4.6) and (3.4.4).

3.12.15 Show that the random vector within the brackets in the proof of Lemma 3.6.2 is bounded in probability.

3.12.16 Show that difference between the numerators of the two F-statistics, (3.6.12) and (3.6.14), converges to 0 in probability under the null hypothesis.

3.12.17 Show that the difference between F_φ, (3.6.12), and A_φ, (3.6.17), converges to 0 in probability under the null hypothesis.

3.12.18 By showing the following results, establish the asymptotic distribution of the least-squares test statistic, F_{LS}, under the sequence of models (3.6.24) with the additional assumption that the random errors have finite variance σ^2.

(a) First show that

$$\frac{1}{\sqrt{n}} \mathbf{X}'\mathbf{Y} \xrightarrow{\mathcal{D}} N\left(\begin{bmatrix} \mathbf{B}\boldsymbol{\theta} \\ \mathbf{A}_2\boldsymbol{\theta} \end{bmatrix}, \sigma^2 \mathbf{I} \right), \tag{3.12.2}$$

where the matrices $\mathbf{A}_2$ and $\mathbf{B}$ are defined in the proof of Theorem 3.6.4. This can be established by using the Lindeberg–Feller central limit theorem (Theorem A.1.1 of the Appendix), to show that an arbitrary linear combination of the components of the random vector on the left-hand side converges in distribution to a random variable with a normal distribution.

(b) Based on (a), show that

$$[-\mathbf{B}'\mathbf{A}_1^{-1}; \mathbf{I}] \frac{1}{\sqrt{n}} \mathbf{X}'\mathbf{Y} \xrightarrow{\mathcal{D}} N(\mathbf{W}^{-1}\boldsymbol{\theta}, \sigma^2 \mathbf{W}^{-1}), \tag{3.12.3}$$

where the matrices $\mathbf{A}_1$ and $\mathbf{W}$ are defined in the proof of Theorem 3.6.4.

(c) Let $F_{LS}(\sigma^2)$ denote the LS F-test statistic with the true value of σ^2 replacing the estimate $\widehat{\sigma}^2$. Show that

$$F_{LS}(\sigma^2) = \left\{ [-\mathbf{B}'\mathbf{A}_1^{-1}; \mathbf{I}] \frac{1}{\sqrt{n}} \mathbf{X}'\mathbf{Y} \right\}' \sigma^{-2} \mathbf{W} \left\{ [-\mathbf{B}'\mathbf{A}_1^{-1}; \mathbf{I}] \frac{1}{\sqrt{n}} \mathbf{X}'\mathbf{Y} \right\}. \tag{3.12.4}$$

(d) Based on (3.12.3) amd (3.12.4), show that $F_{LS}(\sigma^2)$ has a limiting noncentral χ^2 distribution with noncentrality parameter given by (3.6.29).

(e) Obtain the final result by showing that $\widehat{\sigma}^2$ is a consistent estimate of σ^2 under the sequence of models (3.6.24).

Exercises 229

3.12.19 Show that $\overline{D}_e$, (3.6.30), is a scale parameter, i.e. $\overline{D}_e(F_{ae+b}) = |a|\overline{D}_e(F_e)$.

3.12.20 Establish expression (3.6.35).

3.12.21 Suppose Wilcoxon scores are used.
(a) Establish expressions (3.6.36) and (3.6.37).
(b) Similarly, for sign scores establish (3.6.38) and (3.6.39).

3.12.22 Consider the model (3.2.1) and hypotheses (3.6.9). Suppose the errors have a double exponential distribution with density $f(t) = (2b)^{-1} \exp\{-|t|/b\}$. Assume b is known. Show that the likelihood ratio test is equivalent to the drop in dispersion test based on sign scores.

3.12.23 Establish expressions (3.9.8) and (3.9.9).

3.12.24 Let X be a random variable with distribution function $F_X(x)$ and let $Y = aX + b$. Define the quantile function of X as $q_X(p) = F_X^{-1}(p)$. Show that $q_X(p)$ is a linear function of $q_Y(p)$.

3.12.25 Verify expression (3.9.17).

3.12.26 Assume that the errors have a normal distribution. Show that $\widehat{K}_2$, (3.9.25), converges in probability to 1.

3.12.27 Verify expression (3.9.34).

3.12.28 Proceeding as in Theorem 3.9.5, show that the first-order representation of the fitted value $\widehat{\mathbf{Y}}_R$ is given by (3.9.36). Next show that the approximate variance of the ith fitted case is given by (3.9.38).

3.12.29 Consider the mean shift model, (3.9.32). Show that the estimator of θ_i given by the numerator of expression (3.9.35) is based on the inversion of an aligned rank statistic to test the hypotheses (3.9.33).

3.12.30 Assume that the errors have a symmetric distribution. Verify expressions (3.9.41) and (3.9.42).

3.12.31 Assume that the errors have the distribution $GF(2m_1, 2m_2)$.
(a) Show that the optimal rank score function is given by expression (3.10.6).
(b) Show that the asymptotic relative efficiency between the Wilcoxon analysis and the rank-based analysis based on the optimal scores for the distribution $GF(2m_1, 2m_2)$ is given by expression (3.10.8).

3.12.32 Suppose the errors have density function

$$f_{m_2}(x) = e^x(1 + m_2^{-1}e^x)^{-(m_2+1)}, \quad m_2 > 0, \quad -\infty < x < \infty. \quad (3.12.5)$$

(a) Show that the optimal scores are given by expression (3.10.7).
(b) Show that the asymptotic efficiency of the Wilcoxon analysis relative to the rank analysis based on the optimal rank-score function for the density (3.12.5) is given by expression (3.10.9).

3.12.33 The definition of the modulus of a matrix $\mathbf{A}$ is given in expression (3.11.6). Verify the three properties concerning the modulus of a matrix listed in the text following this definition.

3.12.34 Consider Example 3.11.1. If Wilcoxon scores are used, show that $\overline{D}_y = \sqrt{3/4}E|Y_1 - Y_2|$, where Y_1, Y_2 are iid with distribution function G, and that $\overline{D}_e = \sqrt{3/4}E|e_1 - e_2|$, where e_1, e_2 are iid with distribution function F. Next

230   Linear Models

assume that sign scores are used. Show that $\overline{D}_y = E|Y - \text{med } Y|$ where med Y denotes the median of Y. Likewise $\overline{D}_e = E|e - \text{med } e|$.

3.12.25 In Example 3.11.3, show that coefficients of multiple determination $\overline{R}_1$ and $\overline{R}_2$ given by expressions (3.11.27) and (3.11.28) respectively, are one-to-one functions of $\overline{R}^2$ given by expression (3.11.11).

3.12.36 At the end of Example 3.11.3, verify, for Wilcoxon scores and sign scores, that $1/(2T^2) = \pi/6$ and $1/(2T^2) = \pi/4$, respectively.

3.12.37 In Example 3.11.4, show that the density of Y is given by

$$g(y) = \frac{1-\varepsilon}{\sqrt{2}} \phi\left(\frac{y}{\sqrt{2}}\right) + \frac{\varepsilon}{\sqrt{1+\sigma_c^2}} \phi\left(\frac{y}{\sqrt{1+\sigma_c^2}}\right).$$

Using this, verify the expressions for $\overline{D}_y$, $\overline{D}_e$, and τ_φ found in the example.

3.12.38 For the baseball data given in Exercise 1.13.25, consider the variables height and weight.
(a) Obtain the scatterplot of height versus weight.
(b) Obtain the CMDs: R^2, R_1, R_2, R_1^{*2}, and R_2^{*2}.

3.12.39 Consider a linear model of the form

$$\mathbf{Y} = \mathbf{X}^* \boldsymbol{\beta}^* + \mathbf{e}, \tag{3.12.6}$$

where $\mathbf{X}^*$ is $n \times p$ whose column space Ω_F^* does not include $\mathbf{1}$. This model is often called **regression through the origin**. Note for the pseudo-norm $\|\cdot\|_\varphi$ that

$$\|\mathbf{Y} - \mathbf{X}^*\boldsymbol{\beta}^*\|_\varphi = \sum_{i=1}^n a(R(y_i - \mathbf{x}_i^{*'}\boldsymbol{\beta}))(y_i - \mathbf{x}_i^{*'}\boldsymbol{\beta})$$

$$= \sum_{i=1}^n a(R(y_i - (\mathbf{x}_i^* - \overline{\mathbf{x}}^*)'\boldsymbol{\beta}^*))(y_i - (\mathbf{x}_i^* - \overline{\mathbf{x}}^*)'\boldsymbol{\beta}^*)$$

$$= \sum_{i=1}^n a(R(y_i - \alpha - (\mathbf{x}_i^* - \overline{\mathbf{x}}^*)'\boldsymbol{\beta}^*))(y_i - \alpha - (\mathbf{x}_i^* - \overline{\mathbf{x}}^*)'\boldsymbol{\beta}^*), \tag{3.12.7}$$

where $\mathbf{x}_i^*$ is the ith row of $\mathbf{X}^*$ and $\overline{\mathbf{x}}^*$ is the vector of column averages of $\mathbf{X}^*$. Based on this result, the estimate of the regression coefficients based on the R fit of model (3.12.6) is estimating the regression coefficients of the centered model, i.e. the model with the design matrix $\mathbf{X} = \mathbf{X}^* - \mathbf{H}_1\mathbf{X}^*$. Hence, in general, the parameter $\boldsymbol{\beta}$ is not estimated. This problem also occurs in a weighted regression model. Dixon and McKean (1996) proposed the following solution. Assume that (3.12.6) is the true model, but obtain the R fit of the model:

$$\mathbf{Y} = \mathbf{1}\alpha_1 + \mathbf{X}^*\boldsymbol{\beta}_1^* + \mathbf{e} = [\mathbf{1} \ \mathbf{X}^*]\begin{bmatrix}\alpha_1 \\ \boldsymbol{\beta}_1^*\end{bmatrix} + \mathbf{e}, \tag{3.12.8}$$

where the true α_1 is 0. Let $\mathbf{X}_1 = [\mathbf{1} \ \mathbf{X}^*]$ and let Ω_1 denote the column space of $\mathbf{X}_1$. Let $\widehat{\mathbf{Y}}_1 = \mathbf{1}\widehat{\alpha}_1 + \mathbf{X}^*\widehat{\boldsymbol{\beta}}_1^*$ denote the R fitted value based on the fit of model (3.12.8). Note that $\Omega^* \subset \Omega_1$. Let $\widehat{\mathbf{Y}}^* = \mathbf{H}_{\Omega^*}\widehat{\mathbf{Y}}_1$ be the projection of this fitted value onto the desired space Ω^*. Finally, estimate $\boldsymbol{\beta}^*$ by solving the equation

$$\mathbf{X}^*\widehat{\boldsymbol{\beta}}^* = \widehat{\mathbf{Y}}^*. \qquad (3.12.9)$$

(a) Show that $\widehat{\boldsymbol{\beta}}^* = (\mathbf{X}^{*\prime}\mathbf{X}^*)^{-1}\mathbf{X}^{*\prime}\widehat{\mathbf{Y}}_1$ is the solution of (3.12.9).

(b) Assume that the density function of the errors is symmetric, that the R-score function is odd about $1/2$ and that the intercept α_1 is estimated by solving the equation $T^+(\widehat{\mathbf{e}}_R - \alpha) \doteq 0$, as discussed in Section 3.5.2. Under these assumptions show that

$$\widehat{\boldsymbol{\beta}}^* \text{ has an approximate } N(\boldsymbol{\beta}^*, \tau^2(\mathbf{X}^{*\prime}\mathbf{X}^*)^{-1}) \text{ distribution}. \qquad (3.12.10)$$

(c) Establish the asymptotic distribution of $\widehat{\boldsymbol{\beta}}^*$ if the intercept is estimated by median of the residuals from the R fit of (3.12.8).

(d) Show that the invariance to $\overline{\mathbf{x}}^*$ as shown in (3.12.7) is true for any pseudo-norm.

3.12.40 The data in Table 3.12.1 are presented in Graybill and Iyer (1994). The dependent variable is the weight (in grams) of a crystalline form of a certain chemical compound, while the independent variable is the length of time (in hours) that the crystal was allowed to grow. A model of interest is the regression through the origin model (3.12.6). Obtain the R-estimate of β^* for these data using the procedure described in (3.12.9). Compare this fit with the R-fit of the intercept model.

Table 3.12.1: Crystal data for Exercise 3.12.40

Time (hours)	2	4	6	8	10	12	14
Weight (grams)	0.08	1.12	4.43	4.98	4.92	7.18	5.57
Time (hours)	16	18	20	22	24	26	28
Weight (grams)	8.40	8.881	10.81	11.16	10.12	13.12	15.04

4

Experimental Designs

4.1 Introduction

In this chapter we will discuss rank-based inference for experimental designs based on the theory developed in Chapter 3. We will concentrate on factorial-type designs and analysis of covariance designs but, based on our discussion, it will be clear how to extend the rank-based analysis for any fixed effects design. For example, based on this rank-based inference, Vidmar and McKean (1996) developed a response surface methodology which is quite analogous to the traditional response surface methods. We will discuss estimation of effects, tests of linear hypotheses concerning effects, and multiple comparison procedures. We illustrate this rank-based inference with numerous examples. One purpose of our discussion is to show how this rank-based analysis is analogous to the traditional analysis based on least squares. In Section 4.2.5 we will introduce pseudo-observations which are based on an R fit of the full model. We show that the rank-based analysis (Wald type) can be obtained by substituting these pseudo-observations in place of the responses in a package that obtains the traditional analysis. We begin with the one way design. For additional discussion and review of rank methods in designed experiments, see Draper (1988).

In our development we apply rank scores to residuals. In this sense our methods are not pure rank statistics; but they do provide consistent and highly efficient tests for traditional linear hypotheses. The rank-transform method is a pure rank test and is discussed in Section 4.7, where we describe various drawbacks to the approach for testing traditional linear hypotheses in linear models. Brunner and his colleagues have successfully developed a general approach to testing in designed experiments based on pure ranks; however, the hypotheses of their approach are generally not linear hypotheses. Brunner and Puri (1996) provide an excellent survey of these pure rank tests. We will not pursue them further in this book.

While we will only consider linear models in this chapter, there have been extensions of robust inference to other models. For example, Vidmar, McKean, and Hettmansperger (1992) extended this robust inference to generalized linear models for quantal responses in the context of drug combination problems, and Li (1991) discussed rank procedures for a logistic model.

Stefanski, Carroll, and Ruppert (1986) discussed generalized M-estimates for generalized linear models.

4.2 One-way Design

Suppose we want to determine the effect that a single factor A has on a response of interest over a specified population. Assume that A has k levels, each level being referred to as a treatment group. In this situation, the completely randomized design is often used to investigate the effect of A. For this design n subjects are selected at random from the population of interest and n_i of these subjects are randomly assigned to level i of A, for $i = 1, \ldots, k$. Let Y_{ij} denote the response of the jth subject in the ith level of A. We will assume that the responses are independent of one another and that the distributions among levels differ by at most shifts in location. Although the randomization gives some credence to the assumption of independence, after fitting the model a residual analysis should be conducted to check this assumption and the assumption that the level distributions differ by at most a shift in locations.

Under these assumptions, the **full model** can be written as

$$Y_{ij} = \mu_i + e_{ij}, \quad j = 1, \ldots, n_i, \ i = 1, \ldots, k, \qquad (4.2.1)$$

where the e_{ij}s are iid random variables with density $f(x)$ and distribution function $F(x)$ and the parameter μ_i is a convenient location parameter, (for example, the mean or median). Let $T(F)$ denote the location functional. Assume, without loss of generality, that $T(F) = 0$. Let $\Delta_{ii'}$ denote the shift between the distributions of Y_{ij} and $Y_{i'l}$. Recall from Chapter 2 that the parameters $\Delta_{ii'}$ are invariant to the choice of locational functional and that $\Delta_{ii'} = \mu_i - \mu_{i'}$. If μ_i is the mean of the Y_{ij} then Hocking (1985) calls this the **means model**. If μ_i is the median of the Y_{ij} then we will call it the **medians model**; see Section 4.2.4 below.

Observational studies can also be modeled this way. Suppose k independent samples are drawn from k different populations. If we assume further that the distributions for the different populations differ by at most a shift in locations then model (4.2.1) is appropriate. But as in all observational studies, care must be taken in the interpretation of the results of the analyses.

While the parameters μ_i fix the locations, the parameters of interest in this chapter are **contrasts** of the form $h = \sum_{i=1}^{k} c_i \mu_i$, where $\sum_{i=1}^{k} c_i = 0$. Similar to the shift parameters, contrasts are invariant to the choice of location functional. In fact contrasts are linear functions of these shifts, i.e.

$$h = \sum_{i=1}^{k} c_i \mu_i = \sum_{i=1}^{k} c_i(\mu_i - \mu_1) = \sum_{i=2}^{k} c_i \Delta_{i1} = \mathbf{c}_1' \Delta_1, \qquad (4.2.2)$$

where $\mathbf{c}_1' = (c_2, \ldots, c_k)$ and

$$\Delta_1' = (\Delta_{21}, \ldots, \Delta_{k1}) \qquad (4.2.3)$$

is the vector of location shifts from the first cell. In order easily to reference the theory of Chapter 3, we will often use Δ_1 which references cell 1. But picking cell 1 is only for convenience and similar results hold for the selection of any other cell.

As in Chapter 2, we can write this model in terms of a linear model as follows. Let $\mathbf{Z}' = (Y_{11}, \ldots, Y_{1n_1}, \ldots, Y_{k1}, \ldots, Y_{kn_k})$ denote the vector of all observations, $\boldsymbol{\mu}' = (\mu_1, \ldots, \mu_k)$ denote the vector of locations, and $n = \sum n_i$ denote the total sample size. The model can then be expressed as a linear model of the form

$$\mathbf{Z} = \mathbf{W}\boldsymbol{\mu} + \mathbf{e}, \tag{4.2.4}$$

where $\mathbf{e}$ denotes the $n \times 1$ vector of random errors e_{ij} and the $n \times k$ design matrix $\mathbf{W}$ denotes the appropriate **incidence matrix** of 0s and 1s, i.e.

$$\mathbf{W} = \begin{bmatrix} \mathbf{1}_{n_1} & 0 & \cdots & 0 \\ 0 & \mathbf{1}_{n_2} & \cdots & 0 \\ \vdots & \vdots & \cdots & \vdots \\ 0 & 0 & \cdots & \mathbf{1}_{n_k} \end{bmatrix}. \tag{4.2.5}$$

Note that the vector $\mathbf{1}_n$ is in the column space of $\mathbf{W}$; hence, the theory derived in Chapter 3 is valid for this model.

At times it will be convenient to reparameterize the model in terms of a vector of shift parameters. For the vector $\boldsymbol{\Delta}_1$, let $\mathbf{W}_1$ denote the last $k-1$ columns of $\mathbf{W}$ and let $\mathbf{X}$ be the centered $\mathbf{W}_1$, i.e. $\mathbf{X} = (\mathbf{I} - \mathbf{H}_1)\mathbf{W}_1$, where $\mathbf{H}_1 = \mathbf{1}(\mathbf{1}'\mathbf{1})^{-1}\mathbf{1}' = n^{-1}\mathbf{1}\mathbf{1}'$ and $\mathbf{1}' = (1, \ldots, 1)$. Then we can write model (4.2.4) as

$$\mathbf{Z} = \alpha \mathbf{1} + \mathbf{X}\boldsymbol{\Delta}_1 + \mathbf{e}, \tag{4.2.6}$$

where $\boldsymbol{\Delta}_1$ is as given in (4.2.3). It is easy to show for any matrix $[\mathbf{1}|\mathbf{X}^*]$, having the same column space as $\mathbf{W}$, that its corresponding nonintercept parameters are linear functions of the shifts and, hence, are invariant to the selected location functional. The relationship between models (4.2.4) and (4.2.6) will be explored further in Section 4.2.4.

4.2.1 R Fit of the One-way Design

Note that the sum of all the column vectors of $\mathbf{W}$ equals the vector of ones $\mathbf{1}_n$. Thus $\mathbf{1}_n$ is in the column space of $\mathbf{W}$ and we can fit model (4.2.4) by using the R-estimates discussed in Chapter 3. In this chapter we assume that a specified score function, $a(i) = \varphi(i/(n+1))$, has been chosen which, without loss of generality, has been standardized; recall (S.1), (3.4.10). A convenient way of obtaining the fit is by the QR decomposition algorithm on the incidence matrix $\mathbf{W}$; see Section 3.7.3.

For the R fits used in the examples of this chapter, we will use the cell median model; that is, model (4.2.4) with $T(F) = 0$, where T denotes the median functional and F denotes the distribution of the random errors e_i. We will use the score function $\varphi(u)$ to obtain the R fit of this model. Let $\mathbf{X}\widehat{\boldsymbol{\Delta}}_1$ denote the fitted value. As discussed in Chapter 3, $\mathbf{X}\widehat{\boldsymbol{\Delta}}_1$ lies in the column space of the centered matrix $\mathbf{X} = (\mathbf{I} - \mathbf{H}_1)\mathbf{W}_1$. We then estimate the intercept as

$$\widehat{\alpha}_S = \mathrm{med}_{1 \leqslant i \leqslant n}\{Z_i - \mathbf{x}_i'\widehat{\mathbf{\Delta}}_1\}, \qquad (4.2.7)$$

where $\mathbf{x}_i$ is the ith row of $\mathbf{X}$. The final fitted value and the residuals are, respectively,

$$\widehat{\mathbf{Z}} = \widehat{\alpha}_S \mathbf{1} + \mathbf{X}\widehat{\mathbf{\Delta}}_1 \qquad (4.2.8)$$

$$\widehat{\mathbf{e}} = \mathbf{Z} - \widehat{\mathbf{Z}}. \qquad (4.2.9)$$

Note that $\widehat{\mathbf{Z}}$ lies in the column space of $\mathbf{W}$ and that, further, $T(F_n) = 0$, where F_n denotes the empirical distribution function of the residuals and T is the median location functional. Denote the fitted value of the response Y_{ij} as $\widehat{Y}_{ij}$. Given $\widehat{\mathbf{Z}}$, we find from (4.2.4) that $\widehat{\boldsymbol{\mu}} = (\mathbf{W}'\mathbf{W})^{-1}\mathbf{W}'\widehat{\mathbf{Z}}$. Because $\mathbf{W}$ is an incidence matrix, the estimate of μ_i is the common fitted value of the ith cell which, for future reference, is given by

$$\widehat{\mu}_i = \widehat{Y}_{ij}, \qquad (4.2.10)$$

for any $j = 1, \ldots, n_i$. In the examples below, we shall denote the R fit described in this paragraph by stating that the model was fit using Wilcoxon scores and the residuals were **adjusted to have median zero**.

It follows from Section 3.5.2 that $\widehat{\boldsymbol{\mu}}$ is asymptotically normal with mean $\boldsymbol{\mu}$. To do inference based on these estimates of μ_i we need their standard errors, but these can be obtained immediately from the variance of the fitted values given by (3.9.38). First note that the leverage value for an observation in the ith cell is $1/n_i$ and, hence, the leverage value for the centered design is $h_{ci} = h_i - n^{-1} = (n - n_i)/nn_i$. Therefore, by (3.9.38), the approximate variance of $\widehat{\mu}_i$ is given by

$$\mathrm{Var}(\widehat{\mu}_i) \doteq \frac{1}{n_i}\tau_\varphi^2 + \frac{1}{n}(\tau_S^2 - \tau_\varphi^2), \quad i = 1, \ldots, k; \qquad (4.2.11)$$

see Exercise 4.8.18.

Let $\widehat{\tau}_\varphi$ and $\widehat{\tau}_S$ denote respectively the estimates of τ_φ and τ_S presented in Section 3.7.1. The estimated approximate variance of $\widehat{\mu}_i$ is expression (4.2.11) with these estimates in place of τ_φ and τ_S. Define the minimum value of the dispersion function as DE, i.e,

$$DE = D(\widehat{\mathbf{e}}) = \sum_{i=1}^{k}\sum_{j=1}^{n_i} a(R(\widehat{e}_{ij}))\widehat{e}_{ij}. \qquad (4.2.12)$$

The symbol DE stands for the dispersion of the errors and is analogous to LS sums of squared errors, SSE. Upon fitting such a model, a residual analysis as discussed in Section 3.9 should be conducted to assess the goodness of fit of the model.

Example 4.2.1 *LDL Cholesterol in Quail*

Thirty-nine quail were randomly assigned to four diets, each diet containing a different drug compound, which, hopefully, would reduce LDL cholesterol.

Table 4.2.1: Data for Example 4.2.1

Drug	LDL Cholesterol								
I	52	67	54	69	116	79	68	47	120	73
II	36	34	47	125	30	31	30	59	33	98
III	52	55	66	50	58	176	91	66	61	63
IV	62	71	41	118	48	82	65	72	49	

The drug compounds are labeled I, II, III, and IV. At the end of the prescribed experimental time the LDL cholesterol of each quail was measured. The data are displayed in Table 4.2.1.

From the boxplot, Figure 4.2.1(a), it appears that drug compound II was more effective than the other three in lowering LDL. The data appear to be positively skewed with a long right tail. We fitted the data using Wilcoxon scores, $\varphi(u) = \sqrt{12}(u - \frac{1}{2})$, and adjusted the residuals to have median 0. Figure 4.2.1(b) displays the Wilcoxon residuals versus fitted values. The long right tail of the error distribution is apparent from this plot. Figure 4.2.1(c,d) involves the internal R studentized residuals, (3.9.31), with the benchmarks ± 2. The internal R-studentized residuals detected six outlying data points while the normal $q-q$ plot of these residuals clearly shows the skewness.

The estimates of τ_φ and τ_S are 19.19 and 21.96, respectively. For comparison, the LS estimate of σ was 30.49. Table 4.2.2 displays the Wilcoxon and LS estimates of the cell locations along with their standard errors. The Wilcoxon and LS estimates of the location levels are quite different, as they should be since they estimate different functionals under asymmetric errors. The long right tail has drawn out the LS estimates. The standard errors of the Wilcoxon estimates are much smaller than their LS counterparts.

This data set was taken from a much larger study discussed in McKean, Vidmar, and Sievers (1989). Most of the data in that study exhibited long right tails. The left tails were also long; hence, transformations such as logarithms were not effective. Scores more appropriate for positively skewed data were used with considerable success in this study. These scores are briefly discussed in Example 2.5.1.

Table 4.2.2: Estimates of Location Levels for the Quail Data

Drug Compound	Wilcoxon Fit		LS Fit	
	Est.	SE	Est.	SE
I	67.0	6.3	74.5	9.6
II	42.0	6.3	52.3	9.6
III	63.0	6.3	73.8	9.6
IV	62.0	6.6	67.6	10.1

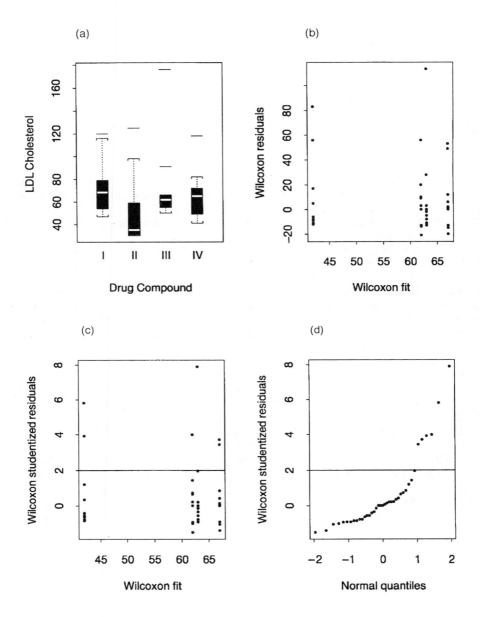

Figure 4.2.1: (a) Comparison boxplots for data of Example 4.2.1; (b) Wilcoxon residual plot; (c) Wilcoxon internal R studentized residual plot; (d) Wilcoxon internal R studentized residual normal q−q plot.

4.2.2 Rank-Based Tests of $H_0 : \mu_1 = \cdots = \mu_k$

Consider model (4.2.4). A hypothesis of interest in the one-way design is that there are no differences in the levels of A, i.e.

$$H_0 : \mu_1 = \cdots = \mu_k \quad \text{versus} \quad H_1 : \mu_i \neq \mu_{i'} \text{ for some } i \neq i'. \quad (4.2.13)$$

Define the $k \times (k-1)$ matrix $\mathbf{M}$ as

$$\mathbf{M} = \begin{bmatrix} 1 & -1 & 0 & 0 & \cdots & 0 \\ 1 & 0 & -1 & 0 & \cdots & 0 \\ \vdots & \vdots & \vdots & \vdots & \vdots & \vdots \\ 1 & 0 & 0 & 0 & \cdots & -1 \end{bmatrix}. \quad (4.2.14)$$

Then $\mathbf{M}\boldsymbol{\mu} = \boldsymbol{\Delta}_1$, (4.2.3), and, hence, H_0 is equivalent to $\mathbf{M}\boldsymbol{\mu} = \mathbf{0}$. Note that the rows of $\mathbf{M}$ form $k-1$ linearly independent contrasts in the vector $\boldsymbol{\mu}$. If the design matrix given in (4.2.6) is used then the null hypothesis is simply $H_0 : \mathbf{I}_{k-1}\boldsymbol{\Delta}_1 = \mathbf{0}$; that is, all the regression coefficients are zero. We shall discuss two rank-based tests for this hypothesis.

One appropriate test statistic is the gradient test statistic, (3.5.8), which is given by

$$T = \sigma_a^{-2} \mathbf{S}(\mathbf{Z})'(\mathbf{X}'\mathbf{X})^{-1}\mathbf{S}(\mathbf{Z}), \quad (4.2.15)$$

where $\mathbf{S}(\mathbf{Z})' = (S_2(\mathbf{Z}), \ldots, S_k(\mathbf{Z}))$ for

$$S_i(\mathbf{Z}) = \sum_{j=1}^{n_i} a(R(Z_{ij})), \quad (4.2.16)$$

and, as defined in Theorem 3.5.1,

$$\sigma_a^2 = (n-1)^{-1} \sum_{i=1}^{n} a^2(i). \quad (4.2.17)$$

Based on Theorem 3.5.2, a level α test for H_0 versus H_1 is:

$$\text{Reject } H_0 \text{ in favor of } H_1 \text{ if } T \geq \chi^2(\alpha, k-1), \quad (4.2.18)$$

where $\chi^2(\alpha, k-1)$ denotes the upper level α critical value of the χ^2 distribution with $k-1$ degrees of freedom. Because the design matrix $\mathbf{X}$ of model (4.2.6) is an incidence matrix, the gradient test simplifies. First note that

$$(\mathbf{X}'\mathbf{X})^{-1} = \frac{1}{n_1}\mathbf{J} + \text{diag}\left\{\frac{1}{n_2}, \ldots, \frac{1}{n_k}\right\}, \quad (4.2.19)$$

where $\mathbf{J}$ is a $(k-1) \times (k-1)$ matrix of ones; see Exercise 4.8.1. Since the scores sum to 0, we have that $\mathbf{S}(\mathbf{Z})'\mathbf{1}_{k-1} = -S_1(\mathbf{Z})$. Upon combining these results, the gradient test statistic simplifies to

$$T_\varphi = \sigma_a^{-2}\mathbf{S}(\mathbf{Z})'(\mathbf{X}'\mathbf{X})^{-1}\mathbf{S}(\mathbf{Z}) = \sigma_a^{-2} \sum_{i=1}^{k} \frac{1}{n_i} S_i^2(\mathbf{Z}). \quad (4.2.20)$$

For Wilcoxon scores further simplification is possible. In this case

$$S_i(Y) = \sum_{j=1}^{n_i} \sqrt{12}\left(\frac{R(Y_{ij})}{n+1} - \frac{1}{2}\right)$$

$$= \frac{\sqrt{12}}{n+1} n_i \left(\overline{R}_i - \frac{n+1}{2}\right), \qquad (4.2.21)$$

where $\overline{R}_i$ denotes the average of the ranks from sample i. Also for Wilcoxon scores $\sigma_a^2 = n/(n+1)$. Thus the test statistic for Wilcoxon scores is given by

$$H_W = \frac{12}{n(n+1)} \sum_{i=1}^{k} n_i \left(\overline{R}_i - \frac{n+1}{2}\right)^2. \qquad (4.2.22)$$

This is the **Kruskal–Wallis** test statistic (Kruskal and Wallis, 1952). It is distribution-free under H_0. In the case of two levels, the Kruskal–Wallis test is equivalent to the MWW test discussed in Chapter 2; see Exercise 4.8.2. From the discussion on efficiency, Section 3.5, the efficiency results for the Kruskal–Wallis test are the same as for the MWW.

As a second rank-based test, we briefly discuss the drop in dispersion test for H_0 versus H_1 given by expression (3.6.12). Under the null hypothesis, the underlying distributions of the k levels of A are the same; hence, the reduced model is

$$Y_{ij} = \mu + e_{ij}, \qquad (4.2.23)$$

where μ is a common location functional. Thus there are no parameters to fit in this case and the reduced model dispersion is

$$DT = D(Y) = \sum_{i=1}^{k} \sum_{j=1}^{n_i} a(R(Y_{ij})) Y_{ij}. \qquad (4.2.24)$$

The symbol DT denotes total dispersion in the problem which is analogous to the classical LS total variation, SST. Hence the reduction in dispersion is $RD = DT - DE$, where DE is defined in (4.2.12), and the drop in dispersion test is given by $F_\varphi = (RD/(k-1))/(\hat{\tau}_\varphi/2)$. As discussed in Section 3.6, this should be compared with F critical values having $k-1$ and $n-k$ degrees of freedom. The analysis can be summarized in an analysis of dispersion table of the form given in Table 4.2.3.

Because the Kruskal–Wallis test is a gradient test, the drop in dispersion test and the Kruskal–Wallis test have the same asymptotic efficiency; see Section 3.6. The third test discussed in that section, the Wald-type test, will be discussed, for this hypothesis, in Section 4.2.5.

Table 4.2.3: Analysis of dispersion table for the hypotheses (4.2.13)

Source	D = Dispersion	df	MD	F
A	RD	$k-1$	$RD/(k-1)$	F_φ
Error		$n-k$	$\hat{\tau}_\varphi/2$	

One-way Design 241

Table 4.2.4: Tests of hypotheses (4.2.13) for the quail data

Procedure	Test statistic	Scale $\widehat{\sigma}$ or $\widehat{\tau}_\varphi$	df	p-value
LS, F_{LS}	1.14	30.5	(3,35)	0.35
Drop Disp., F_φ	3.77	19.2	(3,35)	0.02
Kruskal–Wallis	7.18		3	0.067

Example 4.2.2 *LDL Cholesterol of Quail, Example 4.2.1 continued*

For the hypothesis (4.2.13) of no difference among the locations of the cholesterol levels of the drug compounds, the results of the LS F-test, the Kruskal–Wallis test, and the drop in dispersion test can be found in Table 4.2.4. The long right tail of the errors spoiled the LS test statistic. Using it, one would conclude that there is no significant difference among the drug compounds, which is inconsistent with the boxplots in Figure 4.2.1. On the other hand, both robust procedures detect the differences among the drug compounds, especially the drop in dispersion test statistic.

4.2.3 Tests of General Contrasts

As discussed above, the parameters and hypotheses of interest for model (4.2.4) can usually be defined in terms of contrasts. In this section we discuss R-estimates and tests of contrasts. We will apply these results to more complicated designs in the remainder of the chapter.

For model (4.2.4), consider general linear hypotheses of the form

$$H_0 : \mathbf{M}\boldsymbol{\mu} = \mathbf{0} \text{ versus } H_A : \mathbf{M}\boldsymbol{\mu} \neq \mathbf{0}, \quad (4.2.25)$$

where $\mathbf{M}$ is a $q \times k$ matrix of contrasts (rows sum to 0) of full row rank. Since $\mathbf{M}$ is a matrix of contrasts, the hypothesis H_0 is invariant to the intercept and, hence, can be tested by the R test statistic discussed in Section 3.6. To obtain the test based on the reduction of dispersion, F_φ, discussed in Section 3.6, we need to fit the reduced model ω which is model (4.2.4) subject to H_0. Let $D(\omega)$ denote the minimum value of the dispersion function for the reduced model fit and let $RD = D(\omega) - DE$ denote the reduction in dispersion. Note that RD is analogous to the reduction in sums of squares of the traditional LS analysis. The test statistic is given by $F_\varphi = (RD/q)/\widehat{\tau}_\varphi/2$. As discussed in Chapter 3, this statistic should be compared with F critical values having q and $n-k$ degrees of freedom. The test can be summarized in the analysis of dispersion table found in Table 4.2.5, which is analogous to the traditional analysis of variance table for summarizing a least-squares analysis.

Table 4.2.5: Analysis of dispersion table for $H_0 : \mathbf{M}\boldsymbol{\mu} = \mathbf{0}$

Source	D = Dispersion	df	MD	F
$\mathbf{M}\widetilde{\mu} = \mathbf{0}$	RD	q	$MRD = RD/q$	F_φ
Error		$n-k$	$\widehat{\tau}_\varphi/2$	

242 *Experimental Designs*

Table 4.2.6: Birth weights of Poland China pigs by litter

Litter	Birth weight								
1	2.0	2.8	3.3	3.2	4.4	3.6	1.9	3.3	2.8	1.1
2	3.5	2.8	3.2	3.5	2.3	2.4	2.0	1.6		
3	3.3	3.6	2.6	3.1	3.2	3.3	2.9	3.4	3.2	3.2
4	3.2	3.3	3.2	2.9	3.3	2.5	2.6	2.8		
5	2.6	2.6	2.9	2.0	2.0	2.1				
6	3.1	2.9	3.1	2.5						
7	2.6	2.2	2.2	2.5	1.2	1.2				
8	2.5	2.4	3.0	1.5						

Example 4.2.3 *Poland China Pigs*

This data set, presented by Scheffé (1959, p. 87), concerns the birth weights of Poland China pigs in eight litters. For convenience we have tabulated that data in Table 4.2.6. There are 56 pigs in the eight litters. The sample sizes of the litters vary from 4 to 10.

In Exercise 4.8.3 a residual analysis is conducted of this data set and the hypothesis (4.2.13) is tested. Here we are only concerned with the following contrast suggested by Scheffé. Assume that litters 1, 3, and 4 were sired by one boar while the other litters were sired by another boar. The contrast of interest is that the average litter birthweights of the pigs sired by the boars are the same, i.e. $H_0 : h = 0$, where

$$h = \tfrac{1}{3}(\mu_1 + \mu_3 + \mu_4) - \tfrac{1}{5}(\mu_2 + \mu_5 + \mu_6 + \mu_7 + \mu_8). \qquad (4.2.26)$$

For this hypothesis, the matrix **M** of expression (4.2.25) is given by [5 −3 5 5 −3 −3 −3 −3]. The value of the LS F-test statistic is 11.19, while $F_\varphi = 15.65$. There are 1 and 48 degrees of freedom for this hypothesis, so both tests are highly significant. Hence both tests indicate a difference in average litter birthweights of the boars. The reason F_φ is more significant than F_{LS} is clear from the residual analysis found in Exercise 4.8.3.

4.2.4 More on Estimation of Contrasts and Location

In this section we further explore the relationship between models (4.2.4) and (4.2.6). This will enable us to formulate the contrast procedure based on pseudo-observations discussed in Section 4.2.5. Recall that the design matrix **X** of (4.2.6) is a centered design matrix based on the last $k-1$ columns of the design matrix **W** of (4.2.4). To determine the relationship between the parameters of these models, we simply match them by location parameter for each level. For model (4.2.4) the location parameter for level i is of course μ_i. In terms of model (4.2.6), the location parameter for the first level is $\alpha - \sum_{j=2}^{k}(n_j/n)\Delta_j$ and that of the ith level is $\alpha - \sum_{j=2}^{k}(n_j/n)\Delta_j + \Delta_i$. Hence, letting $\delta = \sum_{j=2}^{k}(n_j/n)\Delta_j$, we can write the vector of level locations as

$$\boldsymbol{\mu} = (\alpha - \delta)\mathbf{1} + (0, \boldsymbol{\Delta}_1)', \qquad (4.2.27)$$

where $\boldsymbol{\Delta}_1$ is defined in (4.2.3).

Let $\mathbf{h} = \mathbf{M}\boldsymbol{\mu}$ be a $q \times 1$ vector of contrasts of interest (i.e. rows of $\mathbf{M}$ sum to 0). Write $\mathbf{M}$ as $[\mathbf{m}\ \mathbf{M}_1]$. Then by (4.2.27) we have

$$\mathbf{h} = \mathbf{M}\boldsymbol{\mu} = \mathbf{M}_1 \boldsymbol{\Delta}_1. \tag{4.2.28}$$

By Corollary 3.5.6, $\widehat{\boldsymbol{\Delta}}_1$ has an asymptotic $N(\boldsymbol{\Delta}, \tau_\varphi^2 (\mathbf{X}'\mathbf{X})^{-1})$ distribution. Hence, based on (4.2.28), the asymptotic variance-covariance matrix of the estimate $\mathbf{M}\widehat{\boldsymbol{\mu}}$ is

$$\boldsymbol{\Sigma}_\mathbf{h} = \tau_\varphi^2 \mathbf{M}_1 (\mathbf{X}'\mathbf{X})^{-1} \mathbf{M}_1'. \tag{4.2.29}$$

Note that the only difference for the LS fit is that σ^2 would be substituted for τ_φ^2. Expressions (4.2.28) and (4.2.29) are the basic relationships used by pseudo-observations discussed in Section 4.2.5.

To illustrate these relationships, suppose we want a confidence interval for $\mu_i - \mu_{i'}$. Based on (4.2.29), an asymptotic $(1-\alpha)100\%$ confidence interval is given by,

$$\widehat{\mu}_i - \widehat{\mu}_{i'} \pm t_{(\alpha/2, n-k)} \widehat{\tau}_\varphi \sqrt{\frac{1}{n_i} + \frac{1}{n_{i'}}}; \tag{4.2.30}$$

that is the same as for LS except that $\widehat{\tau}_\varphi$ replaces $\widehat{\sigma}$.

Example 4.2.4 *LDL Cholesterol of Quail, Example 4.2.1 continued*

To illustrate the above confidence intervals, Table 4.2.7 displays the six pairwise confidence intervals among the four drug compounds. On the basis of these intervals drug compound II seems best. This conclusion, however, is based on six simultaneous confidence intervals and the problem of overall confidence in these intervals needs to be addressed. This is discussed in some detail in Section 4.3, at which time we will return to this example.

Medians Model

Suppose we are interested in estimates of the level locations themselves. We first need to select a location functional. For the discussion we will use the median; for any other functional, only a change of the scale parameter τ_S is necessary. Assume, then, that the R residuals have been adjusted so that their median is zero. As discussed above, (4.2.10), the estimate of μ_i is $\widehat{Y}_{ij}$, for any $j = 1, \ldots, n_i$, where $\widehat{Y}_{ij}$ is the fitted value of Y_{ij}. Let $\widehat{\boldsymbol{\mu}} = (\widehat{\mu}_1, \ldots, \widehat{\mu}_k)'$. Further,

Table 4.2.7: All pairwise 95% confidence intervals for the quail data based on the Wilcoxon fit

Difference	Estimate	Confidence Interval
$\mu_2 - \mu_1$	−25.0	(−42.7, −7.8)
$\mu_2 - \mu_3$	−21.0	(−38.6, −3.8)
$\mu_2 - \mu_4$	−20.0	(−37.8, −2.0)
$\mu_1 - \mu_3$	4.0	(−13.41, 21.41)
$\mu_1 - \mu_4$	5.0	(−12.89, 22.89)
$\mu_3 - \mu_4$	1.0	(−16.89, 18.89)

$\widehat{\boldsymbol{\mu}}$ is asymptotically normal with mean $\boldsymbol{\mu}$ and the asymptotic variance of $\widehat{\mu}_i$ is given in (4.2.11). As Exercise 4.8.4 shows, the asymptotic covariance between estimates of location levels is:

$$\text{Cov}(\widehat{\mu}_i, \widehat{\mu}_{i'}) = (\tau_S^2 - \tau_\varphi^2)/n, \qquad (4.2.31)$$

for $i \neq i'$. As Exercises 4.8.4 and 4.8.18 show, expressions (3.9.38) and (4.2.31) lead to a verification of the confidence interval (4.2.30).

Note that if the scale parameters are the same, say $\tau_S = \tau_\varphi = \kappa$, then the approximate variance reduces to κ^2/n_i and the covariances are 0. Hence, in this case, the estimates $\widehat{\mu}_i$ are asymptotically independent. This occurs in the following two ways:

1. For the fit of model (4.2.4) use a score function $\varphi(u)$ which satisfies (S.2) and use the location functional based on the corresponding signed-rank score function $\varphi^+(u) = \varphi((u+1)/2)$. The asymptotic theory, however, requires the assumption of symmetric errors. If the Wilcoxon score function is used then the location functional would result in the residuals being adjusted so that the median of the Walsh averages of the adjusted residuals is 0.

2. Use the L_1 score function $\varphi_S(u) = \text{sgn}(u - \frac{1}{2})$ to fit model (4.2.4) and use the median as the location functional. This of course is equivalent to using an L_1 fit on model (4.2.4). The estimate of μ_i is then the cell median.

4.2.5 Pseudo-observations

We next discuss a convenient way to estimate and test contrasts once an R fit of model (4.2.4) is obtained. Let $\widehat{\mathbf{Z}}$ denote the R fit of this model, let $\widehat{\mathbf{e}}$ denote the vector of residuals, let $\mathbf{a}(R(\widehat{\mathbf{e}}))$ denote the vector of scored residuals, and let $\widehat{\tau}_\varphi$ be the estimate of τ_φ. Let $\mathbf{H}_W$ denote the projection matrix onto the column space of the incidence matrix $\mathbf{W}$. Because of (3.2.13), the fact that $\mathbf{1}_n$ is in the column space of $\mathbf{W}$, and that the scores sum to 0, we obtain

$$\mathbf{H}_W \mathbf{a}(R(\widehat{\mathbf{e}})) = \mathbf{0}. \qquad (4.2.32)$$

Define the constant ζ_φ by

$$\zeta_\varphi^2 = \frac{n-k}{\sum_{i=1}^n a^2(i)}. \qquad (4.2.33)$$

Because $n^{-1} \sum a^2(i) \doteq 1$, $\zeta \doteq 1 - (k/n)$. Then the vector of **pseudo-observations** is defined by

$$\widetilde{\mathbf{Z}} = \widehat{\mathbf{Z}} + \widehat{\tau}_\varphi \zeta_\varphi \mathbf{a}(R(\widehat{\mathbf{e}})); \qquad (4.2.34)$$

see Bickel (1976) for a discussion of the pseudo-observations. For the Wilcoxon scores, we obtain

$$\zeta_W^2 = \frac{(n-k)(n+1)}{n(n-1)}. \qquad (4.2.35)$$

Let $\widetilde{\widehat{Z}}$ and $\widetilde{\widehat{e}}$ denote the LS fit and residuals, respectively, of the pseudo-observations, (4.2.34). By (4.2.32) we have,

$$\widetilde{\widehat{Z}} = \widehat{Z}, \qquad (4.2.36)$$

and, hence,

$$\widetilde{\widehat{e}} = \widehat{\tau}_\varphi \zeta_\varphi \mathbf{a}(R(\widehat{e})). \qquad (4.2.37)$$

From this last expression and the definition of ζ_φ in (4.2.33), we obtain

$$\frac{1}{n-k}\widetilde{\widehat{e}}'\widetilde{\widehat{e}} = \widehat{\tau}_\varphi^2. \qquad (4.2.38)$$

Therefore the LS fit of the pseudo-observations results in the R fit of model (4.2.4) and, further, the LS estimator MSE is $\widehat{\tau}_\varphi^2$.

The pseudo-observations can be used to compute the R inference on a given contrast, say $\mathbf{h} = \mathbf{M}\boldsymbol{\mu}$. If the pseudo-observations are used in place of the observations in an LS algorithm, then based on the variance-covariance of $\widehat{\mathbf{h}}$, given in (4.2.29), expressions (4.2.36) and (4.2.38) imply that the resulting LS estimate of $\mathbf{h}$ and the LS estimate of the corresponding variance-covariance matrix of $\widehat{\mathbf{h}}$ will be the R-estimate of $\mathbf{h}$ and the R-estimate of the corresponding variance-covariance matrix of $\widehat{\mathbf{h}}$. Similarly for testing the hypotheses (4.2.25), the LS test using the pseudo-observations will result in the Wald-type R-test, $F_{\varphi,Q}$, of these hypotheses, given by expression (3.6.14). Pseudo-observations will be used in many of the subsequent examples of this chapter.

The pseudo-observations are easy to obtain. For example, the package rglm returns the pseudo-observations directly in the output data set of fits and residuals. These pseudo-observations can then be read into Minitab or another package for further analyses. In Minitab itself, for Wilcoxon scores the robust regression command, RREGR, has the subcommand PSEUDO which returns the pseudo-observations. Then the pseudo-observations can be used in place of the observations in Minitab commands to obtain the R inference on contrasts.

Example 4.2.5 *LDL Cholesterol of Quail, Example 4.2.1 continued*

To demonstrate how easy it is to use the pseudo-observations with Minitab, reconsider Example 4.2.1 concerning LDL cholesterol levels of quail under the treatment of four different drug compounds. Suppose we want the Wald-type R-test of the hypotheses that there is no effect due to the different drug compounds. The pseudo-observations were obtained based on the full model R fit and placed in column 10 and the corresponding levels were placed in column 11. The Wald $F_{\varphi,Q}$ statistic is obtained by using the following Minitab command:

oneway c10 c11

The execution of this command returned the value of the $F_{\varphi,Q} = 3.45$ with a p-value 0.027, which is close to the result based on the F_φ-statistic.

4.3 Multiple Comparison Procedures

Our basic model for this section is model (4.2.4), although much of what we do here pertains to the rest of this chapter also. We will discuss methods based on the R fit of this model as described in Section 4.2.1. In particular, we shall use the same notation to describe the fit, i.e. the R residuals and fitted values are respectively $\widehat{\mathbf{e}}$ and $\widehat{\mathbf{Z}}$, the estimates of $\boldsymbol{\mu}$ and τ_φ are $\widehat{\boldsymbol{\mu}}$ and $\widehat{\tau}_\varphi$, and the vector of pseudo-observation is $\widehat{\mathbf{Z}}$. We also denote the pseudo-observation corresponding to the observation Y_{ij} as $\widehat{Z}_{ij}$.

Besides tests of contrasts of level locations, we often want to make comparisons among the location levels, for instance, all pairwise comparisons among the levels. With so many comparisons to make, overall confidence becomes a problem. Multiple comparison procedures, (MCPs) have been developed to offset this problem. In this section we will explore several of these methods in terms of robust estimation. These procedures can often be directly robustified. It is our intent to show this for several popular methods, including the Tukey T-method. We will also discuss simultaneous, rank-based tests among levels. We will show how simple Minitab code, based on the pseudo-observations, suffices to compute these procedures. It is not our purpose to give a full discussion of MCPs. Such discussions can be found, for example, in Miller (1981) and Hsu (1996).

We will focus on the problem of simultaneous inference for all $\binom{k}{2}$ comparisons $\mu_i - \mu_{i'}$ based on an R fit of model (4.2.4). Recall, (4.2.30), that a $(1-\alpha)100\%$ asymptotic confidence interval for $\mu_i - \mu_{i'}$ based on the R-fit of model (4.2.4) is given by

$$\widehat{\mu}_i - \widehat{\mu}_{i'} \pm t_{(\alpha/2, n-k)} \widehat{\tau}_\varphi \sqrt{\frac{1}{n_i} + \frac{1}{n_{i'}}}. \tag{4.3.1}$$

In this section we say that this confidence interval has **experiment error rate** α. As Exercise 4.8.8 illustrates, simultaneous confidence for several such intervals can easily slip well below $1 - \alpha$. The error rate for a simultaneous confidence procedure will be called its **family error rate**.

We next describe six robust multiple comparison procedures for the problem of all pairwise comparisons. The error rates for them are based on asymptotics. But note that the same is true for MCPs based on least squares when the normality assumption is not valid. Sufficient Minitab code is given to demonstrate how easily these procedures can be performed.

1. *Bonferroni Procedure.* This is the simplest of all the MCPs. Suppose we are interested in making l comparisons of the form $\mu_i - \mu_{i'}$. If each individual confidence interval, (4.3.1), has confidence $1 - \alpha/l$, then the family error rate for these l simultaneous confidence intervals is at most α; see Exercise 4.8.8. To do all comparisons just select $l = \binom{k}{2}$. Hence the R Bonferroni procedure declares

$$\text{levels } i \text{ and } i' \text{ differ if } |\widehat{\mu}_i - \widehat{\mu}_{i'}| \geq t_{(\alpha/(2\binom{k}{2})), n-k)} \widehat{\tau}_\varphi \sqrt{\frac{1}{n_i} + \frac{1}{n_{i'}}}. \tag{4.3.2}$$

The asymptotic family error rate for this procedure is at most α.

To obtain these Bonferroni intervals by Minitab assume that pseudo-observations, $\widetilde{Y}_{ij}$, are in column 10, the corresponding levels, i, are in column 11, and the constant $\alpha/\binom{k}{2}$ is in k1. Then the following two lines of Minitab code will obtain the intervals:

```
oneway c10 c11;
bonferroni k1.
```

2. *Protected LSD Procedure of Fisher.* First use the test statistic F_φ to test the hypotheses that all the level locations are the same, (4.2.13), at level α. If H_0 is rejected then the usual level $1 - \alpha$ confidence intervals, (4.2.30), are used to make the comparisons. If we fail to reject H_0 then either no comparisons are made or the comparisons are made using the Bonferroni procedure. In summary, this procedure declares

levels i and i' differ if $F_\varphi \geq F_{\alpha, k-1, n-k}$ and

$$|\widehat{\mu}_i - \widehat{\mu}_{i'}| \geq t_{(\alpha/2, n-k)} \widehat{\tau}_\varphi \sqrt{\frac{1}{n_i} + \frac{1}{n_{i'}}}. \qquad (4.3.3)$$

This MCP has no family error rate but the initial test does offer protection. In a large simulation study conducted by Carmer and Swanson (1973) this procedure, based on LS estimates, performed quite well in terms of power and level. In fact, it was one of the two procedures recommended. In a moderate-sized simulation study conducted by McKean, Vidmar and Sievers (1989) the robust version of the protected LSD discussed here performed similarly to the analogous LS procedure on normal errors and had a considerable gain in power over LS for error distributions with heavy tails.

Upon rejection of the hypotheses (4.2.13) at level α, the following Minitab code will obtain the comparison confidence intervals. Assume that pseudo-observations, $\widetilde{Y}_{ij}$, are in column 10, the corresponding levels, i, are in column 11, and the constant α is in k1.

```
oneway c10 c11;
fisher k1.
```

The F-test that appears in the ANOVA table upon execution of these commands is Wald's test statistic $F_{\varphi, Q}$ for the hypotheses (4.2.13). Recall from Chapter 3 that it is asymptotically equivalent to F_φ under the null and local hypotheses.

3. *Tukey's T Procedure.* This is an MCP for the set of all contrasts, $h = \sum_{i=1}^{k} c_i \mu_i$, where $\sum_{i=1}^{k} c_i = 0$. Assume that the sample sizes for the levels are the same, say, $n_1 = \cdots = n_k = m$. The basic geometric fact for this procedure is the following equivalence due to Tukey (see Miller, 1981): for $t > 0$,

$$\max_{1 \leq i,i' \leq k} |(\widehat{\mu}_i - \mu_i) - (\widehat{\mu}_{i'} - \mu_{i'})| \leq t \iff$$

$$\sum_{i=1}^{k} c_i \widehat{\mu}_i - \tfrac{1}{2} t \sum_{i=1}^{k} |c_i| \leq \sum_{i=1}^{k} c_i \mu_i \leq \sum_{i=1}^{k} c_i \widehat{\mu}_i + \tfrac{1}{2} t \sum_{i=1}^{k} |c_i|, \quad (4.3.4)$$

for all contrasts $\sum_{i=1}^{k} c_i \mu_i$, where $\sum_{i=1}^{k} c_i = 0$. Hence, to obtain simultaneous confidence intervals for the set of all contrasts we need the distribution of the left-hand side of this inequality. But first note that

$$(\widehat{\mu}_i - \mu_i) - (\widehat{\mu}_{i'} - \mu_{i'}) = \{(\widehat{\mu}_i - \mu_i) - (\widehat{\mu}_1 - \mu_1)\} - \{(\widehat{\mu}_{i'} - \mu_{i'}) - (\widehat{\mu}_1 - \mu_1)\}$$

$$= (\widehat{\Delta}_{i1} - \Delta_{i1}) - (\widehat{\Delta}_{i'1} - \Delta_{i'1}).$$

Hence, we need only consider the asymptotic distribution of $\widehat{\boldsymbol{\Delta}}_1$, which by (4.2.19) is $N_{k-1}(\boldsymbol{\Delta}_1, (\tau_\varphi^2/m)[\mathbf{I} + \mathbf{J}])$.

Recall, if $v_1, \ldots, v_k$ are iid $N(0, \sigma^2)$, then $\max_{1 \leq i,i' \leq k} |v_i - v_{i'}|/\sigma$ has the studentized range distribution, with $k - 1$ and ∞ degrees of freedom. But we can write this random variable as

$$\max_{1 \leq i,i' \leq k} |v_i - v_{i'}| = \max_{1 \leq i,i' \leq k} |(v_i - v_1) - (v_{i'} - v_1)|.$$

Hence we need only consider the random vector of shifts $\mathbf{v}'_1 = (v_2 - v_1, \ldots, v_k - v_1)$ to determine the distribution. But $\mathbf{v}_1$ has distribution $N_{k-1}(\mathbf{0}, \sigma^2[\mathbf{I} + \mathbf{J}])$. Based on this, it follows from the asymptotic distribution of $\widehat{\boldsymbol{\Delta}}_1$ that if we substitute $q_{\alpha;k,\infty} \tau_\varphi/\sqrt{m}$ for t in expression (4.3.4), where $q_{\alpha;k,\infty}$ is the upper α critical value of a studentized range distribution with k and ∞ degrees of freedom, then the asymptotic probability of the resulting expression will be $1 - \alpha$.

The parameter τ_φ, however, is unknown and must be replaced by an estimate. In the Tukey T procedure for LS, the parameter is σ. The usual estimate s of σ is such that if the errors are normally distributed then the random variable $(n - k)s^2/\sigma^2$ has a χ^2 distribution and is independent of the LS location estimates. In this case the studentized range distribution with $k - 1$ and $n - k$ degrees of freedom is used. If the errors are not normally distributed then this distribution leads to an approximate simultaneous confidence procedure. We proceed similarly for the procedure based on the robust estimates. Replacing t in expression (4.3.4) by $q_{\alpha;k,n-k} \widehat{\tau}_\varphi/\sqrt{m}$, where $q_{\alpha;k,n-k}$ is the upper α critical value of a studentized range distribution with k and $n - k$ degrees of freedom, yields an approximate simultaneous confidence procedure for the set of all contrasts. As discussed before, however, small-sample studies have shown that the Student t-distribution works well for inference based on the robust estimates. Hopefully these small-sample properties carry over to the approximation based on the studentized range distribution. Further research is needed in this area.

Tukey's procedure requires that the level sample sizes are the same, which is frequently not the case in practice. A simple adjustment due to Kramer (1956) results in the simultaneous confidence intervals,

$$\widehat{\mu}_i - \widehat{\mu}_{i'} \pm \frac{1}{\sqrt{2}} q_{\alpha;k,n-k} \widehat{\tau}_\varphi \sqrt{\frac{1}{n_i} + \frac{1}{n_{i'}}}. \tag{4.3.5}$$

These intervals have approximate family error rate α. This approximation is often called the **Tukey–Kramer procedure**.

In summary, the R Tukey–Kramer procedure declares

$$\text{levels } i \text{ and } i' \text{ differ if } \quad |\widehat{\mu}_i - \widehat{\mu}_{i'}| \geq \frac{1}{\sqrt{2}} q_{\alpha;k,n-k} \widehat{\tau}_\varphi \sqrt{\frac{1}{n_i} + \frac{1}{n_{i'}}}. \tag{4.3.6}$$

The asymptotic family error rate for this procedure is approximately α.

To obtain these R Tukey intervals by Minitab assume that pseudo-observations, $\widetilde{Y}_{ij}$, are in column 10, the corresponding levels, i, are in column 11, and the constant α is in k1. Then the following two lines of Minitab code will obtain the intervals:

```
oneway c10 c11;
tukey k1.
```

4. *Pairwise Tests Based on Joint Rankings.* The above methods were concerned with estimation and simultaneous confidence intervals for effects. Traditionally, simultaneous nonparametric inference has dealt with comparison tests. The first such procedure we will discuss is based on the combined rankings of all levels, i.e. the rankings that are used by the Kruskal–Wallis test. We will discuss this procedure using the Wilcoxon score function; see Exercise 4.8.10 for the analogous procedure based on a selected score function. Assume a common level sample size m. Denote the average of the ranks for the ith level by $\overline{R}_{i\cdot}$ and let $\mathbf{R}'_1 = (\overline{R}_{2\cdot} - \overline{R}_{1\cdot}, \ldots, \overline{R}_{k\cdot} - \overline{R}_{1\cdot})$. Using the results of Chapter 3, under $H_0 : \mu_1 = \cdots = \mu_k$, $\mathbf{R}_1$ is asymptotically $N_{k-1}(\mathbf{0}, k(n+1)/12(\mathbf{I}_{k-1} + \mathbf{J}_{k-1}))$; see Exercise 4.8.9. Hence, as in the development of the Tukey procedure above, we have the asymptotic result

$$P_{H_0}\left[\max_{1 \leq i,i' \leq k} |\overline{R}_{i\cdot} - \overline{R}_{i'\cdot}| \leq \sqrt{\frac{k(n+1)}{12}} q_{\alpha;k,\infty}\right] \doteq 1 - \alpha. \tag{4.3.7}$$

Hence, the joint ranking procedure declares

$$\text{levels } i \text{ and } i' \text{ differ if } \quad |\overline{R}_{i\cdot} - \overline{R}_{i'\cdot}| \geq \sqrt{\frac{k(n+1)}{12}} q_{\alpha;k,\infty}. \tag{4.3.8}$$

This procedure has an approximate family error rate of α. It is not easy to invert for simultaneous confidence intervals for the effects. We would recommend the Tukey procedure, (2), with Wilcoxon scores for corresponding simultaneous inference on the effects.

An approximate level α test of the hypotheses (4.2.13) is given by

$$\text{Reject } H_0 \text{ if } \max_{1 \leq i,i' \leq k} |\overline{R}_{i\cdot} - \overline{R}_{i'\cdot}| \geq \sqrt{\frac{k(n+1)}{12}}\, q_{\alpha;k,\infty}. \qquad (4.3.9)$$

However, the Kruskal–Wallis test is the usual choice in practice.

The joint ranking procedure, (4.3.9), is approximate for the unequal sample size case. Miller (1981, p. 166) describes a procedure similar to the Scheffé procedure in LS which is valid for the unequal sample size case, but which is also much more conservative; see Exercise 4.8.6. A Tukey–Kramer type rule, (4.3.6), for procedure (4.3.9) is

$$\text{levels } i \text{ and } i' \text{ differ if; } |\overline{R}_{i\cdot} - \overline{R}_{i'\cdot}| \geq \sqrt{\frac{n(n+1)}{24}}\sqrt{\frac{1}{n_i} + \frac{1}{n_{i'}}}\, q_{\alpha;k,\infty}.$$

$$(4.3.10)$$

The small-sample properties of this approximation need to be studied.

5. *Pairwise Tests Based on Separate Rankings.* For this procedure we compare levels i and i' by ranking the combined ith and i'th samples. Let $R_{i\cdot}^{(i')}$ denote the sum of the ranks of the ith level when it is compared with the i'th level. Assume that the sample sizes are the same, $n_1 = \cdots = n_k = m$. For $0 < \alpha < 1$, define the critical value $c_{\alpha;m,k}$ by

$$P_{H_0}\left[\max_{1 \leq i,i' \leq k} R_{i\cdot}^{(i')} \geq c_{\alpha;m,k}\right] = \alpha. \qquad (4.3.11)$$

Tables for this critical value at the 5% and 1% levels are provided in Miller (1981). The separate ranking procedure declares

$$\text{levels } i \text{ and } i' \text{ differ if } R_{i\cdot}^{(i')} \geq c_{\alpha;m,k} \text{ or } R_{i'\cdot}^{(i)} \geq c_{\alpha;m,k}. \qquad (4.3.12)$$

This procedure has an approximate family error rate of α and was developed independently by Steel (1960) and Dwass (1960).

An approximate level α test of the hypotheses (4.2.13) is given by

$$\text{Reject } H_0 \text{ if } \max_{1 \leq i,i' \leq k} R_{i\cdot}^{(i')} \geq c_{\alpha;m,k}, \qquad (4.3.13)$$

although as noted the Kruskal–Wallis test is the usual choice in practice.

Corresponding simultaneous confidence intervals can be constructed similar to the confidence intervals developed in Chapter 2 for a shift in locations based on the MWW statistic. For the confidence interval for the ith and i'th samples corresponding to the test (4.3.12), first form the differences between the two samples, say

$$D_{kl}^{ii'} = Y_{ik} - Y_{i'l}, \quad 1 \leq k, l \leq m.$$

Let $D_{(1)}, \ldots, D_{(m^2)}$ denote the ordered differences. Note here that the critical value $c_{\alpha;m,k}$ is for the sum of the ranks and not statistics of the form S_R^+, (2.4.2). But recall that these versions of the Wilcoxon statistic differ by the constant $m(m+1)/2$. Hence the confidence interval is

$$(D_{(c_{\alpha;m,k}-\frac{m(m+1)}{2}+1)},\, D_{(m^2-c_{\alpha;m,k}+\frac{m(m+1)}{2})}). \qquad (4.3.14)$$

It follows that this set of confidence intervals, over all pairs of levels i and i', forms a set of simultaneous $1-\alpha$ confidence intervals. Using the iterative algorithm discussed in Section 3.7.2, the differences need not be formed.

6. **Procedures Based on Pairwise Distribution-Free Confidence Intervals.** Simple pairwise (separate ranking) MCPs can be easily formulated based on the MWW confidence intervals discussed in Section 2.4.2. Such procedures do not depend on equal sample sizes. As an illustration, we describe a **Bonferroni-type procedure** for the situation of all $l = \binom{k}{2}$ comparisons. For the levels (i, i'), let $[D^{ii'}_{(c_{\alpha/(2l)}+1)}, D^{ii'}_{(n_i n_{i'}-c_{\alpha/(2l)})}]$ denote the $(1-(\alpha/l))100\%$ confidence interval discussed in Section 2.4.2 based on the $n_i n_{i'}$ differences between the ith and i'th samples. This procedure declares

levels i and i' differ if 0 is not in $[D^{ii'}_{(c_{\alpha/(2l)}+1)}, D^{ii'}_{(n_i n_{i'}-c_{\alpha/(2l)})}]$. $\qquad (4.3.15)$

This Bonferroni-type procedure has family error rate at most α. Note that the asymptotic value for $c_{\alpha/(2l)}$ is given by

$$c_{\alpha/(2l)} \doteq \frac{n_i n_{i'}}{2} - z_{\alpha/(2l)}\sqrt{\frac{n_i n_{i'}(n_i+n_{i'}+1)}{12}} - 0.5, \qquad (4.3.16)$$

see (2.4.13). A **Protected LSD-type procedure** can be constructed in the same way, using as the overall test either the Kruskal–Wallis test or the test based on F_φ; see Exercise 4.8.12.

Example 4.3.1 *LDL Cholesterol of Quail, Example 4.2.1 continued*

Reconsider the data on the LDL levels of quail subject to four different drug compounds. The full model fit returned the estimate $\widehat{\boldsymbol{\mu}} = (67, 42, 63, 62)$. We set $\alpha = 0.05$ and ran the first five MCPs on this data set. We used the Minitab code based on pseudo-observations to compute the first three procedures and we obtained the next two by Minitab commands. A table that helps for the separate rank procedure can be found in Lehmann (1975, p. 242) which links the tables in Miller (1981) with a table of family error α values for this procedure. Based on these values, the Minitab MANN command can then be used to obtain the confidence intervals (4.3.14). For each procedure, Table 4.3.1 displays the drug compounds that were declared significantly different by the procedure. The first three procedures, based on effects, declared drug compounds I and II different. Fisher's Protected LSD also declared drug compound II different from drug compounds III and IV. The usual summary schematic based on Fisher's is

2	4	3	1

Table 4.3.1: Drug Compounds Declared Significantly Different by MCPs

Procedure	Compunds Declared Different	Respective Confidence Interval
Bonferroni	(I, II)	(1.25, 49.23)
Fisher	(I, II), (II, III), (II, IV)	(7.83, 42.65) (−38.57, −3.75) (−37.79, −2.01)
Tukey–Kramer	(I, II)	(2.13, 48.35)
Joint Ranking	None	
Separate Ranking	None	

which shows the separation of the second drug compound from the other three compounds. On the other hand, the schematic for either the Bonferroni or Tukey–Kramer procedure is

$$\begin{array}{cccc} 2 & 4 & 3 & 1 \\ \hline \end{array}$$

which shows that though treatment II is significantly different from treatment I it does not differ significantly from either treatments IV or III. The joint ranking procedure came close to declaring drug compounds I and II different because the difference in average rankings between these levels was 12.85, slightly less than the critical value of 13.10. The separate-ranking procedure declared none different. Its interval, (4.3.14), for compounds I and II is (−29, 68.99). In comparison, the corresponding confidence interval for the Tukey procedure based on LS is (−14.5, 58.9). Hence, the separate ranking procedure was impaired more by the outliers than least squares.

4.3.1 Discussion

We have presented robust analogs to three of the most popular MCPs: the Bonferroni, Fisher's protected least significant difference, and the Tukey T method. These procedures provide the user with estimates of the most interesting parameters in these experiments, namely the simple contrasts between treatment effects, and estimates of standard errors with which to assess these contrasts. The robust analogs are straightforward. Replace the LS estimates of the effects by the robust estimates and replace the estimate of σ by the estimate of τ_φ. Furthermore, these robust procedures can easily be obtained by using the pseudo-observations as discussed in Section 4.2.5. Hence, the asymptotic relative efficiency between the LS-based MCP and its robust analog is the same as the ARE between the LS estimator and robust estimator, as discussed in Chapters 1–3. In particular, if Wilcoxon scores are used, then the ARE of the Wilcoxon MCP to that of the LS MCP is 0.955 provided the errors are normally distributed. For error distributions with longer tails than the normal, the Wilcoxon MCP is generally much more efficient than its LS MCP counterpart.

The theory behind the robust MCPs is asymptotic, hence, the error rates are approximate. But this is true also for the LS MCPs when the errors are not normally distributed. Verification of the validity and power of both LS and robust MCPs is based on small-sample studies. The small-sample study by McKean et al. (1989) demonstrated that the Wilcoxon Fisher protected LSD had the same validity as its LS counterpart over a variety of error distributions for a one-way design. For normal errors, the LS MCP had slightly more empirical power than the Wilcoxon. Under error distributions with heavier tails than the normal, though, the empirical power of the Wilcoxon MCP was larger than the empirical power of the LS MCP.

The decision as to which MCP to use has long been debated in the literature. It is not our purpose here to discuss these issues. We refer the reader to books devoted to MCPs for discussions on this topic; see, for example, Miller (1981) and Hsu (1996). We do note that, besides τ_φ replacing σ, the error part of the robust MCP is the same as that of LS; hence, arguments that one procedure dominates another in a certain situation will hold for the robust MCP as well as for LS.

There has been some controversy on the two simultaneous rank-based testing procedures that we presented: pairwise tests based on joint rankings and pairwise tests based on separate rankings. Miller (1981) and Hsu (1996) both favor the tests based on separate rankings because in the separate rankings procedure the comparison between two levels is not influenced by any information from the other levels, which is not the case for the procedure based on joint rankings. They point out that this is also true of the LS procedure, since the comparison between two levels is based only on the difference in sample means for those two levels, except for the estimate of scale. However, Lehmann (1975) points out that the joint ranking makes use of all the information in the experiment while the separate ranking procedure does not. The spacings between all the points is information that is utilized by the joint ranking procedure but lost in the separate ranking procedure. The quail data, Example 4.3.1, are illustrative. The separate ranking procedure did quite poorly on this data set. The sample sizes are moderate and in the comparisons when half of the information is lost, the outliers impaired the procedure. In contrast, the joint ranking procedure came close to declaring drug compounds I and II different. Consider also the LS procedure on this data set. It is true that the ouliers impaired the sample means, but the estimated variance, being a weighted average of the level sample variances, was drawn down somewhat over all the information; for example, instead of using $s_3 = 37.7$ in the comparisons with the third level, the LS procedure uses a pooled standard deviation $s = 30.5$. There is no way to make a similar correction to the separate ranking procedure. Also, the separate rankings procedure can lead to inconsistencies in that it could declare treatment A superior to treatment B, treatment B superior to treatment C, while not declaring treatment A superior to treatment C; see Lehmann (1975, p. 245) for a simple illustration.

4.4 Two-way Crossed Factorial

For this design we have two factors, say A at a levels and B at b levels, that may have an effect on the response. Each combination of the ab factor settings is a treatment. For a completely randomized design, n subjects are selected at random from the reference population and then n_{ij} of these subjects are randomly assigned to the (i, j)th treatment combination; hence, $n = \sum\sum n_{ij}$. Let Y_{ijk} denote the response for the kth subject and the (i, j)th treatment combination, let F_{ij} denote the distribution function of Y_{ijk}, and let $\mu_{ij} = T(F_{ij})$. Then the unstructured or **full model** is

$$Y_{ijk} = \mu_{ij} + e_{ijk}, \qquad (4.4.1)$$

where e_{ijk} are iid with distribution and density functions F and f, respectively. Let T denote the location functional of interest and assume without loss of generality that $T(F) = 0$. The submodels described below utilize the two-way structure of the design.

Model (4.4.1) is the same as the one-way design model (4.2.1) of Section 4.2. Using the scores $a(i) = \varphi(i/(n+1))$, the R fit of this model can be obtained as described in that section. We will use the same notation as in Section 4.2: i.e. $\widehat{e}$ denotes the residuals from the fit adjusted so that $T(F_n) = 0$, where F_n is the empirical distribution function of the residuals; $\widehat{\mu}$ denotes the R-estimate of μ, the $ab \times 1$ vector of the μ_{ij}s; and $\widehat{\tau}_\varphi$ denotes the estimate of τ_φ. For the examples discussed in this section, Wilcoxon scores are used and the residuals are adjusted so that their median is 0.

An interesting submodel is the **additive model**, which is given by

$$\mu_{ij} = \overline{\mu} + (\overline{\mu}_{i.} - \overline{\mu}) + (\overline{\mu}_{.j} - \overline{\mu}). \qquad (4.4.22)$$

For the additive model, the **profile plots** (μ_{ij} versus i or j) are parallel. A diagnostic check for the additive model is to plot the **sample profile plots** ($\widehat{\mu}_{ij}$ versus i or j) and see how close the profiles are to parallel. The null hypotheses of interest for this model are the **main effect hypotheses** given by

$$H_{0A} : \overline{\mu}_{i.} = \overline{\mu}_{i'.} \quad \text{for all } i, i' = 1, \ldots a \text{ and} \qquad (4.4.3)$$
$$H_{0B} : \overline{\mu}_{.j} = \overline{\mu}_{.j'} \quad \text{for all } j, j' = 1, \ldots b. \qquad (4.4.4)$$

Note that there are $a - 1$ and $b - 1$ free constraints for H_{0A} and H_{0B}, respectively. Under H_{0A}, the levels of A have no effect on the response.

The **interaction parameters** are defined as the differences between the full model parameters and the additive model parameters, i.e.

$$\gamma_{ij} = \mu_{ij} - [\overline{\mu} + (\overline{\mu}_{i.} - \overline{\mu}) + (\overline{\mu}_{.j} - \overline{\mu})] = \mu_{ij} - \overline{\mu}_{i.} - \overline{\mu}_{.j} + \overline{\mu} \qquad (4.4.5)$$

The hypothesis of **no interaction** is given by

$$H_{0AB} = \gamma_{ij} = 0, \quad i = 1, \ldots, a, \quad j = 1, \ldots, b. \qquad (4.4.6)$$

Note that are $(a-1)(b-1)$ free constraints for H_{0AB}. Under H_{0AB} the additive model holds.

Historically nonparametric tests for interaction were developed in an *ad hoc* fashion. They generally do not appear in nonparametric texts and this has been a shortcoming of the area. Sawilowsky (1990) provides an excellent review of nonparametric approaches to testing for interaction. The methods we present are simply part of the general R theory in testing general linear hypotheses in linear models and they are analogous to the traditional LS tests for interactions.

All these hypotheses are contrasts in the parameters μ_{ij} of the one-way model, (4.4.1); hence they can easily be tested with the rank-based analysis as described in Section 4.2.3. Usually the interaction hypothesis is tested first. If H_{0AB} is rejected then there is difficulty in interpretation of the main effect hypotheses, H_{0A} and H_{0B}. In the presence of interaction H_{0A} concerns the cell mean averaged over factor B, which may have little practical significance. In this case multiple comparisons (see below) between cells may be of more practical significance. If H_{0AB} is not rejected then there are two schools of thought. The 'pooling' school would take the additive model, (4.4.2), as the new full model to test main effects. The 'nonpoolers' would stick with the unstructured model, (4.4.1), as the full model. In either case with little evidence of interaction present, the main effect hypotheses are more interpretable.

Since model (4.4.1) is a one-way design, the multiple comparison procedures discussed in Section 4.3 can be used. The crossed structure of the design makes for several interesting families of contrasts. When interaction is present in the model, it is often of interest to consider simple contrasts between cell locations. Here, we will only mention all $\binom{ab}{2}$ pairwise comparisons. Among others, the Bonferroni, Fisher, and Tukey T procedures described in Section 4.3 can be used. The rule for the Tukey–Kramer procedure is:

$$\text{cells } (i,j) \text{ and } (i',j') \text{ differ if } \quad |\widehat{\mu}_{ij} - \widehat{\mu}_{i'j'}| \geq \frac{1}{\sqrt{2}} q_{\alpha; ab, n-ab} \, \widehat{\tau}_\varphi \sqrt{\frac{1}{n_{ij}} + \frac{1}{n_{i'j'}}}.$$

(4.4.7)

The asymptotic family error rate for this procedure is approximately α.

The pseudo-observations discussed in Section 4.2.5 can be used in order easily to obtain the Wald test statistic, $F_{\varphi, Q}$, (3.6.14), for tests of hypotheses; similarly, they can be used to obtain multiple comparison procedures for families of contrasts. Simply obtain the R fit of model (4.4.1), form the pseudo-observations, (4.2.34), and input these pseudo-observations in a LS package. The resulting analysis of variance table will contain the Wald-type R tests of the main effect hypotheses, (H_{0A} and H_{0B}), and the interaction hypothesis, (H_{0AB}). As with an LS analysis, one has to know what main hypotheses are being tested by the LS package. For instance, the main effect hypothesis H_{0A}, (4.4.3), is a Type III sums of squares hypothesis in SAS; see Speed, Hocking, and Hackney (1978).

Example 4.4.1 *Lifetime of Motors*

This problem is an unbalanced two-way design discussed by Nelson (1982, p. 471); see also McKean and Sievers (1989) for a discussion on R-analyses of this data set. The responses are lifetimes of three motor insulations (1, 2, and 3), tested at three different temperatures (200°F, 225°F, and 250°F). The design is an unbalanced 3 × 3 factorial with five replicates in six of the cells and three replicates in the others. The data are displayed in Table 4.4.1. Following

Table 4.4.1: Data for Example 4.4.1, lifetimes of motors, (hours)

Temp.	Insulation		
	1	2	3
200°F	1176	2856	3528
	1512	3192	3528
	1512	2520	3528
	1512	3192	
	3528	3528	
225°F	624	816	720
	624	912	1296
	624	1296	1488
	816	1392	
	1296	1488	
250°F	204	300	252
	228	324	300
	252	372	324
	300	372	
	324	444	

Nelson, as the response variable we considered the logs of the lifetimes. Let Y_{ijk} denote the log of the lifetime of the kth replicate at temperature level i with motor insulation j. As a full model we will use model (4.4.1). The results found in Tables 4.4.2 and 4.4.3 are for the R analysis based on Wilcoxon scores with the intercept estimated by the median of the residuals. Hence, the R-estimates of μ_{ij} estimate the true cell medians.

The cell median profile plot based on the Wilcoxon estimates, Figure 4.4.1(a), indicates that some interaction is present. Figure 4.4.1(b) is a plot of the internal Wilcoxon studentized residuals, (3.9.31), versus fitted values. It indicates randomness but also shows several outlying data points which are also quite evident in the $q-q$ plot, Figure 4.4.1(c), of the Wilcoxon studentized residuals versus logistic population quantiles. This plot indicates that score functions for distributions with heavier right tails than the logistic would be more appropriate for these data; see McKean and Sievers (1989) for more discussion on score selection for this example. Figure 4.4.1(d) readily identifies the outliers as the fifth observation in cell (1, 1), the fifth observation in cell (2, 1), and the first observation in cell (2, 3).

Table 4.4.2 is an ANOVA table for the R analysis. Since $F(0.05, 4, 30) = 2.69$, the test of interaction is significant at the 0.05 level. This confirms the

Table 4.4.2: Analysis of dispersion table for lifetime of motors data

Source	RD	df	MRD	F_φ
Temperature (T)	26.40	2	13.20	121.7
Motor Insulation (I)	3.72	2	1.86	17.2
T × I	1.24	4	0.310	2.86
Error		30	0.108	

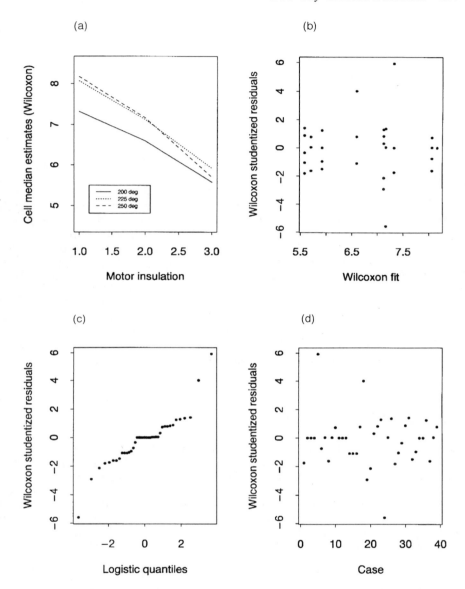

Figure 4.4.1: (a) Cell median profile plot for data of Example 4.4.1, cell medians based on the Wilcoxon fit; (b) internal Wilcoxon studentized residual plot; (c) logistic $q-q$ plot based on internal Wilcoxon studentized residuals; (d) casewise plot of the Wilcoxon studentized residuals

profile plot, Figure 4.4.1(a). It is interesting to note that the least-squares F-test statistic for interaction was 1.30 and, hence, was not significant. The LS analysis was impaired because of the outliers. The row effect hypothesis is that the average row effects are the same. The column effect hypothesis is similarly

Table 4.4.3: Contrasts for differences in insulations at temperature 200°F

Contrast	Estimate	Confidence
$\mu_{11} - \mu_{12}$	−0.76	(−1.22, −0.30)
$\mu_{11} - \mu_{13}$	−0.84	(−1.37, −0.32)
$\mu_{12} - \mu_{13}$	−0.09	(−0.62, 0.44)

defined. Both main effects are significant. In the presence of interaction, though, we have interpretation difficulties with main effects.

In Nelson's discussion of this problem it was of interest to estimate the simple contrasts of mean lifetimes of insulations at the temperature setting of 200°F. Since this is the first temperature setting, these contrasts are $\mu_{1j} - \mu_{1j'}$. Table 4.4.3 displays the estimates of these contrasts along with corresponding confidence intervals formed under the Tukey–Kramer procedure as discussed above, (4.3.5). It seems that insulations 2 and 3 are better than insulation 1 at the temperature of 200°F, but between insulations 2 and 3 there is no discernible difference.

In this example, the number of observations per parameter was less than five. To offset uneasiness over the use of the rank analysis for such small samples, McKean and Sievers (1989) conducted a a Monte Carlo study on this design. The empirical levels and powers of the R analysis were good over situations similar to those suggested by these data.

4.5 Analysis of Covariance

Often there are extraneous variables available besides the response variable. Hopefully these variables explain some of the noise in the data. They are called **covariates** or **concomitant variables** and the traditional analysis of such data is called **analysis of covariance**.

As an example, consider the one-way model (4.2.1) with k levels, and suppose we have a single covariate, say x_{ij}. A first-order model is $y_{ij} = \mu_i + \beta x_{ij} + e_{ij}$. This model, however, assumes that the covariate behaves the same within each treatment combination. A more general model is

$$y_{ij} = \mu_i + \beta x_{ij} + \gamma_i x_{ij} + e_{ij}, \quad j = 1, \ldots, n_i, \ i = 1, \ldots, k. \tag{4.5.1}$$

Hence the slope at the ith level is $\beta_i = \beta + \gamma_i$, and thus each treatment combination has its own linear model. There are two natural hypotheses for this model: $H_{0C} : \beta_1 = \cdots = \beta_k$ and $H_{0L} : \mu_1 = \cdots = \mu_k$. If H_{0C} is true then the differences between the levels of factor A are just the differences in the location parameters μ_i for a given value of the covariate. In this case, contrasts in these parameters are often of interest as well as the hypothesis H_{0L}. If H_{0C} is not true then the covariate and the treatment combinations interact. For example, whether one treatment combination is better than another may depend on where in factor space the responses are measured. Thus, as in crossed factorial designs, the interpretation of main effect hypotheses may not be clear; for more discussion on this point, see Huitema (1980).

The above example is easily generalized. Consider a designed experiment with k treatment combinations. This may be a one-way model with a factor at k levels, a two-way crossed factorial design model with $k = ab$ treatment combinations, or some other design. Suppose we have n_i observations at treatment level i. Let $n = \sum n_i$ denote the total sample size. Denote by $\mathbf{W}$ the full model incidence matrix and by $\boldsymbol{\mu}$ the $k \times 1$ vector of location parameters. Suppose we have p covariates. Let $\mathbf{U}$ be the $n \times p$ matrix of covariates and let $\mathbf{Z}$ denote the $n \times 1$ vector of responses. Let $\boldsymbol{\beta}$ denote the corresponding $p \times 1$ vector of regression coefficients. Then the general **covariate model** is given by

$$\mathbf{Z} = \mathbf{W}\boldsymbol{\mu} + \mathbf{U}\boldsymbol{\beta} + \mathbf{V}\boldsymbol{\gamma} + \mathbf{e}, \tag{4.5.2}$$

where $\mathbf{V}$ is the $n \times pk$ matrix consisting of all column products of $\mathbf{W}$ and $\mathbf{U}$ and the $pk \times 1$ vector $\boldsymbol{\gamma}$ is the vector of interaction parameters between the design and the covariates.

The first hypothesis of interest is

$$H_{0C} : \gamma_{11} = \cdots = \gamma_{pk,pk} \text{ versus } H_{AC} : \gamma_{ij} \neq \gamma_{i'j'} \text{ for some } (i,j) \neq (i',j').$$
$$\tag{4.5.3}$$

Other hypotheses of interest consist of contrasts in the the μ_{ij}. In general, let $\mathbf{M}$ be a $q \times k$ matrix of contrasts and consider the hypotheses

$$H_0 : \mathbf{M}\boldsymbol{\mu} = \mathbf{0} \text{ versus } H_A : \mathbf{M}\boldsymbol{\mu} \neq \mathbf{0}. \tag{4.5.4}$$

Matrices $\mathbf{M}$ of interest are related to the design. For a one-way design $\mathbf{M}$ may be a $(k-1) \times k$ matrix that tests all the location levels to be the same, while for a two-way design it may be used to test that all interactions between the two factors are zero. But as noted above, the hypothesis H_{0C} concerns interaction between the covariate and design spaces. While the interpretation of these later hypotheses, (4.5.4), is clear under H_{0C} it may not be if H_{0C} is false.

The rank-based fit of the full model (4.5.2) proceeds as described in Chapter 3, after a score function is chosen. Once the fitted values and residuals have been obtained, the diagnostic procedures described in Section 3.9 can be used to assess the fit. With a good fit, the model estimates of the parameters and their standard errors can be used to form confidence intervals and regions and multiple comparison procedures can be used for simultaneous inference. Reduced models appropriate for the hypotheses of interest can be obtained and the values of the test statistic F_φ can be used to test them. This analysis can be conducted by the package rglm. It can also be conducted by fitting the full model and obtaining the pseudo-observations. These in turn can be substituted for the responses in a package which performs the traditional LS analysis of covariance in order to obtain the R analysis.

Example 4.5.1 *Snake Data*

As an example of an analysis of covariance problem consider the data set discussed by Afifi and Azen (1972) and reproduced below in Table 4.5.1. It involves four methods, three of which are intended to reduce a human's fear of

Table 4.5.1: Snake data

| Placebo | | Treatment 2 | | Treatment 3 | | Treatment 4 | |
Initial dist.	Final dist.	Initial dist.	Final dist.	Initial dist.	Final dist.	Initial dist.	Final dist.
25	25	17	11	32	24	10	8
13	25	9	9	30	18	29	17
10	12	19	16	12	2	7	8
25	30	25	17	30	24	17	12
10	37	6	1	10	2	8	7
17	25	23	12	8	0	30	26
9	31	7	4	5	0	5	8
18	26	5	3	11	1	29	29
27	28	30	26	5	1	5	29
17	29	19	20	25	10	13	9

snakes. Forty subjects were given a behavior approach test to determine how close they could walk to a snake without feeling uncomfortable. This score was taken as the covariate. Next they were randomly assigned to one of the four treatments, with 10 subjects assigned to a treatment. The first treatment was a control (placebo) while the other three treatments were different methods intended to reduce a human's fear of snakes. The response was a subject's score on the behavior approach test after treatment. Hence, the sample size is 40 and the number of independent variables in model (4.5.2) is 8. Wilcoxon scores were used to conduct the analysis of covariance described above with the residuals adjusted to have median 0.

The plots of the response variable versus the covariate for each treatment are found in Figure 4.5.1(a)–(d). It is clear from the plots that the relationship between the response and the covariate varies with the treatment, from virtually no relationship for the first treatment (placebo) to a fairly strong linear relationship for the third treatment. Outliers are also apparent in these plots. These plots are overlaid with Wilcoxon and LS fits of the full model, model (4.5.1). Figure 4.5.1(e)–(f) show the internal Wilcoxon studentized residual plot and the internal Wilcoxon studentized logistic $q-q$ plot. The outliers stand out in these plots. From the residual plot, the data appear to be heteroscedastic and, as Exercise 4.8.14 shows, the square root transformation of the response does lead to a better fit.

Table 4.5.2 displays the Wilcoxon and LS estimates of the linear models for each treatment. As this table and Figure 4.5.1 show, the larger discrepancies between the Wilcoxon and LS estimates occur for those treatments which have large outliers. The estimates of τ_φ and σ are 3.92 and 5.82, respectively;

Table 4.5.2: Wilcoxon and LS estimates of the linear models for each treatment

| Treatment | Wilcoxon Estimates | | | | LS Estimates | | | |
	Int.	(SE)	Slope	(SE)	Int.	(SE)	Slope	(SE)
1	27.3	(3.6)	−0.02	(0.20)	25.6	(5.3)	0.07	(0.29)
2	−1.78	(2.8)	0.83	(0.15)	−1.39	(4.0)	0.83	(0.22)
3	−6.7	(2.4)	0.87	(0.12)	−6.4	(3.5)	0.87	(0.17)
4	2.9	(2.4)	0.66	(0.13)	7.8	(3.4)	0.49	(0.19)

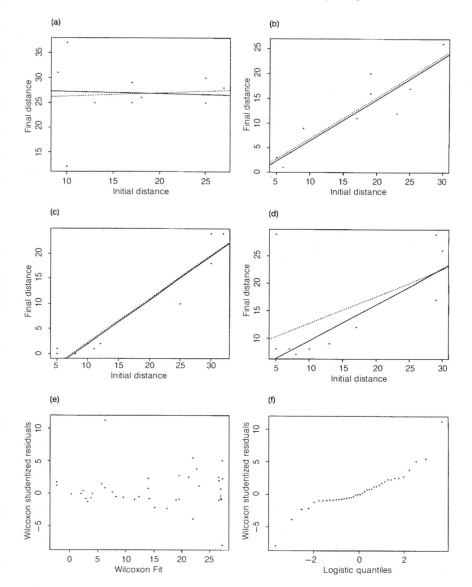

Figure 4.5.1: (a)–(d) For the snake data, scatterplots of final distance versus initial distance for the placebo and treatments 2–4, overlaid with the Wilcoxon fit (solid line) and the LS fit (dashed line); (e) internal Wilcoxon studentized residual plot; (f) Wilcoxon studentized logistic $q-q$ plot.

hence, as the table shows, the estimated standard errors of the Wilcoxon estimates are lower than their LS counterparts.

Table 4.5.3 displays the analysis of dispersion table for these data. Note that F_φ strongly rejects H_{0C}, (p-value is 0.015). This confirms the discussion

Table 4.5.3: Analysis of dispersion (Wilcoxon) for the snake data

Source	D = Dispersion	df	MD	F_φ
H_{OC}	24.06	3	8.021	4.09
Treatment	74.89	3	24.96	12.7
Error		32	1.96	

above based on Figure 4.5.1. The second hypothesis tested is no treatment effect, $H_0 : \mu_1 = \cdots = \mu_4$. Although F_φ strongly rejects this hypothesis, in lieu of the results for H_{0C}, the practical interpretation of such a decision is not obvious. The value of the LS F-test for H_{0C} is 2.34 (p-value is 0.078). If H_{0C} is not rejected then the LS analysis could lead to an invalid interpretation. The outliers spoiled the LS analysis of this data set. As shown in Exercise 4.8.15, both the R analysis and the LS analysis strongly reject H_{0C} for the square root transformation of the response.

4.6 Further Examples

In this section we present two further data examples. Our main purpose in this section is to show how easy it is to use the rank-based analysis on more complicated models. Each example is a three-way crossed factorial design. The first has replicates while the second involves a covariate. Besides displaying tests of the effects, we also consider estimates and standard errors of contrasts of interest.

Example 4.6.1 *Marketing Data*

This data set is drawn from Neter *et al.* (1996, p. 953). A marketing firm research consultant studied the effects that three factors have on the quality of work performed under contract by independent marketing research agencies. The three factors and their levels are: fee level (1, high; 2, average; 3, low); scope (1, all contract work performed in house; 2, some subcontracted out); supervision (1, local supervision; 2, traveling supervisors). The respose was the quality of work performed, as measured by an index. Four agencies were chosen for each level combination. For convenience the data are displayed in Table 4.6.1.

The design is a $3 \times 2 \times 2$ crossed factorial with four replications, which we shall write as

$$y_{ijkl} = \mu_{ijk} + e_{ijkl}, \quad i = 1, \ldots, 3; \; j, k = 1, 2; \; l = 1, \ldots 4, \quad (4.6.1)$$

where y_{ijkl} denotes the response for the lth replicate, at fee i, scope j, and supervision k. Wilcoxon scores were selected for the fit with residuals adjusted to have median 0. Figure 4.6.1(a,b) show the residual and normal $q-q$ plots for the internal R studentized residuals, (3.9.31), based on this fit. The scatter in the residual plot is fairly random and flat. There do not appear to be any outliers. The main trend in the normal $q-q$ plot indicates tails lighter than

Table 4.6.1: Marketing data for Example 4.6.1

	Supervision			
	Local supervision		Traveling supervision	
Fee Level	In House	Sub-out	In House	Sub-out
	124.3	115.1	112.7	88.2
High	120.6	119.9	110.2	96.0
	120.7	115.4	113.5	96.4
	122.6	117.3	108.6	90.1
	119.3	117.2	113.6	92.7
Average	188.9	114.4	109.1	91.1
	125.3	113.4	108.9	90.7
	121.4	120.0	112.3	87.9
	90.9	89.9	78.6	58.6
Low	95.3	83.0	80.6	63.5
	88.8	86.5	83.5	59.8
	92.0	82.7	77.1	62.3

those of a normal distribution. Hence, the fit is good and we proceed with the analysis.

Table 4.6.2 displays the tests of the effects based on the LS and Wilcoxon fits. The Wald-type $F_{\varphi,Q}$ statistic based on the pseudo-observations is also given. The LS and Wilcoxon analyses agree, which is not surprising based on the residual plot. The main effects are highly significant and the only significant interaction is that between scope and supervision.

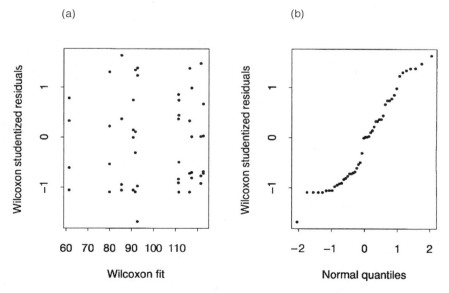

Figure 4.6.1: (a) Wilcoxon studentized residual plot for data of Example 4.6.1; (b) Wilcoxon studentized residual normal $q-q$ plot

As a subsequent analysis, we shall consider nine contrasts of interest. We will use the Bonferroni method based on the pseudo-observations as discussed in Section 4.3. We used Minitab to obtain the results that follow. Because the fee factor does not interact with the other two factors, the contrasts of interest for this factor are: $\mu_{1..} - \mu_{2..}$, $\mu_{1..} - \mu_{3..}$, and $\mu_{2..} - \mu_{3..}$. Table 4.6.3 presents the estimates of these contrasts and the 95% Bonferroni confidence intervals which are given by the estimates of the contrast $\pm t_{(0.05/18;36)} \hat{\tau}_\varphi \sqrt{2/16} \doteq \pm 2.64$. From these results, quality of work significantly improves for either higher or average fees over low fees. The results for high or average fees are insignificant.

Since the scope and supervision factors interact, but do not interact separately or jointly with the fee factor, the parameters of interest are the simple contrasts among $\mu_{.11}, \mu_{.12}, \mu_{.21}$ and $\mu_{.22}$. Table 4.6.4 displays the estimates of these parameters. Using $\alpha = 0.05$, the Bonferroni bound for a simple contrast here is $t_{(0.05/18;36)} \hat{\tau}_\varphi \sqrt{2/12} \doteq 3.04$. Hence all six simple pairwise contrasts among these parameters are significantly different from 0. In particular, averaging over fees, the best quality of work occurs when all contract work is done in house and under local supervision. The source of the interaction between scope and supervision is also clear from these estimates.

Table 4.6.2: Tests of effects for the market data

Effect	df	F_{LS}	F_φ	$F_{\varphi,Q}$
Fee	2	679.0	207.0	793.0
Scope	1	248.0	160.0	290.0
Supervision	1	518.0	252.0	596.0
Fee × Scope	2	0.108	0.098	0.103
Fee × Super.	2	0.053	0.004	0.002
Scope × Super.	1	77.7	70.2	89.6
Fee × Scope × Super.	2	0.266	0.532	0.362
$\hat{\sigma}$ or $\hat{\tau}_\varphi$	36	2.72	2.53	2.53

Table 4.6.3: Contrasts of interest for the market data

Contrast	Estimate	Confidence Interval
$\mu_{1..} - \mu_{2..}$	1.05	(−1.59, 3.69)
$\mu_{1..} - \mu_{3..}$	31.34	(28.70, 33.98)
$\mu_{2..} - \mu_{3..}$	30.28	(27.64, 32.92)

Table 4.6.4: Parameters of interest for the market data

Parameter	$\mu_{.11}$	$\mu_{.12}$	$\mu_{.21}$	$\mu_{.22}$
Estimate	111.9	101.0	106.4	81.64

Example 4.6.2 *Pigs and Diets*

This data set is discussed in Rao (1973, p. 291). It concerns the effect of diets on the growth rate of pigs. There are three diets, called A, B and C. Besides the diet classification, the pigs were classified according to their pens (five levels) and sex (two levels). Their initial weight was also recored as a covariate. The data are displayed in Table 4.6.5.

The design is a $5 \times 3 \times 2$ crossed factorial with only one replication. For comparison purposes, we will use the same model that Rao used which is a fixed effects model with main effects and the two-way interaction between the factors diets and sex. Letting y_{ijk} and x_{ijk} denote, respectively, the growth rate in pounds per week and the initial weight of the pig in pen i, on diet j and of sex k, this model is given by

$$y_{ijk} = \mu + \alpha_i + \beta_j + \gamma_k + (\beta\gamma)_{jk} + \delta x_{ijk} + e_{ijk}, \quad i = 1, \ldots, 5;\ j = 1, 2, 3;\ k = 1, 2.$$
(4.6.2)

Table 4.6.5: Data for Example 4.6.2

Pen	Diet	Sex	Initial wt.	Growth rate
1	A	G	48	9.94
	B	G	48	10.00
	C	G	48	9.75
	C	H	48	9.11
	B	H	39	8.51
	A	H	38	9.52
2	B	G	32	9.24
	C	G	28	8.66
	A	G	32	9.48
	C	H	37	8.50
	A	H	35	8.21
	B	H	38	9.95
3	C	G	33	7.63
	A	G	35	9.32
	B	G	41	9.34
	B	H	46	8.43
	C	H	42	8.90
	A	H	41	9.32
4	C	G	50	10.37
	A	H	48	10.56
	B	G	46	9.68
	A	G	46	10.98
	B	H	40	8.86
	C	H	42	9.51
5	B	G	37	9.67
	A	G	32	8.82
	C	G	30	8.57
	B	H	40	9.20
	C	H	40	8.76
	A	H	43	10.42

266 *Experimental Designs*

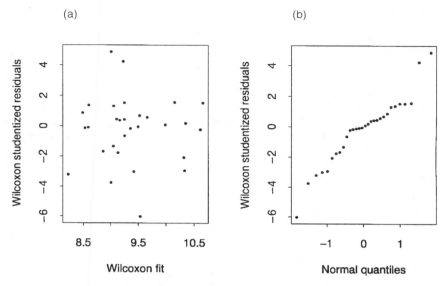

Figure 4.6.2: (a) Internal Wilcoxon studentized residual plot for data of Example 4.6.2; (b) internal Wilcoxon studentized residual normal $q-q$ plot

For convenience we have written the model as an over parameterized model; although, we could have expressed it as a cell means model with constraints for the interaction effects which are assumed to be 0. The effects of interest are the diet effects, β_j.

We fit the model using the Wilcoxon scores. The analysis could also be carried out using pseudo-observations and Minitab. Figure 4.6.2(a) and (b) display the residual plot and normal $q-q$ plot of the internal R studentized residuals based on the Wilcoxon fit. The residual plot shows the three outliers. The outliers are prominent in the $q-q$ plot, but note that even the remaining plotted points indicate an error distribution with heavier tails than the normal. Not surprisingly the estimate of τ_φ is smaller than that of σ, 0.413 and 0.506, respectively. The largest outlier corresponds to the 6th pig which had the lowest initial weight (recall that the internal R studentized residuals account for position in factor space), but its response was above the first quartile. The second largest outlier corresponds to the pig which had the lowest response.

Table 4.6.6 displays the results of the tests for the effects for the LS and Wilcoxon fits. The pseudo-observations were obtained based on the Wilcoxon fit and were inputed as the responses in SAS to obtain $F_{\varphi,Q}$ using Type III sums of squares. The Wilcoxon analyses based on F_φ and $F_{\varphi,Q}$ are quite similar. All three tests indicate no interaction between the factors diet and sex, which clarifies the interpretation of the main effects. Also all three agree on the need for the covariate. Diet has a significant effect on weight gain, as does sex. The robust analyses indicate that pens are also a contributing factor.

Table 4.6.7 displays the results of the analyses when the covariate is not taken into account. It is interesting to note, here, that the factor diet is not

Table 4.6.6: Test statistics for the effects of pigs and diets data with initial weight as a covariate

Effect	df	F_{LS}	F_φ	$F_{\varphi,Q}$
Pen	4	2.35	3.65*	3.48*
Diet	2	4.67*	7.98*	8.70*
Sex	1	5.05*	8.08*	8.02*
Diet × Sex	2	0.17	1.12	0.81
Initial wt.	1	13.7*	19.2*	19.6*
$\hat{\sigma}$ or $\hat{\tau}_\varphi$	19	0.507	0.413	0.413

*Denotes significance at the 0.05 level

Table 4.6.7: Test statistics for the effects of pigs and diets data with no covariate

Effect	df	F_{LS}	F_φ	$F_{\varphi,Q}$
Pen	4	2.95*	4.20*	5.87*
Diet	2	2.77	4.80*	5.54*
Sex	1	1.08	3.01	3.83
Diet × Sex	2	0.55	1.28	1.46
$\hat{\sigma}$ or $\hat{\tau}_\varphi$	20	0.648	0.499	0.501

*Denotes significance at the 0.05 level

significant based on the LS fit, while it is for the Wilcoxon analyses. The heavy tails of the error distribution, as evident in the residual plots, have foiled the LS analysis.

4.7 Rank Transform

In this section we present a short comparison between the rank-based analysis of this chapter and the rank-transform (RT) analysis. Much of this discussion is drawn from McKean and Vidmar (1994). The main point of this section is to show that often the RT test does not work well for testing hypotheses in factorial designs and more complicated models. Hence, we do not recommend using RT methods. On the other hand, Akritas, Arnold and Brunner (1997) develop a unique approach in which factorial hypotheses are replaced by corresponding nonparametric hypotheses based on cdfs. They then show that RT-type methods are appropriate for these nonparametric hypotheses.

As we have pointed out, the rank-based analysis is quite analogous to the LS-based traditional analysis. It is based on R-estimates, while the traditional analysis is based on LS estimates. The only difference in the geometry of estimation is that that the R-estimates are based on the pseudo-norm (3.2.6) while the LS estimates are based on the Euclidean pseudo-norm. The rank-based analysis produces confidence intervals and regions and tests of general linear hypotheses. The diagnostic procedures of Chapter 3 can be used to check the adequacy of the fit of the model and determine outliers and influential points. Furthermore, the efficiency properties discussed for the simple location nonparametric procedures carry over to this analysis. The rank-based analysis offers the user a complete and highly efficient analysis of a linear model as an alternative to the traditional analysis. Further, there are computational algorithms available for these procedures.

Proposed by Conover and Iman (1981), the RT has become a very popular procedure. The RT test of a linear hypothesis consists generally of ranking the dependent variable and then performing the LS test on these ranks. Although in general the RT offers no estimation, and hence no model checking, it is a simple procedure for testing.

Some basic differences between the rank-based analysis and RT are readily apparent. In linear models the Y_is are independent but not identically distributed. Hence when the RT is applied indiscriminately to a linear model, the ranking is performed on nonidentically distributed items. The rankings in the RT are not 'free' of the xs. In contrast, the residuals based on the R-estimates, under Wilcoxon scores, satisfy

$$\sum_{i=1}^n x_{ij} R(Y_i - \mathbf{x}_i' \widehat{\boldsymbol{\beta}}_R) \doteq 0, \quad j = 1, \ldots, p. \quad (4.7.1)$$

Hence the R residuals have been adjusted by the fit so that the ranks are orthogonal to the x-space, i.e. the ranks are 'free' of the xs. These are the ranks that are used in the R test statistic F_φ, for the full model. Under H_0 this would also be true of the expected ranks of the residuals in the R fit of the reduced model. Note also that the statistic F_φ is invariant to the values of the parameters of the reduced model.

Unlike the rank-based analysis, there is no general supporting theory for the RT. Hora and Conover (1984) presented asymptotic null theory on the RT for treatment effect in a randomized block design with no interaction. Thompson and Ammann (1989) explored the efficiency of this RT, showing, however, that this efficiency depends on the block parameters. RT theory for repeated measures designs has been developed by Akritas (1991; 1993) and Thompson (1991b). These extensions also have the unpleasant trait that their efficiencies depend on nuisance parameters.

Many of these theoretical studies on the RT have raised serious questions concerning the validity of the RT for simple two-way and more complicated designs. For a two-way crossed factorial design, Brunner and Neumann (1986) showed that the RT statistics are not reasonable for testing main effects in the presence of interaction for designs larger than 2×2 designs. This was echoed by Akritas (1990) who stated further that RT statistics are not reasonable test statistics for interaction or most other common hypotheses in either two-way crossed or nested classifications. In several of these articles, (see Akritas, 1990; Thompson, 1991a; 1993), the nonlinear nature of the RT is faulted. For a given model the hypotheses of interest are linear contrasts in model parameters. The RT, though, is nonlinear; hence, often the original hypothesis is no longer tested by the rank-transformed data. The same issue was raised earlier by Fligner (1981) in a discussion of the article by Conover and Iman (1981).

In terms of small-sample properties, initial simulations of the RT analysis on certain models (see, for example, Iman, 1974) did appear promising. Now there has been ample evidence based on simulation studies questioning the wisdom of doing RTs on designs as simple as two-way factorial designs with interaction; see, for example, Blair, Sawilowsky, and Higgins (1987). We discuss one such study next and then present an analysis of covariance example where the use of the RT results in a poor analysis.

4.7.1 Monte Carlo Study

Another major Monte Carlo study on the RT was performed by Sawilowsky, Blair, and Higgins (1989), which investigated the behavior of the RT over a three-way factorial design with interaction. In many of their situations, the RT gave severely inflated empirical levels and severely deflated empirical powers. We present the results of a small Monte Carlo study discussed in McKean and Vidmar (1994), which is based on the study of Sawilowsky et al. The model for the study is a $2 \times 2 \times 2$ three-way factorial design. The shortcomings of the RT as discussed in the two-way models above seem to become worse for such models. Letting A, B, and C denote the factors, the model is

$$Y_{ijkl} = \mu + a_i + b_j + c_k + (ab)_{ij} + (ac)_{ik} + (bc)_{jk} + (abc)_{ijk} + e_{ijkl},$$

$$i, j, k = 1, 2; \quad l = 1, \ldots, r,$$

where r is the number of replicates per cell. In the study by Sawilowsky et al., r was set at 2, 5, 10 or 20. Several distributions were considered for the errors e_{ijkl}, including the normal. They considered the usual seven hypotheses (3 main effects, 3 two-way, and 1 three-way) and eight patterns of alternatives. The nonnull effects were set at $\pm c$, where c was a multiple of σ; see also McKean and Vidmar (1992) for further discussion. The study of Sawilowsky et al. found that the RT test for interaction 'is dramatically nonrobust at times and that it has poor power properties in many cases'.

In order to compare the behavior of the rank-based analysis and the RT, on this design, we performed part of their simulation study. We considered standard normal errors and contaminated normal errors, which had 10% contamination from a normal distribution with mean 0 and standard deviation 8. The normal variates were generated as discussed in Marsaglia and Bray (1964) using uniform variates which were generated by a portable Fortran generator written by Kahaner, Moler, and Nash (1989). There were five replications per cell and the nonnull constant of proportionality c was set at 0.75. The simulation size was 1000.

Tables 4.7.1 and 4.7.2 summarize the results of our study for the following two situations: the two-way interaction $A \times C$ and the three-way interaction effect $A \times B \times C$. The alternative for the $A \times C$ situation had all main effects and all two-way interactions in, while the alternative for the $A \times B \times C$ situation had all main effects and two-way interactions besides the three-way alternative in. These were poor situations for the RT in the study conducted by Sawilowsky et al. and, as Tables 4.7.1 and 4.7.2 indicate, the RT behaves poorly for these situations in our study also. Its empirical α levels are deplorable. For instance, at the nominal 0.10 level for the three-way interaction test under normal errors, the RT has an empirical level of 0.777, while the level is 0.511 for the contaminated normal. In contrast, the

Table 4.7.1: Empirical levels and power for test of $A \times C$

	Error distribution											
	Normal errors						Contaminated normal errors					
	Model						Model					
	Null			Alternative			Null			Alternative		
	Nominal α			Nominal α			Nominal α			Nominal α		
Method	0.10	0.05	0.01	0.10	0.05	0.01	0.10	0.05	0.01	0.10	0.05	0.01
LS	0.095	0.040	0.009	0.998	0.995	0.977	0.087	0.029	0.001	0.602	0.505	0.336
Wilcoxon	0.104	0.060	0.006	0.997	0.992	0.970	0.079	0.032	0.004	0.934	0.887	0.713
RT	0.369	0.243	0.076	0.847	0.770	0.521	0.221	0.128	0.039	0.677	0.576	0.319

Table 4.7.2: Empirical levels and power for test of $A \times B \times C$

	Error distribution											
	Normal errors						Contaminated normal errors					
	Model						Model					
	Null			Alternative			Null			Alternative		
	Nominal α			Nominal α			Nominal α			Nominal α		
Method	0.10	0.05	0.01	0.10	0.05	0.01	0.10	0.05	0.01	0.10	0.05	0.01
LS	0.094	0.050	0.005	1.00	0.998	0.980	0.102	0.041	0.001	0.598	0.485	0.301
Wilcoxon	0.101	0.060	0.004	0.997	0.992	0.970	0.085	0.039	0.006	0.948	0.887	0.713
RT	0.777	0.644	0.381	0.484	0.343	0.144	0.511	0.377	0.174	0.398	0.276	0.105

levels of the rank-based analysis were quite close to the nominal levels under normal errors and slightly conservative under the contaminated normal errors. In terms of power, note that the empirical power of the rank-based analysis is slightly less than the empirical power of LS under normal errors while it is substantially greater than the power of LS under contaminated normal errors. For the three-way interaction test, the empirical power of the RT falls below its empirical level.

Example 4.7.1 *The Rat Data*

The following example, taken from Shirley (1981), contrasts the rank-based methods, the rank-transformed methods, and least-squares methods in an analysis of covariance setting. The response is the time it takes a rat to enter a chamber after receiving a treatment designed to delay the time of entry. There were 30 rats in the experiment and they were divided evenly into three groups. The rats in groups 2 and 3 received an antidote to the treatment. The covariate is the time taken by the rat to enter the chamber prior to its treatment. The data are presented in Table 4.7.3 and are displayed in Figure 4.7.1(a).

Table 4.7.3: Rat data

Group 1		Group 2		Group 3	
Initial time	Final time	Initial time	Final time	Initial time	Final time
1.8	79.1	1.6	10.2	1.3	14.8
1.3	47.6	0.9	3.4	2.3	30.7
1.8	64.4	1.5	9.9	0.9	7.7
1.1	68.7	1.6	3.7	1.9	63.9
2.5	180.0	2.6	39.3	1.2	3.5
1.0	27.3	1.4	34.0	1.3	10.0
1.1	56.4	2.0	40.7	1.2	6.9
2.3	163.3	0.9	10.5	2.4	22.5
2.4	180.0	1.6	0.8	1.4	11.4
2.8	132.4	1.2	4.9	0.8	3.3

We considered as a full model,

$$y_{ij} = \alpha_j + \beta_j x_{ij} + e_{ij}, \; j = 1, \ldots, 3; \; i = 1, \ldots, 10 \,, \quad (4.7.2)$$

where y_{ij} denotes the response for the ith rat in group j and x_{ij} denotes the corresponding covariate. There is a slight quadratic aspect to the Wilcoxon residual plot, Figure 4.7.1(b), which is investigated in Exercise 4.8.16.

Figure 4.7.1(c) displays a plot of the internal Wilcoxon studentized residuals by case. Note that there are several outliers. These also can be seen in the plots of the data for groups 2 and 3, Figure 4.7.1(e,f). Note that the outliers have an effect on the LS fits, drawing the fits toward the outliers in each group. In particular, for group 3, it only took one outlier to spoil the LS fit. On the other hand, the Wilcoxon fit is not affected by the outliers. The estimates are given in Table 4.7.4. As the plots indicate, the LS and Wilcoxon estimates differ numerically. Further evidence of the more precise R fits relative to the LS fits is given by the estimates of the scale parameters σ and τ_φ found in Table 4.7.4.

We first test for homogeneity of slopes for the groups; i.e. $H_0 : \beta_1 = \beta_2 = \beta_3$. As clearly shown in Figure 4.7.1(a), this does not appear to be true for these data. While the slopes for groups 2 and 3 seem to be about the same (the Wilcoxon 95% confidence interval for $\beta_2 - \beta_3$ is 3.9 ± 27.2), the slope for group 1 appears to differ from the other two. To confirm this statistically, the value of the F_φ statistic to test homogeneity of slopes, H_0, has the value 9.88 with 2 and 24 degrees of freedom, which is highly significant ($p < 0.001$). This says that group 1, the group that did not receive the antidote, does differ significantly from the other two groups in terms of how the groups interact with the

Table 4.7.4: LS and Wilcoxon estimates (standard errors) for the rat data

Procedure	Group 1		Group 2		Group 3		σ or τ_φ
	α	β	α	β	α	β	
LS	−39.1 (20.0)	76.8 (10.0)	−15.6 (22.0)	20.5 (14.0)	−14.7 (19.0)	21.9 (12.0)	20.5
Wilcoxon	−54.3 (16.0)	84.2 (8.6)	−19.3 (18.0)	21.0 (11.0)	−11.6 (16.0)	17.4 (10.0)	17.0

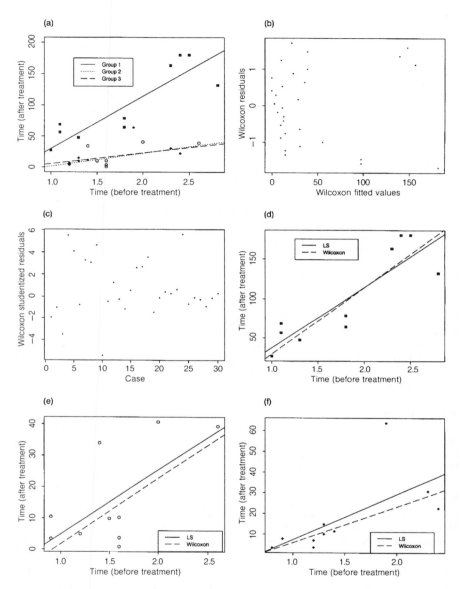

Figure 4.7.1: (a) Wilcoxon fits of all groups; (b) internal Wilcoxon studentized residual plot; (c) internal Wilcoxon studentized residuals by case; (d) LS (solid line) and Wilcoxon (dashed line) fits for group 1; (e) LS (solid line) and Wilcoxon (dashed line) fits for group 2; (f) LS (solid line) and Wilcoxon (dashed line) fits for group 3.

covariate. In particular, the estimated slope of post-treatment time over pretreatment time for the rats in group 1 is about 4 times as large as the slope for the rats in the two groups which received the antidote. Because there is interaction between the groups and the covariate, we did not proceed with the second test on average group effects, i.e. testing $\alpha_1 = \alpha_2 = \alpha_3$.

Shirley (1981) performed a rank-transform on this data by ranking the response and then applying standard least-squares analysis. It is clear from Figure 4.7.1(a) that this nonlinear transform will result in homogeneous slopes for the ranked problem, as confirmed by Shirley's analysis. But the RT is a nonlinear transform and the subsequent analysis based on the rank-transformed data does not test homogeneity of slopes in model (4.7.2). The RT analysis is misleading in this case.

Note that, using the rank-based analysis, we performed an overall analysis of this data set, including a residual analysis for model criticism. Hypotheses of interest were readily tested and estimates of contrasts, along with standard errors, were easily obtained.

4.8 Exercises

4.8.1 *Derive expression (4.2.19).*

4.8.2 *In Section 4.2.2 when we have only two levels, show that the Kruskal–Wallis test is equivalent to the MWW test discussed in Chapter 2.*

4.8.3 *Consider a one-way design for the data in Example 4.2.3. Fit the model using Wilcoxon estimates and conduct a residual analysis, including residual and q–q plots of standardized residuals. Identify any outliers. Next test the hypothesis (4.2.13) using the Kruskal–Wallis test and the test based on F_φ.*

4.8.4 *Using the notation of Section 4.2.4, show that the asymptotic covariance between $\widehat{\mu}_i$ and $\widehat{\mu}_{i'}$ is given by expression (4.2.31). Next show that expressions (3.9.38) and (4.2.31) lead to a verification of the confidence interval (4.2.30).*

4.8.5 *Show that the asymptotic covariance between estimates of location levels is given by expression (4.2.31).*

4.8.6 *Suppose $\mathbf{D}$ is a symmteric, positive definite matrix. Prove that*

$$\sup_{\mathbf{h}} \frac{\mathbf{h}'\mathbf{y}}{\sqrt{\mathbf{h}'\mathbf{D}^{-1}\mathbf{h}}} = \sqrt{\mathbf{y}'\mathbf{D}\mathbf{y}}. \qquad (4.8.1)$$

Refer to the Kruskal–Wallis statistic H_W, given in expression (4.2.22). Let $\mathbf{y}' = ((\overline{R}_1 - (n+1)/2, \ldots, \overline{R}_k - (n+1)/2)$ and $\mathbf{D} = 12/[n(n+1)]\mathrm{diag}(n_1, \ldots, n_k)$. Then, using (4.8.1), show that

$$H_W \leq \chi^2_\alpha(k-1) \text{ if and only if } \sum_{i=1}^{k} \frac{h_i\left(\overline{R}_i - \frac{n+1}{2}\right)}{\sqrt{\frac{n(n+1)}{12} \sum_{j=1}^{k} \frac{1}{n_j} h_j^2}} \leq \sqrt{\chi^2(k-1)},$$

for all vectors $\mathbf{h}$ such that $\sum h_i = 0$.

Hence, if the Kruskal–Wallis test rejects H_0 at level α then there must be at least one contrast in the rank averages that exceeds the critical value $\sqrt{\chi^2(k-1)}$.

This provides Scheffé-type multiple contrast tests with family error rate approximately equal to α.

4.8.7 Apply the procedure presented in Exercise 4.6.8 to the quail data of Example 4.2.1. Use $\alpha = 0.10$.

4.8.8 Let I_1 and I_2 be $(1-\alpha)100\%$ confidence intervals for parameters θ_1 and θ_2, respectively. Show that

$$P[\{\theta_1 \in I_1\} \cap \{\theta_2 \in I_2\}] \geq 1 - 2\alpha. \qquad (4.8.2)$$

(a) Suppose the confidence intervals I_1 and I_2 are independent. Show that

$$1 - 2\alpha \leq P[\{\theta_1 \in I_1\} \cap \{\theta_2 \in I_2\}] \leq 1 - \alpha.$$

(b) Generalize expression (4.8.2) to k confidence intervals and derive the Bonferroni procedure described in (4.3.2).

4.8.9 In the notation of the pairwise tests based on joint rankings procedure of Section 4.7, show that $\overline{\mathbf{R}}_1$ is asymptotically $N_{k-1}(\mathbf{0}, (k(n+1))/12) \times (\mathbf{I}_{k-1} + \mathbf{J}_{k-1}))$ under $H_0 : \mu_1 = \cdots = \mu_k$. (Hint: The asymptotic normality follows as in Theorem 3.5.2. In order to determine the covariance matrix of $\overline{\mathbf{R}}_1$, first obtain the covariance matrix of the random vector $\overline{\mathbf{R}}' = (\overline{R}_1., \ldots, \overline{R}_k.)$ and then obtain the covariance matrix of $\overline{\mathbf{R}}_1$ by using the transformation $[-\mathbf{1}_{k-1} \; \mathbf{I}_{k-1}]$.)

4.8.10 In Section 4.3, the **pairwise tests based on joint rankings** procedure was discussed based on Wilcoxon scores. Generalize this procedure for an arbitrary score function $\varphi(u)$.

4.8.11 For the baseball data in Exercise 1.13.25, consider the following one-way problem. The response of interest is the hitter's average and the three groups are left-handed hitters, right-handed hitters, and switch hitters. Using either Minitab or rglm, obtain the following analyses based on Wilcoxon scores:

(a) Using the test statistic F_φ, test for an overall group effect. Obtain the p-value and conclude at the 5% level.

(b) Use the protected LSD procedure of Fisher to compare the groups at the 5% level.

4.8.12 Consider the Bonferroni-type procedure described in item 6 of Section 4.3. Formulate a similar protected LSD-type procedure based on the test statistic F_φ. Use these procedures to make the comparisons discussed in Exercise 4.8.11.

4.8.13 Consider the baseball data in Exercise 1.13.25. In Exercise 3.12.38, we investigated the linear relationship between a player's height and his weight. For this problem, consider the simple linear model

$$\text{height} = \alpha + \text{weight}\,\beta + e.$$

Using Wilcoxon scores and either Minitab or rglm, investigate whether or not the same simple linear model can be used for both the pitchers and hitters. Obtain the p-value for the test of this hypothesis based on the statistic F_φ.

4.8.14 In Example 4.5.1 obtain the square root of the response and fit it to the full model. Perform a residual analysis on the resulting fit. In particular, identify any outliers and compare the heteroscedasticity in the plot of the residuals versus the fitted values with the analogous plot in Example 4.5.1.

4.8.15 For Example 4.5.1, overlay the Wilcoxon and LS fits for the four treatments based on the square root transformation of the response. Then obtain an analysis of covariance for both the Wilcoxon and LS analyses for the transformed data. Comment on the plots and the results of the analyses.

4.8.16 Consider Example 4.7.1. Investigate whether a model which also includes quadratic terms in the covariates is more appropriate for the rat data than model (4.7.2).

4.8.17 Consider Example 4.7.1. Eliminate the placebo group, Group 1, and perform an analysis of covariance on groups 2 and 3. Use the linear model, (4.7.2). Is there any difference between these groups?

4.8.18 Let $\mathbf{H_W} = \mathbf{W(W'W)^{-1}W'}$ be the projection matrix based on the incidence matrix, (4.2.5). Show that $\mathbf{H_W}$ is a block diagonal matrix with the ith block an $n_i \times n_i$ matrix of all ones. Recall $\mathbf{X} = (\mathbf{I} - \mathbf{H_1})\mathbf{W_1}$ in Section 4.2.1. Let $\mathbf{H_X} = \mathbf{X(X'X)^{-1}X'}$ be the projection matrix. Then argue that $\mathbf{H_W} = \mathbf{H_1} + \mathbf{H_X}$ and, hence, $\mathbf{H_X} = \mathbf{H_W} - \mathbf{H_1}$ is easy to find. Using (4.2.8), show that, for the one-way design, $\operatorname{Cov}(\widehat{\mathbf{Z}}) \doteq \tau_S^2 \mathbf{H_1} + \tau_\varphi^2 \mathbf{H_X}$ and, hence, show that $\operatorname{Var}(\widehat{\mu}_i)$ is given by (4.2.11) and that $\operatorname{cov}(\widehat{\mu}_i, \widehat{\mu}_{i'})$ is given by (4.2.31).

4.8.19 Suppose we have k treatments of interest and we employ a block design consisting of a blocks. Within each block, we randomly assign mk subjects to the treatments so that each treatment receives m subjects. Suppose we model the responses Y_{ijl} as

$$Y_{ijl} = \mu + \alpha_i + \beta_j + e_{ijl}, \quad i = 1, \ldots, a; \ j = 1, \ldots, k; \ l = 1, \ldots, m,$$

where e_{ijl} are iid with cdf $F(t)$. We want to test

$$H_0: \beta_1 = \cdots = \beta_k \text{ versus } H_A: \beta_j \neq \beta_{j'} \text{ for some } j \neq j'.$$

Suppose we rank the data in the ith block from 1 to mk for $i = 1, \ldots, a$. Let R_j be the sum of the ranks for the jth treatment. Show that

$$E(R_j) = \frac{am(mk+1)}{2}$$

$$\operatorname{Var}(R_j) = \frac{am^2(mk+1)(k-1)}{12}$$

$$\operatorname{Cov}(R_j, R_l) = -\frac{am^2(mk+1)}{12}.$$

Further, argue that

$$K_m = \sum_{j=1}^{k} \left(\frac{k-1}{k}\right) \left[\frac{R_j - E(R_j)}{\sqrt{\operatorname{Var}(R_j)}}\right]^2$$

$$= \left[\frac{12}{akm^2(mk+1)} \sum_{j=1}^{k} R_j^2\right] - 3a(mk+1)$$

276   Experimental Designs

is asymptotically χ^2 with $k-1$ degrees of freedom. Note if $m=1$ then K_1 is the **Friedman statistic**. Show that the efficiency of the Friedman test relative to the two-way LS F-test is $12\sigma^2[\int f^2(x)\,dx]^2 k/(k+1)$. Plot the efficiency as a function of k when f is $N(0, 1)$.

4.8.20 The data in Table 4.8.1 are the results of a 3×4 design discussed in Box and Cox (1964). Forty-eight animals were exposed to three different poisons and four different treatments. The response was the survival time of the animal. The design was balanced. Use (4.4.1) as the full model to answer the questions below.

(a) Using Wilcoxon scores obtain the fit of the full model. Sketch the cell median profile plot based on this fit and discuss whether or not interaction between poison and treatments is present.

(b) Based on the Wilcoxon fit, plot the residuals versus the fitted values.

Table 4.8.1: Box–Cox Data, Exercise 4.8.20

Poisons	Treatments			
	1	2	3	4
1	0.31	0.82	0.43	0.45
	0.45	1.10	0.45	0.71
	0.46	0.88	0.63	0.66
	0.43	0.72	0.76	0.62
2	0.36	0.92	0.44	0.56
	0.29	0.61	0.35	1.02
	0.40	0.49	0.31	0.71
	0.23	1.24	0.40	0.38
3	0.22	0.30	0.23	0.30
	0.21	0.37	0.25	0.36
	0.18	0.38	0.24	0.31
	0.23	0.29	0.22	0.33

Comment on the appropriateness of the model. Also obtain the internal Wilcoxon studentized residuals and identify any outliers.

(c) Using the statistic F_φ, obtain the robust ANOVA table (main effects and interaction) for these data. Conclude in terms of the p-values.

(d) Note that the hypothesis matrix for interaction defines six interaction contrasts. Use the Bonferroni and protected LSD multiple comparison procedures, (4.3.2) and (4.3.3), to investigate these contrasts. Determine which, if any, are significant.

(e) Repeat the analysis in (c) and (d) (Bonferroni analysis) using LS. Compare the Wilcoxon and LS results.

4.8.21 For testing the ordered alternative

$$H_0: \mu_1 = \cdots = \mu_k \text{ versus } H_A: \mu_1 \leq \cdots \leq \mu_k,$$

with at least one strict inequality, let

$$J = \sum_{s<t} S_{st}^+,$$

where $S_{st}^+ = \#(Y_{tj} > Y_{si})$ for $i = 1, \ldots, n_s$ and $j = 1, \ldots, n_t$; see (2.2.20). This test for ordered alternatives was proposed independently by Jonckheere (1954) and Terpstra (1952). Under H_0, show the following:

(a)
$$E(J) = \frac{n^2 - \sum n_t^2}{4}.$$

(b)
$$\text{Var}(J) = \frac{n^2(2n+3) - \sum n_t^2(2n_t+3)}{72}.$$

(c) $z = (J - E(J))/\sqrt{\text{Var}(J)}$ is approximately $N(0, 1)$.

Hence, based on (a)–(c) an asymptotic test for H_0 versus H_A, is to reject H_0 if $z \geq z_\alpha$.

5
Bounded Influence and High-Breakdown Methods

From (3.5.17) the influence function of the R-estimate is unbounded in the x-space. While in a designed experiment this is of little consequence, for non-designed experiments where there are widely dispersed xs (i.e. outliers in factor space), this is of some concern. In this chapter we present R-estimates which have influence functions bounded in both spaces and which have positive breakdown. Further, we derive diagnostics which differentiate between fits based on these estimates and fits based on regular R-estimates. Tableman (1990) provides an alternative development of bounded influence R-estimates.

5.1 Geometry of the GR-Estimates

Consider the linear model (3.2.3). In Chapter 3, estimation and testing are based on the pseudo-norm, (3.2.6). Here we shall consider the function

$$\|\mathbf{u}\|_{GR} = \sum_{i<j} b_{ij}|u_i - u_j|, \qquad (5.1.1)$$

where the b_{ij} are specified functions of the xs. We will assume that these weights, b_{ij} are positive and symmetric, i.e. $b_{ij} = b_{ji}$. It is then easy to show (see Exercise 5.10.1) that the function (5.1.1) is a pseudo-norm. As noted in Section 2.2.2, if the weights $b_{ij} \equiv 1$, then this pseudo-norm is proportional to the pseudo-norm based on Wilcoxon scores. Hence, we will refer to this as a generalized R (GR) pseudo-norm. In Section 5.8.1, we will let the weights depend also on the residuals.

Since this is a pseudo-norm we can develop estimation and testing procedures using the same geometry as in the previous chapter. Briefly, the GR-estimate of $\boldsymbol{\beta}$ in model (3.2.3) is a vector $\widehat{\boldsymbol{\beta}}_{GR}$ which minimizes $\|\mathbf{Y} - \mathbf{X}\boldsymbol{\beta}\|_{GR}$. Equivalently, we can define the dispersion function

$$D_{GR}(\boldsymbol{\beta}) = \|\mathbf{Y} - \mathbf{X}\boldsymbol{\beta}\|_{GR}. \qquad (5.1.2)$$

Since it is based on a pseudo-norm, D_{GR} is a continuous, nonnegative, convex function of $\boldsymbol{\beta}$. The negative of its gradient is given by

$$S_{GR}(\boldsymbol{\beta}) = \sum_{i<j} b_{ij}(\mathbf{x}_i - \mathbf{x}_j)\operatorname{sgn}((Y_i - Y_j) - (\mathbf{x}_i - \mathbf{x}_j)'\boldsymbol{\beta}). \tag{5.1.3}$$

Thus the GR-estimate solves the equation

$$S_{GR}(\boldsymbol{\beta}) \doteq \mathbf{0}.$$

The GR-estimates were proposed by Sievers (1983). Many of their properties were further developed by Naranjo and Hettmansperger (1994).

In terms of testing, a gradient test for the null hypothesis that $\boldsymbol{\beta} = \mathbf{0}$ can be based on the statistic $S_{GR}(\mathbf{0})$. Its asymptotic null distribution is given in Corollary 5.2.4. For the general linear hypotheses (3.2.5), a test statistic can be constructed based on the drop in dispersion between the reduced and the full model. Using the notation in Chapter 3, let

$$RD_{GR} = D_{GR}(\mathbf{Y}, \omega) - D_{GR}(\mathbf{Y}, \Omega) \tag{5.1.4}$$

denote this drop in dispersion. Small values of RD_{GR} favor the null hypothesis while large values indicate the alternative. Proper standardization is discussed in Theorem 5.2.12 which gives the null asymptotic distribution of this statistic. Wald-type tests as well as gradient tests for this hypothesis can also be developed; see Exercise 5.10.2.

5.2 Asymptotic Theory

We next develop the asymptotic theory for the GR-estimates and tests. We shall make use of the theory found in Section A.3 of the Appendix; hence, the theoretical development in this chapter is much like that of Chapter 3 and, thus, our discussion will be brief. Our first goal is to derive an asymptotic linearity result from which theory for the GR-estimates and tests will follow.

It will be convenient to represent the weights as follows:

$$w_{ij} = \begin{cases} -(1/n)b_{ij} & i \neq j \\ (1/n)\sum_{k \neq i} b_{ik} & i = j. \end{cases} \tag{5.2.1}$$

Let $\mathbf{W}$ be the $n \times n$ matrix of elements w_{ij}. Note that $\mathbf{W}$ is symmetric and that its rows sum to zero.

Besides the assumptions (E.1), (D.2), and (D.3) of Section 3.4, we will need some assumptions on the weights. As in Chapter 3, $\mathbf{X}$ will denote a centered design matrix. Assume that

(G.1) $\quad \lim_{n \to \infty} n^{-1}\mathbf{X}'\mathbf{W}\mathbf{X} = \mathbf{C}$ \hfill (5.2.2)

(G.2) $\quad \lim_{n \to \infty} n^{-1}\mathbf{X}'\mathbf{W}^2\mathbf{X} = \mathbf{V}$ \hfill (5.2.3)

(G.3) $\quad \mathbf{W}\mathbf{X}$ satisfies condition (D.2), \hfill (5.2.4)

where $\mathbf{C}$ and $\mathbf{V}$ are positive definite matrices.

The linearity result is given in Theorem 5.2.9. Some of the lemmas leading to this result are of independent interest. For example, Lemmas 5.2.2 and 5.2.4 give the projection and asymptotic distribution of the gradient $S_{GR}(0)$.

Our first step is to obtain an analog of Theorem A.3.8 of the Appendix. Thus we need to consider only the simple linear model

$$Y_i = \alpha + x_i\beta + e_i. \tag{5.2.5}$$

Although we will consider this model, we will state vector results when appropriate. For example, we will use both $\mathbf{x}$ and $\mathbf{X}$ to denote the independent variable.

Let $z_i = Y_i - x_i\beta$. For this model, the negative of the gradient of D_{GR} is the process

$$S(\beta) = \sum\sum_{i<j} b_{ij}(x_i - x_j)\operatorname{sgn}(z_i - z_j). \tag{5.2.6}$$

Without loss of generality assume that the true β is zero. We begin with the following lemma concerning the moments of $S(0)$.

Lemma 5.2.1 *Under the assumption (E.1), (3.4.1),*

$$E_0[S(0)] = 0$$

$$\operatorname{Var}_0[S(0)] = \frac{n^2}{3}\mathbf{x'W^2 x} + \frac{n^2}{2}\sum_{i=1}^{n}\sum_{j=1}^{n} w_{ij}^2(x_i - x_j)^2.$$

Proof. Because $\beta = 0$, the random variables $Y_1, \ldots, Y_n$ are iid. Hence, the result for the mean is immediate. For the variance, note that we can write $S(0)$ as

$$S(0) = \frac{n}{2}\sum_{i,j} w_{ij}(x_i - x_j)\operatorname{sgn}(Y_i - Y_j);$$

hence,

$$\operatorname{Var}[S(0)] = \frac{n^2}{4}\sum_{i,j}\sum_{l,k} w_{ij}(x_i - x_j)w_{lk}(x_l - x_k)\operatorname{Cov}(\operatorname{sgn}(Y_i - Y_j), \operatorname{sgn}(Y_l - Y_k)).$$

The covariance is zero unless there is at least one tie among the subscripts, excluding, of course, $i = j$ and $l = k$. This leads to the six cases listed next, along with the values of the covariance function for each case. We derive the result for the first case. The other cases are similar and are left as Exercise 5.10.3. We repeatedly use the fact that $F(Y_i)$ has a uniform $(0, 1)$ distribution.

1. $i = l, j \neq k$: The covariance in this case is

$$\operatorname{Cov} = E[\operatorname{sgn}(Y_1 - Y_2)\operatorname{sgn}(Y_1 - Y_3)]$$
$$= 2E[F^2(Y_1)] - 2E[F(Y_1)(1 - F(Y_1))] = \tfrac{1}{3}.$$

2. $i = l, j = k$: Cov $= 1$.
3. $i = k, j \neq l$: Cov $= -\tfrac{1}{3}$.
4. $i = k, j = l$: Cov $= -1$.

5. $i \neq l, j = k$: Cov $= \frac{1}{3}$.
6. $i \neq k, j = l$: Cov $= -\frac{1}{3}$.

Using the fact that the weights w_{ij} are symmetric, the variance simplifies to

$$\text{Var}[S(0)] = \frac{n^2}{4} \frac{4}{3} \sum_{i=1}^{n} \sum_{j=1}^{n} \sum_{k=1}^{n} w_{ij}(x_i - x_j)w_{ik}(x_i - x_k)$$

$$+ \frac{n^2}{4} 2 \sum_{i=1}^{n} \sum_{j=1}^{n} w_{ij}^2 (x_i - x_j)^2. \quad (5.2.7)$$

Since the rows of the matrix $\mathbf{W}$ sum to zero, the first term is $\mathbf{x}'\mathbf{W}^2\mathbf{x}$, hence the result.

Consider the variance of $n^{-3/2}S(0)$. Exercise 5.10.4 shows, under the above assumptions, that the second standardized term of the variance goes to 0 as $n \to \infty$; hence, we have the result

$$\lim_{n \to \infty} \text{Var}[n^{-3/2}S(0)] = \lim_{n \to \infty} \frac{1}{3}\left(\frac{1}{n}\mathbf{x}'\mathbf{W}^2\mathbf{x}\right). \quad (5.2.8)$$

Although we derived the above results for a simple linear regression model, it follows immediately that for a p-variate model the mean of the vector $n^{-3/2}\mathbf{S(0)}$ is $\mathbf{0}$ and the asymptotic variance-covariance matrix is $\Sigma_{\mathbf{W}}$, where

$$\Sigma_{\mathbf{W}} = \lim_{n \to \infty} \frac{1}{3}\left(\frac{1}{n}\mathbf{X}'\mathbf{W}^2\mathbf{X}\right) = \frac{1}{3}\mathbf{V} \quad (5.2.9)$$

and $\mathbf{V}$ is given by (5.2.3).

We next derive the projection of $S(0)$.

Lemma 5.2.2 *Let* $\overline{S(0)} = n^{-3/2}S(0)$. *Then the projection of* $\overline{S(0)}$ *is given by*

$$\overline{T(0)} = \frac{2n}{n^{3/2}}T(0) = \frac{2n}{n^{3/2}} \sum_{k=1}^{n}\left(\sum_{i=1}^{n} w_{ki}x_i\right)F(Y_k). \quad (5.2.10)$$

Furthermore, $\text{Var}(\overline{T(0)}) = (\frac{1}{3})n^{-1}\mathbf{x}'\mathbf{W}^2\mathbf{x}$ *and* $\sqrt{n}\overline{T(0)}$ *is asymptotically normal.*

Proof. To obtain the projection, we first need $E[S(0)|Y_k = y]$. Using independence of the Y_is, symmetry of the weights, and the definition of w_{ij}, we obtain

$$E[S(0)|Y_k = y] = \sum_{i<j} b_{ij}(x_i - x_j)E[\text{sgn}(Y_i - Y_j)|Y_k = y]$$

$$= \sum_{i=1}^{k-1} b_{ik}(x_i - x_k)(1 - 2F(y)) + \sum_{j=k+1}^{n} b_{kj}(2F(y) - 1)$$

$$= \sum_{i=1}^{n} b_{ki}(x_k - x_i)(2F(y) - 1)$$

$$= n(2F(y) - 1)\sum_{i=1}^{n} w_{ki}x_i.$$

Asymptotic Theory 283

From this last line, we have the result for the projection. Since $F(Y_k)$ has a uniform distribution, the result on the variance is immediate. In order to obtain the asymptotic normality of $\sqrt{n}\overline{T(0)}$, note first that it is a sum of independent random variables of the form $c_k F(Y_k)$, $k = 1, \ldots, n$, where c_k is the kth row of the matrix $\mathbf{WX}$ which satisfies Huber's condition, assumption (G.3), (5.2.4). Using this along with assumption (G.2), (5.2.3), on the variance of $\sqrt{n}\overline{T(0)}$, asymptotic normality can be argued as in the proof of Theorem 3.5.2.

From this last result and the result on the asymptotic variance of $S(0)$, we have the following lemma:

Lemma 5.2.3 *Under assumptions (E.1), (3.4.1), (G.1), (5.2.2), (G.2), (5.2.3), and (G.3), (5.2.4), $\overline{T(0)} - \overline{S(0)} \xrightarrow{P} 0$.*

These results also give us the asymptotic distribution of $S(0)$. We state the multiple regression result which follows easily from the univariate result.

Corollary 5.2.4 *Under the assumptions (E.1), (3.4.1), (G.1), (5.2.2), (G.2), (5.2.3), and (G.3), (5.2.4), and assuming that $\boldsymbol{\beta} = \boldsymbol{0}$,*

$$\frac{1}{n^{3/2}}\mathbf{S}(0) \text{ has an asymptotic } N_p(\mathbf{0}, \Sigma_W) \text{ distribution,}$$

where Σ_W is given by (5.2.9).

From this result a gradient test for $H_0 : \boldsymbol{\beta} = \boldsymbol{0}$ can be based on the test statistic $3n^{-2}\mathbf{S}'(\mathbf{X}'\mathbf{W}^2\mathbf{X})^{-1}\mathbf{S}$. By the above corollary, under H_0 this test statistic has a central χ^2 distribution with p degrees of freedom.

Define the shifted projection $T(\Delta)$ as

$$T(\Delta) = 2n \sum_{k=1}^{n}\left(\sum_{i=1}^{n} w_{ki} x_i\right) F(Y_k + \Delta d_k), \quad (5.2.11)$$

where the sequence d_k satisfies $\max_{1 \leq k \leq n} |d_k| \to 0$; see Section A.2.2 of the Appendix.

Lemma 5.2.5 *Under the assumptions (E.1), (3.4.1), (G.1), (5.2.2), (G.2), (5.2.3), and (G.3), (5.2.4), $n^{-3/2}(T(0) - (T(\Delta) - E_0(T(\Delta)))) \xrightarrow{P} 0$.*

Proof. Since the mean of $T(0)$ is 0, we need only show that $n^{-3}V(T(0) - T(\Delta)) \to 0$. Using independence, we first get the bound

$$\text{Var}(T(0) - T(\Delta)) \leq 4n^2 \sum_{k=1}^{n}\left(\sum_{i=1}^{n} w_{ki} x_i\right)^2 E[(F(Y_k) - F(Y_k + \Delta d_k))^2]$$

$$= 4n^2 \sum_{k=1}^{n}\left(\sum_{i=1}^{n} w_{ki} x_i\right)^2 \int (F(y) - F(y + \Delta d_k))^2 f(y) dy.$$

From assumption (E.1), (3.4.1), we know that f is uniformly bounded, i.e.

$f(x) \leq M_f$ for some M_f for all x. Hence, by an application of the mean value theorem for integrals, we obtain,

$$\max_k |F(y) - F(y + \Delta d_k)| \leq |\Delta| M_f \max_k |d_k|.$$

Based on these results and after some algebraic simplification, we have

$$n^{-3}[\text{Var}(T(0) - T(\Delta))] \leq 4\Delta^2 M_f^2 \left(\max_k |d_k|\right)^2 \frac{1}{n} \mathbf{x'W^2 x}.$$

From $\max_k |d_k| \to 0$ and (5.2.8), the desired result follows.

We need the next result on the expectation of the shifted projection. First for $i = 1, \ldots, n$, let $d_k = -x_k/\sqrt{n}$. By assumption (D.2), (3.4.7), note that $\max_k |d_k| \to 0$ as $n \to \infty$.

Lemma 5.2.6 *Under the assumptions* (E.1), (3.4.1), (G.1), (5.2.2), (G.2), (5.2.3), *and* (G.3), (5.2.4),

$$\lim_{n \to \infty} E_0 \left[\frac{1}{n^{3/2}} T(\Delta) \right] = -2\Delta \int f^2(x) dx \, c_0,$$

where $c_0 = \lim_{n \to \infty} n^{-1} \mathbf{x'Wx}$; *see* (5.2.2).

Proof. Using the mean value theorem we obtain

$$E_0\left[\frac{1}{n^{3/2}} T(\Delta)\right] = \frac{2n}{n^{3/2}} \sum_{k=1}^{n} \left(\sum_{i=1}^{n} w_{ki} x_i \right) \int F(y + \Delta d_k) f(y) dy$$

$$= \frac{2n}{n^{3/2}} \sum_{k=1}^{n} \left(\sum_{i=1}^{n} w_{ki} x_i \right) \left[\int F(y) f(y) dy + \int \Delta d_k f(y_k^*) f(y) dy \right]$$

$$= \frac{2n}{n^{3/2}} \sum_{k=1}^{n} \left(\sum_{i=1}^{n} w_{ki} x_i \right) \int F(y) f(y) dy + \frac{2n}{n^{3/2}} \sum_{k=1}^{n} \left(\sum_{i=1}^{n} w_{ki} x_i \right) \int \Delta d_k f^2(y) dy$$

$$+ \frac{2n}{n^{3/2}} \sum_{k=1}^{n} \left(\sum_{i=1}^{n} w_{ki} x_i \right) \int \Delta d_k (f(y_k^*) - f(y)) f(y) dy,$$

where, because $\max_k |d_k| \to 0$, $y_k^* \to y$ uniformly in $k = 1, \ldots, n$ as $n \to \infty$. By interchanging the order of summation, the first term is 0 since the columns of $\mathbf{W}$ sum to 0. For the third term, let $\varepsilon > 0$ be given. Since f is uniformly continuous we can find δ such that $|f(x) - f(y)| < \varepsilon$ for $|x - y| < \delta$. Hence, choose n_0 so large that $|\Delta d_k| \leq |\Delta| \max_j |d_j| < \delta$, for all $k = 1, \ldots, n$ and for all $n \geq n_0$. Thus for $n \geq n_0$, the absolute value of the third term is less than or equal to

$$\frac{2n\varepsilon|\Delta|}{n^{3/2}} \left| \sum_{k=1}^{n} \left(\sum_{i=1}^{n} w_{ki} x_i \right) d_k \right| \leq \frac{2n\varepsilon|\Delta|}{n^{3/2}} \left[\sum_{k=1}^{n} \left(\sum_{i=1}^{n} w_{ki} x_i \right)^2 \right]^{1/2} \left[\sum_{l=1}^{n} d_l^2 \right]^{1/2}$$

$$= 2\varepsilon|\Delta| \left[\frac{1}{n} \mathbf{x'W^2 x} \right]^{1/2} \left[\sum_{l=1}^{n} d_l^2 \right]^{1/2}.$$

Note that the last two factors in the last expression converge to finite limits as $n \to \infty$; therefore, the third term goes to 0 as $n \to \infty$. Using the fact that $d_k = -x_k/\sqrt{n}$, some simple algebra shows that the second term is equal to

$$\frac{-2n\Delta \int f^2}{n^2} \sum_{k=1}^{n} \sum_{i=1}^{n} w_{ki} x_i x_k = -2\Delta \int f^2(t)\, dt \left(\frac{1}{n} \mathbf{x}' \mathbf{W} \mathbf{x}\right),$$

which converges to the desired limit.

Next define the process $S(\Delta)$ as

$$S(\Delta) = \sum_{i<j} b_{ij} \operatorname{sgn}((Y_i - Y_j) - \Delta(x_i - x_j)/\sqrt{n}). \tag{5.2.12}$$

Denote the likelihood of $Y_1, \ldots, Y_n$ as $p = p(\mathbf{y}) = \Pi_{i=1}^{n} f(y_i)$. The likelihood of the random variables $Y_i - \Delta x_i/\sqrt{n}$ is $q = q(\mathbf{y}) = \Pi_{i=1}^{n} f(t + \Delta x_i/\sqrt{n})$. It follows from assumption (D.2), (3.4.7), that the likelihood q is contiguous to p; see Section A.2.2 of the Appendix. Since $n^{-3/2}(T(0) - S(0))$ converges to 0 in probability under p it follows that it converges to 0 in probability under q. By a change of variables,

$$\lim_{n \to \infty} P_p[|T(\Delta) - S(\Delta)| \geq n^{3/2}\varepsilon] = \lim_{n \to \infty} P_q[|T(0) - S(0)| \geq n^{3/2}\varepsilon] = 0,$$

for any $\varepsilon > 0$. Hence, we have proved the following lemma:

Lemma 5.2.7 *Under the assumptions (E.1), (3.4.1), (G.2), (5.2.2), (G.2), (5.2.3), and (G.3), (5.2.4),*

$$n^{-3/2}(T(\Delta) - S(\Delta)) \xrightarrow{P} 0,$$

as $n \to \infty$.

The above series of lemmas leads us to the following key result.

Lemma 5.2.8 *Under the assumptions (E.1), (3.4.1), (G.1), (5.2.2), (G.2), (5.2.3), and (G.3), (5.2.4), for any fixed* Δ,

$$n^{-3/2} S(\Delta) - \left(n^{-3/2} S(0) - 2\Delta \int f^2(x)\, dx\, c_0\right) = o_p(1),$$

where c_0 *is defined in Lemma 5.2.6.*

Proof. Simply write the left-hand side of the above expression as the sum

$$n^{-3/2}\{S(\Delta) - T(\Delta)\} + n^{-3/2}\{T(\Delta) - E_0(T(\Delta)) - T(0)\}$$
$$+ \left\{n^{-3/2} E_0(T(\Delta)) - \left(-2\Delta \int f^2(x)\, dx\, c_0\right)\right\} + n^{-3/2}\{T(0) - S(0)\}.$$

From the above lemmas the first, second, and fourth terms converge to 0 in probability while the third term converges to 0. Hence we are done.

Based on the last result, as discussed in Section A.3 of the Appendix, lemma 5.2.8 holds for the multiple regression problem. Because D_{GR} is a convex function of $\boldsymbol{\beta}$, asymptotic linearity and quadraticity follow in the same way as in the Appendix, Section A.3. For reference we next state the asymptotic linearity result, where, in the notation of Chapter 3, $\tau = (\sqrt{12}\int f^2(x)dx)^{-1}$.

Theorem 5.2.9 *Under the assumptions (E.1), (3.4.1), (G.1), (5.2.2), (G.2), (5.2.3), and (G.3), (5.2.4),*

$$n^{-3/2}\mathbf{S}(\boldsymbol{\beta}) = n^{-3/2}\mathbf{S}(\mathbf{0}) - \sqrt{n}(\sqrt{3}\tau)^{-1}\mathbf{C}\boldsymbol{\beta} + o_p(1),$$

uniformly for all $\boldsymbol{\beta}$ such that $\sqrt{n}\|\boldsymbol{\beta}\| \leq d$, for any $d > 0$.

To obtain the asymptotic distribution of $\widehat{\boldsymbol{\beta}}_{GR}$, we proceed as in Chapter 3. The approximating quadratic in this case is

$$Q(\boldsymbol{\beta}) = \frac{n}{2\sqrt{3}\tau}\boldsymbol{\beta}'\mathbf{X}'\mathbf{W}\mathbf{X}\boldsymbol{\beta} - \boldsymbol{\beta}'\mathbf{S}(\mathbf{0}) + D_{GR}(\mathbf{0}). \tag{5.2.13}$$

The value of $\boldsymbol{\beta}$ which minimizes this quadratic is the pseudo-estimate

$$\widetilde{\boldsymbol{\beta}}_{GR} = \frac{\sqrt{3}\tau}{n}(\mathbf{X}'\mathbf{W}\mathbf{X})^{-1}\mathbf{S}(\mathbf{0}). \tag{5.2.14}$$

By Corollary 5.2.4, we immediately have the result that

$\widetilde{\boldsymbol{\beta}}_{GR}$ is approximately $N_p\left(\mathbf{0}, \tau^2(\mathbf{X}'\mathbf{W}\mathbf{X})^{-1}\mathbf{X}'\mathbf{W}^2\mathbf{X}(\mathbf{X}'\mathbf{W}\mathbf{X})^{-1}\right)$ distributed.

$$\tag{5.2.15}$$

Thus we need only show that

$$\sqrt{n}(\widetilde{\boldsymbol{\beta}}_{GR} - \widehat{\boldsymbol{\beta}}_{GR}) \overset{P}{\to} 0. \tag{5.2.16}$$

As we mentioned above, asymptotic quadraticity holds, i.e.

$$\max_{\sqrt{n}\|\boldsymbol{\beta}\| \leq c} |Q_{GR}(\boldsymbol{\beta}) - D_{GR}(\boldsymbol{\beta})| \overset{P}{\to} 0, \tag{5.2.17}$$

for all $c > 0$. This, along with the convexity of $D_{GR}(\boldsymbol{\beta})$, implies that the analog of Theorem A.3.9 of the Appendix holds; hence, (5.2.16) is true. Thus we have established the asymptotic distribution of the GR-estimate which we now state for reference:

Theorem 5.2.10 *Under the assumptions (E.1), (3.4.1), (G.1), (5.2.2), (G.2), (5.2.3), and (G.3), (5.2.4),*

$\widehat{\boldsymbol{\beta}}_{GR}$ *has an approximate* $N_p\left(\boldsymbol{\beta}, \tau^2(\mathbf{X}'\mathbf{W}\mathbf{X})^{-1}\mathbf{X}'\mathbf{W}^2\mathbf{X}(\mathbf{X}'\mathbf{W}\mathbf{X})^{-1}\right)$ *distribution.*

$$\tag{5.2.18}$$

The following lemma will be helpful in a comparison of the efficiency between the R- and the GR-estimates.

Lemma 5.2.11 *The matrix*

$$\mathbf{B} = (\mathbf{X}'\mathbf{W}\mathbf{X})^{-1}\mathbf{X}'\mathbf{W}^2\mathbf{X}(\mathbf{X}'\mathbf{W}\mathbf{X})^{-1} - (\mathbf{X}'\mathbf{X})^{-1}$$

is positive semi-definite.

Proof. Let $\mathbf{v}$ be any vector in $\mathcal{R}^p$. Since $\mathbf{X}'\mathbf{W}\mathbf{X}$ is non singular, there exists a vector $\mathbf{u}$ such that $\mathbf{v} = \mathbf{X}'\mathbf{W}\mathbf{X}\mathbf{u}$. Hence, by the Pythagoras's theorem,

$$\mathbf{v}'\mathbf{B}\mathbf{v} = \|\mathbf{W}\mathbf{X}\mathbf{u}\|^2 - \|\mathbf{H}\mathbf{W}\mathbf{X}\mathbf{u}\|^2 \geq 0,$$

where $\mathbf{H}$ is the projection matrix onto the column space of $\mathbf{X}$.

Based on this lemma, it is easy to see that there is always a loss of efficiency when using the GR-estimates. As the examples below show, this loss can be severe. If the design matrix, however, has clusters of outlying points then this downweighting may be necessary.

We next obtain the asymptotic null distribution of the test based on the drop in dispersion of the general linear hypothesis, (3.2.5). As in Chapter 3, without loss in generality we can write the model as

$$\mathbf{Y} = \mathbf{1}\alpha + \mathbf{X}_1\boldsymbol{\beta}_1 + \mathbf{X}_2\boldsymbol{\beta}_2 + \mathbf{e}, \tag{5.2.19}$$

so that the hypotheses become $H_0 : \boldsymbol{\beta}_2 = \mathbf{0}$ versus $H_A : \boldsymbol{\beta}_2 \neq \mathbf{0}$. Let $\mathbf{C}_{11}$ denote the analog of $\mathbf{C}$, (G.1), (5.2.2), for the reduced model design matrix $\mathbf{X}_1$ and define the matrix

$$\mathbf{C}_r^+ = \begin{bmatrix} \mathbf{C}_{11}^{-1} & \mathbf{0} \\ \mathbf{0} & \mathbf{0} \end{bmatrix}. \tag{5.2.20}$$

Theorem 5.2.12 *Under the assumptions of this section, if H_0 is true then*

$$\frac{\sqrt{12}}{n\tau} RD_{GR} \xrightarrow{\mathcal{D}} \sum_{i=1}^{q} \lambda_i \chi_i^2(1), \tag{5.2.21}$$

where $\lambda_1, \ldots, \lambda_q$ are the q positive eigenvalues of the matrix $\mathbf{V}(\mathbf{C}^{-1} - \mathbf{C}_r^+)$ and $\chi_1^2(1), \ldots, \chi_q^2(1)$ are iid χ^2 random variables each with one degree of freedom.

The proof proceeds as that of Theorem 3.6.4, i.e. the drop is written as a sum of five differences. Four of these differences converge to 0 in probability under H_0 while the remaining difference involving the approximating quadratic yields the result after algebraic simplification; see Exercise 5.10.5 or Witt, Naranjo, and McKean (1994) for details.

Unlike the result for the drop in dispersion test statistic for R-estimates the asymptotic distribution is not a central χ^2 distribution but the distribution of a linear combination of $\chi^2(1)$ random variables. For future use, we will still standardize the test statistic as an F random variable, i.e.

$$F_{GR} = \frac{\sqrt{12}}{n} \frac{RD_{GR}/q}{\hat{\tau}}. \tag{5.2.22}$$

For general designs we do not recommend comparing this with F critical values. Boot strapping the p-value of F_{GR} would be one possibility; see Exercise 5.20.6 or Schrader and McKean (1987). Wald-type test statistics, quadratic forms in $\widehat{\boldsymbol{\beta}}_{GR}$, do have null asypmptotic χ^2 distributions; see Exercise 5.10.8.

For another possibility, note that the weights depend only on the design matrix $\mathbf{X}$. In practice, we can compute the matrix $\mathbf{V}(\mathbf{C}^{-1} - \mathbf{C}_r^+)$ and obtain its eigenvalues. Using these, it is straightforward to simulate the null distribution of the random variable $\sum_{i=1}^{q} \lambda_i \chi_i^2(1)$. This null distribution can then be used to find the p-value of the test statistic qF_{GR}; see Exercise 5.10.7.

5.3 Implementation

To implement the GR-estimate we need to discuss an estimate of the intercept parameter, an estimate of τ, and the weights. Clearly the estimate of τ given in (3.7.8) of Chapter 3 can be used because for consistency of the estimate the only condition placed on the residuals is that $\sqrt{n}$ times the estimate of $\boldsymbol{\beta}$ is bounded in probability; see Koul et al. (1987). As in Chapter 3, the estimate of the intercept parameter α is the median of the residuals, i.e.

$$\widehat{\alpha}_{GR} = \mathrm{med}_{1 \leqslant i \leqslant n}\{y_i - \mathbf{x}_i'\widehat{\boldsymbol{\beta}}_{GR}\}. \tag{5.3.1}$$

Under the above assumptions, a version of Theorem 3.5.11 also holds for the joint asymptotic distribution of the GR-estimates of α and $\boldsymbol{\beta}$. For completeness we state the result for the **uncentered design matrix** $\mathbf{X}_1$. Also, as in that section, assume that the median of the random errors is 0. Let $\widehat{\alpha}_{GR}^* = \widehat{\alpha}_{GR} - \overline{\mathbf{x}}'\widehat{\boldsymbol{\beta}}_{GR}$, where $\overline{\mathbf{x}}$ is the vector of column averages of $\mathbf{X}_1$, denote the intercept parameter for the uncentered design, then under assumptions (E.1), (3.4.1), (E.2), (3.4.3), (G.1), (5.2.2), (G.2), (5.2.3), and (G.3), (5.2.4),

$$\begin{pmatrix} \widehat{\alpha}_{GR}^* \\ \widehat{\boldsymbol{\beta}}_{GR} \end{pmatrix} \text{ has an approximate } N_{p+1}\left(\begin{pmatrix} \alpha \\ \boldsymbol{\beta} \end{pmatrix}, \begin{bmatrix} \kappa_{GRn} & -\tau^2\overline{\mathbf{x}}'\mathbf{A} \\ -\tau^2\mathbf{A}'\overline{\mathbf{x}} & \tau^2\mathbf{A} \end{bmatrix}\right) \text{ distribution,}$$

where $\mathbf{A} = (\mathbf{X}'\mathbf{W}\mathbf{X})^{-1}\mathbf{X}'\mathbf{W}^2\mathbf{X}(\mathbf{X}'\mathbf{W}\mathbf{X})^{-1}$, $\kappa_{GRn} = n^{-1}\tau_S^2 + \tau^2\overline{\mathbf{x}}'\mathbf{A}\overline{\mathbf{x}}$, and $\tau_S = (2f(0))^{-1}$; see (3.5.30) and Exercise 3.12.14. Also, we used the fact that $\mathbf{X}_1'\mathbf{W}\mathbf{X}_1 = \mathbf{X}'\mathbf{W}\mathbf{X}$, which follows because $\mathbf{W}$ is a contrast matrix (its rows sum to zero).

The weight for the ith case should be a measure of the distance of $\mathbf{x}_i$ from the center of factor space. Traditionally this can be measured in terms of the ith leverage value, h_{ii}, of the design matrix. This yields weights of the form $b_{ij} = b_i b_j$, where

$$b_i = \min\left\{1, c_1/\sqrt{h_{ii}}\right\}^{\alpha_1}. \tag{5.3.2}$$

In all the examples of GR-estimates in this chapter, for weights of the form (5.3.2) we set the parameter α_1 at 2 and the parameter c_1 at the 30th percentile of the leverage values, $h_{(0.3n)}$; see McKean, Sheather, and Hettmansperger (1994) for some discussion on the setting of these tuning constants. This measure of distance, however, is based on Euclidean distance and, hence, may be sensitive to outliers in factor space. To avoid this we can consider weights based on robust measures of center and scatter. Consider the weights given by

$$b_i = \min\left\{1, \; c_2/\sqrt{(\mathbf{x}_i - \mathbf{v})'\mathbf{V}^{-1}(\mathbf{x}_i - \mathbf{v})}\right\}^{\alpha_2}, \qquad (5.3.3)$$

where $\mathbf{v}$ is the center of the minimal volume ellipsoid which covers at least $(n + p)/2$ of the rows of $\mathbf{X}$, and $\mathbf{V}$ is the positive definite matrix corresponding to the ellipsoid. Taken together $(\mathbf{v}, \mathbf{V})$ are the minimum volume ellipsoid (MVE) measures of location and scatter and have 50% breakdown; see Rousseeuw (1984) and Rousseeuw and van Zomeren (1990). For weights of the form (5.3.3), we set c_2 at the 95th percentile of the χ^2 distribution with p degrees of freedom and set α_2 at 1 (Simpson, Ruppert, and Carroll, 1992). In addition, we used the inflated form of the MVE matrix $\mathbf{V}$ as discussed by Rousseeuw and van Zomeren (1991). For these parameterizations conditions (G.1)–(G.3) hold. Other estimates of scatter that can be considered are those suggested by Ammann (1993) and Hadi and Simonoff (1993).

For the actual computation of the GR fit, we used the algorithm proposed by Simonoff and Hawkins (1993) to compute the MVE estimates of location $\mathbf{v}$ and scatter $\mathbf{V}$. The criterion function for the GR-estimates, D_{GR}, is a convex function of $\boldsymbol{\beta}$. Hence, we used a Gauss–Newton type of algorithm similar to the one outlined in Section 3.7.2 to compute the GR-estimates.

Example 5.3.1 *Stars Data*

This data set is drawn from an astronomy study on the star cluster CYG OB1 which contains 47 stars; see Rousseeuw and Leroy (1987) for a discussion on the history of the data set. The response is the logarithm of the light intensity of the star, while the independent variable is the logarithm of the temperature of the star. The data are tabled in Table 5.3.1 and are shown in Figure 5.3.1(a). Note that four of the stars, called giants, are outliers in factor space while the rest of the stars fall in a point cloud. The three fits LS, Wilcoxon, and GR are overlaid on this plot of the data. Note that the GR fit falls through the point cloud whereas the other two fits are drawn towards the outliers. This illustrates the GR-estimate's insensitivity to outliers in factor space. The actual estimates of the regression coefficients and their standard errors are found in Table 5.3.2. Note that there is a difference in sign on the estimate of the slope parameter between the GR-estimate and the other two estimates.

Figure 5.3.1(b,c) show the internal studentized residuals plots for the Wilcoxon and GR fits; see Section 5.4.3. The GR fit has fitted the point cloud well and the four outlying stars stand out in the plot. Figure 5.3.1(d) is a plot of the standardized GR residuals, as discussed in Section 4.5, versus the case number. The four giant stars are indeed outliers.

Example 5.3.2 *Hawkins Data*

This is an artificial data set proposed by Hawkins, Bradu, and Kass (1984) involving three independent variables. There are a total of 75 data points in

Table 5.3.1: Stars data

Star	log Temp	log Intensity	Star	log Temp	log Intensity
1	4.37	5.23	25	4.38	5.02
2	4.56	5.74	26	4.42	4.66
3	4.26	4.93	27	4.29	4.66
4	4.56	5.74	28	4.38	4.90
5	4.30	5.19	29	4.22	4.39
6	4.46	5.46	30	3.48	6.05
7	3.84	4.65	31	4.38	4.42
8	4.57	5.27	32	4.56	5.10
9	4.26	5.57	33	4.45	5.22
10	4.37	5.12	34	3.49	6.29
11	3.49	5.73	35	4.23	4.34
12	4.43	5.45	36	4.62	5.62
13	4.48	5.42	37	4.53	5.10
14	4.01	4.05	38	4.45	5.22
15	4.29	4.26	39	4.53	5.18
16	4.42	4.58	40	4.43	5.57
17	4.23	3.94	41	4.38	4.62
18	4.42	4.18	42	4.45	5.06
19	4.23	4.18	43	4.50	5.34
20	3.49	5.89	44	4.45	5.34
21	4.29	4.38	45	4.55	5.54
22	4.29	4.22	46	4.45	4.98
23	4.42	4.42	47	4.42	4.50
24	4.49	4.85			

the set and the first 14 of them are outlying in factor space. The other 61 points follow a linear model. Of the 14 outlying points, the first 10 points do not follow the model, while points 11 through 14 do; hence, the first 10 cases are referred to as bad points of high leverage while the next four cases are referred to as good points of high leverage.

Figure 5.3.2(a)–(c) show the unstandardized residual plots from the LS, Wilcoxon and GR fits, respectively. Note that the LS and Wilcoxon fits are fooled by the bad outliers. Their fits are drawn by the bad points of high leverage while they both flag the four good points of high leverage. The last plot, Figure 5.3.2(d), shows the internal GR studentized residuals plotted versus case number. The bad points of high leverage are declared as outliers. Table 5.3.3 displays the estimates and their standard errors.

Table 5.3.2: Estimates of regression coefficients for the stars data

	α	(se)	β	(se)
LS	6.79	(1.2)	−0.413	(0.29)
Wilcoxon	7.20	(1.3)	−0.477	(0.31)
GR	−6.08	(2.2)	2.53	(0.50)

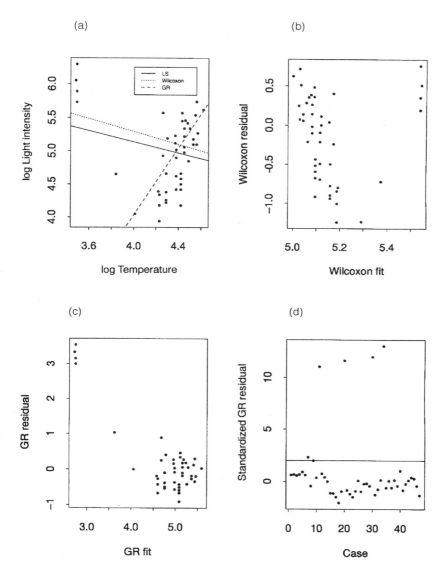

Figure 5.3.1: (a) Stars data overlaid with LS, Wilcoxon and GR fits; (b) Wilcoxon residual plot; (c) GR residual plot; (d) Internal GR studentized residuals by case

Table 5.3.3: Estimates of regression coefficients for the Hawkins data

	α	(se)	β_1	(se)	β_2	(se)	β_3	(se)
LS	−0.387	(0.42)	0.239	(0.26)	−0.334	(0.15)	0.383	(0.13)
Wilcoxon	−0.772	(0.21)	0.167	(0.11)	0.017	(0.06)	0.269	(0.05)
GR	−0.177	(0.26)	0.100	(0.10)	0.053	(0.09)	−0.044	(0.07)

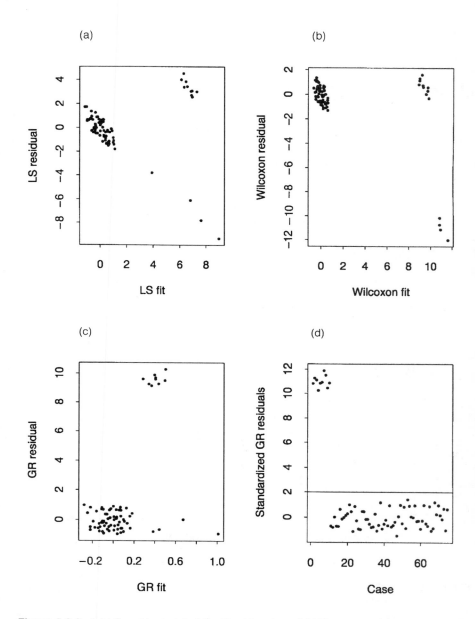

Figure 5.3.2: (a) LS residual plot of the Hawkins data; (b) Wilcoxon residual plot; (c) GR residual plot; (d) Internal GR-studentized residuals by case

5.4 Diagnostics

In Section 3.9, we presented diagnostic techniques for R-estimates. In this chapter we shall continue this discussion for the GR-estimates. Recall that the principal uses of diagnostics are to check on quality of the fit and to detect outliers and points of influence. By their very nature, GR-estimates offer protection against influential points so we will concentrate only on proper standardization for detection of outliers and criticism of fit.

5.4.1 Interpretation of GR Residuals

To gain an understanding of GR residuals we consider their behavior under the misspecified model

$$\mathbf{Y} = \mathbf{1}\alpha + \mathbf{X}\boldsymbol{\beta} + \mathbf{Z}\boldsymbol{\gamma} + \mathbf{e}, \qquad (5.4.1)$$

where, as in Section 3.9, $\mathbf{Z}$ is an $n \times q$ centered matrix of constants and $\boldsymbol{\gamma} = \boldsymbol{\theta}/\sqrt{n}$, for $\boldsymbol{\theta} \neq \mathbf{0}$. As in the previous section, let $\widehat{\boldsymbol{\beta}}_{GR}$ denote the GR-estimate under the model $\mathbf{Y} = \mathbf{X}\boldsymbol{\beta} + \mathbf{e}$. The asymptotic representation for the R-estimate is given by (3.9.5). Its analog for the GR-estimate is given by the following:

Theorem 5.4.1 *Suppose the conditions of Section 5.2 are true for the augmented matrix* [$\mathbf{X}$ $\mathbf{Z}$]; *then under model (5.4.1)*,

$$\widehat{\boldsymbol{\beta}}_{GR} = \boldsymbol{\beta} + \frac{\sqrt{3}\tau}{n}(\mathbf{X}'\mathbf{W}\mathbf{X})^{-1}\mathbf{S}(\boldsymbol{\beta}) + (\mathbf{X}'\mathbf{W}\mathbf{X})^{-1}\mathbf{X}'\mathbf{W}\mathbf{Z}\boldsymbol{\gamma} + o_p(n^{-1/2}). \qquad (5.4.2)$$

The proof of this theorem is similar to the proof of Theorem 3.9.1. Details can be found in Naranjo *et al.* (1994).

Based on this last theorem we obtain the following first-order representations of the GR-residuals and fitted values.

$$\widehat{\mathbf{Y}}_{GR} \doteq \alpha\mathbf{1} + \mathbf{X}\boldsymbol{\beta} + \frac{\sqrt{3}}{n\tau}\mathbf{X}(\mathbf{X}'\mathbf{W}\mathbf{X})^{-1}\mathbf{S}(\boldsymbol{\beta}) + \mathbf{K}_W\mathbf{Z}\boldsymbol{\gamma} \qquad (5.4.3)$$

$$\widehat{\mathbf{e}}_{GR} \doteq \mathbf{e} - \frac{\sqrt{3}}{n\tau}\mathbf{X}(\mathbf{X}'\mathbf{W}\mathbf{X})^{-1}\mathbf{S}(\boldsymbol{\beta}) + (\mathbf{I} - \mathbf{K}_W)\mathbf{Z}\boldsymbol{\gamma}, \qquad (5.4.4)$$

where $\mathbf{K}_W = \mathbf{X}(\mathbf{X}'\mathbf{W}\mathbf{X})^{-1}\mathbf{X}'\mathbf{W}$. Similar to the discussion in Section 3.9 for the R-estimates, in terms of model misspecification, the vector of parameters of interest is the vector of regression coefficients; hence, as in Chapter 3, we will consider the fitted value based on $\widehat{\boldsymbol{\beta}}_{GR}$ only. Note that unlike the results for the R-estimates, the GR residuals and fitted values are not the same functions of the misspecified part of the model as the LS estimates. Instead the expressions are functions of the weights also. As the examples illustrate below, this can lead to difficulties in interpretation of the GR residual plots.

More light is shed on this interpretation by considering the parameter, (3.9.10), that measures the expected departure from orthogonality of the residuals and fitted values, i.e.

$$v_{GR} = E\left[\widehat{\mathbf{e}}'_{GR}\widehat{\mathbf{Y}}_{GR}\right]. \qquad (5.4.5)$$

Based on the above representations, we have the following:

Theorem 5.4.2 *Suppose the conditions of the last theorem are true and $E[e] = 0$. Then the representation for v_{GR} under the misspecified model (5.4.1) is*

$$v_{GR,\gamma} \doteq \frac{\xi\sqrt{3}}{\tau}(p-1) - \tau^{-2}\text{tr}\mathbf{K}'_W\mathbf{K}_W + \gamma'\mathbf{Z}'[\mathbf{I} - \mathbf{K}_W][\mathbf{X}\beta + \mathbf{K}_W\mathbf{Z}\gamma], \quad (5.4.6)$$

where $\xi = E[e_1 \text{sgn}(e_1 - e_2)]$. Note that if the correct model is fit, i.e. $\gamma = 0$, then v_{GR} reduces to

$$v_{GR,0} \doteq \frac{\xi\sqrt{3}}{\tau}(p-1) - \tau^{-2}\text{tr}\mathbf{K}'_W\mathbf{K}_W. \quad (5.4.7)$$

Hence, unlike the case for the R-estimates, the value of v_{GR} changes for the misspecified model which may hinder interpretation of GR residual plots. Furthermore, the weights tend to produce a negative correlation between residuals and fitted values. For instance, if Wilcoxon scores are used for $\widehat{\beta}_R$ then it follows from expression (3.9.11) and the identity $E[F(e)e] = \xi/2$, that

$$v_{GR,0} \doteq v_R + \tau^2 \text{tr}(\mathbf{H} - \mathbf{K}_W\mathbf{K}'_W). \quad (5.4.8)$$

The second term can be shown to be less than or equal to zero; see Exercise 5.10.9. This negative correlation is consistent with the observations in McKean *et al.* (1993) about the behavior of residual plots based on high-breakdown fits. Illustrations of this negative correlation are found in the next two examples, in which the message is that bounded influence methods cannot generally be trusted to diagnose model misspecification.

Example 5.4.1 *Synthetic Rubber Data.*

This data set consists of 37 observations from a synthetic rubber process taken from Mason, Gunst, and Hess (1989, p. 488). The response variable is the percentage weight of a solvent, while the independent variable is the corresponding production rate of the rubber process. For computational convenience we have divided the predictor variable by 100 and then centered it by subtracting the mean. The data are given in Table 5.4.1.

A plot of the data is given in Figure 5.4.1(a) which is overlaid with the Wilcoxon and GR fits of a quadratic model. Note how poor the GR fit is. The downweighting of the outlying points has spoiled this fit causing it to turn up much too soon on the left. The corresponding residual plot, Figure 5.4.1(b), gives little insight on how poor the fit is. The negative correlation in this plot is an illustration of what can occur for generalized robust estimates; see (5.4.8) above. In contrast, the Wilcoxon estimate fits the point cloud well. Its accompanying residual plot, Figure 5.4.1(c), shows that the quadratic model is appropriate. It does show some heteroscedasticity, which Mason *et al.* (1989) also point out on the basis of the LS fit. The plot of the internal Wilcoxon studentized residuals by case, Figure 5.4.1(d), shows that two of the points are outlying. The estimates appear in Table 5.4.2.

Table 5.4.1: Synthetic rubber data

Production rate centered	Weight of solvent	Production rate centered	Weight of solvent
−7.31351	0.02	0.88649	0.17
−6.81351	0.03	1.48649	0.12
−6.71351	0.06	1.58649	0.12
−6.51351	0.01	1.58649	0.15
−6.51351	0.03	1.58649	0.19
−4.91351	0.03	1.88649	0.13
−3.71351	0.06	1.88649	0.16
−2.41351	0.09	1.88649	0.21
−1.11351	0.16	1.98649	0.25
−0.41351	0.16	2.08649	0.13
−0.31351	0.12	2.18649	0.20
−0.31351	0.10	2.48649	0.16
−0.11351	0.14	2.48649	0.23
0.08649	0.12	2.78649	0.30
0.08649	0.14	2.88649	0.43
0.18649	0.11	4.78649	0.41
0.08649	0.13	5.28649	0.40
0.18649	0.13	6.18649	0.26
0.58649	0.19		

Table 5.4.2: Estimates of regression coefficients for the synthetic rubber data

	α	(se)	β_1	(se)	β_2	(se)
Wilcoxon	0.1329	(0.01)	0.0298	(0.003)	0.00215	(0.0006)
GR	0.1194	(0.05)	0.00179	(0.014)	0.00655	(0.0046)

McKean, Sheather, and Hettmansperger (1994) discuss robust fitting of polynomial models for M, GM, and least median squares (LMS) estimates. The GM estimates are a class of high-breakdown estimates which use the same weights as the GR estimates. As discussed in this article, the GM and LMS estimates also fit this data set poorly. The article includes a discussion of the poor efficiency results for the GM-estimates. The efficiency results of the GR-estimates are similar; see Exercise 5.10.10. The discussion shows the folly of basing weights on elliptical contours in the space $(x_i, x_i^2, \ldots, x_i^k)$ for most polynomial-type designs.

Example 5.4.2 *Analysis of Covariance Data*

This is an artificial data set which illustrates the difficulties in interpreting the GR residual plots discussed above; see Theorem 5.4.1. These data consist of one covariate and two groups. Thirty observations were drawn from each group so the total sample size is 60. The model can be expressed as

$$y_i = \beta_0 + \beta_1 d_i + \beta_2 x_i + \beta_3 x_i d_i + e_i, \quad i = 1, \ldots, 60, \qquad (5.4.9)$$

where the indicator variable d_i is 0 for the first group and 1 for the second group. The true values of the parameters were: $\beta_0 = \beta_1 = 0$, $\beta_2 = 0.5$ and

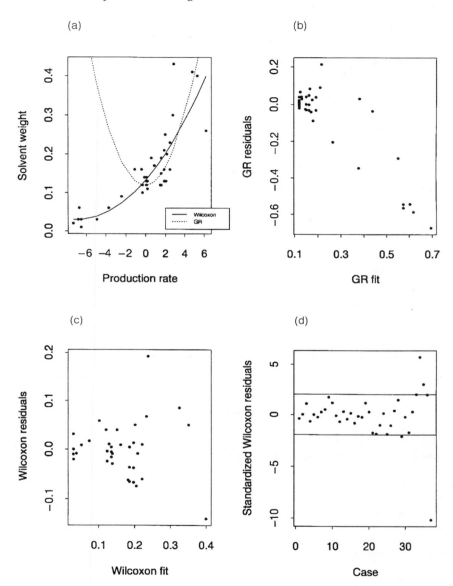

Figure 5.4.1: (a) Scatterplot of the synthetic rubber data overlaid with the Wilcoxon and GR fits; (b) GR residual plot; (c) Wilcoxon residual plot; (d) internal Wilcoxon studentized residuals by case

$\beta_3 = -0.1$. The errors, e_{ij}, were generated $N(0, 1)$ variates using the generator in S-Plus. For each sample, 20 of the xs were uniformly spaced between -0.5 and 0.5, while 10 xs were placed between 4 and 8 and between -8 and -4 in jumps of 1. Thus there are a total of 20 points outlying in factor space. These were severely downweighted by the weights for the GR-estimates.

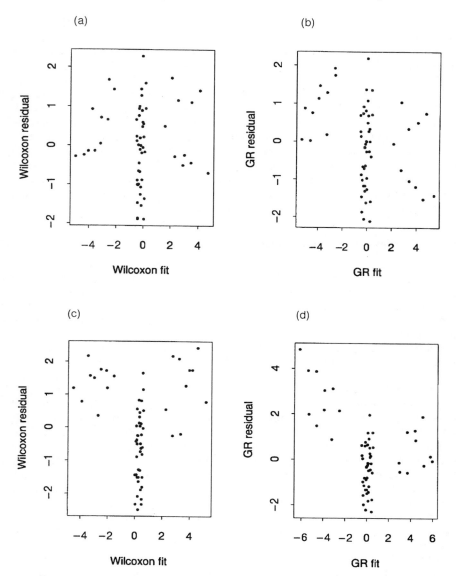

Figure 5.4.2: (a,b) Wilcoxon and GR residual plots, respectively, for the analysis of covariance data, following model (5.4.9); (c,d) Wilcoxon and GR residual plots, respectively, for the fit of the linear model, (5.4.9), when the data follow the quadratic model discussed in the text

Figure 5.4.2(a,b) displays the residual plots for the Wilcoxon and GR fits. The plot based on the Wilcoxon fit shows the group of xs with high leverage, but shows no overall pattern. These are desired plots in this case since the correct model was fit. On the other hand, the GR residual plot shows a strong negative correlation between the residuals and fitted values. With such a strong pattern, the usual interpretation of this plot is model misspecification, which is not true here. Thus in this instance the GR residual plot cannot be interpreted as containing information about model misspecification.

We next inserted a small bias by adding $0.05x_i^2 + 0.025x_i^2 d_i$ to the original observations. The original model (5.4.9), however, was fitted; hence there is a real bias in the fitted model which is correlated with the design matrix fitted. Ideally residual plots would detect this curvature. The residual plots for the Wilcoxon and GR fits are found in Figure 5.4.2(c,d). The Wilcoxon fit detects the curvature, indicating correctly that a quadratic model in the xs needs to be considered. The GR residual plot, though, does not detect the curvature. As in the original data, it indicates that a linear effect in the model is missing. Its ability to detect the curvature is obscured by the relationship between the residuals and fitted values as suggested by Theorem 5.4.1.

5.4.2 Studies for Curvature Detection

The first-order approximations of the residuals and fitted values based on GR fits and the parameter v_{GR} indicate that their corresponding residual plots may have trouble in correctly identifying misspecified models. Another indication is that the parameter v_{GR} is a function of the weights and the misspecified model. In this section, we discuss two Monte Carlo studies which verify these indications.

Naranjo et al. (1994) conducted a simulation study to investigate the relationship between the residuals and fitted values and the interpretation of their residual plots for R, GR, and LS fits. The R-estimate was based on Wilcoxon scores and the GR-estimate employed the weights given in (5.3.3) with $\alpha_2 = 1$. The investigation was over several different models, (both correct and misspecified) and error distributions. The relationships between the residuals and fitted values were measured by the correlation coefficients: Pearson's r, Spearman's ρ and Kendall's τ. These measurements show that the negative correlation patterns predicted by the above theoretical properties are generally true for the GR-estimates; whereas for the Wilcoxon estimates there was much less of a relationship between their residuals and fitted values. In fact, for the nonparametric correlation coefficients, the correlations based on the Wilcoxon estimates were quite similar to correlations between the LS residuals and fitted values. As theory predicts for the GR-estimates, the negative correlation pattern is worse in the case of misspecified models than for true models.

The residual plots of the different fits were assessed by runs tests on the randomness of the residuals and a quadrant count test, which is a χ^2 test based on the counts of points in the four quadrants determined by the median of the residuals and the median of the fitted values. The tests based on LS and Wilcoxon fits exhibited validity for their respective residual plots when the correct models were fitted and exhibited power for nonrandomness when the misspecified models were fitted. The empirical powers for the Wilcoxon and LS results were similar. The empirical power of the tests based on the GR fits was much less than the power for the tests based on the Wilcoxon fits.

Next we present the results of a small simulation study of Hettmansperger, McKean, and Sheather (1997) which compared the curvature detection

Table 5.4.3: Top row contains counts of times a minimum was identified out of 500 simulations. Bottom row contains counts of times a saddle point was identified

Distribution	Type Ident.	LS	Wil.	$\alpha_2 = 1$	$\alpha_2 = 0.75$	$\alpha_2 = 0.5$	$\alpha_2 = 0.25$
$6 \cdot N(0, 1)$	Minimum	478	474	162	261	385	459
	Saddle point	22	26	322	239	115	41
$6 \cdot CN(0.1, 16)$	Minimum	406	427	154	238	346	426
	Saddle point	94	73	331	258	154	74
$6 \cdot CN(0.1, 64)$	Minimum	320	389	141	226	332	401
	Saddle point	179	111	333	270	167	99
$6 \cdot CN(0.25, 16)$	Minimum	328	378	124	190	284	355
	Saddle point	172	122	351	308	216	145

abilities of the rank-based methods for several error distributions over a central composite type design; see also McKean, Sheather, and Hettmansperger (1994). Such a design contains the bulk of the design points in the center and an inner shell of factor space, while several points are in an outer shell. Our data set consists of 32 points, 12 of which were generated uniformly within the circle of radius 0.25, 12 in the ring with radii 1 and 1.25, and 8 in the ring with radii 2 and 2.25. As a model we considered

$$y = 14x_1 + 10x_2 + 3x_1^2 + 3x_2^2 + 2x_1 x_2 + e, \quad (5.4.10)$$

where e is distributed as a $N(0, 6^2)$ variate. The surface $E(y)$ has a minimum at the point $x_1 = -2$ and $x_2 = -1$ which is within the design space. This design was used in a similar study performed by McKean, Sheather, and Hettmansperger (1994) for M-estimates. In the present study we considered LS, Wilcoxon R-, and GR-estimates. We used the weights given by (5.3.3) and $\alpha_2 = 1, 0.75, 0.5,$ and 0.25. We ran 500 simulations of this design over each of the four error distributions: an $N(0, 6^2)$ distribution and the three contaminated normal error distributions listed in Table 5.4.3.

To measure how well an estimation procedure detected curvature, we counted the times its fitted surface correctly predicted a minimum; that is, the times when both eigenvalues of the Jacobian of the fitted surface are positive. These counts are displayed on the top line for each distribution in Table 5.4.3. We also tallied the number of times the surface predicted a saddle point. These are the values in the second line. Note that the Wilcoxon and LS results are quite similar for the normal, while over the contaminated normal error distributions the Wilcoxon performed somewhat better than LS. The GR-estimates with the most severe weights, $\alpha_2 = 1$, did poorly across all the distributions. Note that as α_2 decreases, the resulting GR-estimate does better. The results are indistinguishable from the Wilcoxon at the lowest value of $\alpha_2 = 0.25$.

5.4.3 Standardization of GR Residuals

For this section we additionally assume that the variance of the random errors is finite and is given by σ^2. In order properly to standardize the GR residuals, the variation in the estimation of the intercept parameter α must be taken into account. The only estimate of α that we will consider in this section is the median of the GR residuals, (5.3.1). As we mentioned in Section 5.3, the asymptotic theory for $\widehat{\alpha}_{GR}$ is similar to that of the estimate based on the R residuals. In particular, the representation (3.5.23) also holds for $\widehat{\alpha}_{GR}$. Using this and the above representations of the GR residuals and fitted values, (5.4.3) and (5.4.4), a first-order representation to the variance-covariance matrix for $\widehat{\mathbf{e}}_{GR}$ is given by

$$\mathrm{Var}(\widehat{\mathbf{e}}_{GR}) \doteq \sigma^2 \mathbf{I} - K_3 \mathbf{J} - (K_4 \mathbf{I} - K_5 \mathbf{J})\mathbf{K}'_W + \tau^{-2}\mathbf{K}_W \mathbf{K}'_W, \qquad (5.4.11)$$

where

$$\mathbf{J} = \mathbf{1}\mathbf{1}',$$

$$\mathbf{K}_W = \mathbf{X}(\mathbf{X}'\mathbf{W}\mathbf{X})^{-1}\mathbf{X}'\mathbf{W},$$

$$K_3 = 2\tau_S \delta_S - \tau_S^2,$$

$$K_4 = \frac{\sqrt{12}\xi}{\tau},$$

$$K_5 = \frac{\sqrt{12}\tau_S \delta_S}{\tau},$$

$$\xi = E[e_1 \mathrm{sgn}(e_1 - e_2)],$$

$$\delta_S = E[\mathrm{sgn}(e_1)\mathrm{sgn}(e_1 - e_2)].$$

The derivation can be found in Exercise 5.10.11. Estimates of the last two parameters are given by

$$\widehat{\xi} = n^{-2} \sum_i \sum_j \widehat{e}_{GRi} \mathrm{sgn}(\widehat{e}_{GRi} - \widehat{e}_{GRj})$$

$$\widehat{\delta}_S = n^{-2} \sum_i \sum_j \mathrm{sgn}(\widehat{e}_{GRi})\mathrm{sgn}(\widehat{e}_{GRi} - \widehat{e}_{GRj}),$$

while estimates of the other parameters are discussed in Section 3.9.2.

As in Chapter 3, the standard errors of the GR residuals are of a similar form to their R and LS counterparts. They depend on the underlying density of the errors and location in factor space. The standard errors of the GR residuals are also functions of the weights. This is as it should be since the GR estimates weigh cases according to their position in factor space. Example 5.3.2 illustrates the importance of correctly standardizing in outlier identification. The standardized residuals used here are the **internal GR studentized residuals** similar to (3.9.31).

5.5 Diagnostics that Detect Differences between Fits

Based on the discussion and examples of the previous section, fits based on R- and GR-estimates can differ on data sets where there are clusters of outliers in the x-space and/or points where curvature is more extreme. In exploring a messy data set these are precisely the points that should be determined; for example, these are the points that should be 'clicked' on in rotational (spin) plots in order visually to explore a given data set. It seems natural, then, to derive diagnostics which detect such points. McKean, Naranjo, and Sheather (1996a, 1996b) proposed such diagnostics based on R and GR fits. In this section we discuss these diagnostics and explore several data sets with them.

The idea behind the diagnostics is to choose fits which can substantially differ on such messy data sets. With this in mind we have selected the highly efficient Wilcoxon fit as the R fit and the GR fit based on the high-breakdown weights given in expression (5.3.3), with $\alpha_2 = 1$. Denote the Wilcoxon estimate by $\widehat{\boldsymbol{\beta}}_R$ and the GR-estimate by $\widehat{\boldsymbol{\beta}}_{GR}$. A measure of overall change in parameter estimates due to the downweighting of high-leverage points is given by the statistic $\|\widehat{\boldsymbol{\beta}}_R - \widehat{\boldsymbol{\beta}}_{GR}\|$. In order to be useful, this difference must be assessed relative to some scale. The following theorem gives the asymptotic distribution of the difference. Clearly $\widehat{\boldsymbol{\beta}}_R - \widehat{\boldsymbol{\beta}}_{GR} = \mathbf{0}$ when no observations are downweighted, in which case the distribution is degenerate. We thus assume that the weights have made enough difference so that $\widehat{\boldsymbol{\beta}}_R - \widehat{\boldsymbol{\beta}}_{GR}$ converges in law to a proper distribution, i.e. assume that

$$(\mathbf{X'WX})^{-1}\mathbf{X'W^2X}(\mathbf{X'WX})^{-1} - (\mathbf{X'X})^{-1} \text{ is positive definite.} \quad (5.5.1)$$

Theorem 5.5.1 *Under regularity conditions (E.1), (D.2), and (D.3) of Section 3.4, (G.1)–(G.3) of Section 4.2, and (5.5.1),*

$$\widehat{\boldsymbol{\beta}}_R - \widehat{\boldsymbol{\beta}}_{GR} \text{ is approximately distributed}$$

$$N\left(\mathbf{0},\ \tau^2[(\mathbf{X'WX})^{-1}\mathbf{X'W^2X}(\mathbf{X'WX})^{-1} - (\mathbf{X'X})^{-1}]\right), \quad (5.5.2)$$

where $\tau = (\sqrt{12} \int f^2(x)\,dx)^{-2}$ *and* $f(x)$ *is the pdf of the errors.*

Proof. Recall from expressions (3.5.16) and (5.4.2), respectively, the asymptotic representations for $\widehat{\boldsymbol{\beta}}_R$ and $\widehat{\boldsymbol{\beta}}_{GR}$:

$$\widehat{\boldsymbol{\beta}}_R = \boldsymbol{\beta}_0 + \sqrt{12}\tau(\mathbf{X'X})^{-1}\mathbf{X'F}_c(\mathbf{e}) + o_p(n^{-1/2})$$

$$\widehat{\boldsymbol{\beta}}_{GR} = \boldsymbol{\beta}_0 + (\sqrt{3}\tau/n)(\mathbf{X'WX})^{-1}\mathbf{S}(\boldsymbol{\beta}_0) + o_p(n^{-1/2}),$$

where $\mathbf{F}_c(\mathbf{e}) = (F(e_1) - \frac{1}{2}, \ldots, F(e_n) - \frac{1}{2})'$ is an $n \times 1$ vector of independent random variables, and $\mathbf{S}(\boldsymbol{\beta}_0) = n\sum\sum_{i<j} w_i w_j (x_i - x_j)\text{sgn}(e_i - e_j)$ is $p \times 1$. The projection of $\mathbf{S}(\boldsymbol{\beta}_0)$ is of the form $\mathbf{S}^*(\boldsymbol{\beta}_0) = \sum_i \mathbf{a}_i^*[F(e_i) - \frac{1}{2}]$, where the coefficients $\{\mathbf{a}_i^*\}$ are a function of $\{w_i\}$ and $\mathbf{x}_i$; see Exercise 5.10.12. Hence $\widehat{\boldsymbol{\beta}}_R - \widehat{\boldsymbol{\beta}}_{GR}$ is asymptotically a linear combination of $F(e_1), \ldots, F(e_n)$, from which asymptotic normality follows; see Exercise 5.10.13 for details.

Next consider the covariance matrix

$$\text{Cov}(\widehat{\boldsymbol{\beta}}_R - \widehat{\boldsymbol{\beta}}_{GR}) = \text{Cov}(\widehat{\boldsymbol{\beta}}_R) + \text{Cov}(\widehat{\boldsymbol{\beta}}_{GR}) - 2E[(\widehat{\boldsymbol{\beta}}_R - \boldsymbol{\beta}_0)(\widehat{\boldsymbol{\beta}}_{GR} - \boldsymbol{\beta}_0)'].$$

Based on the asymptotic distributions of $\widehat{\boldsymbol{\beta}}_R$ and $\widehat{\boldsymbol{\beta}}_{GR}$, (3.5.6) and (5.2.18) respectively, we need only show that $E(\widehat{\boldsymbol{\beta}}_R - \boldsymbol{\beta}_0)(\widehat{\boldsymbol{\beta}}_{GR} - \boldsymbol{\beta}_0)' = \tau^2(\mathbf{X}'\mathbf{X})^{-1}$. It can be shown (Exercise 5.10.14) that $E[\mathbf{F}_c(\mathbf{e})\mathbf{S}'(\boldsymbol{\beta}_0)] = (n/6)\mathbf{W}\mathbf{X}$, so that

$$E(\widehat{\boldsymbol{\beta}}_R - \boldsymbol{\beta}_0)(\widehat{\boldsymbol{\beta}}_{GR} - \boldsymbol{\beta}_0)' \doteq (6\tau^2/n)(\mathbf{X}'\mathbf{X})^{-1}\mathbf{X}'E[\mathbf{F}_c(\mathbf{e})\mathbf{S}'(\boldsymbol{\beta}_0)](\mathbf{X}'\mathbf{W}\mathbf{X})^{-1}$$
$$= \tau^2(\mathbf{X}'\mathbf{X})^{-1},$$

which proves the theorem.

For a messy data set, the R- and GR-estimates of the intercept parameter can also differ; hence, in order adequately to assess differences in fits the intercepts should be included. For both we shall use the medians of their residuals. Denote these estimates by

$$\widehat{\alpha}_{S,R} = \text{med}_i\{Y_i - \mathbf{x}_i\widehat{\boldsymbol{\beta}}_R\}$$

$$\widehat{\alpha}_{S,GR} = \text{med}_i\{Y_i - \mathbf{x}_i\widehat{\boldsymbol{\beta}}_{GR}\}.$$

Let $\widehat{\mathbf{b}}_R = (\widehat{\alpha}_{S,R}, \widehat{\boldsymbol{\beta}}_R')'$ and $\widehat{\mathbf{b}}_{GR} = (\widehat{\alpha}_{S,GR}, \widehat{\boldsymbol{\beta}}_{GR}')'$ denote the vectors of R- and GR-estimates, respectively.

Let $\mathbf{A}_D$ denote the asymptotic variance-covariance of the statistic $\widehat{\mathbf{b}}_R - \widehat{\mathbf{b}}_{GR}$. The matrix $\mathbf{A}_D$ is obtained in Exercise 5.10.15. It seems natural to standardize $\widehat{\mathbf{b}}_R - \widehat{\mathbf{b}}_{GR}$ by the inverse of $\mathbf{A}_D$. Under the same regularity conditions as Theorem 5.5.1, however, the asymptotic representations of $\widehat{\alpha}_{S,R}$ and $\widehat{\alpha}_{S,GR}$ are the same, (3.5.23). Because of this the asymptotic variance-covariance matrix of $\widehat{\mathbf{b}}_R - \widehat{\mathbf{b}}_{GR}$ is singular; see Exercise 5.10.16. A solution is to standardize $\widehat{\mathbf{b}}_R - \widehat{\mathbf{b}}_{GR}$ by the standard error of $\widehat{\mathbf{b}}_R$ instead of the standard error of the numerator. This produces the following statistic, which measures the total difference in the fits of $\widehat{\mathbf{b}}_R$ and $\widehat{\mathbf{b}}_{GR}$,

$$TDBETAS_R = (\widehat{\mathbf{b}}_R - \widehat{\mathbf{b}}_{GR})'\mathbf{A}_R^{-1}(\widehat{\mathbf{b}}_R - \widehat{\mathbf{b}}_{GR}), \qquad (5.5.3)$$

where

$$\mathbf{A}_R = \text{Cov}\begin{pmatrix}\widehat{\alpha}_R \\ \widehat{\boldsymbol{\beta}}_R\end{pmatrix} = \begin{bmatrix}\tau_S^2/n & 0 \\ 0 & \tau^2(\mathbf{X}'\mathbf{X})^{-1}\end{bmatrix};$$

see Theorem 3.5.11. We leave discussion of a benchmark for $TDBETAS_R$ until later in this section.

The diagnostic $TDBETAS_R$ measures the overall difference in the estimates. If it is large then usually we want to determine the cases causing this discrepancy in the fits. As we mentioned above, these would be the cases to investigate to see whether they are outliers in factor space or cases of extreme curvature. Let $\widehat{y}_{R,i} = \widehat{\alpha}_{S,R} + \mathbf{x}'\widehat{\boldsymbol{\beta}}_R$ and $\widehat{y}_{GR,i} = \widehat{\alpha}_{S,GR} + \mathbf{x}'\widehat{\boldsymbol{\beta}}_{GR}$ denote the respective fitted values for the ith case. A statistic which detects the observations that are fitted differently by R and GR is

$$CFITS_{D,i} = \frac{\widehat{y}_{R,i} - \widehat{y}_{GR,i}}{\{[1, \ \mathbf{x}_i'] \mathbf{A}_D[1, \ \mathbf{x}_i']'\}^{1/2}}, \qquad (5.5.4)$$

where $\mathbf{A}_D$ is the covariance of the difference discussed previously. Since $\mathbf{A}_D$ is singular, (the denominator of (5.5.4) is 0 at $\mathbf{x}_i = \bar{\mathbf{x}}$), we propose standardizing (5.5.4) by the standard error of the R fitted value $\widehat{y}_{R,i}$. Recall that this is the standardization given in expression (3.9.40). This gives

$$CFITS_{R,i} = \frac{\widehat{y}_{R,i} - \widehat{y}_{GR,i}}{(n^{-1}\widehat{\tau}_S^2 + h_{c,i}\widehat{\tau}^2)^{1/2}}. \tag{5.5.5}$$

This type of standardization is similar to $RDFFITS$, (3.9.39), and we propose the same benchmark $2\sqrt{(p+1)/n}$. We should note here that the objective of the diagnostic $CFITS$ is *not* outlier deletion. Rather the intent is to identify the *critical few* data points for closer study, because these critical few points often largely determine the outcome of the analysis or the direction that further analyses should take. This closer study may involve subject matter expertise, or rotation and spin plots, or data-collection-site investigation for accuracy of measurements, among other things. In this regard, the proposed benchmark of $2\sqrt{(p+1)/n}$ is meant as a heuristic aid, not a boundary to some formal critical region.

Using the benchmark for $CFITS$, we can derive an analogous benchmark for $TDBETAS$. In doing so, we will replace τ_S with τ. We realize that this may be a crude approximation but we are deriving a benchmark. Let $\mathbf{X}_1 = [\mathbf{1} : \mathbf{X}]$ and denote the projection matrix by $\mathbf{H} = \mathbf{X}_1(\mathbf{X}_1'\mathbf{X}_1)^{-1}\mathbf{X}_1'$. Replacing τ_S with τ, we have $\text{Cov}(\widehat{\mathbf{b}}_R) \doteq \tau^2(\mathbf{X}_1'\mathbf{X}_1)^{-1}$ and $\text{SE}(\widehat{y}_{R,i}) \doteq \tau\sqrt{h_{ii}}$. Under this approximation, it follows from (5.5.5) that an observation is flagged by the diagnostic $CFITS_{R,i}$ whenever

$$\frac{|\widehat{y}_{R,i} - \widehat{y}_{GR,i}|}{\widehat{\tau}\sqrt{h_{ii}}} > 2\sqrt{(p+1)/n}. \tag{5.5.6}$$

We use this expression to obtain a benchmark for the diagnostic $TDBETAS_R$ as follows:

$$\begin{aligned} TDBETAS_R &= (\widehat{\mathbf{b}}_R - \widehat{\mathbf{b}}_{GR})'[\widehat{\tau}^2(\mathbf{X}_1'\mathbf{X}_1)^{-1}]^{-1}(\widehat{\mathbf{b}}_R - \widehat{\mathbf{b}}_{GR}) \\ &= (1/\widehat{\tau}^2)[\mathbf{X}_1(\widehat{\mathbf{b}}_R - \widehat{\mathbf{b}}_{GR})]'[\mathbf{X}_1(\widehat{\mathbf{b}}_R - \widehat{\mathbf{b}}_{GR})] \\ &= (1/\widehat{\tau}^2)\sum_i (\widehat{y}_{R,i} - \widehat{y}_{GR,i})^2 \\ &= (p+1)\frac{1}{n}\sum_i \left(\frac{\widehat{y}_{R,i} - \widehat{y}_{GR,i}}{\widehat{\tau}\sqrt{(p+1)/n}}\right)^2. \end{aligned}$$

Since h_{ii} has the average value $(p+1)/n$, (5.5.6) suggests flagging $TDBETAS_R$ as large whenever $TDBETAS_R > (p+1)(2\sqrt{(p+1)/n})^2$, or

$$TDBETAS_R > \frac{4(p+1)^2}{n}. \tag{5.5.7}$$

In order to assess the performance of the benchmarks under both null and alternative situations, McKean et al. (1996a) conducted a simulation over two situations. For each situation, 500 simulations were performed.

Each simulation generated a sample of size $n = 20$ from the model $y = \alpha + \beta x + e$. The situations were:

(1) $\alpha = 0$, $\beta = 0.5$, e_i iid $N(0, 1)$, and x_i iid Unif$(0,1)$.
(2) Same as Situation (1) except that the last two x's were changed to $x_{19} = 1.5$, $x_{20} = 1.6$. The model at these two points was changed to $\alpha = \beta = 0$.

This study showed that the benchmarks for *CFITS* and *TDBETAS* were quite conservative under situation (1). Both diagnostics exceeded their benchmarks less than 1% of the time when no high-leverage outliers are present. On the other hand, for situation (2), when high-leverage outliers are present, the diagnostic *TDBETAS* had large values, often exceeding the benchmark by a substantial amount. In situation (2), the diagnostic *CFITS* had quite large values for the influential cases, sometimes exceeding the benchmark more than a hundredfold. In light of this, along with the benchmarks, we have found it useful in many examples to simply to look for *gaps* in size that separate large *CFITS* from small *CFITS*. This is illustrated in the examples below.

The following procedure describes one way that we have made use of these diagnostics in analyzing data sets. It separates the data set into two sets, A and B. On set A the fits agree, while the addition of any point in Set B causes disagreement. This is not an expert system. As with other diagnostic tools, we strongly recommend that the procedure described below be used in conjunction with knowledge of the subject matter, graphical methods and common sense.

1. Using R- and GR-estimates, make initial fits on all the data. Obtain the diagnostic *TDBETAS*.
2. If *TDBETAS* is less than its benchmark, (5.5.7), this gives evidence that there are no serious influential points in the **x**-space. This in no way says that the model is adequate. But it does give some confidence in using the highly efficient fit to check the adequacy of the fit of the model. This includes the use of residual plots, $q-q$ plots, and standard robust diagnostics, as discussed in Chapter 3. Such a check may very well lead to the consideration of other models. For example, there may be a curvature pattern in the residual plot or an indication of heteroscedasticity in the residual plot.
3. If *TDBETAS* exceeds its benchmark then there is evidence that R and GR fits substantially differ. The casewise diagnostics, *CFITS*, are next examined to determine what cases were the main contributors to this overall difference. We have found it best to consider the ordered $|CFITS_i|$ and look at the larger values. For instance, gaps between groups of values are worthy of inspection because a cluster of outliers in the **x**-space will often have similar $CFITS_i$ values. We then place the cases with largest values into a group labeled group B and place all other cases in group A.

The use of dynamic plots such as rotation plots cannot be overemphasized, here. In such plots, the cases in group B are the ones to click on.

Often by rotating and changing the axes for the various predictor variables, clusters of points are brought to one's attention; see Cook and Weisberg (1994). This may be the most important contribution of these diagnostics, leading to an investigation of the subject matter for the rationale behind such clusters.

4. If the preceding analysis or subject matter indicates that the discrepancies are due to curvature then we may not want to proceed with the next step which involves refitting the model using only the points in group A. In these circumstances, we may feel comfortable with the highly efficient robust fit. The second example below serves as an illustration of this.
5. If we do decide to refit the cases in group A, using both R- and GR-estimates, we proceed as follows:

    (a) If, upon refitting, $TDBETAS$ is still large we proceed through step 3 once more. This will lead to an enlargement of group B.
    (b) If $TDBETAS$ is small or step 5(a) was performed, the standard R diagnostics are run on the cases that were set aside in group B. This is performed by adding these points one at a time to Group A and then obtaining the R diagnostics. Hence, the **masking effect** that a cluster of outliers in the x-space has on the R-estimate is avoided. The diagnostics to use here are $RDFFITS_i$, (3.9.39), and the external t residual, (3.9.35), to see if an excluded case follows the same model as group A does.

The next two examples illustrate the use of the diagnostics $TDBETAS$ and $CFITS$. Other examples are given in the exercises.

Example 5.5.1 *Wood Data*

This is a real data set (Draper and Smith, 1966, p. 227) consisting of measurements on 20 mature wood samples of slash pine cross-sections. The responses are the specific gravities of the sections while the predictors are the measurements of five anatomical factors of the sections. The data set was modified by Rousseeuw (1984) to contain four outliers and this is the data that is displayed in Table 5.5.1.

The diagnostic $TDBETAS$ has the value 280 for this data set, implying a major difference in the R and GR fits. Cases 4, 6, 8, and 19 have $|CFITS_i|$ values of 14.7, 15.7, 15.3, and 16.2, respectively. These points were clicked on in a rotational plot and upon rotating the axes they clearly showed as a cluster of points in factor space; see Figure 5.5.1 for one such rotation obtained from R code, (Cook and Weisberg, 1994). These are the four cases which were modified by Rousseeuw. The next largest value of $|CFITS_i|$ was 4.80 for case 11. Hence, we let group B consist of the four largest $|CFITS_i|$ values. Our second fit was based on group A. For the second fit, $TDBETAS$ had the value 2.74 which is less than the benchmark value of 9. Table 5.5.2 displays the diagnostics $RDFFITS_i$, (3.9.39), and the external t residual, (3.9.35), for these points when they were included in set A, one at a time. Based on these values it is clear that these cases do not conform to the first-order model which is

Table 5.5.1: The modified wood data as presented in Rousseeuw (1984)

x_1	x_2	x_3	x_4	x_5	y
0.5730	0.1059	0.4650	0.5380	0.8410	0.5340
0.6510	0.1356	0.5270	0.5450	0.8870	0.5350
0.6060	0.1273	0.4940	0.5210	0.9200	0.5700
0.4370	0.1591	0.4460	0.4230	0.9920	0.4500
0.5470	0.1135	0.5310	0.5190	0.9150	0.5480
0.4440	0.1628	0.4290	0.4110	0.9840	0.4310
0.4890	0.1231	0.5620	0.4550	0.8240	0.4810
0.4130	0.1673	0.4180	0.4300	0.9780	0.4230
0.5360	0.1182	0.5920	0.4640	0.8540	0.4750
0.6850	0.1564	0.6310	0.5640	0.9140	0.4860
0.6640	0.1588	0.5060	0.4810	0.8670	0.5540
0.7030	0.1335	0.5190	0.4840	0.8120	0.5190
0.6530	0.1395	0.6250	0.5190	0.8920	0.4920
0.5860	0.1114	0.5050	0.5650	0.8890	0.5170
0.5340	0.1143	0.5210	0.5700	0.8890	0.5020
0.5230	0.1320	0.5050	0.6120	0.9190	0.5080
0.5800	0.1249	0.5460	0.6080	0.9540	0.5200
0.4480	0.1028	0.5220	0.5340	0.9180	0.5060
0.4170	0.1687	0.4050	0.4150	0.9810	0.4010
0.5280	0.1057	0.4240	0.5660	0.9090	0.5680

Table 5.5.2: One-at-a-time diagnostics for outliers of the wood data

	Case			
Diagnostic	4	8	6	19
External t residual	−8.0	−10.2	−9.1	−6.9
RDFFITS<$_i$	−17.8	−24.5	−22.9	−30.4

followed by the majority of the cases. Thus the procedure has isolated the points which caused the discrepancy. In practice, hopefully subject knowledge would provide a rationale for why these four cases are outliers.

The purpose of the next example is to illustrate the use of the diagnostics *TDBETAS* and *CFITS* on a data set that contains curvature.

Example 5.5.2 *Simulated Response Surface Data*

This is an artificially generated data set. For two predictor variables, say x_1 and x_2, we chose a response surface design. Six points were chosen within the circle of radius 1, five points were chosen in the ring with radii 1 and 2, and four ponts were chosen in the ring with radii 2 and 3. There were two replicates at each of these points; hence, 30 points altogether. As a model we chose

$$y = -9x_1 - 9x_2 + 3x_1^2 + 3x_2^2 - 2x_1x_2 + e, \qquad (5.5.8)$$

where the errors are simulated iid $N(0, 100)$. $E(Y)$ has a minimum at the point (2.25, 2.25). The data are displayed in Table 5.5.3.

Diagnostics that Detect Differences between Fits 307

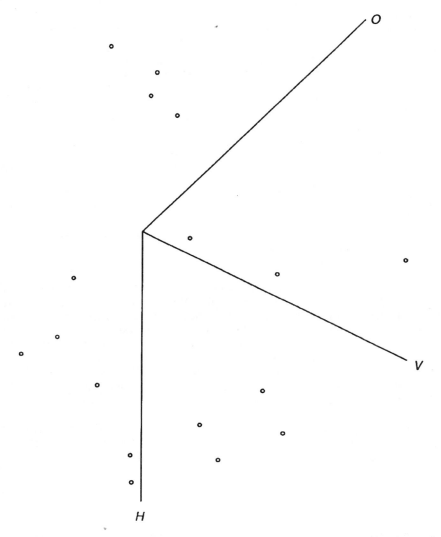

Figure 5.5.1: Rotational plot of wood data

As a start, we fit the first-order model with x_1 and x_2 as the predictors. Based on the R and GR fits of the linear model, $TDBETAS$ had the value 0.22, while the benchmark is 1.2. Hence, the R and GR fits are similar. This is further confirmed by very similar residual plots for the Wilcoxon and GR fits of this first-order model, Figure 5.5.2(a,c). Note, however, that both of these plots indicate curvature. Although not shown, this was also true of the LS residual plot. Hence, we would proceed to fit at least a quadratic (second-order) model.

Table 5.5.3: Simulated response surface data

x_1	x_2	y	x_1	x_2	y
0.33524	−0.14966	8.2262	0.33524	−0.14966	−9.5881
−0.70350	0.05567	9.5947	−0.70350	0.05567	3.4073
0.28335	0.84827	−8.0016	0.28335	0.84827	−11.2798
−0.55313	0.15834	0.5192	−0.55313	0.15834	−5.8145
0.36275	−0.41027	5.0605	0.36275	−0.41027	−0.7018
−0.85961	−0.26525	23.3692	−0.85961	−0.26525	21.5892
−0.18148	1.03942	1.4821	−0.18148	1.03942	−9.8291
−1.43606	0.52026	−6.2537	−1.43606	0.52026	−1.5601
−1.09385	1.31888	−7.3118	−1.09385	1.31888	11.5337
0.82857	−1.33497	15.2159	0.82857	−1.33497	10.7931
−0.80846	−1.78504	37.2183	−0.80846	−1.78504	39.8418
1.64558	2.28647	−10.0508	1.64558	2.28647	−20.9083
2.53973	−1.31050	26.9540	2.53973	−1.31050	11.8045
−0.30379	−2.47121	37.9522	−0.30379	−2.47121	45.5355
−0.32892	−2.12256	27.9355	−0.32892	−2.12256	33.9767

The residual plots of the full second-order model are presented in Figure 5.5.2(b,d). The random scatter in the Wilcoxon plot is an improvement over its respective plot for the first-order model. It indicates a better fit with the second-order model. On the other hand, the GR residual plot is poor. It shows a negative linear pattern which is similar to the patterns discussed in Section 5.4 and McKean *et al.* (1993; 1994) for high-breakdown fits in the presence of curvature. The LS plot was similar to the Wilcoxon plot. Both LS and R fitted response surfaces indicate a minimum value while the GR fit indicates a saddle point; hence, the GR-estimate has misread the curvature in this data set.

In light of the residual plots, it is not surprising that the diagnostic *TDBETAS* for the quadratic model is high (1140). For this model, though, the series of residual plots show that the Wilcoxon fit results in a much better fit to the data than the GR fit.

5.6 Coefficients of Multiple Determination

For this and the next section on robustness properties, the underlying model will be the correlation model (3.11.1), in which **x** is stochastic. This was discussed for R-estimation in Section 3.11. In this brief section, we discuss coefficients of multiple determination (CMDs) based on GR-estimates. These coefficients are much less sensitive to outliers in factor space than their R counterparts but they are more difficult to interpret. Much of the material in this section was discussed by Witt, Naranjo, and McKean (1995).

Consider, then, the correlation model (3.11.1). Recall for this model that Y and **x** are stochastically independent if and only if the vector of regression

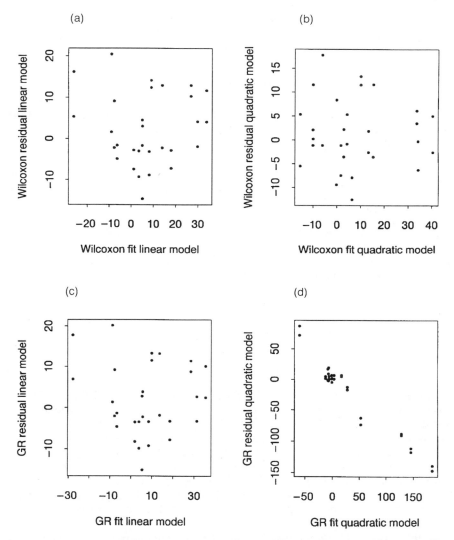

Figure 5.5.2: (a,c) Wilcoxon and GR residual plots for the simulated response surface data based on the fits of the first-order model; (b,d) Wilcoxon and GR residual plots for the simulated response surface data based on the fits of the second-order model

coefficients $\boldsymbol{\beta} = \mathbf{0}$. A natural CMD is the percentage of dispersion accounted for by fitting $\mathbf{x}$, i.e.

$$R_{GR1} = \frac{RD_{GR}}{D_{GR}(\mathbf{0})}, \qquad (5.6.1)$$

where the reduction in dispersion is $RD_{GR} = D_{GR}(\mathbf{0}) - D_{GR}(\widehat{\boldsymbol{\beta}}_{GR})$. The influence function, though, of the denominator of R_{GR1} is unbounded in the Y-space, similar to its R counterpart. As with the R-estimates, we define a second

CMD based on the drop in dispersion test statistic F_{GR}, (5.2.22), for $H_0 : \boldsymbol{\beta} = \mathbf{0}$, which is given by

$$R_{GR2} = \frac{RD_{GR}}{RD_{GR} + \frac{n(n-p-1)}{\sqrt{12}}\hat{\tau}}. \qquad (5.6.2)$$

As we note in Section 5.7, the influence function of R_{GR2} is bounded in both the x- and the Y-spaces. We assume throughout that $b_{ij} = b_i b_j$.

The functionals corresponding to $D_{GR}(\mathbf{0})$ and $D_{GR}(\hat{\boldsymbol{\beta}}_{GR})$ for the bounded influence R-estimates are

$$\overline{D}_{GRy} = \frac{\sqrt{3}}{2} \iint b(\mathbf{x}_1) b(\mathbf{x}_2) |y_1 - y_2| \, dH(\mathbf{x}_1, y_1) \, dH(\mathbf{x}_2, y_2) \qquad (5.6.3)$$

$$\overline{D}_{GRe} = \frac{\sqrt{3}}{2} \int b(\mathbf{x}_1) b(\mathbf{x}_2) \, dM(\mathbf{x}_1) \, dM(\mathbf{x}_2) \, E|e_1 - e_2|. \qquad (5.6.4)$$

Let $\overline{RD}_{GR} = \overline{D}_{GRy} - \overline{D}_{GRe}$. In the same way as for regular R-estimates, we define population CMDs $\overline{R}_{GR1}$ and $\overline{R}_{GR2}$ as

$$\overline{R}_{GR1} = \frac{\overline{RD}_{GR}}{\overline{RD}_{GRy}} \qquad (5.6.5)$$

$$\overline{R}_{GR2} = \frac{\overline{RD}_{GR}}{\overline{RD}_{GR} + \frac{\tau}{\sqrt{12}}}. \qquad (5.6.6)$$

Because of the weight function these will not simplify as did the CMDs for R-estimates based on Wilcoxon scores. Furthermore, for the classical multivariate normal model the relationships between the Wilcoxon CMDs and the classical $\overline{R}^2$, expressions (3.11.27) and (3.11.28), will not hold.

It is clear from their definitions that both $\overline{D}_{GRy}$ and $\overline{D}_{GRe}$ are strictly positive. Similarly, we can show that $\overline{RD}_{GR}(\boldsymbol{\beta})$ is a strictly convex function of $\boldsymbol{\beta}$ with a minimum value of 0 at $\boldsymbol{\beta} = \mathbf{0}$; see Exercise 5.10.17. Thus $0 \leqslant \overline{R}_{GRi} \leqslant 1$ for $i = 1, 2$ and $\overline{R}_{GRi} = 0$ if and only if Y and $\mathbf{x}$ are independent. Due to the weights, however, simple expressions for these CMDs are not obtainable.

Example 5.6.1 *A Contaminated Normal Model, Example 3.11.4 continued*

In Example 3.11.4 we presented the results of a small study to see how sensitive the population CMDs were to contamination in the distribution of errors. It seems fitting here to extend this study to contamination in the distribution of x. So, as an additional situation to that discussed in Section 3.9.4, we consider the case where the random error e has an $N(0, 1)$ distribution while the random variable x has a contaminated normal distribution with proportion of contamination ε and ratio of contaminated variance to uncontaminated σ_c^2. The model is $Y = x + e$; hence, $\beta = 1$ and Y and x are dependent. We used the weight function $b(x) = 1$ or $2/|x|$ depending on whether $|x| < 2$ or $|x| > 2$. The population CMDs for the GR-estimates were obtained by numerical integration routines in NAG (Numerical Algorithms Group, 1983). We also

Table 5.6.1: Coefficients of multiple determination under contaminated **x**

	$x \sim CN(\varepsilon, \sigma_c^2 = 9)$						$x \sim CN(\varepsilon, \sigma_c^2 = 100)$					
CMD	0.00	0.01	0.02	ε 0.05	0.10	0.15	0.00	0.01	0.02	ε 0.05	0.10	0.15
$\overline{R}^2$	0.50	0.52	0.54	0.58	0.64	0.69	0.50	0.67	0.75	0.86	0.92	0.94
$\overline{R}_1^*$	0.50	0.51	0.52	0.56	0.62	0.66	0.50	0.58	0.63	0.75	0.85	0.90
$\overline{R}_2^*$	0.50	0.51	0.53	0.57	0.62	0.66	0.50	0.57	0.63	0.74	0.85	0.90
$\overline{R}_{GR1}^{*2}$	0.50	0.50	0.51	0.52	0.56	0.59	0.50	0.51	0.52	0.55	0.62	0.66
$\overline{R}_{GR2}^{*2}$	0.49	0.50	0.50	0.52	0.55	0.57	0.49	0.50	0.51	0.53	0.58	0.62

formulated GR analogs to $\overline{R}_1^{*2}$ and $\overline{R}_2^{*2}$, (3.11.33) and (3.11.34), respectively, which we will denote as $\overline{R}_{GR1}^{*2}$ and $\overline{R}_{GR2}^{*2}$. Recall, when x and e are normal random variables, that $\overline{R}_1^{*2} = \overline{R}_2^{*2} = \overline{R}^2$. While this is not true for the GR counterparts, it does put them on a similar scale.

Table 5.6.1 presents the values of the CMDs for ε between 0 and 0.15 and for $\sigma_c^2 = 9$ and 100. In this situation, all the coefficients except $\overline{R}_{GR1}^{*2}$ and $\overline{R}_{GR2}^{*2}$ inflate, rapidly for $\sigma_c^2 = 100$, as ε increases. In terms of fitting data, this corresponds to least-squares and regular R fits sensitivity to outliers in the x-space. Such points draw these fits toward them, resulting in larger R^2, R_1, and R_2. On the other hand, the bounded influence R-estimate is not as sensitive to these outliers.

5.7 Robustness Properties

In this section we obtain the robust properties of the GR-estimate, showing that its influence function is bounded in both the x- and Y-spaces and that it has positive breakdown. Our underlying model will be the correlation model of the previous section.

5.7.1 Influence Functions

As in Chapter 3, let H denote the joint distribution of **x** and Y and let M denote the distribution function of **x**. Let $b(x_1, x_2)$ denote the weight function. Because **x** is stochastic the assumptions listed in Section 5.2 need to be rephrased. In particular, the limits in (G.1), (5.2.2), and (G.2), (5.2.3), are now convergences in probability and condition (G.3), (5.2.4), is replaced by Theorem 3.11.2. Upon expanding the right-hand sides of (G.1) and (G.2), for this stochastic model, we obtain,

$$\mathbf{C} = \tfrac{1}{2}\int\int (\mathbf{x}_2 - \mathbf{x}_1)(\mathbf{x}_2 - \mathbf{x}_1)' b(\mathbf{x}_1, \mathbf{x}_2) dM(\mathbf{x}_1) dM(\mathbf{x}_2) \qquad (5.7.1)$$

$$\mathbf{V} = \int\left\{\int (\mathbf{x}_2 - \mathbf{x}_1) b(\mathbf{x}_2, \mathbf{x}_1) dM(\mathbf{x}_2)\right\}\left\{\int (\mathbf{x}_2 - \mathbf{x}_1) b(\mathbf{x}_2, \mathbf{x}_1) dM(\mathbf{x}_2)\right\}' dM(\mathbf{x}_1). \quad (5.7.2)$$

As defined in expression (3.11.5), let Σ be the variance-covariance matrix of $\mathbf{x}$.

Let $T(H)$ denote the functional corresponding to the estimator $\widehat{\boldsymbol{\beta}}_{GR}$. Based on the estimating equation for $\widehat{\boldsymbol{\beta}}_{GR}$, namely $S_{GR}(\widehat{\boldsymbol{\beta}}_{GR}) \doteq \mathbf{0}$, the functional $T(H)$ satisfies the implicit equation

$$\int\int b(\mathbf{x}_1, \mathbf{x}_2)(\mathbf{x}_2 - \mathbf{x}_1)\mathrm{sgn}\{(y_2 - \mathbf{x}_2'T(H))$$

$$- (y_1 - \mathbf{x}_1'T(H))\}dH(\mathbf{x}_2, y_2)dH(\mathbf{x}_1, y_1) = 0. \quad (5.7.3)$$

Denote by $H_s = (1-s)H + s\Delta_{\mathbf{x}_0, y_0}$ the distribution function H contaminated by a point mass at $(\mathbf{x}_0, y_0)$.

The influence function of the GR-estimate was derived by Naranjo and Hettmansperger (1994).

Theorem 5.7.1 *The influence function of $\widehat{\boldsymbol{\beta}}_{GR}$ is*

$$IF(\mathbf{x}_0, y_0; \widehat{\boldsymbol{\beta}}_{GR}) = \tau\sqrt{12}(F(y_0 - \alpha - \mathbf{x}_0'\boldsymbol{\beta}) - \tfrac{1}{2})\mathbf{C}^{-1}\int (\mathbf{x}_0 - \mathbf{x})b(\mathbf{x}_0, \mathbf{x})dM(\mathbf{x}).$$

(5.7.4)

This result follows immediately from Theorem 5.8.6 of Section 5.8.3. Note that the influence function is bounded in the Y-space and by a proper choice of weight function it is also bounded in the x-space. The weights given in Section 5.3 are examples of such weight functions. As Exercise 5.10.18 shows, the asymptotic distribution of $\widehat{\boldsymbol{\beta}}_{GR}$ is read from the influence function. If the weights are identically equal to 1, i.e. Wilcoxon R-estimates, then the influence function simplifies to expression (3.5.17) with Wilcoxon scores.

As in Chapter 3, we also give the influence function for the test statistic based on the drop in dispersion, more precisely the square root of the test statistic. As there, the proof of this result is rather technical and we will not give it here. Details of the proof can be found in Witt et al. (1995). We shall use the notation of Theorem 5.2.12.

Theorem 5.7.2 *Consider the hypotheses of Theorem 5.2.12 and the test statistic T_{GR}, (5.2.21). Under the assumptions of Theorem 5.71,*

$$IF\left(\mathbf{x}_0, y_0; \sqrt{T_{GR}}\right) = \sqrt{12}[F(y_0 - \mathbf{x}_0'\boldsymbol{\beta}) - \tfrac{1}{2}]$$

$$\cdot \{[\int(\mathbf{x}_0 - \mathbf{x})b(\mathbf{x}_0, \mathbf{x})dM(\mathbf{x})]'(\mathbf{C}^{-1} - \mathbf{C}_r^+)[\int(\mathbf{x}_0 - \mathbf{x})b(\mathbf{x}_0, \mathbf{x})dM(\mathbf{x})]\}^{1/2},$$

where $\mathbf{C}_r^+$ is defined in expression (5.2.20).

As in Chapter 3, the asymptotic distribution of the test statistic T_{GR} can be read from this result. Details are given in Exercise 5.10.19.

As with the influence function for the estimate $\widehat{\boldsymbol{\beta}}_{GR}$, a proper choice of weights results in a test statistic whose influence is bounded in both the x- and the Y-spaces. The results of the study in the previous section on the sensitivity of the CMD based on T_{GR} was a verification of this robustness of T_{GR}.

5.7.2 Breakdown

In order to obtain the breakdown of the GR-estimates, consider the contaminated distribution function $H_s = (1 - s)H + s\Delta_{(\mathbf{x}_0, y_0)}$, where $0 < s < 1$ and $\Delta_{(\mathbf{x}_0, y_0)}$ denotes a point mass cdf at $(\mathbf{x}_0, y_0)$. Denote the bias of GR functional as

$$\text{bias}(s) = \|T(H_s) - T(H)\|.$$

For this section we will define the breakdown point as

$$\varepsilon^* = \sup\{s < \tfrac{1}{2} : \max_{(\mathbf{x}_0, y_0)}\{\text{bias}(s) < \infty\}\}, \tag{5.7.5}$$

see Hampel *et al.* (1986). We have the following result:

Theorem 5.7.3 *Let $\mathbf{x}_1$ and $\mathbf{x}_2$ be independent random vectors with distribution function M. The estimate $\widehat{\boldsymbol{\beta}}_{GR}$ has gross error breakdown*

$$\varepsilon^* = \inf_{\|\lambda\|=1}\left\{\frac{\tfrac{1}{2}E|\lambda'(\mathbf{x}_1 - \mathbf{x}_2)|b(\mathbf{x}_1, \mathbf{x}_2)}{\tfrac{1}{2}E|\lambda'(\mathbf{x}_1 - \mathbf{x}_2)|b(\mathbf{x}_1, \mathbf{x}_2) + \sup_{\mathbf{x}_0 \in \chi} E|\lambda'(\mathbf{x}_0 - \mathbf{x})|b(\mathbf{x}_0, \mathbf{x})}\right\}. \tag{5.7.6}$$

The proof of this theorem can be found in Naranjo and Hettmansperger (1994). Since $E[|\lambda'(\mathbf{x}_1 - \mathbf{x}_2|b(\mathbf{x}_1, \mathbf{x}_2)] \leq \sup_{\mathbf{x}_0} E[|\lambda'(\mathbf{x}_0 - \mathbf{x}_2|b(\mathbf{x}_0, \mathbf{x}_2)]$, it follows that $\varepsilon^* \leq \tfrac{1}{3}$. Further, it can be quite small under certain circumstances; see McKean, Naranjo, and Sheather (1996a) for an investigation of ε^*.

5.8 High-Breakdown (HBR) Estimates

While the GR-estimates discussed in the previous sections have bounded influence and positive breakdown, they do not have 50% breakdown. In this section, we present GR-estimates which attain 50% breakdown. We shall call them the HBR-estimates. They were proposed by Chang (1995) and were further developed by Chang *et al.* (1997). These estimates are obtained by allowing the weights used by the GR-estimates to be a function of response (Y) space in addition to factor (x) space. This gives us a large class of weighted Wilcoxon estimates. They range from the highly efficient Wilcoxon estimates, when all the weights equal one, to bounded influence estimates and, now, to 50% breakdown estimates. The HBR-estimates can differ from the R-estimates on a given data set but the diagnostics presented in Section 5.5 to detect differences between robust fits are easily extended to the HBR-estimates; see Exercise 5.10.20. This wide class of estimates and their accompanying diagnostics offer the user powerful tools with which to investigate messy data sets. Hössjer (1994) presents another class of high-breakdown rank-based estimates.

5.8.1 Definition of the HBR-Estimates

The estimates proposed below are generalized R-estimates but, unlike the weights of the GR-estimates of Section 5.1, they use weights which also depend on the responses. Define the function $\psi(t)$ by $\psi(t) = 1, t,$ or -1 according as $t \geq 1, -1 < t < 1,$ or $t \leq -1$. Let $m_i = \psi[b/(\mathbf{x}_i - \mathbf{v})'\mathbf{V}^{-1}(\mathbf{x}_i - \mathbf{v})]$, where $\mathbf{v}$ and $\mathbf{V}$ are robust estimates of location and scatter in factor space and b is a tuning constant. The minimum volume ellipsoid estimates discussed in Section 5.3 could serve as estimates of $\mathbf{v}$ and $\mathbf{V}$. Let $\hat{\boldsymbol{\beta}}^{(0)}$ be an initial estimate of the vector of regression parameters $\boldsymbol{\beta}$ and denote the residuals based on this initial fit by $\hat{e}_i^{(0)} = Y_i - \mathbf{x}_i'\hat{\boldsymbol{\beta}}^{(0)}$. Consider the weights defined by

$$b_{ij} = \psi\left[\left|\frac{c\, m_i m_j}{(\hat{e}_i^{(0)}/\hat{\sigma})(\hat{e}_j^{(0)}/\hat{\sigma})}\right|\right], \tag{5.8.1}$$

where c is a tuning constant and $\hat{\sigma}$ is an initial scaling estimate such as the MAD which is given by

$$\text{MAD} = 1.483\, \text{med}_i|\hat{e}_i^{(0)} - \text{med}_j\{\hat{e}_j^{(0)}\}|. \tag{5.8.2}$$

Letting $Q_i = (\mathbf{x}_i - \mathbf{v})'\mathbf{V}^{-1}(\mathbf{x}_i - \mathbf{v})$, we can write

$$m_i = \psi\left(\frac{b}{Q_i}\right) = \min\left\{1, \frac{b}{Q_i}\right\},$$

where b is a tuning constant. Hence the weights can be estimated by

$$\hat{b}_{ij} = \min\left\{1, \frac{c\hat{\sigma}}{|\hat{e}_i|}\frac{\hat{\sigma}}{|\hat{e}_j|}\min\left\{1, \frac{b}{Q_i}\right\}\min\left\{1, \frac{b}{Q_j}\right\}\right\}. \tag{5.8.3}$$

From this point of view, it is clear that these weights downweight both outlying points in factor space and outlying responses. More discussion on the tuning parameters and initial estimates is given below.

The criterion function of interest is the same as that used to define the GR-estimates, namely (5.1.1), except now the weights are given by (5.8.1). Once the weights are given, this function is a pseudo-norm which we will denote by $\|\cdot\|_{HBR}$ and the corresponding dispersion function by

$$D_{HBR}(\boldsymbol{\beta}) = \|\mathbf{Y} - \mathbf{X}\boldsymbol{\beta}\|_{HBR}. \tag{5.8.4}$$

Since it is based on a pseudo-norm, D_{HBR} is a continuous, nonnegative, convex function of $\boldsymbol{\beta}$. We define the HBR-estimate of $\boldsymbol{\beta}$ by

$$\hat{\boldsymbol{\beta}}_{HBR} = \text{Argmin}\, D_{HBR}(\boldsymbol{\beta}). \tag{5.8.5}$$

The negative of the gradient of $D_{HBR}(\boldsymbol{\beta})$ is given by

$$\mathbf{S}_{HBR}(\boldsymbol{\beta}) = \sum_{i<j} b_{ij}(\mathbf{x}_i - \mathbf{x}_j)\text{sgn}((Y_i - Y_j) - (\mathbf{x}_i - \mathbf{x}_j)'\boldsymbol{\beta}); \tag{5.8.6}$$

hence, the HBR-estimate also solves the equation

$$\mathbf{S}_{HBR}(\boldsymbol{\beta}) \doteq \mathbf{0}.$$

5.8.2 Asymptotic Normality of $\widehat{\boldsymbol{\beta}}_{HBR}$

The asymptotic normality of the HBR-estimates was developed by Chang (1995) and Chang et al. (1997). Much of our development is in Appendix A.6, taken from Chang et al. (1997). In order to establish asymptotic normality of $\widehat{\boldsymbol{\beta}}_{HBR}$, we need some further notation and assumptions. Define the parameters

$$\gamma_{ij} = B'_{ij}(0)/E_{\boldsymbol{\beta}}(b_{ij}), \quad \text{for } 1 \leq i,j \leq n, \tag{5.8.7}$$

where

$$B_{ij}(t) = E_{\boldsymbol{\beta}}[b_{ij}I(0 < y_i - y_j < t)]. \tag{5.8.8}$$

Consider the symmetric $n \times n$ matrix $\mathbf{A}_n = [a_{ij}]$ defined by

$$a_{ij} = \begin{cases} -\gamma_{ij}b_{ij} & \text{if } i \neq j \\ \sum_{k \neq i} \gamma_{ik}b_{ik} & \text{if } i = j. \end{cases} \tag{5.8.9}$$

Define the $p \times p$ matrix $\mathbf{C}_n$ as

$$\mathbf{C}_n = \mathbf{X}' \mathbf{A}_n \mathbf{X}. \tag{5.8.10}$$

Since the rows and columns of $\mathbf{A}_n$ sum to zero, it can be shown that

$$\mathbf{C}_n = \sum_{i<j} \gamma_{ij}b_{ij}(\mathbf{x}_j - \mathbf{x}_i)(\mathbf{x}_j - \mathbf{x}_i)'; \tag{5.8.11}$$

see Exercise 5.10.21. Let

$$\mathbf{U}_i = (1/n) \sum_{j=1}^{n} (\mathbf{x}_j - \mathbf{x}_i) E(b_{ij}\text{sgn}(y_j - y_i)|y_i). \tag{5.8.12}$$

Besides assumptions (E.1), (3.4.1), (D.2), (3.4.7), and (D.3), (3.4.8) of Chapter 3, we need to make the following additional assumptions:

(H.1) There exists a matrix $\mathbf{C}_H$ such that $n^{-2}\mathbf{C}_n = n^{-2}\mathbf{X}'\mathbf{A}_n\mathbf{X} \xrightarrow{P} \mathbf{C}_H$.
$$\tag{5.8.13}$$

(H.2) There exists a $p \times p$ matrix Σ_H, $(1/n) \sum_{i=1}^{n} \text{Var}(\mathbf{U}_i) \to \Sigma_H$. (5.8.14)

(H.3) $\sqrt{n}(\widehat{\boldsymbol{\beta}}^{(0)} - \boldsymbol{\beta}) \xrightarrow{D} N(\mathbf{0}, \Xi)$ where $\widehat{\boldsymbol{\beta}}^{(0)}$ is the initial estimator and Ξ is a positive definite matrix. (5.8.15)

(H.4) The weight function $b_{ij} = g(\mathbf{x}_i, \mathbf{x}_j, y_i, y_j, \widehat{\boldsymbol{\beta}}^{(0)}) \equiv g_{ij}(\widehat{\boldsymbol{\beta}}^{(0)})$ is continuous and the gradient ∇g_{ij} is bounded uniformly in i and j.
$$\tag{5.8.16}$$

For the correlation model, an explicit expression can be given for the matrix $\mathbf{C}_H$ assumed in (H.1); see (5.8.26) and also Lemma 5.8.3.

As our theory will show, the HBR-estimate attains 50% breakdown (Section 5.8.3) and asymptotic normality, at rate $\sqrt{n}$, provided the initial estimates of regression coefficients have these qualities. One such estimate is the **least trimmed squares**, (LTS), which is given by,

$$\widehat{\boldsymbol{\beta}}^{(0)} = \text{Argmin} \sum_{i=1}^{h} \{y - \alpha - \mathbf{x}'\boldsymbol{\beta}\}_{i:n}^{2}, \qquad (5.8.17)$$

where $\{y - \alpha - \mathbf{x}'\boldsymbol{\beta}\}_{i:n}$ are the ordered residuals and h is an integer between 1 and n. If $h = [n/2] + 1$ then the LTS estimates possess 50% breakdown. The LTS estimates are also asymptotically normal at rate $\sqrt{n}$. See Rousseeuw and Leroy (1987) for discussion. Another class of such estimates are the rank-based estimates proposed by Hössjer (1994); see also Croux, Rousseeuw and Hössjer (1994).

The development of the theory for $\widehat{\boldsymbol{\beta}}_{HBR}$ proceeds similarly to that of the GR-estimates presented in Section 5.2, except that now we have the further complication of stochastic weights. The theory is sketched in the appendix, Section A.6, and here we present only the two main results: the asymptotic distribution of the gradient and the asymptotic distribution of the estimate.

Theorem 5.8.1 *Under assumptions (E.1), (3.4.1), and (H.1)–(H.4), (5.8.13)–(5.8.16),*

$$n^{-3/2} \mathbf{S}_{HBR}(\mathbf{0}) \xrightarrow{\mathcal{D}} N(\mathbf{0}, \Sigma_H).$$

The proof of this theorem proceeds along the same lines of theory as used to obtain the null distribution of the gradients of the R- and GR-estimates. The projection of $\mathbf{S}_{HBR}(\mathbf{0})$ is first determined and its asymptotic distribution is established as $N(\mathbf{0}, \Sigma_H)$. The result follows then upon showing that the difference between $\mathbf{S}_{HBR}(\mathbf{0})$ and its projection goes to zero in second mean; see the proof of Theorem A.6.9 for details. The following theorem gives the asymptotic distribution of $\widehat{\boldsymbol{\beta}}_{HBR}$.

Theorem 5.8.2 *Under assumptions (E.1), (3.4.1), and (H.1)–(H.4), (5.8.13)–(5.8.16),*

$$\sqrt{n}(\widehat{\boldsymbol{\beta}}_{HBR} - \boldsymbol{\beta}) \xrightarrow{\mathcal{D}} N(\mathbf{0}, \tfrac{1}{4} \mathbf{C}_H^{-1} \Sigma_H \mathbf{C}_H^{-1}).$$

The proof of this theorem is similar to that of the R- and GR-estimates. First asymptotic linearity and quadraticity are established. These results are then combined with Theorem 5.8.1 to yield the result; see Theorem A.6.1 for details.

The following lemma derives another representation of the limiting matrix $\mathbf{C}_H$, which will prove useful in the derivation of the influence function of $\widehat{\boldsymbol{\beta}}_{HBR}$ found in Sections 5.8.3 and 5.9 which concerns the implementation of these high-breakdown estimates. For what follows, assume without loss of generality that the true parameter value $\boldsymbol{\beta} = \mathbf{0}$. Let $g_{ij}(\widehat{\boldsymbol{\beta}}^{(0)}) \equiv b(\mathbf{x}_i, \mathbf{x}_j, y_i, y_j, \widehat{\boldsymbol{\beta}}^{(0)})$ denote the weights as a function of the initial estimator. Let $g_{ij}(\mathbf{0}) \equiv$

$b(\mathbf{x}_i, \mathbf{x}_j, y_i, y_j)$ denote the weight function evaluated at the true value $\boldsymbol{\beta} = \mathbf{0}$. The following result is proved in Lemma A.6.2:

$$B'_{ij}(t) = \int_{-\infty}^{\infty} \cdots \int_{-\infty}^{\infty} b\left(x_i, x_j, y_j + t, y_j, \widehat{\boldsymbol{\beta}}^{(0)}\right) f(y_j + t) f(y_j) \prod_{k \neq i,j} f(y_k)\, dy_1 \cdots dy_n. \quad (5.8.18)$$

It is further shown that $B'_{ij}(t)$ is continuous in t. The representation we want is:

Lemma 5.8.3 *Under assumptions (E.1), (3.4.1), and (H.1)–(H.4), (5.8.13)–(5.8.16),*

$$E\left[\frac{1}{2}\frac{1}{n^2}\sum_{i=1}^{n}\sum_{j=1}^{n}\int_{-\infty}^{\infty} b(\mathbf{x}_i, \mathbf{x}_j, y_i, y_j) f^2(y_j)\, dy_j (\mathbf{x}_j - \mathbf{x}_i)(\mathbf{x}_j - \mathbf{x}_i)'\right] \rightarrow C_H. \quad (5.8.19)$$

Proof. By (5.8.7), (5.8.8), (5.8.11), and (5.8.18),

$$E\left[\frac{1}{n^2}\mathbf{C}_n\right] = \frac{1}{2}\frac{1}{n^2}\sum_{i=1}^{n}\sum_{j=1}^{n} B'_{ij}(0)(\mathbf{x}_j - \mathbf{x}_i)(\mathbf{x}_j - \mathbf{x}_i)'. \quad (5.8.20)$$

Because $B'_{ij}(0)$ is uniformly bounded over all i and j, and the matrix $(1/n^2)\sum_i \sum_j (\mathbf{x}_j - \mathbf{x}_i)(\mathbf{x}_j - \mathbf{x}_i)'$ converges to a positive definite matrix, the right-hand side of (5.8.20) also converges. By Lemmas A.6.2 and A.6.4, we have

$$B'_{ij}(0) = \int b(\mathbf{x}_i, \mathbf{x}_j, y_j, y_j) f^2(y_j)\, dy_j + o(1), \quad (5.8.21)$$

where the remainder term is uniformly small over all i and j. Under assumption (H.1), (5.8.13), the result follows.

5.8.3 Robustness Properties of the HBR-Estimate

In this section we show that the HBR-estimate can attain 50% breakdown and derive its influence function. We show that its influence function is bounded in both the **x**- and the *Y*-spaces. The argument for breakdown is taken from Chang (1995) while the influence function derivation is taken from Chang et al. (1997).

Breakdown of the HBR-Estimate

Let $\mathbf{Z} = \{\mathbf{z}_i\} = \{(\mathbf{x}_i, y_i)\}$, $i = 1, \ldots, n$ denote the sample of data points and $\|\cdot\|$ the Euclidean norm. Define the **breakdown point** of the estimator at sample **Z** as

$$\varepsilon_n^*(\widehat{\boldsymbol{\beta}}, \mathbf{Z}) = \max\left\{\frac{m}{n}; \sup_{\mathbf{Z}'} \|\widehat{\boldsymbol{\beta}}(\mathbf{Z}') - \widehat{\boldsymbol{\beta}}(\mathbf{Z})\| < \infty\right\},$$

where the supremum is taken over all samples $\mathbf{Z}'$ that can result from replacing m observations in **Z** by arbitrary values. See also Definition 1.6.1.

We now state conditions under which the HBR-estimate remains bounded.

Lemma 5.8.4 *Suppose there exist finite constants $M_1 > 0$ and $M_2 > 0$ such that the following conditions hold:*

(B1) $\inf_{\|\beta\|=1} \sup_{ij} \{b_{ij}(\mathbf{x}_j - \mathbf{x}_i)'\beta\} = M_1$;

(B2) $\sup_{ij}\{b_{ij}|y_j - y_i|\} = M_2$.

Then
$$\|\widehat{\beta}_{HBR}\| < \frac{1}{M_1}\left[1 + 2\binom{n}{2}\right]M_2.$$

Proof. Note that

$$D_{HBR}(\beta) \geq \sup_{ij}\{b_{ij}|y_j - y_i - (\mathbf{x}_j - \mathbf{x}_i)'\beta|\} \geq \|\beta\|M_1 - M_2$$

$$\geq 2\binom{n}{2}M_2$$

whenever $\|\beta\| \geq 1/M_1[1 + 2\binom{n}{2}]M_2$. Since $D_{HBR}(0) = \sum\sum_{i<j} b_{ij}|y_j - y_i| \leq \binom{n}{2}M_2$ and D_{HBR} is a convex function of β, it follows that $\widehat{\beta}_{HBR} = \operatorname{Argmin} D_{HBR}(\beta)$ satisfies

$$\|\widehat{\beta}_{HBR}\| < \frac{1}{M_1}\left[1 + 2\binom{n}{2}\right]M_2.$$

The lemma follows.

For our result, we need further to assume that the data points $\mathbf{Z}$ are in **general position**; that is, any subset of $p + 1$ of these points determines a unique solution β. In particular, this implies that neither all of the $\mathbf{x}_i$s are the same nor all of the y_is are the same; hence, provided the weights have not broken down, this implies that both constants M_1 and M_2 of Lemma 5.8.4 are positive.

Theorem 5.8.5 *Assume that the data points $\mathbf{Z}$ are in general position. Let $\mathbf{v}$, $\mathbf{V}$, and $\widehat{\beta}^{(0)}$ denote the initial estimates of location, scatter, and β. Let $\varepsilon_n^*(\mathbf{v}, \mathbf{Z})$, $\varepsilon_n^*(\mathbf{V}, \mathbf{Z})$, and $\varepsilon_n^*(\widehat{\beta}^{(0)}, \mathbf{Z})$ denote their corresponding breakdown points. Then breakdown point of the HBR estimator is*

$$\varepsilon_n^*(\widehat{\beta}_{HBR}, \mathbf{Z}) = \min\{\varepsilon_n^*(\mathbf{v}, \mathbf{Z}), \varepsilon_n^*(\mathbf{V}, \mathbf{Z}), \varepsilon_n^*(\widehat{\beta}^{(0)}, \mathbf{Z}), 1/2\}. \quad (5.8.22)$$

Proof. Corrupt m points in the data set $\mathbf{Z}$ and let $\mathbf{Z}'$ be the sample consisting of these corrupt points and the remaining $n - m$ points. Assume that $\mathbf{Z}'$ is in general position. Assume that $\mathbf{v}(\mathbf{Z}')$, $\mathbf{V}(\mathbf{Z}')$, and $\widehat{\beta}^{(0)}(\mathbf{Z}')$ have not broken down. Then the constants M_1 and M_2 of Lemma 5.8.4 are positive and finite. Hence, by Lemma 5.8.4, $\|\widehat{\beta}_{HBR}(\mathbf{Z}')\| < \infty$ and the theorem follows.

Based on this last result, the HBR-estimate has 50% breakdown provided the initial estimates $\mathbf{v}$, $\mathbf{V}$, and $\widehat{\beta}^{(0)}$ all have 50% breakdown. Assuming that the data points are in general position, the MVE estimates of location and scatter defined in Section 5.3 have 50% breakdown; see Rousseeuw and Leroy (1987). For initial estimates of the regression coefficients, again assuming that the data points are in general position, the LTS estimates, (5.8.17), and the LMS

High-Breakdown (HBR) Estimates   319

estimates, (5.9.7), have 50% breakdown; see, also, Hössjer (1994). The HBR estimates used in the examples of Section 5.9 employ the MVE estimates of location and scatter and the LMS estimate of the regression coefficients.

Influence Function of the HBR-Estimate

In order to derive the influence function, we start with the gradient equation $S(\beta) \doteq 0$, written as

$$0 \doteq \frac{1}{n^2} \sum_{i=1}^{n} \sum_{j=1}^{n} b_{ij} \mathrm{sgn}(z_j - z_i)(\mathbf{x}_j - \mathbf{x}_i).$$

Note, by Lemma A.6.4, that $b_{ij} = g_{ij}(0) + O_p(1/\sqrt{n})$ so that the defining equation may be written as

$$0 \doteq \frac{1}{n^2} \sum_{i=1}^{n} \sum_{j=1}^{n} g_{ij}(0) \mathrm{sgn}(z_j - z_i)(\mathbf{x}_j - \mathbf{x}_i), \tag{5.8.23}$$

ignoring a remainder term of magnitude $O_p(1/\sqrt{n})$.

Influence functions are derived for the model where both $\mathbf{x}$ and y are stochastic; hence, consider the correlation model of Section 5.6,

$$y = \mathbf{x}'\boldsymbol{\beta} + e, \tag{5.8.24}$$

where e has density f, $\mathbf{x}$ is a $p \times 1$ random vector with density function m, and e and $\mathbf{x}$ are independent. Let F and M denote the corresponding distribution functions of e and $\mathbf{x}$. Let H and h denote the joint distribution function and density of y and $\mathbf{x}$. It then follows that

$$h(\mathbf{x}, y) = f(y - \mathbf{x}'\boldsymbol{\beta}) m(\mathbf{x}). \tag{5.8.25}$$

If we rewrite (5.8.23) using the Stieltjes integral notation of the empirical distribution of $(\mathbf{x}_i, y_i)$, for $i = 1, \ldots, n$, we see that the functional $\boldsymbol{\beta}(H)$ solves the equation

$$0 = \iint b(\mathbf{x}_1, \mathbf{x}_2, y_1, y_2) \mathrm{sgn}\{y_2 - y_1 - (\mathbf{x}_2 - \mathbf{x}_1)'\boldsymbol{\beta}(H)\}(\mathbf{x}_2 - \mathbf{x}_1) dH(\mathbf{x}_1, y_1) dH(\mathbf{x}_2, y_2).$$

Let $I(a < b) = 1$ or 0, depending on whether $a < b$ or $a > b$. Then using the fact that the sign function is odd and the symmetry of the weight function in its $\mathbf{x}$ and y arguments we can write the defining equation of the functional $\boldsymbol{\beta}(H)$ as

$$0 = \iint \mathbf{x}_1 b(\mathbf{x}_1, \mathbf{x}_2, y_1, y_2) [I(y_2 - y_1 < (\mathbf{x}_2 - \mathbf{x}_1)'\boldsymbol{\beta}(H)) - \tfrac{1}{2}] dH(\mathbf{x}_1, y_1) dH(\mathbf{x}_2, y_2).$$

Define the matrix $\mathbf{C}_H$ by

$$\mathbf{C}_H = \left\{ \tfrac{1}{2} \iint \int (\mathbf{x}_2 - \mathbf{x}_1) b(\mathbf{x}_1, \mathbf{x}_2, y_1, y_1)(\mathbf{x}_2 - \mathbf{x}_1)' f^2(y_1) dy_1 dM(\mathbf{x}_1) dM(\mathbf{x}_2) \right\}.$$
$$\tag{5.8.26}$$

Note that under the correlation model $\mathbf{C}_H$ is the assumed limiting matrix of assumption (H.1), (5.8.13); see Lemma 5.8.3.

The next theorem gives the result for the influence function of $\widehat{\boldsymbol{\beta}}_{HBR}$. Its proof is given in Theorem A.5.2.

Theorem 5.8.6 *The influence function for the estimate* $\widehat{\boldsymbol{\beta}}_{HBR}$ *is given by*

$$\Omega(\mathbf{x}_0, y_0, \widehat{\boldsymbol{\beta}}_{HBR}) = \mathbf{C}_H^{-1} \frac{1}{2} \int\int (\mathbf{x}_0 - \mathbf{x}_1) b(\mathbf{x}_1, \mathbf{x}_0, y_1, y_0) \operatorname{sgn}\{y_0 - y_1\} \, dF(y_1) dM(\mathbf{x}_1),$$

(5.8.27)

where $\mathbf{C}_H$ is given by expression (5.8.26).

In order to show that the influence function correctly identifies the asymptotic distribution of the estimator, define W_i as

$$W_i = \int\int (\mathbf{x}_i - \mathbf{x}_1) b(\mathbf{x}_1, \mathbf{x}_i, y_1, y_i) \operatorname{sgn}(y_i - y_1) \, dF(y_1) dM(\mathbf{x}_1). \quad (5.8.28)$$

Next write W_i in terms of a Stieltjes integral over the empirical distribution of $(\mathbf{x}_j, y_j)$ as

$$W_i^* = \frac{1}{n} \sum_{j=1}^{n} (\mathbf{x}_i - \mathbf{x}_j) b(\mathbf{x}_j, \mathbf{x}_i, y_j, y_i) \operatorname{sgn}(y_i - y_j). \quad (5.8.29)$$

If we can show that $(1/\sqrt{n}) \sum_{j=1}^{n} W_i^* \xrightarrow{D} N(\mathbf{0}, \Sigma_H)$, then we are done. From the proof of Theorem A.6.9, it will suffice to show that

$$\frac{1}{\sqrt{n}} \sum_{i=1}^{n} (U_i - W_i^*) \xrightarrow{P} 0, \quad (5.8.30)$$

where $U_i = (1/n) \sum_{j=1}^{n} (\mathbf{x}_i - \mathbf{x}_j) E[b_{ij} \operatorname{sgn}(y_i - y_j) | y_i]$. Writing the left-hand side of (5.8.30) as

$$(1/n^{3/2}) \sum_{i=1}^{n} \sum_{j=1}^{n} (\mathbf{x}_i - \mathbf{x}_j) \{ E[b_{ij} \operatorname{sgn}(y_i - y_j) | y_i] - g_{ij}(0) \operatorname{sgn}(y_i - y_j) \},$$

where $g_{ij}(0) \equiv b(\mathbf{x}_j, \mathbf{x}_i, y_j, y_i)$, the proof is analogous to the proof of Theorem A.6.9.

5.8.4 Discussion

The influence function, $\Omega(\mathbf{x}_0, y_0, \widehat{\boldsymbol{\beta}}_{HBR})$, for the HBR-estimate is a continuous function of $\mathbf{x}_0$ and y_0. With a proper choice of a weight function it is bounded in both the x- and Y-spaces. This is true for the weights given by (5.8.1); furthermore, for these weights $\Omega(\mathbf{x}_0, y_0, \widehat{\boldsymbol{\beta}}_{HBR})$ goes to zero as $\mathbf{x}_0$ and y_0 get large in any direction.

The influence function $\Omega(\mathbf{x}_0, y_0, \widehat{\boldsymbol{\beta}})$ is a generalization of the influence functions for the Wilcoxon and GR-estimates; see Exercise 5.10.22. Figure 5.8.1(a) shows the influence function of the HBR-estimate for the special case where $(\mathbf{x}, Y)$ has a bivariate normal distribution with mean **0** and the identity matrix as the variance-covariance matrix. For this plot we used the weights given by (5.8.1) where $m_i = \psi(b/x_i^2)$ with the constants $b = c = 4$.

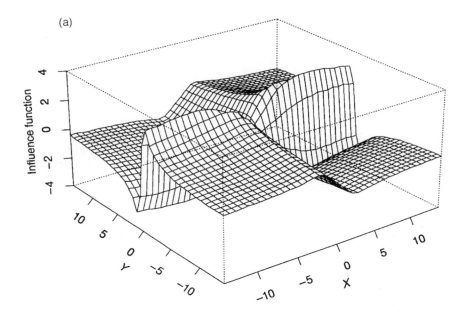

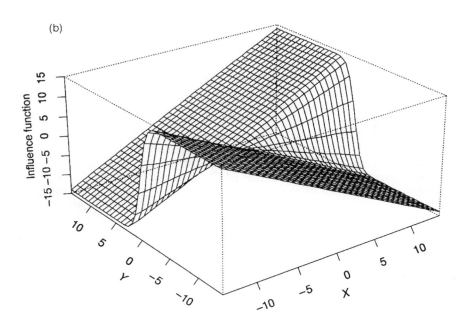

Figure 5.8.1: Influence function for: (a) the HBR-estimate; (b) the Wilcoxon estimate; (c) the GR-estimate (see over).

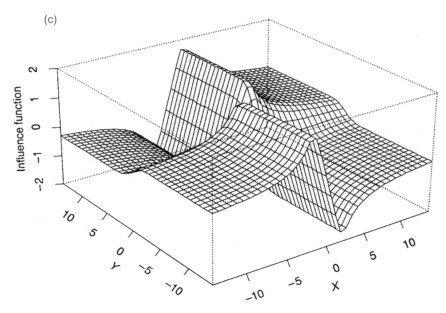

Figure 5.8.1: (c) the GR-estimate.

For comparison purposes, the influence functions of the Wilcoxon and GR-estimates are also shown in Figure 5.8.1(b,c). The Wilcoxon influence function is bounded in the Y-space but is unbounded in the x-space while the GR-estimate is bounded in both spaces. Note that because the weights of the GR-estimate do not depend on Y, it does not taper to 0 as $y_0 \to \infty$, as the influence function of the HBR-estimate does. For all three plots, we used the method of Monte Carlo, (10 000 simulations for each of 1600 grid points), to perform the numerical integration. The plot of the Wilcoxon influence function is an easily verifiable check on the Monte Carlo because of its closed form, (3.5.17).

High-breakdown estimates can have unbounded influence functions. Such estimates can have instability problems, as discussed in Sheather, McKean, and Hettmansperger (1997) for the LMS estimate which has unbounded influence in the x-space at the quartiles of Y. The generalized S estimators discussed in Croux et al. (1994) also have unbounded influence functions in the x-space for particular values of Y. In contrast, the influence function of the HBR-estimate is bounded everywhere. This helps to explain its more stable behavior than the LMS in the stability study discussed in Chang et al. (1997).

5.9 Implementation and Examples of High-Breakdown Fits

In this section, we discuss how to estimate the standard errors of the HBR estimates and how properly to standardize the residuals. We then consider two examples.

5.9.1 Standard Errors of $\widehat{\boldsymbol{\beta}}_{HBR}$ and Internal Studentized Residuals

Using the asymptotic distribution of the HBR-estimate as a guideline and upon substituting the estimated weights for the true weights, we can estimate the asymptotic standard errors for these estimates. The asymptotic variance-covariance matrix of $\widehat{\boldsymbol{\beta}}_{HBR}$ is a function of the two matrices Σ_H and $\mathbf{C}_H$, given in (5.8.14) and (5.8.13), respectively. The matrix Σ_H is the variance-covariance matrix of the random vector $\mathbf{U}_i$, (5.8.12). We can approximate $\mathbf{U}_i$ by the expression

$$\widehat{\mathbf{U}}_i = \frac{1}{n}\sum_{j=1}^{n}(\mathbf{x}_j - \mathbf{x}_i)\widehat{b}_{ij}(1 - 2F_n(\widehat{e}_i)), \tag{5.9.1}$$

where $\widehat{b}_{ij}$ are the estimated weights, $\widehat{e}_i$ are the HBR residuals and F_n is the empirical distribution function of the residuals. Our estimate of Σ_H is then the sample variance-covariance matrix of $\widehat{\mathbf{U}}_1, \ldots, \widehat{\mathbf{U}}_n$, i.e.

$$\widehat{\Sigma}_H = \frac{1}{n-1}\sum_{i=1}^{n}\left(\widehat{\mathbf{U}}_i - \overline{\widehat{\mathbf{U}}}\right)\left(\widehat{\mathbf{U}}_i - \overline{\widehat{\mathbf{U}}}\right)'. \tag{5.9.2}$$

For the matrix $\mathbf{C}_H$, consider the results in Lemma 5.8.3. Upon substituting the estimated weights for the weights, expression (5.8.21) simplifies to

$$B'_{ij}(0) \doteq \widehat{b}_{ij}\int f^2(t)\,dt = \widehat{b}_{ij}\frac{1}{\sqrt{12\tau}}, \tag{5.9.3}$$

where τ is the scale parameter when the Wilcoxon score function $\varphi(u) = \sqrt{12}(u - \frac{1}{2})$ is used. To estimate τ, we will use the estimator $\widehat{\tau}$ given in expression (3.7.8). Now approximating b_{ij} in $\mathbf{C}_n$ using (5.8.3) leads to the estimate

$$n^{-2}\widehat{\mathbf{C}}_n = \frac{1}{4\sqrt{3}}(\widehat{\tau}n)^{-2}\sum_{i=1}^{n}\sum_{j=1}^{n}\widehat{b}_{ij}(\mathbf{x}_j - \mathbf{x}_i)(\mathbf{x}_j - \mathbf{x}_i)'. \tag{5.9.4}$$

Similar to the R- and GR-estimates, we estimate the intercept by

$$\widehat{\alpha}_{HBR} = \text{med}_{1\leq i\leq n}\{y_i - \mathbf{x}'_i\widehat{\boldsymbol{\beta}}_{HBR}\}. \tag{5.9.5}$$

Similar to the GR-estimates, a version of Theorem 3.5.11 holds for the joint asymptotic distribution of $\widehat{\alpha}_{HBR}$ and $\widehat{\boldsymbol{\beta}}_{HBR}$. Using (5.9.4) and (5.9.2) as estimates of $\mathbf{C}_H$ and Σ_H, we can obtain an estimate of the asymptotic covariance matrix of $(\widehat{\alpha}_{HBR}, \widehat{\boldsymbol{\beta}}'_{HBR})'$. Furthermore, using Theorems 5.8.2 and A.6.7, a first-order approximation of the standard errors of the residuals, $\widehat{e}_i$, is obtained following the development of Sections 3.9.2 and 5.4.3. Let $\widehat{\sigma}_{\widehat{e}_i}$ denote this standard error of $\widehat{e}_i$. Define the **internal HBR studentized residuals** by

$$\widehat{e}_i^* = \frac{\widehat{e}_i}{\widehat{\sigma}_{\widehat{e}_i}}. \tag{5.9.6}$$

For flagging outliers, appropriate benchmarks for these residuals are ± 2; see Section 3.9.2 for discussion.

5.9.2 Examples

Similar to the discussion in Section 5.4 on the difference between GR-estimates and R-estimates, high-breakdown estimates and highly efficient estimates often give conflicting results. High-breakdown estimates are less sensitive to outliers and clusters of outliers in the x-space; hence, for data sets where this is a problem high-breakdown estimates often give better fits than highly efficient fits. On the other hand, similar to the GR-estimates, the HBR estimates are hampered in fitting and detecting curvature, while this is not true of the highly efficient estimates. We choose two examples which illustrate these disagreements between the high-breakdown fit and the highly efficient Wilcoxon fit.

To obtain the HBR-estimates we used the weights given by (5.8.1). As initial estimates of location and scatter we chose the MVE estimates discussed in Section 5.3. They were computed by the algorithm proposed by Hadi and Simonoff (1993). For initial estimates of the regression coefficients we chose the LMS estimates given by

$$\begin{pmatrix} \widehat{\alpha}^{(0)} \\ \widehat{\boldsymbol{\beta}}^{(0)} \end{pmatrix} = \text{Argmin med}_{1 \leqslant i \leqslant n} \{y_i - \alpha - \mathbf{x}'_i \boldsymbol{\beta}\}^2, \quad (5.9.7)$$

see Rousseeuw and Leroy (1987). These estimates have 50% breakdown. We computed them with the algorithm written by Stromberg (1993). The Mallows tuning constant b was set at the 95th percentile of $\chi^2(p)$. The tuning constant c was set at $[\text{med}\{a_i\} + 3 \text{ MAD}\{a_i\}]^2$ where $a_i = e_i(\widehat{\boldsymbol{\beta}}_0)/(\widehat{\sigma} \ m_i)$ and $\widehat{\sigma} = \text{MAD}$, (5.8.2). Once the weights are computed, a Gauss-Newton type algorithm similar to that used by rglm (see Kapenga et al., 1988) is used to obtain the HBR estimates.

Example 5.9.1 *Stars Data, Example 5.3.1 continued*

In Example 5.3.1 we compared the GR and Wilcoxon fits for a data set consisting of measurements of stars. Recall that there is a cluster of outliers in the x-space which greatly influences the Wilcoxon fit but has little effect on the GR fit. The HBR-estimates of the intercept and slope are -6.91 and 2.71, respectively, which are quite close to the GR-estimates as displayed in Table 5.3.2. Hence similar to the GR-estimates, the outlying cluster of giant stars has little effect on the HBR fit.

Example 5.9.2 *Quadratic Data*

In order to demonstrate the problems that the high-breakdown estimates have in fitting curvature, we simulated data from the following quadratic model:

$$Y_i = 5.5|x_i| - 0.6x_i^2 + e_i, \quad (5.9.8)$$

where the e_is were simulated iid $N(0, 1)$ variates and the x_is were simulated contaminated normal variates with the contamination proportion set at 0.25 and the ratio of the variance of the contaminated part to the noncontaminated part set at 16. Figure 5.9.1(a) displays a scatterplot of the data overlaid with

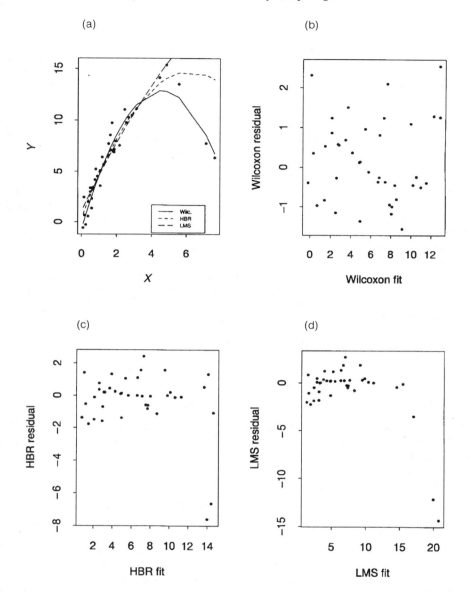

Figure 5.9.1: (a) For the quadratic data of Example 5.9.2, scatterplot of data overlaid by Wilcoxon, HBR and LMS fits; (b) studentized residual plot based on the Wilcoxon fit; (c) studentized residual plot based on the HBR fit; (d) residual plot based on the LMS fit

the Wilcoxon, HBR, and LMS fits. The estimated coefficients for these fits are in Table 5.9.1. As shown, the Wilcoxon fit is quite good in fitting the curvature of the data. Its estimates are close to the true values. On the other hand, the high-breakdown fits are quite poor. The LMS fit missed the

Table 5.9.1: Estimates of coefficients for the quadratic data

Fit	Intercept	Quadratic data Linear	Quadratic
Wilcoxon	−0.665	5.95	−0.652
HBR	0.422	4.64	−0.375
LMS	1.12	3.65	−0.141

curvature in the data. This is true too for the HBR fit, although the fit did correct itself somewhat from the poor LMS starting values. Figure 5.9.1(b,c) contain the internal studentized residual plots based on the Wilcoxon and the HBR fits, respectively. Based on the Wilcoxon residual plot, no further models would be considered. The HBR residual plot shows as outliers the two points which were fitted poorly. It also has a mild linear trend in it, which is not helpful since a linear term was fit. This trend is true for the LMS residual plot, Figure 5.9.1(d), although it gives an overall impression of the lack of a quadratic term in the model. In such cases in practice, a higher-degree polynomial may be fitted, which in this case would be incorrect. Difficulties in reading residual plots from high-breakdown fits, as encountered here, were discussed in Section 5.4; see also McKean *et al.* (1993).

5.10 Exercises

5.10.1 Show that the function (5.1.1) is a pseudo-norm.

5.10.2 Consider the hypotheses $H_0 : \boldsymbol{\beta} = \mathbf{0}$ versus $H_A : \boldsymbol{\beta} \neq \mathbf{0}$.
(a) Using Theorem 5.2.18, derive the Wald-type test based on the GR-estimate of $\boldsymbol{\beta}$.
(b) Using Theorem 5.2.4, derive the gradient test.

5.10.3 In the derivation of Lemma 5.2.1 show the results for cases 2–6.

5.10.4 Show that the second standardized term of the variance of $n^{-3/2}S(0)$ in expression (5.2.7) goes to 0 as $n \to \infty$.

5.10.5 Consider Theorem 5.2.12.
(a) If the weights are identically equal to 1, show that the random variable $(\sqrt{12}/n\tau)RD_{GR}$ reduces to the random variable $RD_\varphi/(\tau_\varphi/2)$ of Theorem 3.6.4 provided $\varphi(u) = \sqrt{12}(u - \frac{1}{2})$, (i.e. Wilcoxon scores).
(b) Complete the proof of Theorem 5.2.12.

5.10.6 Keeping in mind that the weights for the GR-estimates depend only on the xs, discuss how the bootstrap described in Section 3.8 for the L_1 estimate can be used to bootstrap the p-value of the test based on the statistic F_{GR}.

5.10.7
(a) Write an algorithm which obtains the simulated distribution of $\sum_{i=1}^{q} \lambda_i \chi_i^2(1)$, where $\lambda_1, \ldots, \lambda_q$ are specified and $\chi_i^2(1)$ are iid χ^2 random variables with one degree of freedom.

(b) Write a second algorithm which uses the algorithm in (a) to obtain the p-value of the test statistic qF_{GR}.

5.10.8 For general linear hypotheses, $H_0: \mathbf{M}\boldsymbol{\beta} = \mathbf{0}$ versus $H_A: \mathbf{M}\boldsymbol{\beta} \neq \mathbf{0}$, discuss the Wald-type test based on the GR-estimates.

5.10.9 Show that the second term in expression (5.4.8) is less than or equal to zero.

5.10.10 For the data in Example 5.4.1 consider polynomial models of degree p,

$$y_i = \alpha + \sum_{j=1}^{p} \beta_j (x_i - \bar{x})^j.$$

Once a good fit has been established, a hypothesis of interest is $H_0: \beta_p = 0$. Suppose for the GR-estimates we use the weights given by expression (5.3.2) parameterized by the exponent α_1. For $\alpha_1 = 0.5, 1,$ and 2, obtain the asymptotic relative efficiency between the GR-estimate and the Wilcoxon estimate of β_p for polynomial models of degree $p = 1, \ldots, 4$. If software is available, obtain these efficiencies for the weights given by expression (5.3.3) using the tuning constants recommended.

5.10.11 Obtain the details of the derivation of the $\text{Var}(\widehat{\mathbf{e}}_{GR})$ given in expression (5.4.11).

5.10.12 Obtain the projection that is used in the proof of Theorem 5.5.1.

5.10.13 In the proof of Theorem 5.5.1, show that $\sqrt{n}\left(\widehat{\boldsymbol{\beta}}_R - \widehat{\boldsymbol{\beta}}_{GR}\right)$ is asymptotically normal.

5.10.14 In the proof of Theorem 5.5.1, show that $E[\mathbf{F}_c(\mathbf{e})\mathbf{S}'(\boldsymbol{\beta}_0)] = (n/6)\mathbf{WX}$.

5.10.15 By completing the following steps, obtain the asymptotic variance-covariance matrix of $\widehat{\mathbf{b}}_R^* - \widehat{\mathbf{b}}_{GR}^*$, where $\widehat{\mathbf{b}}_R^* = (\widehat{\alpha}_{S,R}^*, \widehat{\boldsymbol{\beta}}_R')'$ and $\widehat{\mathbf{b}}_{GR}^* = (\widehat{\alpha}_{S,GR}^*, \widehat{\boldsymbol{\beta}}_{GR}')'$.

(a) Show that $\widehat{\mathbf{b}}_R^* = \mathbf{T}\widehat{\mathbf{b}}_R$, where

$$\mathbf{T} = \begin{bmatrix} 1 & -\bar{\mathbf{x}}' \\ \mathbf{0} & \mathbf{I} \end{bmatrix}.$$

(b) Show that

$$\text{AV}\left[\begin{pmatrix} \widehat{\alpha}_{S,R}^* \\ \widehat{\boldsymbol{\beta}}_R \end{pmatrix} - \begin{pmatrix} \widehat{\alpha}_{S,GR}^* \\ \widehat{\boldsymbol{\beta}}_{GR} \end{pmatrix}\right] = \mathbf{T}\text{AV}\left[\begin{pmatrix} \widehat{\alpha}_{S,R} - \widehat{\alpha}_{S,GR} \\ 0 \end{pmatrix} + \begin{pmatrix} 0 \\ \widehat{\boldsymbol{\beta}}_{GR} - \widehat{\boldsymbol{\beta}}_{GR} \end{pmatrix}\right]\mathbf{T}',$$

where AV represents asymptotic variance.

(c) Since the asymptotic representation (3.5.23) holds for both $\widehat{\alpha}_{S,R}^*$ and $\widehat{\alpha}_{S,GR}^*$ and since the centered intercept is asymptotically independent of the regression estimates, use (5.5.1) to conclude that

$$\text{AV}\left[\begin{pmatrix} \widehat{\alpha}_{S,R}^* \\ \widehat{\boldsymbol{\beta}}_R \end{pmatrix} - \begin{pmatrix} \widehat{\alpha}_{S,GR}^* \\ \widehat{\boldsymbol{\beta}}_{GR} \end{pmatrix}\right] = \tau^2 \begin{bmatrix} \bar{\mathbf{x}}'\mathbf{D}\bar{\mathbf{x}} & -\bar{\mathbf{x}}'\mathbf{D} \\ -\mathbf{D}\bar{\mathbf{x}} & \mathbf{D} \end{bmatrix},$$

where $\mathbf{D} = (\mathbf{X}'\mathbf{WX})^{-1}\mathbf{X}'\mathbf{W}^2\mathbf{X}(\mathbf{X}'\mathbf{WX})^{-1} - (\mathbf{X}'\mathbf{X})^{-1}$.

5.10.16 Show that the asymptotic variance-covariance matrix derived in Exercise 5.10.15 is singular by showing that the vector $(1, \bar{\mathbf{x}})'$ is in its kernel.

5.10.17 Show that $\overline{RD}_{GR}(\beta) = \overline{D}_{GRy} - \overline{D}_{GRe}$, (5.6.3) and (5.6.4), is a strictly convex function of β with a minimum value of 0 at $\beta = 0$.

5.10.18 Using the influence function, (5.7.4), of the GR-estimate, obtain the asymptotic distribution of $\widehat{\beta}_{GR}$.

5.10.19 The influence function of the test statistic T_{GR} is given in Theorem 5.7.2. Use it to obtain the asymptotic distribution of T_{GR}.

5.10.20 Obtain the approximate distribution of $\widehat{\beta}_{HBR} - \widehat{\beta}_R$, where $\widehat{\beta}_R$ is the Wilcoxon estimate. Use it to obtain HBR analogs of the diagnostics statistics $TDBETAS_R$, (5.5.3), and $CFITS_{R,i}$, (5.5.5).

5.10.21 Show that the $p \times p$ matrix $\mathbf{C}_n$ defined in expression (5.8.10) can be written alternately as $\mathbf{C}_n = \sum_{i<j} \gamma_{ij} b_{ij} (\mathbf{x}_j - \mathbf{x}_i)(\mathbf{x}_j - \mathbf{x}_i)'$.

5.10.22 Consider the influence function of the HBR-estimate given in expression (A.5.24).

(a) If the weights for residuals are set at 1, show that the influence function of the HBR estimate simplifies to the influence function of the GR-estimate given in (5.7.4).

(b) If the weights for residuals and the xs are both set at 1, show that the influence function of the HBR-estimate simplifies to the influence function of the Wilcoxon estimate given in (3.5.17).

6
Multivariate Models

6.1 Bivariate Location Model

We now consider a statistical model in which we observe vectors of observations. For example, we may record both the standard aptitude test (SAT) verbal and math scores on students. We then wish to investigate the bivariate distribution of scores. We may wish to test the hypothesis that the vector of population locations has changed over time or to estimate the vector of locations. The framework in which we carry out the statistical inference is the bivariate location model which is similar to the location model of Chapter 1.

Suppose that $\mathbf{X}_1, \ldots, \mathbf{X}_n$ are iid random vectors with $\mathbf{X}_i^T = (X_{i1}, X_{i2})$. In this chapter, T denotes transpose and we reserve prime for differentiation. We assume that $\mathbf{X}$ has an absolutely continuous distribution with cdf $F(s - \theta_1, t - \theta_2)$ and pdf $f(s - \theta_1, t - \theta_2)$. We also assume that the marginal distributions are absolutely continuous. The vector $\boldsymbol{\theta} = (\theta_1, \theta_2)^T$ is the location vector. Throughout this chapter we will emphasize the bivariate model. We will state results for k-dimensional multivariate models and provide references that cover the developments.

Definition 6.1.1 *Distribution models for bivariate data. Let $F(s, t)$ be a prototype cdf, then the underlying model will be a shifted version:* $H(s, t) = F(s - \theta_1, t - \theta_2)$.

The following models will be used throughout this chapter.

1. We say the distribution is **symmetric** when $\mathbf{X}$ and $-\mathbf{X}$ have the same distribution or $f(s, t) = f(-s, -t)$. This is sometimes called **diagonal symmetry**. The vector $(0, 0)^T$ is the center of symmetry of F and the location functionals all equal the center of symmetry. Unless stated otherwise, we will assume symmetry throughout this chapter.
2. The distribution has **spherical symmetry** when $\Gamma\mathbf{X}$ and $\mathbf{X}$ have the same distribution where Γ is an orthogonal matrix. The pdf has the form $g(\|\mathbf{x}\|)$, where $\|\mathbf{x}\| = (\mathbf{x}^T\mathbf{x})^{1/2}$ is the Euclidean norm of $\mathbf{x}$. The contours of the density are circular.
3. In an **elliptical** model the pdf has the form $|\det \Sigma|^{-1/2} g(\mathbf{x}^T \Sigma^{-1} \mathbf{x})$, where det denotes determinant and Σ is a symmetric, positive definite matrix. The contours of the density are ellipses.

330   *Multivariate Models*

In an elliptical model, the contours of the density are elliptical and if Σ is the identity matrix then we have a spherically symmetric distribution. An elliptical distribution can be transformed into a spherical one by a transformation of the form $\mathbf{Y} = \mathbf{DX}$, where $\mathbf{D}$ is a nonsingular matrix. Along with various models, we will encounter various transformations in this chapter. The following definition summarizes the transformations.

Definition 6.1.2 *Data transformations.*

(a) $\mathbf{Y} = \mathbf{\Gamma X}$ *is an* **orthogonal** *transformation when the matrix* $\mathbf{\Gamma}$ *is orthogonal. These transformations include rotations and reflections of the data.*

(b) $\mathbf{Y} = \mathbf{AX} + \mathbf{b}$ *is called an* **affine** *transformation when* $\mathbf{A}$ *is a nonsingular matrix and* $\mathbf{b}$ *is any vector of real numbers.*

(c) When the matrix $\mathbf{A}$ *in (b) is diagonal, we have a special affine transformation called a* **scale and location** *transformation.*

(d) Suppose $\mathbf{t}(\mathbf{X})$ *represents one of the above transformations of the data. Let* $\hat{\boldsymbol{\theta}}(\mathbf{t}(\mathbf{X}))$ *denote the estimator computed from the transformed data. Then we say the estimator is* **equivariant** *if* $\hat{\boldsymbol{\theta}}(\mathbf{t}(\mathbf{X})) = \mathbf{t}(\hat{\boldsymbol{\theta}}(\mathbf{X}))$. *Let* $V(\mathbf{t}(\mathbf{X}))$ *denote a test statistic computed from the transformed data. We say the test statistic is* **invariant** *when* $V(\mathbf{t}(\mathbf{X})) = V(\mathbf{X})$.

In Exercise 6.8.1, the reader is asked to show that the vector of sample means is affine equivariant and **Hotelling's T^2 test** statistic is affine invariant. Recall that

$$T^2 = n(\overline{\mathbf{X}} - \boldsymbol{\mu})^T \mathbf{S}^{-1} (\overline{\mathbf{X}} - \boldsymbol{\mu})$$

is Hotelling's statistic, where $\mathbf{S}$ is the sample covariance matrix. Table 6.1.1 provides a map of some of the statistical methods, classified by invariance and type of statistic. The examples refer to one-sample, two-sample, one-way layout, and linear models.

Table 6.1.1: Some of the statistical methods discussed in the chapter classified by invariance and type of statistic

Invariance	Statistic	Formula number	Example number
Location and Scale	Component sign	(6.1.8)	6.2.1
	Component rank	(6.2.13)–(6.2.15)	6.2.1, 6.6.1, 6.6.3, 6.6.4
Rotation	Spatial sign	(6.1.5), (6.3.2)	6.3.1
	Spatial rank		
	Ranking distances	(6.3.11)	
	Rank vector	(6.3.16), (6.3.20)	6.3.2
Affine	Sign		
	Directions	(6.4.2), (6.4.3)	
	Interdirections	(6.4.4)	
	Oja	(6.4.5), (6.4.7)	
	Rank		
	Ranking distances	(6.4.8)	
	Oja	(6.4.10), (6.4.11)	6.4.1, 6.6.2

As in the earlier chapters, we begin with a criterion function or with a set of estimating equations. To fix the ideas, suppose that we wish to estimate $\boldsymbol{\theta}$ or test the hypothesis $H_0 : \boldsymbol{\theta} = \mathbf{0}$ and we are given a pair of estimating equations:

$$\mathbf{S}(\boldsymbol{\theta}) = \begin{pmatrix} S_1(\boldsymbol{\theta}) \\ S_2(\boldsymbol{\theta}) \end{pmatrix} = \mathbf{0}, \tag{6.1.1}$$

see Example 6.1.1 for three criterion functions. We now list the usual set of assumptions that we have been using throughout the book. These assumptions guarantee that the estimating equations are Pitman regular in the sense of Definition 1.5.3 so that we can define the estimate and test and develop the necessary asymptotic distribution theory. It will often be convenient to suppose the the true value of $\boldsymbol{\theta}$ is $\mathbf{0}$, which we can do without loss of generality.

Definition 6.1.3 *Pitman regularity conditions.*

(a) The components of $\mathbf{S}(\boldsymbol{\theta})$ should be nonincreasing functions of θ_1 and θ_2.

(b) $E_0(\mathbf{S}(\mathbf{0})) = \mathbf{0}$.

(c) $\frac{1}{\sqrt{n}}\mathbf{S}(\mathbf{0}) \xrightarrow{D_0} \mathbf{Z} \sim N_2(\mathbf{0}, \mathbf{A})$.

(d) $\sup_{\|\mathbf{b}\| \leq B} \left| \frac{1}{\sqrt{n}} \mathbf{S}\left(\frac{1}{\sqrt{n}} \mathbf{b}\right) - \frac{1}{\sqrt{n}} \mathbf{S}(\mathbf{0}) + \mathbf{B}\mathbf{b} \right| \xrightarrow{P} \mathbf{0}$.

The matrix $\mathbf{A}$ in (c) is the asymptotic covariance matrix of $(1/\sqrt{n})\mathbf{S}(\mathbf{0})$ and the matrix $\mathbf{B}$ in (d) can be computed in various ways, depending on when differentiation and expectation can be interchanged. We list the various computations of $\mathbf{B}$ for completeness:

$$\begin{aligned}
\mathbf{B} &= -E_0 \nabla \frac{1}{n} \mathbf{S}(\boldsymbol{\theta}) \mid_{\boldsymbol{\theta}=\mathbf{0}} \\
&= \nabla E_{\boldsymbol{\theta}} \frac{1}{n} \mathbf{S}(\mathbf{0}) \mid_{\boldsymbol{\theta}=\mathbf{0}} \\
&= E_0[(-\nabla \log f(\mathbf{X}))\boldsymbol{\Psi}^T(\mathbf{X})],
\end{aligned} \tag{6.1.2}$$

where ∇ denotes differentiation with respect to the components of $\boldsymbol{\theta}$, $\nabla \log f(\mathbf{X})$ denotes the vector of partial derivatives of $\log f(\mathbf{X})$, and $\boldsymbol{\Psi}(\cdot)$ is such that

$$\frac{1}{\sqrt{n}} \mathbf{S}(\boldsymbol{\theta}) = \frac{1}{\sqrt{n}} \sum_{i=1}^n \boldsymbol{\Psi}(\mathbf{X}_i - \boldsymbol{\theta}) + \mathbf{o}_p(1).$$

Brown (1985) proved a multivariate counterpart to Theorem 1.5.6. We state it next and refer the reader to the paper for the proof.

Theorem 6.1.1 *Suppose conditions (a)–(c) of Definition 6.1.3 hold. Suppose further that $\mathbf{B}$ is given by the second expression in (6.1.2) and is positive* definite. If, for any **b**,

$$\text{tr}\left\{n\text{Cov}\left[\frac{1}{n}\mathbf{S}\left(\frac{1}{\sqrt{n}}\mathbf{b}\right) - \frac{1}{n}\mathbf{S}(0)\right]\right\} \to 0,$$

then (d) of Definition 6.1.3 also holds.

The estimate of $\boldsymbol{\theta}$ is, of course, the solution of the estimating equations, denoted $\widehat{\boldsymbol{\theta}}$. Conditions (a) and (b) make this reasonable. To test the hypothesis $H_0 : \boldsymbol{\theta} = \mathbf{0}$ versus $H_A : \boldsymbol{\theta} \neq \mathbf{0}$, we reject the null hypothesis when $(1/n)\mathbf{S}^T(\mathbf{0})\mathbf{A}^{-1}\mathbf{S}(\mathbf{0}) \geqslant \chi_\alpha^2(2)$, the upper α percentile of a chi-square distribution with two degrees of freedom. Condition (c) implies that this is an asymptotically size α test; however, we may have to estimate the matrix $\mathbf{A}$ in practice.

With condition (d) we can determine the asymptotic distribution of the estimate and the asymptotic local power of the test; hence, asymptotic efficiencies can be computed. We can determine the quantity that corresponds to the efficacy in the univariate case described in Section 1.5.2 of Chapter 1. We do this next before discussing specific estimating equations. The following proposition follows at once from the assumptions.

Theorem 6.1.2 *Suppose conditions (a)–(d) in Definition 6.1.3 are satisfied, $\boldsymbol{\theta} = \mathbf{0}$ is the true parameter value, and $\boldsymbol{\theta}_n = \boldsymbol{\gamma}/\sqrt{n}$ for some fixed vector $\boldsymbol{\gamma}$. Further, $\widehat{\boldsymbol{\theta}}$ is the solution of the estimating equation. Then*

1. $\sqrt{n}\widehat{\boldsymbol{\theta}} = \mathbf{B}^{-1}\frac{1}{\sqrt{n}}\mathbf{S}(\mathbf{0}) + o_p(1) \xrightarrow{D_0} \mathbf{Z} \sim \text{MVN}(\mathbf{0}, \mathbf{B}^{-1}\mathbf{A}\mathbf{B}^{-1})$,
2. $\frac{1}{n}\mathbf{S}^T(\mathbf{0})\mathbf{A}^{-1}\mathbf{S}(\mathbf{0}) \xrightarrow{D_{\theta_n}} \chi^2(2, \boldsymbol{\gamma}^T\mathbf{B}\mathbf{A}^{-1}\mathbf{B}\boldsymbol{\gamma})$,

where $\chi^2(a, b)$ is noncentral chi-square with a degrees of freedom and noncentrality parameter b.

Proof. Part 1 follows immediately from condition (d) and letting $\boldsymbol{\theta}_n = \widehat{\boldsymbol{\theta}} \to \mathbf{0}$ in probability; see Theorem 1.5.8. Part 2 follows by observing (see Theorem 1.5.9) that

$$P_{\boldsymbol{\theta}_n}\left(\frac{1}{n}\mathbf{S}^T(\mathbf{0})\mathbf{A}^{-1}\mathbf{S}(\mathbf{0}) \geqslant \chi_\alpha^2(2)\right) = P_0\left(\frac{1}{n}\mathbf{S}^T\left(-\frac{1}{\sqrt{n}}\boldsymbol{\gamma}\right)\mathbf{A}^{-1}\mathbf{S}\left(-\frac{1}{\sqrt{n}}\boldsymbol{\gamma}\right) \geqslant \chi_\alpha^2(2)\right)$$

and, from (d),

$$\frac{1}{\sqrt{n}}\mathbf{S}\left(-\frac{1}{\sqrt{n}}\boldsymbol{\gamma}\right) = \frac{1}{\sqrt{n}}\mathbf{S}(\mathbf{0}) + \mathbf{B}\boldsymbol{\gamma} + o_p(1) \xrightarrow{D_0} \mathbf{Z} \sim MVN(\mathbf{B}\boldsymbol{\gamma}, \mathbf{A}).$$

Hence, we have a noncentral chi-square limiting distribution for the quadratic form. Note that the influence function of $\widehat{\boldsymbol{\theta}}$ is $\Omega(\mathbf{x}) = \mathbf{B}^{-1}\Psi(\mathbf{x})$ and we say $\widehat{\boldsymbol{\theta}}$ has **bounded influence** provided $\|\Omega(\mathbf{x})\|$ is bounded.

Definition 6.1.4 *Estimation efficiency.* The efficiency of a bivariate estimator can be measured using the **Wilks generalized variance**, defined to be the determinant of the covariance matrix of the estimator: $\sigma_1^2\sigma_2^2(1 - \rho_{12}^2)$ where $((\rho_{ij}\sigma_i\sigma_j))$ is the covariance matrix of the bivariate vector of estimates. The **estimation**

efficiency of $\widehat{\theta}_1$ relative to $\widehat{\theta}_2$ is the square root of the reciprocal ratio of the generalized variances.

This means that the asymptotic covariance matrix given by $\mathbf{B}^{-1}\mathbf{A}\mathbf{B}^{-1}$ of the more efficient estimator will be 'small' in the sense of generalized variance. See Bickel (1964) for further discussion of efficiency in the multivariate case.

Definition 6.1.5 *Test efficiency. When comparing two tests based on* $\mathbf{S}_1$ *and* $\mathbf{S}_2$, *since the asymptotic local power is an increasing function of the noncentrality parameter, we define the* **test efficiency** *as the ratio of the respective noncentrality parameters.*

In the bivariate case, we have $\boldsymbol{\gamma}^T\mathbf{B}_1\mathbf{A}_1^{-1}\mathbf{B}_1\boldsymbol{\gamma}$ divided by $\boldsymbol{\gamma}^T\mathbf{B}_2\mathbf{A}_2^{-1}\mathbf{B}_2\boldsymbol{\gamma}$ and, unlike the estimation case, the test efficiency depends on the direction $\boldsymbol{\gamma}$ along which we approach the origin; see Theorem 6.1.2. Hence, we note that, unlike the univariate case, the testing and estimation efficiencies are not necessarily equal. Bickel (1965) shows that the ratio of noncentrality parameters can be interpreted as the limiting ratio of sample sizes needed for the same asymptotic level and same asymptotic power along the same sequence of alternatives, as in the Pitman efficiency used throughout this book. We can see that $\mathbf{B}\mathbf{A}^{-1}\mathbf{B}$ should be 'large' just as $\mathbf{B}\mathbf{A}^{-1}\mathbf{B}$ should be 'small'. In the next section we consider how to set up the estimating equations and consider what sort of estimates and tests result. We will be in a position to compute the efficiency of the estimates and tests relative to the traditional least-squares estimates and tests. First we list three important criterion functions and their associated estimating equations. Other criterion functions will be introduced in later sections.

Example 6.1.1 *Three criterion functions*

We now introduce three criterion functions that, in turn, produce estimating equations through differentiation. One of the criterion functions will generate the vector of means, the L_2 or least-squares estimates. The other two criterion functions will generate different versions of what may be considered L_1 estimates or bivariate medians. The two types of medians differ in their equivariance properties. See Small (1990) for an excellent review of multidimensional medians. The vector of means is equivariant under affine transformations of the data; see Exercise 6.8.1. The three criterion functions are:

$$D_1(\boldsymbol{\theta}) = \sqrt{\sum_{i=1}^{n}[(x_{i1} - \theta_1)^2 + (x_{i2} - \theta_2)^2]} \qquad (6.1.3)$$

$$D_2(\boldsymbol{\theta}) = \sum_{i=1}^{n}\sqrt{(x_{i1} - \theta_1)^2 + (x_{i2} - \theta_2)^2} \qquad (6.1.4)$$

$$D_3(\boldsymbol{\theta}) = \sum_{i=1}^{n}\{|x_{i1} - \theta_1| + |x_{i2} - \theta_2|\}. \qquad (6.1.5)$$

334 Multivariate Models

In each of these criterion functions we have pushed the square root operation deeper into the expression. As we will see, this produces very different types of estimates. We now take the gradients of these criterion functions and display the corresponding estimating equations.

$$\mathbf{S}_1(\boldsymbol{\theta}) = [D_1(\boldsymbol{\theta})]^{-1} \begin{pmatrix} \sum(x_{i1} - \theta_1) \\ \sum(x_{i2} - \theta_2) \end{pmatrix} \tag{6.1.6}$$

$$\mathbf{S}_2(\boldsymbol{\theta}) = \sum_{i=1}^{n} \|\mathbf{x}_i - \boldsymbol{\theta}_i\|^{-1} \begin{pmatrix} x_{i1} - \theta_1 \\ x_{i2} - \theta_2 \end{pmatrix} \tag{6.1.7}$$

$$\mathbf{S}_3(\boldsymbol{\theta}) = \begin{pmatrix} \sum \text{sgn}(x_{i1} - \theta_1) \\ \sum \text{sgn}(x_{i2} - \theta_2) \end{pmatrix}. \tag{6.1.8}$$

The computation of these gradients is given in Exercise 6.8.2.

In (6.1.8) if the vector is zero, then we take the term in the summation to be zero also. In Exercise 6.8.3 the reader is asked to verify that $\mathbf{S}_2(\boldsymbol{\theta}) = \mathbf{S}_3(\boldsymbol{\theta})$ in the univariate case; hence, we already see something new in the structure of the bivariate location model over the univariate location model. On the other hand, $\mathbf{S}_1(\boldsymbol{\theta})$ and $\mathbf{S}_3(\boldsymbol{\theta})$ are componentwise equations, unlike $\mathbf{S}_2(\boldsymbol{\theta})$ in which the two components are entangled. The solution to (6.1.8) is the vector of medians, and the solution to (6.1.7) is the spatial median which is discussed in Section 6.3. We will begin with an analysis of componentwise estimating equations and then consider other types.

Sections 6.2.3 through 6.4.3 deal with one-sample estimates and tests based on vector signs and ranks. Both rotational and affine invariant/equivariant methods are developed. Two- and several-sample models are treated in Section 6.6 as examples of location models. In Section 6.6 we will be primarily concerned with componentwise methods.

6.2 Componentwise Estimating Equations

Note that $\mathbf{S}_1(\boldsymbol{\theta})$ and $\mathbf{S}_3(\boldsymbol{\theta})$ are of the general form

$$\mathbf{S}(\boldsymbol{\theta}) = \begin{pmatrix} \sum \psi(x_{i1} - \theta_1) \\ \sum \psi(x_{i2} - \theta_2) \end{pmatrix}, \tag{6.2.1}$$

where $\psi(t) = t$ and $\text{sgn}(t)$ for (6.1.6) and (6.1.8), respectively. We need to find the matrices $\mathbf{A}$ and $\mathbf{B}$ in Definition 6.1.3. It is straightforward to verify that, when the true value of $\boldsymbol{\theta}$ is $\mathbf{0}$,

$$\mathbf{A} = \begin{pmatrix} E\psi^2(X_{11}) & E\psi(X_{11})\psi(X_{12}) \\ E\psi(X_{11})\psi(X_{12}) & E\psi^2(X_{22}) \end{pmatrix}, \tag{6.2.2}$$

and, from (6.1.2),

$$\mathbf{B} = \begin{pmatrix} E\psi'(X_{11}) & 0 \\ 0 & E\psi'(X_{12}) \end{pmatrix}. \tag{6.2.3}$$

Provided that **A** is positive definite, the multivariate central limit theorem implies that condition (c) in Definition 6.1.3 is satisfied for the componentwise estimating functions. In the case that $\psi(t) = \text{sgn}(t)$, we use the second representation in (6.1.2).

Example 6.2.1 *Pulmonary Measurements on Workers Exposed to Cotton Dust.*

In this example we extend the discussion to $k = 3$ dimensions. The data consist of $n = 12$ trivariate ($k = 3$) observations on workers exposed to cotton dust. The measurements in Table 6.2.1 are changes in measurements of pulmonary functions: FVC (forced vital capacity), FEV_3 (forced expiratory volume), and CC (closing capacity); see Merchant *et al.* (1975).

Let $\boldsymbol{\theta}^T = (\theta_1, \theta_2, \theta_3)$ and consider $H_0 : \boldsymbol{\theta} = \mathbf{0}$ versus $H_A : \boldsymbol{\theta} \neq \mathbf{0}$. First we compute the componentwise sign test. In (6.2.1) take $\psi(x) = \text{sgn}(x)$, then $n^{-1/2} \mathbf{S}_3^T = n^{-1/2}(-6, -6, -3)$ and the estimate of $\mathbf{A} = \text{Cov}(n^{-1/2}\mathbf{S}_3)$ is

$$\widehat{\mathbf{A}} = \frac{1}{n}\begin{bmatrix} n & \sum \text{sgn}(x_{i1})\text{sgn}(x_{i2}) & \sum \text{sgn}(x_{i1})\text{sgn}(x_{i3}) \\ \sum \text{sgn}(x_{i1})\text{sgn}(x_{i2}) & n & \sum \text{sgn}(x_{i2})\text{sgn}(x_{i3}) \\ \sum \text{sgn}(x_{i1})\text{sgn}(x_{i3}) & \sum \text{sgn}(x_{i2})\text{sgn}(x_{i3}) & n \end{bmatrix}$$

$$= \frac{1}{12}\begin{bmatrix} 12 & 8 & -4 \\ 8 & 12 & 0 \\ -4 & 0 & 12 \end{bmatrix}.$$

Here the diagonal elements are $\sum_i \text{sgn}^2(x_{is}) = n$ and the off-diagonal elements are values of the statistics $\sum_i \text{sgn}(x_{is})\text{sgn}(x_{it})$. Hence, the test statistic $n^{-1}\mathbf{S}_3^T \widehat{\mathbf{A}}^{-1} \mathbf{S}_3 = 4.667$, and using $\chi^2(3)$, the approximate p-value is 0.198; see Section 6.2.2.

Table 6.2.1: Changes in pulmonary function after 6 hours of exposure to cotton dust

Subject	FVC	FEV_3	CC
1	−0.11	−0.12	−4.3
2	0.02	0.08	4.4
3	−0.02	0.03	7.5
4	0.07	0.19	−0.30
5	−0.16	−0.36	−5.8
6	−0.42	−0.49	14.5
7	−0.32	−0.48	−1.9
8	−0.35	−0.30	17.3
9	−0.10	−0.04	2.5
10	0.01	−0.02	−5.6
11	−0.01	−0.17	2.2
12	−0.26	−0.30	5.5

We can also consider the finite-sample conditional distribution in which sign changes are generated with a binomial with $n = 12$ and $p = 0.5$; see the discussion in Section 6.2.2. Again note that the signs of all components of the observation vector are either changed or not. The matrix $\widehat{\mathbf{A}}$ remains unchanged so it is simple to generate many values of $n^{-1}\mathbf{S}_3^T\widehat{\mathbf{A}}^{-1}\mathbf{S}_3$ in Minitab. Out of 2500 values we found 518 greater than or equal to 4.667; hence, the randomization or sign change p-value is approximately $518/2500 = 0.207$, quite close to the asymptotic approximation. At any rate, we fail to reject $H_0 : \boldsymbol{\theta} = \mathbf{0}$ at any reasonable level.

Further, Hotelling's $T^2 = n\overline{\mathbf{X}}^T\widehat{\boldsymbol{\Sigma}}^{-1}\overline{\mathbf{X}} = 2.32$ with a p-value of 0.143, based on the F-distribution. Hence, Hotelling's T^2 also fails to reject the null hypothesis at reasonable levels.

Figure 6.2.1 provides boxplots for the data and componentwise normal $q-q$ plots. These boxplots suggest that there is little evidence to reject the null hypothesis, consistent with the component sign test. The normal $q-q$ plot of the component CC shows two outlying values on the right-hand side. In the case of the componentwise Wilcoxon test (Section 6.2.3), we consider $(n + 1)\mathbf{S}_4(\mathbf{0})$ in (6.2.13) along with $(n + 1)^2\mathbf{A}$, essentially in (6.2.14). For the pulmonary function data $(n + 1)\mathbf{S}_4^T(\mathbf{0}) = (-58, -54, 28)$ and

$$(n+1)^2\widehat{\mathbf{A}} = \frac{1}{n}\begin{bmatrix} 649 & 608 & -259.5 \\ 608 & 649.5 & -143.5 \\ -259.5 & -143.5 & 650 \end{bmatrix}.$$

The diagonal elements are $\sum_i R^2(|x_{is}|)$ which should be $\sum_i i^2 = 650$ but differ for the first two components due to ties among the absolute values. The off-diagonal elements are $\sum_i R(|x_{is}|)R(|x_{it}|)\text{sgn}|x_{is}|\text{sgn}|x_{it}|$. The test statistic is then $n^{-1}\mathbf{S}_4^T(\mathbf{0})\widehat{\mathbf{A}}^{-1}\mathbf{S}_4(\mathbf{0}) = 5.23$. From the $\chi^2(3)$ distribution, the approximate p-value is 0.156. Further, based on 2500 randomizations or sign changes, as in the sign test, we estimated the p-value of the conditional test to be 0.138. Hence, the Wilcoxon test fails to reject.

In the construction of tests we generally must estimate the matrix $\mathbf{A}$. When testing $H_0 : \boldsymbol{\theta} = \mathbf{0}$ the question arises as to whether or not we should center the data using $\widehat{\boldsymbol{\theta}}$. If we do not center then we are using a reduced model estimate of $\mathbf{A}$; otherwise, it is a full model estimate. Reduced model estimates are generally used in randomization tests. In this case, generally, $\widehat{\mathbf{A}}$ must only be computed once in the process of randomizing and recomputing the test statistic $\mathbf{S}^T\widehat{\mathbf{A}}^{-1}\mathbf{S}$. Note also that when $H_0 : \boldsymbol{\theta} = \mathbf{0}$, $\widehat{\boldsymbol{\theta}} \xrightarrow{P} \mathbf{0}$. Hence, the centered $\widehat{\mathbf{A}}$ is valid under H_0. When estimating $\text{Cov}(\widehat{\boldsymbol{\theta}})$, $\mathbf{B}^{-1}\mathbf{A}\mathbf{B}^{-1}$, we should center $\widehat{\mathbf{A}}$ because we no longer assume that H_0 is true.

6.2.1 Estimation

Further, when (6.1.2) holds, the asymptotic covariance matrix in Theorem 6.1.2 is

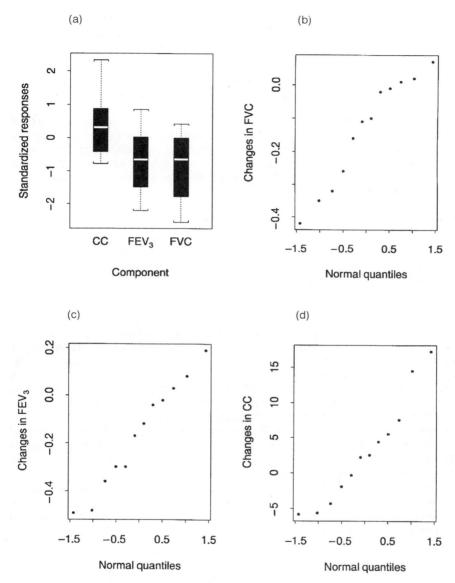

Figure 6.2.1: (a) Boxplots of the changes in pulmonary function for the cotton dust data. Note that the responses have been standardized by componentwise standard deviations; (b) normal q–q plot for the component FVC, original scale; (c) normal q–q plot for the component FEV_3, original scale; (d) normal q–q plot for the component CC, original scale.

$$\mathbf{B}^{-1}\mathbf{A}\mathbf{B}^{-1} = \begin{pmatrix} \dfrac{E\psi^2(X_{11})}{[E\psi'(X_{11})]^2} & \dfrac{E\psi(X_{11})\psi(X_{12})}{E\psi'(X_{11})E\psi'(X_{12})} \\ \dfrac{E\psi(X_{11})\psi(X_{12})}{E\psi'(X_{11})E\psi'(X_{12})} & \dfrac{E\psi^2(X_{12})}{[E\psi'(X_{12})]^2} \end{pmatrix}.$$

Now Theorem 6.1.2 can be applied for various M-estimates to establish asymptotic normality. Our interest is in the comparison of L_2 and L_1 estimates and we now turn to that discussion. In the case of L_2 estimates, corresponding to $\mathbf{S}_1(\boldsymbol{\theta})$, we take $\psi(t) = t$. It is then easy to see that $\mathbf{B}^{-1}\mathbf{A}\mathbf{B}^{-1}$ is equal to the covariance matrix of the underlying model, say Σ_f. The estimate is the vector of component means.

In the first L_1 case corresponding to $\mathbf{S}_3(\boldsymbol{\theta})$, using (6.1.2), we take $\psi(t) = \text{sgn}(t)$ and find, using the second representation in (6.1.2), that

$$\mathbf{B}^{-1}\mathbf{A}\mathbf{B}^{-1} = \begin{pmatrix} \dfrac{1}{4f_1^2(0)} & \dfrac{E\text{sgn}(X_{11})\text{sgn}(X_{12})}{4f_1(0)f_2(0)} \\ \dfrac{E\text{sgn}(X_{11})\text{sgn}(X_{12})}{4f_1(0)f_2(0)} & \dfrac{1}{4f_2^2(0)} \end{pmatrix}, \quad (6.2.4)$$

where f_1 and f_2 denote the marginal pdfs. The estimate is the vector of component medians. Note that in addition to the marginal densities, we must also estimate the quantity $E\text{sgn}(X_{11})\text{sgn}(X_{12})$ if we wish to estimate the asymptotic covariance matrix of the vector of component medians. We recommend a simple moment estimator $n^{-1}\sum \text{sgn}(x_{i1} - \hat{\theta}_1)\text{sgn}(x_{i2} - \hat{\theta}_2)$ for $E\text{sgn}(X_{11})\text{sgn}(X_{12})$, and for estimation of f_1 and f_2 see the discussion in Examples 1.5.5 and 1.5.6 on the estimation of the standard error of the sample median. In the case of the componentwise sign statistics, we do not center the estimate of $\mathbf{A}$ because in that case the null hypothesis specifies $\boldsymbol{\theta} = \mathbf{0}$. We now turn to the efficiency of the vector of sample medians with respect to the vector of sample means.

Let $\delta = \det(\mathbf{B}^{-1}\mathbf{A}\mathbf{B}^{-1}) = \det(\mathbf{A})/[\det(\mathbf{B})]^2$ be Wilk's generalized variance of $\sqrt{n}\hat{\boldsymbol{\theta}}$ in Definition 6.1.4. For the vector of means we have $\delta = \sigma_1^2\sigma_2^2(1 - \rho^2)$, the determinant of the underlying variance-covariance matrix. For the vector of sample medians we have

$$\delta = \frac{1 - [E\text{sgn}(X_{11})\text{sgn}(X_{12})]^2}{16f_1^2(0)f_2^2(0)},$$

and the efficiency of the vector of medians with respect to the vector of means is given by:

$$e(\text{med, mean}) = 4\sigma_1\sigma_2 f_1(0)f_2(0)\sqrt{\frac{1 - \rho^2}{1 - [E\text{sgn}(X_{11})\text{sgn}(X_{12})]^2}}. \quad (6.2.5)$$

Note that $E\text{sgn}(X_{11})\text{sgn}(X_{12}) = 4P(X_{11} < 0, X_{12} < 0) - 1$. When the underlying distribution is bivariate normal with means 0, variances 1, and

Table 6.2.2: Efficiency (6.2.7) of the vector of medians relative to the vector of means when the underlying distribution is bivariate normal

ρ	0	0.1	0.2	0.3	0.4	0.5	0.6	0.7	0.8	0.9	0.99
eff	0.64	0.63	0.63	0.62	0.60	0.58	0.56	0.52	0.47	0.40	0.22

correlation ρ, Exercise 6.8.4 shows that

$$P(X_{11} < 0, X_{12} < 0) = \frac{1}{4} + \frac{1}{2\pi \sin \rho}. \tag{6.2.6}$$

Further, the marginal distributions are standard normal; hence, (6.2.5) becomes

$$e(\text{med, mean}) = \frac{2}{\pi}\sqrt{\frac{1-\rho^2}{1-[(2/\pi)\sin^{-1}\rho]^2}}. \tag{6.2.7}$$

The first factor, $2/\pi \cong 0.637$, is the univariate efficiency of the median relative to the mean when the underlying distribution is normal and also the efficiency of the vector of medians relative to the vector of means when the correlation in the underlying model is zero. The second factor accounts for the bivariate structure of the model and, in general, depends on the correlation ρ. Some values of the efficiency are given in Table 6.2.2.

Clearly, as the elliptical contours of the underlying normal distribution flatten out, the efficiency of the vector of medians decreases. This is the first indication that the vector of medians is not affine (or even rotation) equivariant. The vector of means is affine equivariant and hence the dependency of the efficiency on ρ must be due to the vector of medians. Indeed, Exercise 6.8.5 asks the reader to construct an example showing that when the axes are rotated the vector of means rotates into the new vector of means while the vector of medians fails to do so.

6.2.2 Testing

We now consider the properties of bivariate tests. Recall that we assume the underlying bivariate distribution is symmetric. In addition, we would generally use an odd ψ-function, so that $\psi(t) = -\psi(-t)$. This implies that $\psi(t) = \psi(|t|)\text{sgn}(t)$, which will be useful shortly.

Now referring to Theorem 6.1.2 along with the corresponding matrix $\mathbf{A}$, the test of $H_0 : \boldsymbol{\theta} = \mathbf{0}$ versus $H_A : \boldsymbol{\theta} \neq \mathbf{0}$ rejects the null hypothesis when $(1/n)\mathbf{S}^T(\mathbf{0})\mathbf{A}^{-1}\mathbf{S}(\mathbf{0}) \geq \chi_\alpha^2(2)$. Note that the covariance term in $\mathbf{A}$ is $E\psi(X_{11})\psi(X_{12}) = \iint \psi(s)\psi(t)f(s,t)\,dsdt$ and it depends upon the underlying bivariate distribution f. Hence, even the sign test based on the componentwise sign statistic $\mathbf{S}_3(\mathbf{0})$ is not distribution-free under the null hypothesis as it is in the univariate case. In this case, $E\psi(X_{11})\psi(X_{12}) = 4P(X_{11} < 0, X_{12} < 0) - 1$, as we saw in the discussion of estimation.

To make the test operational we must estimate the components of $\mathbf{A}$. Since they are expectations, we use moment estimates, under the null hypothesis. Now condition (c) in Definition 6.1.3 guarantees that the test with the estimated **A** is asymptotically distribution-free since it has a limiting chi-square distribution, independent of the underlying distribution. What can we say about finite samples?

First note that

$$\mathbf{S}(0) = \begin{pmatrix} \Sigma \psi(|x_{i1}|)\mathrm{sgn}(x_{i1}) \\ \Sigma \psi(|x_{i2}|)\mathrm{sgn}(x_{i2}) \end{pmatrix}. \qquad (6.2.8)$$

Under the assumption of symmetry, $(\mathbf{x}_1, \ldots, \mathbf{x}_n)$ is a realization of $(s_1\mathbf{x}_1, \ldots, s_n\mathbf{x}_n)$, where $(s_1, \ldots, s_n)$ is a vector of independent random variables each equalling ± 1 with probability $\frac{1}{2}, \frac{1}{2}$. Hence $Es_i = 0$ and $Es_i^2 = 1$. Condition on $(\mathbf{x}_1, \ldots, \mathbf{x}_n)$; then, under the null hypothesis, there are 2^n equally likely sign combinations associated with these vectors. Note that the sign changes attach to the entire vector. From (6.2.8), we see that, conditionally, the scores are not affected by the sign changes and $\mathbf{S}(0)$ depends on the sign changes only through the signs of the components of the observation vectors. It follows at once that the conditional mean of $\mathbf{S}(0)$ under the null hypothesis is $\mathbf{0}$. Further, the conditional covariance matrix is given by

$$\begin{pmatrix} \Sigma \psi^2(|x_{i1}|) & \Sigma \psi(|x_{i1}|)\psi(|x_{i2}|)\mathrm{sgn}(x_{i1})\mathrm{sgn}(x_{i2}) \\ \Sigma \psi(|x_{i1}|)\psi(|x_{i2}|)\mathrm{sgn}(x_{i1})\mathrm{sgn}(x_{i2}) & \Sigma \psi^2(|x_{i2}|) \end{pmatrix}. \qquad (6.2.9)$$

Note that, conditionally, n^{-1} times this matrix is an estimate of the matrix **A** above. Thus we have a conditionally distribution-free sign-change distribution. For small to moderate n the test statistic (quadratic form) can be computed for each combination of signs and a conditional p-value of the test is the number of values (divided by 2^n) of the test statistic at least as large as the observed value of the test statistic. In the first chapter on univariate methods this argument also leads to unconditionally distribution-free tests in the case of the univariate sign and rank tests since in those cases the signs and the ranks do not depend on the values of the conditioning variables. Again, the situation is different in the bivariate case due to the matrix **A** which must be estimated since it depends on the unknown underlying distribution. In Exercise 6.8.6 the reader is asked to construct the sign-change distributions for some examples.

We now turn to a more detailed analysis of the tests based on $\mathbf{S}_1 = \mathbf{S}_1(\mathbf{0})$ and $\mathbf{S}_3 = \mathbf{S}_3(\mathbf{0})$. Recall that $\mathbf{S}_1$ is the vector of sample means. The matrix **A** is the covariance matrix of the underlying distribution and we take the sample covariance matrix as the natural estimate. The resulting test statistic is $n\overline{\mathbf{X}}^T\widehat{\mathbf{A}}^{-1}\overline{\mathbf{X}}$, which is Hotelling's T^2 statistic. Note that, for T^2, we typically use a centered estimate of **A**. If we want the randomization distribution then we use the uncentered estimate. Since $\mathbf{B}\mathbf{A}^{-1}\mathbf{B} = \Sigma_f^{-1}$, the covariance matrix of the underlying distribution, the asymptotic noncentrality parameter for Hotelling's test is $\boldsymbol{\gamma}^T\Sigma_f^{-1}\boldsymbol{\gamma}$. The vector $\mathbf{S}_3$ is the vector of component sign statistics. By inverting (6.2.4) we can write down the noncentrality parameter for the bivariate componentwise sign test.

To illustrate the efficiency of the bivariate sign test relative to Hotelling's test we simplify the structure as follows: assume that the marginal distributions are identical. Let $\xi = 4P(X_{11} < 0, X_{12} < 0) - 1$ and let ρ denote the underlying correlation, as usual. Then Hotelling's noncentrality parameter is

$$\frac{1}{\sigma^2(1-\rho^2)} \gamma^T \begin{pmatrix} 1 & -\rho \\ -\rho & 1 \end{pmatrix} \gamma = \frac{\gamma_1^2 - 2\rho\gamma_1\gamma_2 + \gamma_2^2}{\sigma^2(1-\rho^2)}. \tag{6.2.10}$$

Likewise the noncentrality parameter for the bivariate sign test is

$$\frac{4f^2(0)}{(1-\xi^2)} \gamma^T \begin{pmatrix} 1 & -\xi \\ -\xi & 1 \end{pmatrix} \gamma = \frac{4f^2(0)(\gamma_1^2 - 2\xi\gamma_1\gamma_2 + \gamma_2^2)}{1-\xi^2}. \tag{6.2.11}$$

The efficiency of the bivariate sign test relative to Hotelling's test is the ratio of their respective noncentrality parameters:

$$\frac{4f^2(0)\sigma^2(1-\rho^2)(\gamma_1^2 - 2\xi\gamma_1\gamma_2 + \gamma_2^2)}{(1-\xi^2)(\gamma_1^2 - 2\rho\gamma_1\gamma_2 + \gamma_2^2)}. \tag{6.2.12}$$

There are three contributing factors in this efficiency: $4f^2(0)\sigma^2$, which is the univariate efficiency of the sign test relative to the t-test; $(1-\rho^2)/(1-\xi^2)$, due to the dependence structure in the bivariate distribution; and the final factor, which reflects the direction of approach of the sequence of alternatives. It is this last factor which separates the testing efficiency from the estimation efficiency. In order to see the effect of direction on the efficiency we will use the following result from matrix theory; see Graybill (1983).

Lemma 6.2.1 Suppose $\mathbf{D}$ is a nonsingular, square matrix and $\mathbf{C}$ is any square matrix, and suppose λ_1 and λ_2 are the minimum and maximum eigenvalues of $\mathbf{CD}^{-1}$; then

$$\lambda_1 \leq \frac{\gamma^T \mathbf{C} \gamma}{\gamma^T \mathbf{D} \gamma} \leq \lambda_2.$$

The following proposition is left as Exercise 6.8.7.

Theorem 6.2.2 The efficiency $e(\mathbf{S}_3, \mathbf{S}_1)$ is bounded between the minimum and maximum of $4f^2(0)\sigma^2(1-\rho)/(1-\xi)$ and $4f^2(0)\sigma^2(1+\rho)/(1+\xi)$.

In Table 6.2.3 we give some values of the maximum and minimum efficiencies when the underlying distribution is bivariate normal with means 0, variances 1 and correlation ρ. This table can be compared to Table 6.2.2, which contains the corresponding estimation efficiencies. We have

Table 6.2.3: Minimum and maximum efficiencies of the bivariate sign test relative to Hotelling's T^2 when the underlying distribution is bivariate normal

ρ	0	0.2	0.4	0.6	0.8	0.9	0.99
min	0.64	0.58	0.52	0.43	0.31	0.22	0.07
max	0.64	0.68	0.71	0.72	0.72	0.71	0.66

$f^2(0) = (2\pi)^{-1}$ and $\xi = (2/\pi)\sin^{-1}\rho$. Hence, the dependence of the efficiency on direction determined by γ is apparent. The examples involving the bivariate normal distribution also show the superiority of the vector of means over the vector of medians and of Hotelling's test over the bivariate sign test, as expected. Bickel (1964; 1965) gives a more thorough analysis of the efficiency for general models. He points out that when heavy-tailed models are expected then the medians and sign test will be much better provided ρ is not too close to ± 1.

In the exercises the reader is asked to show that Hotelling's T^2 statistic is affine invariant. Thus the efficiency properties of this statistic do not depend on ρ. This means that the bivariate sign test cannot be affine invariant; again, this is developed in the exercises. It is now natural to inquire about the properties of the estimate and test based on S_2. This estimating function cannot be written in the componentwise form that we have been considering. Before we turn to this statistic, we consider estimates and tests based on componentwise ranking.

6.2.3 Componentwise Rank Methods

In this subsection we will sketch the results for the vector of Wilcoxon signed-rank statistics discussed in Section 1.7 for each component. See Example 6.2.1 for an illustration of the calculations. In Section 6.6 we provide a full development of componentwise rank-based methods for location and regression models with examples. We let

$$S_4(\boldsymbol{\theta}) = \begin{pmatrix} \sum \dfrac{R(|x_{i1} - \theta_1|)}{n+1} \operatorname{sgn}(x_{i1} - \theta_1) \\ \sum \dfrac{R(|x_{i2} - \theta_2|)}{n+1} \operatorname{sgn}(x_{i2} - \theta_2) \end{pmatrix}. \quad (6.2.13)$$

Using the projection method (Theorem 2.4.6), we have from Exercise 6.8.8, for the case $\boldsymbol{\theta} = \mathbf{0}$,

$$S_4(\mathbf{0}) = \begin{pmatrix} \sum F_1^+(|x_{i1}|)\operatorname{sgn}(x_{i1}) \\ \sum F_2^+(|x_{i2}|)\operatorname{sgn}(x_{i2}) \end{pmatrix} + o_p(1) = \begin{pmatrix} 2\sum[F_1(x_{i1}) - \tfrac{1}{2}] \\ 2\sum[F_2(x_{i2}) - \tfrac{1}{2}] \end{pmatrix} + o_p(1),$$

where F_j^+ is the marginal distribution of $|X_{1j}|$ for $j = 1, 2$ and F_j is the marginal distribution of X_{1j} for $j = 1, 2$; see also Section A.2.3 of the Appendix. Symmetry of the marginal distributions is used in the computation of the projections. Conditions (a)–(d) of Definition 6.1.3 can now be verified for the projection and then we note that the vector of rank statistics has the same asymptotic properties. We must identify the matrices $\mathbf{A}$ and $\mathbf{B}$ for the purposes of constructing the quadratic form test statistic, the asymptotic distribution of the vector of estimates, and the noncentrality parameter.

The first two conditions, (a) and (b), are easy to check since the multivariate central limit theorem can be applied to the projection. Since under the null hypothesis that $\boldsymbol{\theta} = \mathbf{0}$, $F(X_{i1})$ has a uniform distribution on $(0, 1)$, and introducing $\boldsymbol{\theta}$ and differentiating with respect to θ_1 and θ_2, the matrices $\mathbf{A}$ and $\mathbf{B}$ are

$$\frac{1}{n}\mathbf{A} = \begin{pmatrix} \frac{1}{3} & \delta \\ \delta & \frac{1}{3} \end{pmatrix} \text{ and } \mathbf{B} = \begin{pmatrix} 2\int f_1^2(t)dt & 0 \\ 0 & 2\int f_2^2(t)dt \end{pmatrix}, \qquad (6.2.14)$$

where $\delta = 4 \iint F_1(s)F_2(t)dF(s,t) - 1$. Hence, similar to the vector of sign statistics, the vector of Wilcoxon signed-rank statistics also has a covariance which depends on the underlying bivariate distribution. We could construct a conditionally distribution-free test but not an unconditionally distribution-free one. Of course, the test is asymptotically distribution-free.

A consistent estimate of the parameter δ in $\mathbf{A}$ is given by

$$\widehat{\delta} = \frac{1}{n}\sum_{t=1}^{n} \frac{R_{it}R_{jt}}{(n+1)(n+1)} \, \text{sgn}(X_{it})\text{sgn}(X_{jt}), \qquad (6.2.15)$$

where R_{it} is the rank of $|X_{it}|$ in the tth component among $|X_{1t}|, \ldots, |X_{nt}|$. This estimate is the conditional covariance and can be used in estimating $\mathbf{A}$ in the construction of an asymptotically distribution-free test; when we estimate the asymptotic covariance matrix of $\boldsymbol{\theta}$ we first center the data and then compute (6.2.15).

The estimator that solves $\mathbf{S}_4(\boldsymbol{\theta}) = \mathbf{0}$ is the vector of Hodges–Lehmann (HL) estimates for the two components; that is, the vector of medians of Walsh averages for each component. Like the vector of medians, the vector of HL estimates is not equivariant under orthogonal transformations and the test is not invariant under these transformations. This will show up in the efficiency with respect to the L_2 methods which are an equivariant estimate and an invariant test. Theorem 6.1.2 provides the asymptotic distribution of the estimator and the asymptotic local power of the test.

Suppose the underlying distribution is bivariate normal with means 0, variances 1, and correlation ρ; then the estimation and testing efficiencies are given by

$$e(\text{HL, mean}) = \frac{3}{\pi}\sqrt{\frac{1-\rho^2}{1-9\delta^2}} \qquad (6.2.16)$$

$$e(\text{Wilcoxon, Hotelling}) = \frac{3}{\pi}\frac{1-\rho^2}{1-9\delta^2}\left\{\frac{\gamma_1^2 - 6\delta\gamma_1\gamma_2 + \gamma_2^2}{\gamma_1^2 - 2\rho\gamma_1\gamma_2 + \gamma_2^2}\right\}. \qquad (6.2.17)$$

Exercise 6.8.9 asks the reader to apply Lemma 6.2.1 and show the testing efficiency is bounded between

$$\frac{3(1+\rho)}{2\pi\left[2 - \frac{3}{\pi}\cos^{-1}\left(\frac{\rho}{2}\right)\right]} \text{ and } \frac{3(1-\rho)}{2\pi\left[\frac{3}{\pi}\cos^{-1}\left(\frac{\rho}{2}\right)\right]}. \qquad (6.2.18)$$

In Table 6.2.4 we provide some values of the minimum and maximum efficiencies as well as estimation efficiency. Note how much more stable the rank

Table 6.2.4: Efficiencies of componentwise Wilcoxon methods relative to L_2 methods when the underlying distribution is bivariate normal

ρ	0	0.2	0.4	0.6	0.8	0.9	0.99
min	0.96	0.94	0.93	0.91	0.89	0.88	0.87
max	0.96	0.96	0.97	0.97	0.96	0.96	0.96
est	0.96	0.96	0.95	0.94	0.93	0.92	0.91

methods are than the sign methods. Bickel (1964) points out, however, that when there is heavy contamination and ρ is close to ± 1 the estimation efficiency can be arbitrarily close to 0. Further, this efficiency can be arbitrarily large. This behavior is due to the fact that the sign and rank methods are not invariant and equivariant under orthogonal transformations, unlike the L_2 methods. Hence, we now turn to an analysis of the methods generated by $\mathbf{S}_2(\boldsymbol{\theta})$. Additional material on the componentwise methods can be found in the papers of Bickel (1964; 1965) and the monograph by Puri and Sen (1971). The extension of the results to dimensions higher than 2 is straightforward and the formulas are obvious. One interesting question is how the efficiencies of the sign or rank methods relative to the L_2 methods depend on the dimension.

6.3 Spatial Methods

We are now ready to consider the estimate and test generated by $\mathbf{S}_2(\boldsymbol{\theta})$; recall (6.1.4) and (6.1.7). This estimating function cannot be written in componentwise fashion because $\|\mathbf{x}_i - \boldsymbol{\theta}\|$ appears in both components. Note that $\mathbf{S}_2(\boldsymbol{\theta}) = \sum \|\mathbf{x}_i - \boldsymbol{\theta}\|^{-1}(\mathbf{x}_i - \boldsymbol{\theta})$, a sum of unit vectors, so that the estimating function depends on the data only through the directions and not on the magnitudes of $\mathbf{x}_i - \boldsymbol{\theta}$, $i = 1, \ldots, n$. Hence, the test is called the **angle test** or **angle sign test** and the estimate is called the **spatial median**; see Brown (1983). Milasevic and Ducharme (1987) show that the spatial median is always unique, unlike the univariate median. We will see that the test is invariant under orthogonal transformations and the estimate is equivariant under these transformations. Hence, the methods are rotation invariant and equivariant, properties suitable for methods used on spatial data. However, applications do not have to be confined to spatial data and we will consider these methods as competitors to the other methods already discussed. Gower (1974) calls the estimate the mediancentre and provides a Fortran program for its computation. See Bedall and Zimmerman (1979) for a program in dimensions higher than 2. Further, for higher dimensions see Möttönen and Oja (1995).

Following our pattern above, we first consider the matrices $\mathbf{A}$ and $\mathbf{B}$ in Definition 6.1.3. Suppose $\boldsymbol{\theta} = \mathbf{0}$; then since $\mathbf{S}_2(\mathbf{0})$ is a sum of independent random variables, condition (c) is immediate with $\mathbf{A} = E\|\mathbf{X}\|^{-2}\mathbf{X}\mathbf{X}^T$ and the obvious estimate of $\mathbf{A}$, under H_0, is

$$\widehat{\mathbf{A}} = \frac{1}{n}\sum \|\mathbf{x}_i\|^{-2}\mathbf{x}_i\mathbf{x}_j^T, \qquad (6.3.1)$$

which can be used to construct the quadratic test statistic with

$$\frac{1}{\sqrt{n}} \mathbf{S}_2(\mathbf{0}) \xrightarrow{D} N_2(\mathbf{0}, \mathbf{A}) \text{ and } \frac{1}{n} \mathbf{S}_2^T(\mathbf{0}) \widehat{\mathbf{A}}^{-1} \mathbf{S}_2(\mathbf{0}) \xrightarrow{D} \chi^2(2). \tag{6.3.2}$$

In order to compute $\mathbf{B}$, we first compute the partial derivatives; then we take the expectation. This yields

$$\mathbf{B} = E\left\{\frac{1}{\|\mathbf{X}\|}\left[\mathbf{I} - \frac{1}{\|\mathbf{X}\|^2}(\mathbf{XX}^T)\right]\right\}, \tag{6.3.3}$$

where $\mathbf{I}$ is the identity matrix. Use a moment estimate for $\mathbf{B}$ similar to the estimate of $\mathbf{A}$. Then we have

$$\frac{1}{\sqrt{n}} \widehat{\boldsymbol{\theta}} = \mathbf{B}^{-1} \frac{1}{\sqrt{n}} \mathbf{S}_2(\mathbf{0}) + \mathbf{o}_p(1) \xrightarrow{D} N_2(\mathbf{0}, \mathbf{B}^{-1}\mathbf{A}\mathbf{B}^{-1}).$$

Chaudhuri (1992) provides a sharper analysis for the remainder term in his Theorem 3.2. The consistency of the moment estimate of $\mathbf{A}$ and $\mathbf{B}$ is established rigorously in the linear model setting by Bai et al. (1990). Hence, we would use $\widehat{\mathbf{A}}$ and $\widehat{\mathbf{B}}$ computed from the residuals. Bose and Chaudhuri (1993) develop estimates of $\mathbf{A}$ and $\mathbf{B}$ that converge more quickly than the moment estimates. Bose and Chaudhuri provide a very interesting analysis of why it is easier to estimate the asymptotic covariance matrix of $\widehat{\boldsymbol{\theta}}$ than to estimate the asymptotic variance of the univariate median. Essentially, unlike the univariate case, we do not need to estimate the multivariate density at a point.

Example 6.3.1 *Cork Borings Data*

We consider a well-known example due to Rao (1948) of testing whether the weight of cork borings on trees is independent of the directions: north, south, east and west. In this case we have four measurements on each tree and we wish to test the equality of marginal locations: $H_0 : \theta_N = \theta_S = \theta_E = \theta_W$. This is a common hypothesis in repeated measures designs. See Jan and Randles (1996) for an excellent discussion of issues in repeated measures designs. We reduce the data to trivariate vectors via $N - E, E - S, S - W$. Then we test $\boldsymbol{\delta} = \mathbf{0}$, where $\boldsymbol{\delta}^T = (\theta_N - \theta_S, \theta_S - \theta_E, \theta_E - \theta_W)$. Table 6.3.1 displays the original $n = 28$ four-component data vectors.

In Table 6.3.2 we display the data differences, $N - S, S - E$, and $E - W$ along with the unit spatial vectors $\|\mathbf{x}\|^{-1}\mathbf{x}$, for each data point. Note that, except for rounding error, the sum of squares in each row is 1 for the spatial sign vectors.

We compute the spatial sign vector to be $\mathbf{S}_2^T = (7.78, -4.99, 6.65)$ and, from (6.3.1),

$$\widehat{\mathbf{A}} = \begin{bmatrix} 0.2809 & -0.1321 & -0.0539 \\ -0.1321 & 0.3706 & -0.0648 \\ -0.0539 & -0.0648 & 0.3484 \end{bmatrix}.$$

Then $n^{-1}\mathbf{S}_2^T(\mathbf{0})\widehat{\mathbf{A}}^{-1}\mathbf{S}_2(\mathbf{0}) = 14.74$, which yields an asymptotic p-value of 0.002, using a χ^2 approximation with three degrees of freedom. Hence, we easily reject $H_0 : \boldsymbol{\delta} = \mathbf{0}$ and conclude that boring size depends on direction.

346   *Multivariate Models*

Table 6.3.1: Weight of cork borings (in centigrams) in four directions for 28 trees

N	E	S	W	N	E	S	W
72	66	76	77	91	79	100	75
60	53	66	63	56	68	47	50
56	57	64	58	79	65	70	61
41	29	36	38	81	80	68	58
32	32	35	36	78	55	67	60
30	35	34	26	46	38	37	38
39	39	31	27	39	35	34	37
42	43	31	25	32	30	30	32
37	40	31	25	60	50	67	54
33	29	27	36	35	37	48	39
32	30	34	28	39	36	39	31
63	45	74	63	50	34	37	40
54	46	60	52	43	37	39	50
47	51	52	43	48	54	57	43

For estimation we return to the original component data. Since we have rejected the null hypothesis of equality of locations, we want to estimate the four components of the location vector $\boldsymbol{\theta}^T = (\theta_1, \theta_2, \theta_3, \theta_4)$. The spatial median

Table 6.3.2: Each row is a data vector for N − S, S − E, E − W and also gives the components of the spatial sign vector

Row	N − E	E − S	S − W	S_1	S_2	S_3
1	6	−10	−1	0.51	−0.85	−0.09
2	7	−13	3	0.46	−0.86	0.20
3	−1	−7	6	−0.11	−0.75	0.65
4	12	−7	−2	0.85	−0.50	−0.14
5	0	−3	−1	0.00	−0.95	−0.32
6	−5	1	8	−0.53	0.11	0.84
7	0	8	4	0.00	0.89	0.45
8	−1	12	6	−0.07	0.89	0.45
9	−3	9	6	−0.27	0.80	0.53
10	4	2	−9	0.40	0.19	−0.90
11	2	−4	6	0.27	−0.53	0.80
12	18	−29	11	0.50	−0.80	0.31
13	8	−14	8	0.44	−0.78	0.44
14	−4	−1	9	−0.40	−0.10	0.91
15	12	−21	25	0.34	−0.60	0.71
16	−12	21	−3	−0.49	0.86	−0.12
17	14	−5	9	0.81	−0.29	0.52
18	1	12	10	0.06	0.77	0.64
19	23	−12	7	0.86	−0.44	0.26
20	8	1	−1	0.98	0.12	−0.12
21	4	1	−3	0.78	0.20	−0.59
22	2	0	−2	0.71	0.00	−0.71
23	10	−17	13	0.42	−0.72	0.55
24	−2	−11	9	−0.14	−0.77	0.63
25	3	−3	8	0.33	−0.33	0.88
26	16	−3	−3	0.97	−0.18	−0.18
27	6	−2	−11	0.47	−0.16	−0.87
28	−6	−3	14	−0.39	−0.19	0.90

solves $\mathbf{S}_2(\boldsymbol{\theta}) = \mathbf{0}$, and we find $\widehat{\boldsymbol{\theta}}^T = (45.38, 41.54, 43.91, 41.03)$. For comparison the mean vector is $(50.54, 46.18, 49.68, 45.18)^T$. Computations were carried out using SAS macros provided by Möttönen (1997a) which are available in a Technical Report from the University of Oulu, Finland. The issue of how to apply rank methods in repeated measures designs has an extensive literature. In addition to Jan and Randles (1996), Kepner and Robinson (1988) and Akritas and Arnold (1994) discuss the use of rank transforms and pure ranks for testing hypotheses in repeated measures designs. The Friedman test (Exercise 4.8.19), can also be used for repeated measures designs.

Expressions for $\mathbf{A}$ and $\mathbf{B}$ can be simplified and the computation of efficiencies made easier if we transform to polar coordinates. We write

$$\mathbf{x} = r\begin{pmatrix} \cos\phi \\ \sin\phi \end{pmatrix} = rs\begin{pmatrix} \cos\varphi \\ \sin\varphi \end{pmatrix}, \tag{6.3.4}$$

where $r = \|\mathbf{x}\| \geq 0$, $0 \leq \phi < 2\pi$, and $s = \pm 1$ depending on whether $\mathbf{x}$ is above or below the horizontal axis with $0 < \varphi < \pi$. The second representation is similar to (6.2.8) and is useful in the development of the conditional distribution of the test under the null hypothesis. Hence

$$\mathbf{S}_2(\mathbf{0}) = \sum s_i \begin{pmatrix} \cos\varphi_i \\ \sin\varphi_i \end{pmatrix}, \tag{6.3.5}$$

where φ_i is the angle measured counterclockwise between the positive horizontal axis and the line through $\mathbf{x}_i$ extending indefinitely through the origin and s_i indicates whether the observation is above or below the axis. Under the null hypothesis $\boldsymbol{\theta} = \mathbf{0}$, $s_i = \pm 1$ with probabilities $\frac{1}{2}, \frac{1}{2}$ and $s_1, \ldots, s_n$ are independent. Thus, we can condition on $\varphi_1, \ldots, \varphi_n$ to obtain a conditionally distribution-free test. The conditional covariance matrix is

$$\sum_{i=1}^n \begin{pmatrix} \cos^2\varphi_i & \cos\varphi_i\sin\varphi_i \\ \cos\varphi_i\sin\varphi_i & \sin^2\varphi_i \end{pmatrix}, \tag{6.3.6}$$

and this is used in the quadratic form with $\mathbf{S}_2(\mathbf{0})$ to construct the test statistic; see Möttönen and Oja (1995, Section 2.1).

To consider the asymptotically distribution-free version of this test we use the form

$$\mathbf{S}_2(\mathbf{0}) = \sum \begin{pmatrix} \cos\phi_i \\ \sin\phi_i \end{pmatrix}, \tag{6.3.7}$$

where the reader will recall that $0 \leq \phi < 2\pi$, and the multivariate central limit theorem implies that $(1/\sqrt{n})\mathbf{S}_2(\mathbf{0})$ has a limiting bivariate normal distribution with mean $\mathbf{0}$ and covariance matrix $\mathbf{A}$. We now translate $\mathbf{A}$ and its estimate into polar coordinates:

$$\mathbf{A} = E\begin{pmatrix} \cos^2\phi & \cos\phi\sin\phi \\ \cos\phi\sin\phi & \sin^2\phi \end{pmatrix} \text{ and } \widehat{\mathbf{A}} = \frac{1}{n}\sum_{i=1}^n \begin{pmatrix} \cos^2\phi_i & \cos\phi_i\sin\phi_i \\ \cos\phi_i\sin\phi_i & \sin^2\phi_i \end{pmatrix}. \tag{6.3.8}$$

Hence, $(1/n)\mathbf{S}_2^T(\mathbf{0})\widehat{\mathbf{A}}^{-1}\mathbf{S}_2(\mathbf{0}) \geq \chi_\alpha^2(2)$ is an asymptotically size α test.

We next express $\mathbf{B}$ in terms of polar coordinates:

$$\mathbf{B} = Er^{-1}\left\{\mathbf{I} - \begin{pmatrix} \cos^2\phi & \cos\phi\sin\phi \\ \cos\phi\sin\phi & \sin^2\phi \end{pmatrix}\right\} = Er^{-1}\begin{pmatrix} \sin^2\phi & -\cos\phi\sin\phi \\ -\cos\phi\sin\phi & \cos^2\phi \end{pmatrix}. \tag{6.3.9}$$

Hence, $\sqrt{n}$ times the spatial median is limiting bivariate normal with asymptotic covariance matrix equal to $\mathbf{B}^{-1}\mathbf{A}\mathbf{B}^{-1}$. The corresponding noncentrality parameter of the noncentral chi-square limiting distribution of the test is $\boldsymbol{\gamma}^T\mathbf{B}\mathbf{A}^{-1}\mathbf{B}\boldsymbol{\gamma}$. We are now in a position to evaluate the efficiency of the spatial median and the angle sign test with respect to the mean vector and Hotelling's test under various model assumptions. The following result is basic and is derived in Exercise 6.8.10.

Theorem 6.3.1 *Suppose the underlying distribution is spherically symmetric so that the joint density is of the form $f(\mathbf{x}) = h(\|\mathbf{x}\|)$. Let (r, ϕ) be the polar coordinates. Then r and ϕ are stochastically independent, the pdf of ϕ is uniform on $(0, 2\pi]$ and the pdf of r is $g(r) = 2\pi r f(r)$, for $r > 0$.*

Theorem 6.3.2 *If the underlying distribution is spherically symmetric, then the matrices $\mathbf{A} = (1/2)\mathbf{I}$ and $\mathbf{B} = [(Er^{-1})/2]\mathbf{I}$. Hence, under the null hypothesis, the test statistic $n^{-1}\mathbf{S}_2^T(\mathbf{0})\mathbf{A}^{-1}\mathbf{S}_2(\mathbf{0})$ is distribution-free over the class of spherically symmetric distributions.*

Proof. First note that

$$E\cos\phi\sin\phi = \frac{1}{2\pi}\int \cos\phi\sin\phi\, d\phi = 0.$$

Then note that

$$Er^{-1}\cos\phi\sin\phi = Er^{-1}E\cos\phi\sin\phi = 0.$$

Finally, note that $E\cos^2\phi = E\sin^2\phi = \frac{1}{2}$.

We can then compute $\mathbf{B}^{-1}\mathbf{A}\mathbf{B}^{-1} = [2/(Er^{-1})^2]\mathbf{I}$ and $\mathbf{B}^{-1}\mathbf{B} = [(Er^{-1})^2/2]\mathbf{I}$. This implies that the generalized variance of the spatial median and the noncentrality parameter of the angle sign test are given by $\det \mathbf{B}^{-1}\mathbf{A}\mathbf{B}^{-1} = 2/(Er^{-1})^2$ and $[(Er^{-1})^2/2]\boldsymbol{\gamma}^T\boldsymbol{\gamma}$. Notice that the efficiencies relative to the mean and Hotelling's test are now equal and independent of the direction. This is because both the spatial L_1 methods and the L_2 methods are equivariant and invariant with respect to orthogonal transformations (rotations and reflections). Hence, we see that the efficiency

$$e(\text{spatial } L_1, L_2) = \tfrac{1}{4}Er^2\{Er^{-1}\}^2. \tag{6.3.10}$$

If, in addition, we assume the underlying distribution is spherical normal (bivariate normal with means 0 and identity covariance matrix) then $Er^{-1} = \sqrt{\pi/2}$, $Er^2 = 2$ and $e(\text{spatial } L_1, L_2) = \pi/4 \approx 0.785$. Hence, the efficiency of the spatial L_1 methods based on $\mathbf{S}_2(\boldsymbol{\theta})$ is greater relative to the L_2

Table 6.3.3: Efficiency as a function of dimension for a k-variate spherical normal model

k	2	4	6
e(spatial L_1, L_2)	0.785	0.884	0.920

methods for the spherical normal model than the componentwise L_1 methods (0.637) discussed in Section 6.2.3.

In Exercise 6.8.12 the reader is asked to show that the efficiency of the spatial L_1 methods relative to the L_2 methods with a k-variate spherical model is given by

$$e_k(\text{spatial } L_1, L_2) = \left(\frac{k-1}{k}\right)^2 E(r^2)[E(r^{-1})]^2.$$

When the k-variate spherical model is normal, the exercise shows that $Er^{-1} = \{\Gamma[(k-1)/2]\}/[\sqrt{2}\Gamma(k/2)]$ with $\Gamma(\frac{1}{2}) = \sqrt{\pi}$. Table 6.3.3 gives some values for this efficiency as a function of dimension. Hence, we see that the efficiency increases with dimension. This suggests that the spatial methods are superior to the componentwise L_1 methods, at least for spherical models. We need to consider what happens to the efficiency when the model is elliptical but not spherical. Since the methods that we are considering are equivariant and invariant to rotations, we can eliminate the correlation from the elliptical model with a rotation but then the variances are typically not equal. Hence, we study, without loss of generality, the efficiency when the underlying model has unequal variances but covariance 0. Now the L_2 methods are affine equivariant and invariant but the spatial L_1 methods are not scale equivariant and invariant (hence not affine equivariant and invariant); hence, the efficiency will be a function of the underlying variances.

The computations are now more difficult. To fix the ideas, suppose the underlying model is bivariate normal with means 0, variances 1 and σ^2, and covariance 0. If we let X and Z denote iid $N(0, 1)$ random variables, then the model distribution is that of X and $Y = \sigma Z$. Note that $W^2 = Z^2/X^2$ has a standard Cauchy distribution. Now we are ready to determine the matrices **A** and **B**.

First, by symmetry, we have $E \cos \phi \sin \phi = E[XY/(X^2 + Y^2)] = 0$ and $Er^{-1} \cos \phi \sin \phi = E[XY/(X^2 + Y^2)^{3/2}] = 0$; hence, the matrices **A** and **B** are diagonal. Next, $\cos^2 \phi = X^2/[X^2 + \sigma^2 Z^2] = 1/[1 + \sigma^2 Z^2]$ so we can use the Cauchy density to compute the expectation. Using the method of partial fractions:

$$E \cos^2 \phi = \int \frac{1}{(1 + \sigma^2 w^2)} \frac{1}{\pi(1 + w^2)} dw = \frac{1}{1 + \sigma}.$$

Hence, $E \sin^2 \phi = \sigma/(1 + \sigma)$. The next two formulas are given by Brown (1983) and are derivable by several steps of partial integration:

$$Er^{-1} = \sqrt{\frac{\pi}{2}} \sum_{j=0}^{\infty} \left\{\frac{(2j)!}{2^{2j}(j!)^2}\right\}^2 (1 - \sigma^2)^j,$$

Table 6.3.4: Efficiencies of spatial L_1 methods relative to the L_2 methods for bivariate normal model with means 0, variances 1 and σ^2, and 0 correlation

σ	1	0.8	0.6	0.4	0.2	0.05	0.01
e(spatial L_1, L_2)	0.785	0.783	0.773	0.747	0.678	0.593	0.321

$$Er^{-1}\cos^2\phi = \frac{1}{2}\sqrt{\frac{\pi}{2}}\sum_{j=0}^{\infty}\left\{\frac{(2j+2)!(2j)!}{2^{4j+1}(j!)^2[(j+1)!]^2}\right\}^2 (1-\sigma^2)^j,$$

and

$$Er^{-1}\sin^2\phi = Er^{-1} - Er^{-1}\cos^2\phi.$$

Thus $\mathbf{A} = \text{diag}[(1+\sigma)^{-1}, \sigma(1+\sigma)^{-1}]$ and the distribution of the test statistic, even under the normal model, depends on σ. The formulas can be used to compute the efficiency of the spatial L_1 methods relative to the L_2 methods; numerical values are given in Table 6.3.4. The dependency of the efficiency on σ reflects the dependency of the efficiency on the underlying correlation which is present prior to rotation.

Hence, just as the componentwise L_1 methods have decreasing efficiency as a function of the underlying correlation, the spatial L_1 methods have decreasing efficiency as a function of the ratio of underlying variances. It should be emphasized that the spatial methods are most appropriate for spherical models where they have equivariance and invariance properties. The componentwise methods, although equivariant and invariant under scale transformations of the components, cannot tolerate changes in correlation. See Mardia (1972) and Fisher (1987; 1993) for further discussion of spatial methods. In higher dimensions, Mardia (1972, Section 9.3.1) refers to the angle test as Rayleigh's test. Möttönen and Oja (1995) extend the spatial median and the spatial sign test to higher dimensions. See Table 6.3.7 for efficiencies relative to Hotelling's test for higher dimensions and for a multivariate t underlying distribution. Note that for higher dimensions and lower degrees of freedom, the spatial sign test is superior to Hotelling's T^2.

The question that we must now consider is whether there are L_1-type methods that are affine equivariant and invariant like the L_2 methods, and, if so, what sort of efficiency properties do they enjoy. We take up this topic in the next section. First, however, we consider spatial rank methods.

6.3.1 Spatial Rank Methods

Ranking Lengths

We sketch some results by Hössjer and Croux (1995) in which they introduce rank estimates and tests for spherical models. They also consider elliptical models. However, we will consider only the methods that are developed for spatial models with spherical distributions. We begin with the Wilcoxon criterion function for the univariate case: $D(\theta) = \sum R(|\mathbf{x}_i - \theta|)|\mathbf{x}_i - \theta|$, where $R(|\mathbf{x}_i - \theta|)$ is the rank of $|\mathbf{x}_i - \theta|$ among all the absolute values; see Section 1.7.

Then we extend this to the bivariate case as follows: $D(\boldsymbol{\theta}) = D(\|\mathbf{x}_1 - \boldsymbol{\theta}\|, \ldots, \|\mathbf{x}_n - \boldsymbol{\theta}\|) = \sum R(\|\mathbf{x}_i - \boldsymbol{\theta}\|)\|\mathbf{x}_i - \boldsymbol{\theta}\|$, where $\|\cdot\|$ is the L_2 norm. Now let $\boldsymbol{\theta}$, as usual, be the value that minimizes $D(\boldsymbol{\theta})$. We then compute the gradient to find the corresponding estimating equations:

$$\mathbf{S}_5(\boldsymbol{\theta}) = \sum_{i=1}^{n} \frac{R(\|\mathbf{x}_i - \boldsymbol{\theta}\|)}{n+1} \|\mathbf{x}_i - \boldsymbol{\theta}\|^{-1}(\mathbf{x}_i - \boldsymbol{\theta}). \qquad (6.3.11)$$

Compare this to expression (6.1.7). We have here a weighted version of the angle sign statistic. See Hössjer and Croux (1995) for a discussion of a general scores version of this statistic. It will be convenient to transform to polar coordinates. When $x_{i1} = r_i \cos \phi_i$ and $x_{i2} = r_i \sin \phi_i$ and when $\boldsymbol{\theta} = \mathbf{0}$, (6.3.11) becomes

$$\mathbf{S}_5(\mathbf{0}) = \sum_{i=1}^{n} \frac{R(r_i)}{n+1} \begin{pmatrix} \cos \phi_i \\ \sin \phi_i \end{pmatrix}. \qquad (6.3.12)$$

Under the assumption of spherical symmetry, the radius r and the angle ϕ are independent. Then the asymptotic normality of $n^{-1/2}\mathbf{S}_5(\mathbf{0})$ under the null hypothesis that $\boldsymbol{\theta} = \mathbf{0}$ follows along the same lines as the univariate Wilcoxon signed-rank statistic described in Section 1.7. In fact, we can write (6.3.11) as $\mathbf{S}_5(\mathbf{0}) = \sum [i/(n+1)]\mathbf{u}_{(i)}$, where $\mathbf{u}_i = \|\mathbf{x}_i\|^{-1}\mathbf{x}_i$ and $\mathbf{u}_{(i)}$ corresponds to the direction with the ith largest norm. Under spherical symmetry $\mathbf{u}_{(i)} i = 1, \ldots, n$ are iid and multivariate central limit theory can be used to establish the asymptotic normality. We must, however, display the asymptotic covariance matrix $\mathbf{A}$ required in Definition 6.1.3.

Since the radius and the angle are independent, the vectors of ranks of the radii and the angles are independent. Hence, conditioning on $r_1, \ldots, r_n$, the conditional covariance matrix of $\mathbf{S}_5(\mathbf{0})$ is $[\sum R^2(r_i)/(n+1)^2](\frac{1}{2})\mathbf{I}$ since $E\cos^2\phi = E\sin^2\phi = \frac{1}{2}$ and $E\cos\phi\sin\phi = 0$. Now $\sum R^2(r_i) = \sum i^2 = n(n+1)(2n+1)/6$ and we see that

$$\text{Cov}\mathbf{S}_5(\mathbf{0}) = [n(2n+1)/12(n+1)]\mathbf{I} \qquad (6.3.13)$$

and, with the matrix $\mathbf{A} = \frac{1}{6}\mathbf{I}$,

$$\frac{1}{\sqrt{n}}\mathbf{S}_5(\mathbf{0}) \xrightarrow{D} \mathbf{Z} \sim MVN(\mathbf{0}, \tfrac{1}{6}\mathbf{I}). \qquad (6.3.14)$$

Hence, the test of $H_0 : \boldsymbol{\theta} = \mathbf{0}$ versus $H_A : \boldsymbol{\theta} \neq \mathbf{0}$ rejects the null hypothesis when $6n^{-1}\mathbf{S}_5^T(\mathbf{0})\mathbf{S}_5(\mathbf{0}) \geq \chi_\alpha(2)$, the upper α chi-square percentile with two degrees of freedom. This test is asymptotically size α.

The asymptotics could also be approached via projections. Hössjer and Croux (1995) show that the empirical cdf $G_n(r)$ of the radii can be replaced by the cdf $G(r)$ as in the univariate case to give

$$\frac{1}{\sqrt{n}}\mathbf{S}_5(\mathbf{0}) = \frac{1}{\sqrt{n}}\sum_{i=1}^{n} G(\|\mathbf{X}_i\|)\|\mathbf{X}_i\|^{-1}\mathbf{X}_i + o_p(1), \qquad (6.3.15)$$

where $G(\cdot)$ is the cdf of $r = \|\mathbf{x}\|$, the radius; see Theorem 6.3.1. We will now use (6.1.2) to compute $\mathbf{B}$ in Definition 6.1.3.

352 Multivariate Models

Theorem 6.3.3 *Under mild regularity conditions and assuming that the underlying distribution of* $\mathbf{X}$ *is spherically symmetric with joint pdf* $f(\|\mathbf{x}\|)$*, the matrix* $\mathbf{B}$ *in (6.1.2) is given by*

$$\mathbf{B} = \int_0^\infty -\tfrac{1}{2} \frac{f'(r)}{f(r)} G(r)g(r)dr \mathbf{I},$$

where $g(r) = 2\pi r f(r) r > 0$ *is the pdf of the radius and* $\mathbf{I}$ *is the identity matrix.*

Proof. We compute the expectation in (6.1.2). From (6.3.15) we have that $\Psi(\mathbf{x}) = G(\|\mathbf{x}\|)\|\mathbf{x}\|^{-1}\mathbf{x}$ and hence

$$-E \nabla \log f(\mathbf{X}) \Psi^T(\mathbf{X}) = -E \frac{f'(\mathbf{X})}{f(\mathbf{X})\|\mathbf{X}\|} \mathbf{X} G(\mathbf{X}) \frac{1}{|\mathbf{X}|} \mathbf{X}^T$$

$$= -\iint \frac{f'(\|\mathbf{x}\|)}{f(\|\mathbf{x}\|)} G(\|\mathbf{x}\|) \frac{1}{\|\mathbf{x}\|^2} \mathbf{x}\mathbf{x}^T f(\|\mathbf{x}\|) dx_1 dx_2$$

$$= -\iint f'(r)G(r) \frac{1}{r^2} \begin{pmatrix} r^2 \cos^2 \phi & r^2 \cos \phi \sin \phi \\ r^2 \cos \phi \sin \phi & r^2 \sin^2 \phi \end{pmatrix} r\, dr\, d\phi$$

$$= -\frac{2\pi}{2} \int r f'(r) G(r) dr \mathbf{I}.$$

The formula follows by dividing and multiplying the integrand by $f(r)$ and replacing $2\pi r f(r)$ by $g(r)$.

The spatial R-estimator is found by setting (6.3.11) equal to $\mathbf{0}$. The computation of this estimate is described in detail by Hössjer and Croux (1995). They provide explicit algorithms for carrying out the computation. The limiting distribution of $\sqrt{n}\,\hat{\boldsymbol{\theta}}$ is bivariate normal with mean $\mathbf{0}$ and covariance $\mathbf{B}^{-1}\mathbf{A}\mathbf{B}^{-1}$, with $\mathbf{B}$ given in Theorem 6.3.3. Recall that the vector of $\sqrt{n}$ times the vector of sample means has an asymptotic bivariate normal distribution with mean $\mathbf{0}$ and covariance matrix $\mathbf{I}$ when the underlying distribution is spherically symmetric with variances 1. Hence the efficiency of the spatial R-estimate of Hössjer and Croux relative to the mean vector is given by the reciprocal ratio of Wilk's generalized variances in Definition 6.1.4.

Exercise 6.8.13 shows that when the underlying distribution is spherical normal the efficiency is 0.985. This is in contrast to the spatial median, which has efficiency 0.785. It is interesting to note that in both cases the bivariate efficiency is greater than the corresponding univariate efficiency. When the formulas are generalized to the k-variate spherical normal case, Hössjer and Croux (1995) calculate the efficiencies given in Table 6.3.5.

Table 6.3.5: Efficiency of the signed-rank test relative to Hotelling's test for multivariate normal models

k	1	2	3	4	5
efficiency	0.955	0.985	0.975	0.961	0.949

The efficiency slowly decreases as k increases. Hössjer and Croux show that the efficiency converges to 0.75 as k tends to infinity. This is in contrast to the spatial median case in Table 6.3.2 in which the efficiency increases with k. However, the efficiency is still quite high for the spatial R-estimate for moderate values of k. The testing efficiency will provide a similar picture, but recall that it depends on the direction that the alternative approaches the null hypothesis. We will not pursue the testing efficiency here.

Rank Vectors

Möttönen and Oja (1995) develop the concept of a **rotation invariant rank vector**. Hence, rather than use the univariate concept of rank in the construction of a test, they define a spatial rank vector that has both magnitude and direction. This problem is delicate since there is no inherently natural way to order or rank vectors.

We must first review the relationship between sign, rank, and signed rank. Recall the norm, (1.3.17) and (1.3.21), that was used to generate the Wilcoxon signed-rank statistic. Further, recall that the second term in the norm was the basis, in Section 2.2.2, for the Mann–Whitney–Wilcoxon rank-sum statistic. We reverse this approach here and show how the one-sample signed-rank statistic based on ranks of the absolute values can be developed from the ranks of the data. This will provide the motivation for a one-sample spatial rank statistic.

Let $x_1, \ldots, x_n$ be a univariate sample. Then $2[R_n(x_i) - (n+1)/2] = \sum_j \mathrm{sgn}(x_i - x_j)$. Thus the centered rank is constructed from the signs of the differences. Now to construct a one-sample statistic, we introduce the reflections $-x_1, \ldots, -x_n$ and consider the centered rank of x_i among the combined $2n$ observations and their reflections.

$$2[R_{2n}(x_i) - (2n+1)/2] = \sum_j \mathrm{sgn}(x_i - x_j) + \sum_j \mathrm{sgn}(x_i + x_j) \tag{6.3.16}$$
$$= [2R_n(|x_i|) - 1]\mathrm{sgn}(x_i),$$

see Exercise 6.8.14. The subscript $2n$ indicates that the reflections are included in the ranking. Hence, ranking observations in the combined observations and reflections is essentially equivalent to ranking the absolute values $|x_1|, \ldots, |x_n|$. In this way, one sample methods can be developed from two-sample methods.

Möttönen and Oja (1995) use this approach to develop a **one-sample spatial signed-rank statistic**. The key is the expression $\mathrm{sgn}(x_i - x_j) + \mathrm{sgn}(x_i + x_j)$ which requires only the concept of sign, not rank. Hence, we must find the appropriate extension of sign to two dimensions. In one dimension, $\mathrm{sgn}(x)$ can be thought of as a unit vector pointing in the positive or negative directions toward x.

Likewise $\mathbf{u}(\mathbf{x}) = \|\mathbf{x}\|^{-1}\mathbf{x}$ is a unit vector in the direction of $\mathbf{x}$. Hence, we will take $\mathbf{u}(\mathbf{x})$ to be the **vector spatial sign**. The **vector centered spatial rank** of $\mathbf{x}_i$ is then $\mathbf{R}(\mathbf{x}_i) = \sum_j \mathbf{u}(\mathbf{x}_i - \mathbf{x}_j)$. Thus, the **vector spatial signed-rank statistic** is

$$\mathbf{S}_6(\mathbf{0}) = \sum_i \sum_j \{\mathbf{u}(\mathbf{x}_i - \mathbf{x}_j) + \mathbf{u}(\mathbf{x}_i + \mathbf{x}_j)\}. \tag{6.3.17}$$

This is also the sum of the centered spatial ranks of the observations when ranked in the combined observations and their reflections. Note that $-\mathbf{u}(\mathbf{x}_i - \mathbf{x}_j) = \mathbf{u}(\mathbf{x}_j - \mathbf{x}_i)$ so that $\sum\sum \mathbf{u}(\mathbf{x}_i - \mathbf{x}_j) = \mathbf{0}$ and the statistic can be computed from

$$\mathbf{S}_6(\mathbf{0}) = \sum_i \sum_j \mathbf{u}(\mathbf{x}_i + \mathbf{x}_j), \qquad (6.3.18)$$

which is the direct analog of (1.3.24).

We now develop a conditional test by conditioning on the data $\mathbf{x}_1, \ldots, \mathbf{x}_n$. From (6.3.17) we can write

$$\mathbf{S}_6(\mathbf{0}) = \sum_i \mathbf{r}^+(\mathbf{x}_i), \qquad (6.3.19)$$

where $\mathbf{r}^+(\mathbf{x}) = \sum_j \{\mathbf{u}(\mathbf{x} - \mathbf{x}_j) + \mathbf{u}(\mathbf{x} + \mathbf{x}_j)\}$. Now it is easy to see that $\mathbf{r}^+(-\mathbf{x}) = -\mathbf{r}^+(\mathbf{x})$. Under the null hypothesis of symmetry about $\mathbf{0}$, we can think of $\mathbf{S}_6(\mathbf{0})$ as a realization of $\sum_i b_i \mathbf{r}^+(\mathbf{x}_i)$, where $b_1, \ldots, b_n$ are iid variables with $P(b_i = +1) = P(b_i = -1) = \frac{1}{2}$. Hence, $Eb_i = 0$ and $\text{Var}(b_i) = 1$. This means that, conditional on the data,

$$E\mathbf{S}_6(\mathbf{0}) = \mathbf{0} \text{ and } \widehat{\mathbf{A}} = \text{Cov}\mathbf{S}_6(\mathbf{0}) = \sum_i (\mathbf{r}^+(\mathbf{x}_i))(\mathbf{r}^+(\mathbf{x}_i))^T. \qquad (6.3.20)$$

The approximate size α conditional test of $H_0 : \boldsymbol{\theta} = \mathbf{0}$ versus $H_A : \boldsymbol{\theta} \neq \mathbf{0}$ rejects H_0 when

$$\mathbf{S}_6^T \widehat{\mathbf{A}}^{-1} \mathbf{S}_6 \geq \chi_\alpha^2(2), \qquad (6.3.21)$$

where $\chi_\alpha^2(2)$ is the upper α percentile from a chi-square distribution with two degrees of freedom. Note that the extension to higher dimensions is done in exactly the same way.

Example 6.3.2 *Cork Borings, Example 6.3.1 continued*

We use the spatial signed-rank method (6.3.21) to test the hypothesis. Table 6.3.6 provides the vector signed-ranks, $\mathbf{r}^+(\mathbf{x}_i)$ defined in expression (6.3.19).

Then $\mathbf{S}_6^T(\mathbf{0}) = (4.94, -2.90, 5.17)$,

$$\widehat{\mathbf{A}}^{-1} = \begin{bmatrix} 0.1231 & -0.0655 & 0.0050 \\ -0.0655 & 0.1611 & -0.0373 \\ 0.0050 & -0.0373 & 0.1338 \end{bmatrix},$$

and $\mathbf{S}_6^T(\mathbf{0})\widehat{\mathbf{A}}^{-1}\mathbf{S}_6(\mathbf{0}) = 14.01$, with an approximate p-value of 0.003 based on a χ^2 distribution with 3 degrees of freedom. The Hodges–Lehmann estimate of $\boldsymbol{\theta}$, which solves $\mathbf{S}_6(\boldsymbol{\theta}) \doteq \mathbf{0}$, is computed to be $\widehat{\boldsymbol{\theta}}^T = (49.30, 45.07, 48.90, 44.59)$.

The test in (6.3.21) can be developed from the point of view of asymptotic theory and the efficiency can be computed. The computations are quite involved and are given in Möttönen and Oja (1995). In particular, they compute the efficiency of $\mathbf{S}_6(\mathbf{0})$ relative to Hotelling's T^2 for an underlying multivariate t-distribution. The multivariate t provides both a range of

Table 6.3.6: Each row is a spatial signed-rank vector for the data differences in Table 6.3.2

Row	SR1	SR2	SR3	Row	SR1	SR2	SR3
1	0.28	−0.49	−0.07	15	0.30	−0.54	0.69
2	0.28	−0.58	0.12	16	−0.40	0.73	−0.07
3	−0.09	−0.39	0.31	17	0.60	−0.14	0.39
4	0.58	−0.29	−0.11	18	0.10	0.56	0.49
5	−0.03	−0.20	−0.07	19	0.77	−0.34	0.22
6	−0.28	0.07	0.43	20	0.48	0.10	−0.03
7	0.07	0.43	0.23	21	0.26	0.08	−0.16
8	0.01	0.60	0.32	22	0.12	0.00	−0.11
9	−0.13	0.46	0.34	23	0.32	−0.58	0.48
10	0.23	0.13	−0.49	24	−0.14	−0.53	0.42
11	0.12	−0.20	0.33	25	0.19	−0.12	0.45
12	0.46	−0.76	0.28	26	0.73	−0.07	−0.14
13	0.30	−0.56	0.34	27	0.31	−0.12	−0.58
14	−0.22	−0.05	0.49	28	−0.30	−0.14	0.67

tailweights and a range of dimensions. A summary of these efficiencies is found in Table 6.3.7; see Möttönen, Oja and Tienari (1997).

Note first that, unlike the Hössjer and Croux (1995) test (Table 6.3.5), the Möttönen and Oja (1995) test efficiency increases with the dimension; see, especially, the circular normal case. The efficiency begins at 0.95 and increases! The efficiency also increases with tailweight, as expected. This strongly suggests that the Möttönen and Oja approach is an excellent way to extend the idea of signed rank from the univariate case. See Example 6.6.2 for a discussion of the two-sample spatial rank test.

The estimator derived from $S_6(\theta) \doteq 0$ is the spatial median of the pairwise averages, a spatial Hodges–Lehmann estimator (Hodges and Lehmann, 1963). This estimator is studied in great detail by Chaudhuri (1992). His paper contains a thorough review of multidimensional location estimates. He

Table 6.3.7: The row labeled spatial SR gives the asymptotic efficiencies of the multivariate spatial signed-rank test, (6.3.21), relative to Hotelling's test under the multivariate t distribution; the efficiencies for the spatial sign test, (6.3.2), are given in the rows labeled Spatial sign

Dimension	Test	Degress of Freedom							
		3	4	6	8	10	15	20	∞
1	Spatial SR	1.90	1.40	1.16	1.09	1.05	1.01	1.00	0.95
	Spatial sign	1.62	1.13	0.88	0.80	0.76	0.71	0.60	0.64
2	Spatial SR	1.95	1.43	1.19	1.11	1.07	1.03	1.01	0.97
	Spatial sign	2.00	1.39	1.08	0.98	0.93	0.88	0.85	0.79
3	Spatial SR	1.98	1.45	1.20	1.12	1.08	1.04	1.02	0.97
	Spatial sign	2.16	1.50	1.17	1.06	1.01	0.95	0.92	0.85
4	Spatial SR	2.00	1.46	1.21	1.13	1.09	1.04	1.025	0.98
	Spatial sign	2.25	1.56	1.22	1.11	1.05	0.99	0.96	0.88
6	Spatial SR	2.02	1.48	1.22	1.14	1.10	1.05	1.03	0.98
	Spatial sign	2.34	1.63	1.27	1.15	1.09	1.03	1.00	0.92
10	Spatial SR	2.05	1.49	1.23	1.14	1.10	1.06	1.04	0.99
	Spatial sign	2.42	1.68	1.31	1.19	1.13	1.06	1.03	0.95

develops a Bahadur representation for the estimate. From his Theorem 3.2, we can immediately conclude that

$$\sqrt{n}\widehat{\boldsymbol{\theta}} = \mathbf{B}_2^{-1} \frac{\sqrt{n}}{n(n-1)} \sum_{i=1}^{n}\sum_{j=1}^{n} \mathbf{u}\left(\frac{1}{2}(\mathbf{x}_i + \mathbf{x}_j)\right) + \mathbf{o}_p(1), \qquad (6.3.22)$$

where $\mathbf{B}_2 = E\{\|\mathbf{x}^*\|^{-1}(\mathbf{I} - \|\mathbf{x}^*\|^{-2}\mathbf{x}^*(\mathbf{x}^*)^T)\}$ and $\mathbf{x}^* = \frac{1}{2}(\mathbf{x}_1 + \mathbf{x}_2)$. Hence, the asymptotic distribution of $\sqrt{n}\widehat{\boldsymbol{\theta}}$ is determined by that of $n^{-3/2}\mathbf{S}_6(\mathbf{0})$. This leads to

$$\sqrt{n}\widehat{\boldsymbol{\theta}} \xrightarrow{\mathcal{D}} N_2(\mathbf{0}, \mathbf{B}_2^{-1}\mathbf{A}_2\mathbf{B}_2^{-1}), \qquad (6.3.23)$$

where $\mathbf{A}_2 = E\{\mathbf{u}(\mathbf{x}_1 + \mathbf{x}_2)(\mathbf{u}(\mathbf{x}_1 + \mathbf{x}_2))^T\}$. Moment estimates of $\mathbf{A}_2$ and $\mathbf{B}_2$ can be used. In fact the estimator $\widehat{\mathbf{A}}$, defined in expression (6.3.20), is a consistent estimate of $\mathbf{A}_2$. Bose and Chaudhuri (1993) and Chaudhuri (1992) discuss refinements in the estimation of $\mathbf{A}_2$ and $\mathbf{B}_2$.

Choi and Marden (1997) extend these spatial rank methods to the two-sample model and the one-way layout. They also consider tests for ordered alternatives.

6.4 Affine Equivariant and Invariant L_1 Methods

We begin with yet another representation of the estimating function $\mathbf{S}_2(\boldsymbol{\theta})$, (6.1.7). Let the ordered φ angles be given by $0 \leq \varphi_{(1)} < \varphi_{(2)} < \ldots < \varphi_{(n)} < \pi$ and let $s_{(i)} = \pm 1$ when the observation corresponding to $\varphi_{(i)}$ is above or below the horizontal axis. Then we can write, as in expression (6.3.5),

$$\mathbf{S}_2(\boldsymbol{\theta}) = \sum_{i=1}^{n} s_{(i)} \begin{pmatrix} \cos \varphi_{(i)} \\ \sin \varphi_{(i)} \end{pmatrix}. \qquad (6.4.1)$$

Now under the assumption of spherical symmetry, $\varphi_{(i)}$ is distributed as the ith order statistic from the uniform distribution on $[0, \pi)$ and, hence, $E\varphi_{(i)} = \pi i/(n+1)$, $i = 1, \ldots, n$. Recall, in the univariate case, if we believe that the underlying distribution is normal then we could replace the data by the normal scores (expected values of the order statistics from a normal distribution) in a signed-rank statistic. The result is the distribution-free normal scores test. We will do the same thing here. We replace $\varphi_{(i)}$ by its expected value to construct a scores statistic. Let

$$\mathbf{S}_7(\boldsymbol{\theta}) = \sum_{i=1}^{n} s_{(i)} \begin{pmatrix} \cos \frac{\pi i}{n+1} \\ \sin \frac{\pi i}{n+1} \end{pmatrix} = \sum_{i=1}^{n} s_i \begin{pmatrix} \cos \frac{\pi R_i}{n+1} \\ \sin \frac{\pi R_i}{n+1} \end{pmatrix}, \qquad (6.4.2)$$

where $R_1, \ldots, R_n$ are the ranks of the unordered angles $\varphi_1, \ldots, \varphi_n$. Note that $s_1, \ldots, s_n$ are iid with $P(s_i = 1) = P(s_i = -1) = \frac{1}{2}$ even if the underlying model is elliptical rather than spherical. Since we now have constant vectors in $\mathbf{S}_7(\boldsymbol{\theta})$, it follows that the sign test based on $\mathbf{S}_7(\boldsymbol{\theta})$ is distribution-free over the class of elliptical models. In fact, the test statistic is affine invariant just as Hotellings's statistic is affine invariant. We look at the test in more detail and consider the efficiency of this **affine invariant sign test** relative to Hotelling's test. First, we have immediately, under the null hypothesis, from the distribution of $s_1, \ldots, s_n$, that

$$\text{Cov}\left[\frac{1}{\sqrt{n}}\mathbf{S}_7(\mathbf{0})\right]$$

$$= \begin{pmatrix} \dfrac{\sum \cos^2[\pi i/(n+1)]}{n} & \dfrac{\sum \cos[\pi i/(n+1)]\sin[\pi i/(n+1)]}{n} \\ \dfrac{\sum \cos[\pi i/(n+1)]\sin[\pi i/(n+1)]}{n} & \dfrac{\sum \sin^2[\pi i/(n+1)]}{n} \end{pmatrix} \to \mathbf{A},$$

where

$$\mathbf{A} = \begin{pmatrix} \int_0^1 \cos^2 \pi t\, dt & \int_0^1 \cos \pi t \sin \pi t\, dt \\ \int_0^1 \cos \pi t \sin \pi t\, dt & \int_0^1 \sin^2 \pi t\, dt \end{pmatrix} = \tfrac{1}{2}\mathbf{I},$$

as $n \to \infty$. So reject $H_0 : \boldsymbol{\theta} = \mathbf{0}$ if $(2/n)\mathbf{S}_7^\mathrm{T}(\mathbf{0})\mathbf{S}_7(\mathbf{0}) \geq \chi_\alpha^2(2)$ for the asymptotic size α distribution-free version of the test, where

$$\frac{2}{n}\mathbf{S}_7^\mathrm{T}(\mathbf{0})\mathbf{S}_7(\mathbf{0}) = \frac{2}{n}\left\{\left(\sum s_{(i)} \cos \frac{\pi i}{n+1}\right)^2 + \left(\sum s_{(i)} \sin \frac{\pi i}{n+1}\right)^2\right\} \quad (6.4.3)$$

is the distribution-free **Blumen bivariate sign test** proposed by Blumen (1958). We can think of Blumen's statistic as an elliptical scores version of the angle statistic of Brown (1983).

To compute the efficiency of this test relative to Hotelling's test we must compute the noncentrality parameter of the limiting chi-square distribution. Hence, we must compute $\mathbf{BA}^{-1}\mathbf{B}$ and this leads us to $\mathbf{B}$. Theorem 6.3.2 provides the matrices $\mathbf{A}$ and $\mathbf{B}$ for the angle sign statistic when the underlying distribution is spherically symmetric. The following theorem shows that the affine invariant sign statistic has the same $\mathbf{A}$ and $\mathbf{B}$ matrices as in Theorem 6.3.2 and they hold for all elliptical distributions. We discuss the implications after the proof of the proposition.

Theorem 6.4.1 *If the underlying distribution is elliptical, then corresponding to $\mathbf{S}_7(\mathbf{0})$ we have $\mathbf{A} = \tfrac{1}{2}\mathbf{I}$ and $\mathbf{B} = (Er^{-1}/2)\mathbf{I}$. Hence, the efficiency of Blumen's test relative to Hotelling's test is $e(\mathbf{S}_7, Hotelling) = (Er^{-1})^2/2$ which is the same for all elliptical models.*

Proof. To prove this we show that under a spherical model the angle statistic $\mathbf{S}_2(\mathbf{0})$ and scores statistic $\mathbf{S}_7(\mathbf{0})$ are asymptotically equivalent. Then $\mathbf{S}_7(\mathbf{0})$ will have the same $\mathbf{A}$ and $\mathbf{B}$ matrices as in Theorem 6.3.2. But since $\mathbf{S}_7(\mathbf{0})$ leads to an affine invariant test statistic, it follows that the same $\mathbf{A}$ and $\mathbf{B}$ continue to apply for elliptical models.

Recall that under the spherical model, $s_{(1)}, \ldots, s_{(n)}$ are iid with $P(s_i = 1) = P(s_i = -1) = \frac{1}{2}$ random variables. Then we consider

$$\frac{1}{\sqrt{n}} \sum_{i=1}^{n} s_{(i)} \begin{pmatrix} \cos \dfrac{\pi i}{n+1} \\ \sin \dfrac{\pi i}{n+1} \end{pmatrix} - \frac{1}{\sqrt{n}} \sum_{i=1}^{n} s_{(i)} \begin{pmatrix} \cos \varphi_i \\ \sin \varphi_i \end{pmatrix}$$

$$= \frac{1}{\sqrt{n}} \sum_{i=1}^{n} s_{(i)} \begin{pmatrix} \cos \dfrac{\pi i}{n+1} - \cos \varphi_{(i)} \\ \sin \dfrac{\pi i}{n+1} - \sin \varphi_{(i)} \end{pmatrix}$$

We treat the two components separately. First

$$\left| \frac{1}{\sqrt{n}} \sum s_{(i)} \left(\cos\left(\frac{\pi i}{n+1}\right) - \cos \varphi_{(i)} \right) \right| \leq \max_i \left| \cos\left(\frac{\pi i}{n+1}\right) - \cos \varphi_{(i)} \right| \left| \frac{1}{\sqrt{n}} \sum s_{(i)} \right|.$$

The cdf of the uniform distribution on $[0, \pi)$ is equal to t/π for $0 \leq t < \pi$. Let $G_n(t)$ be the empirical cdf of the angles φ_i, $i = 1, \ldots, n$. Then $G_n^{-1}(i/(n+1)) = \varphi_{(i)}$ and $\max_i |(\pi i/(n+1) - \varphi_{(i)}| \leq \sup_t |G_n^{-1}(t) - t\pi| = \sup_t |G_n(t) - t\pi| \to 0$ with probability one, by the Glivenko–Cantelli lemma. The result now follows by using a linear approximation to $\cos(\pi i/(n+1)) - \cos \varphi_{(i)}$ and noting that the cos and sin are bounded. The same argument applies to the second component. Hence, the difference of the two statistics is $o_p(1)$ and they are asymptotically equivalent. The results for the angle statistic now apply to $\mathbf{S}_7(\mathbf{0})$ for a spherical model. The affine invariance extends the result to an elliptical model.

The main implication of this proposition is that the efficiency of the test based on $\mathbf{S}_7(\mathbf{0})$ relative to Hotelling's test is $\pi/4 \approx 0.785$ for all bivariate normal models, not just the spherical normal model. Recall that the test based on $\mathbf{S}_2(\mathbf{0})$, the angle sign test, has efficiency $\pi/4$ only for the spherical normal and declining efficiency for elliptical normal models. Hence, we not only gain affine invariance but also have a constant, nondecreasing efficiency. We next consider how to extend the test based on $\mathbf{S}_7(\mathbf{0})$ to more than *two* dimensions. The problem is that in higher dimensions we can no longer order the φ-angles. Thus we must next develop a new representation of the statistic due to Randles (1989).

From (6.4.2) we have

$$\frac{2}{n} \mathbf{S}_7'(\mathbf{0}) \mathbf{S}_7(\mathbf{0}) = \frac{2}{n} \left\{ \left(\sum_{i=1}^{n} s_i \cos \frac{\pi R_i}{n+1} \right)^2 + \left(\sum_{i=1}^{n} s_i \sin \frac{\pi R_i}{n+1} \right)^2 \right\}.$$

Now recall from trigonometry that $\cos(a - b) = \cos(a)\cos(b) + \sin(a)\sin(b)$ and apply this identity to the above formula. The test statistic is equivalent to $(2/n) \sum_{i=1}^{n} \sum_{j=1}^{n} s_i s_j \cos[\pi(R_i - R_j)]/(n+1)$, where R_i is the rank of φ_i. We will call a line through $\mathbf{x}_i$ and the origin, extended in both directions, a **data axis**. Given points $\mathbf{x}_i$ and $\mathbf{x}_j$, we will be interested in how many data axes fall between the two points or how many times the two points fall on opposite sides of data axes. This is a measure of how far apart the two points are and is similar to a rank in the univariate case.

Theorem 6.4.2 *Let p_{ij} equal 1 plus the number of data axes between $\mathbf{x}_i$ and $\mathbf{x}_j$ divided by n. Then*

$$\frac{2}{n}\sum_{i=1}^{n}\sum_{j=1}^{n} s_i s_j \cos\frac{\pi(R_i - R_j)}{n+1} \doteq \frac{2}{n}\sum_{i=1}^{n}\sum_{j=1}^{n} \cos\frac{\pi p_{ij}}{n}.$$

Proof. The approximation results because we replaced $n+1$ by n. When s_i and s_j are of the same sign, the two points are on the same side of the horizontal axis and $R_i - R_j$ is the number of data axes that fall between the two points plus 1. We also use $\cos(t) = \cos(-t)$. If the points are on opposite sides of the horizontal axis then $n + 1 - (R_i - R_j)$ is the number we are looking for. The formula follows since $-\cos(t) = \cos(\pi - t)$.

This theorem provides the way to extend the affine invariant test statistic to more than *two* dimensions. Define **Randles' test statistic** for k dimensions:

$$\frac{k}{n}\sum_{i=1}^{n}\sum_{j=1}^{n}\cos(\pi p_{ij}), \qquad (6.4.4)$$

where

$$p_{ij} = (C_{ij} + d_n) \Big/ \binom{n}{k-1} \text{ if } i \neq j \text{ and } 0 \text{ if } i = j \text{ and } d_n = \left(\binom{n}{k-1} - \binom{n-2}{k-1}\right)\Big/2.$$

Here, C_{ij} is the number of hyperplanes formed by the origin and $k-1$ other points (not $\mathbf{x}_i$ or $\mathbf{x}_j$) such that $\mathbf{x}_i$ and $\mathbf{x}_j$ are on opposite sides of the hyperplanes formed. The counts C_{ij} are called **interdirections** by Randles (1989) who introduced and studied this test. The test of $H_0: \boldsymbol{\theta} = \mathbf{0}$ rejects when the statistic in (6.4.4) exceeds a chi-square critical value with k degrees of freedom. The efficiency of this test relative to Hotelling's test for the class of k-variate normal models is given by $[(k-1)/k]^2 E(r^2) E^2(r^{-1})$; see Table 6.3.4. This table was computed for the k-variate angle sign test for spherical normal models but since **Randles' sign test** is affine invariant along with Hotelling's test we have the same table for elliptical normal models. Note, especially that the efficiency increases with dimension. Randles (1989) also studied the efficiency of the test for a power family of elliptical distributions that includes the elliptical normal model, light- and heavy-tailed models. For heavy-tailed models the sign test has efficiency greater than one relative to Hotelling's test; this efficiency decreases as the dimension increases, but generally stays above one. On the other hand, for light-tailed models, the efficiency of the sign test is generally below one but increases with increasing dimension. Randles (1989) also includes a simulation study which confirms the results of the efficiency analysis.

Oja and Nyblom (1989) study a class of sign tests for the bivariate location problem. They show that Blumen's test is locally most powerful invariant for the entire class of elliptical models. Ducharme and Milasevic (1987) define a normalized spatial median as an estimate of location of a spherical distribution. They construct a confidence region for the modal direction. These methods are resistant to outliers.

Estimation methods corresponding to Randles' and Blumen's tests have not yet been developed. Before turning to the question of how to introduce ranks in a multivariate context and maintain invariance and equivariance properties, we introduce one more approach to the extension of L_1 methods to the multivariate model.

6.4.1 The Oja Criterion Function

We will only sketch the results in this section and give references where the more detailed derivations can be found. Recall from the univariate location model that L_1 and L_2 are special cases of methods that are derived from minimizing $\sum |x_i - \theta|^m$, for $m = 1$ and $m = 2$. Oja (1983) proposed the bivariate objective function: $D_8(\theta) = \sum_{i<j} A(\mathbf{x}_i, \mathbf{x}_j, \theta)$, where $A(\mathbf{x}_i, \mathbf{x}_j, \theta)$ is the area of the triangle formed by the three vectors $\mathbf{x}_i, \mathbf{x}_j, \theta$. When $m = 2$, Wilks (1960) showed that $D_8(\theta)$ is proportional to the determinant of the classical scatter matrix and the sample mean vector minimizes this criterion. Thus, by analogy with the univariate case, the $m = 1$ case will be called the L_1 case. The same results carry over to dimensions greater than 2 in which the triangles are replaced by simplices.

We introduce the following notation:

$$\mathbf{A}_{ij} = \frac{1}{2}\begin{pmatrix} 1 & 1 & 1 \\ \theta_1 & x_{i1} & x_{j1} \\ \theta_2 & x_{i2} & x_{j2} \end{pmatrix}.$$

Then $D_8(\theta) = \frac{1}{2}\sum_{i<j} \text{abs}\{\det \mathbf{A}_{ij}\}$, where det stands for determinant and abs stands for absolute value. Now if we differentiate this criterion function with respect to θ_1 and θ_2 we obtain a new set of estimating equations:

$$\mathbf{S}_8(\theta) = \frac{1}{2}\sum_{i=1}^{n-1}\sum_{j=i+1}^{n} \text{sgn}\{\det \mathbf{A}_{ij}\}(\mathbf{x}_j^* - \mathbf{x}_i^*) = \mathbf{0}, \qquad (6.4.5)$$

where $\mathbf{x}_i^*$ is the vector $\mathbf{x}_i$ rotated counterclockwise by $\pi/2$ radians, $\mathbf{x}_i^* = (-x_{i2}, x_{i1})^T$. Note that θ enters only through the $\mathbf{A}_{ij}$. The expression in (6.4.5) is found as follows:

$$\text{sgn}\{(\mathbf{x}_j^* - \mathbf{x}_i^*)^T(\theta - \mathbf{x}_i)\}(\mathbf{x}_j^* - \mathbf{x}_i^*)$$

$$= \begin{cases} \mathbf{x}_j^* - \mathbf{x}_i^* & \text{if } \mathbf{x}_i^* \to \mathbf{x}_j^* \to \theta \text{ is counterclockwise} \\ -(\mathbf{x}_j^* - \mathbf{x}_i^*) & \text{if } \mathbf{x}_i^* \to \mathbf{x}_j^* \to \theta \text{ is clockwise.} \end{cases}$$

The estimator that solves (6.4.5) is called the **Oja median** and we will be interested in its properties. This estimator minimizes the sum of triangular areas formed by all pairs of observations along with θ. Niinimaa, Oja, and Nyblom (1992) provide a fortran program for computing the Oja median and discuss further aspects of its computation. Brown and Hettmansperger (1987a) present a geometric description of the determination of the Oja median. The statistic $S_8(0)$ forms the basis of a sign-type statistic for testing $H_0 : \boldsymbol{\theta} = \mathbf{0}$. We will refer to this test as the **Oja sign test**. In order to study the Oja median and the Oja sign test we need once again to determine the matrices $\mathbf{A}$ and $\mathbf{B}$. Before doing this we will rewrite (6.4.5) in a more convenient form, a form that expresses it as a function of $s_1, \ldots, s_n$. Recall the polar form of $\mathbf{x}$, (6.3.4), that we have been using and at the same time introduce the vector $\mathbf{y}$ as follows:

$$\mathbf{x} = r \begin{pmatrix} \cos \phi \\ \sin \phi \end{pmatrix} = rs \begin{pmatrix} \cos \varphi \\ \sin \varphi \end{pmatrix} = s\mathbf{y}.$$

As usual, $0 \leq \varphi < \pi$, s indicates whether $\mathbf{x}$ is above or below the horizontal axis, and r is the length of $\mathbf{x}$. Hence, if $s = 1$ then $\mathbf{y} = \mathbf{x}$, and if $s = -1$ then $\mathbf{y} = -\mathbf{x}$, so $\mathbf{y}$ is always above the horizontal axis.

Theorem 6.4.3 *The following string of equalities is true:*

$$\frac{1}{n}\mathbf{S}_8(0) = \frac{1}{2n}\sum_{i=1}^{n-1}\sum_{j=i+1}^{n} \operatorname{sgn}\left\{\det\begin{pmatrix} x_{i1} & x_{j1} \\ x_{i2} & x_{j2} \end{pmatrix}\right\}(\mathbf{x}_j^* - \mathbf{x}_i^*)$$

$$= \frac{1}{2n}\sum_{i=1}^{n-1}\sum_{j=i+1}^{n} s_i s_j (s_j \mathbf{y}_j^* - s_i \mathbf{y}_i^*)$$

$$= \frac{1}{2}\sum_{i=1}^{n} s_i \mathbf{z}_i,$$

where

$$\mathbf{z}_i = \frac{1}{n}\sum_{j=1}^{n-1} \mathbf{y}_{i+j}^* \text{ and } \mathbf{y}_{n+i} = -\mathbf{y}_i.$$

Proof. The first formula follows at once from (6.4.5). In the second formula we need to recall the $*$ operation. It entails a counterclockwise rotation of 90 degrees. Suppose, without loss of generality, that $0 \leq \varphi_1 \leq \ldots \leq \varphi_n \leq \pi$. Then

$$\operatorname{sgn}\left\{\det\begin{pmatrix} x_{i1} & x_{j1} \\ x_{i2} & x_{j2} \end{pmatrix}\right\} = \operatorname{sgn}\left\{\det\begin{pmatrix} s_i r_i \cos\varphi_i & s_j r_j \cos\varphi_j \\ s_i r_i \sin\varphi_i & s_j r_j \sin\varphi_j \end{pmatrix}\right\}$$

$$= \operatorname{sgn}\{s_i s_j r_i r_j \cos\varphi_i \sin\varphi_j - \sin\varphi_i \cos\varphi_j\}$$

$$= s_i s_j \operatorname{sgn}\{\sin(\varphi_j - \varphi_i)\}$$

$$= s_i s_j.$$

Now if $\mathbf{x}_i$ is in the first or second quadrant then $\mathbf{y}_i^* = \mathbf{x}_i^* = s_i\mathbf{x}_i^*$ and if $\mathbf{x}_i$ is in the third or fourth quadrant then $\mathbf{y}_i^* = -\mathbf{x}_i^* = s_i\mathbf{x}_i^*$. Hence, in all cases we have $\mathbf{x}_i^* = s_i\mathbf{y}_i^*$. The second formula now follows.

The third formula follows by straightforward algebraic manipulations. We leave those details to the reader (Exercise 6.8.15), and instead point out the following helpful facts:

$$\mathbf{z}_i = \sum_{j=i+1}^{n} \mathbf{y}_j^* - \sum_{j=1}^{i-1} \mathbf{y}_j^*, i = 2, \ldots, n-1, \mathbf{z}_1 = \sum_{j=2}^{n} \mathbf{y}_j^*, \mathbf{z}_n = -\sum_{j=1}^{n-1} \mathbf{y}_j^*. \quad (6.4.6)$$

The third formula shows that we have a sign statistic similar to the ones that we have been studying. Under the null hypothesis $(s_1, \ldots, s_n)$ and $(\mathbf{z}_1, \ldots, \mathbf{z}_n)$ are independent. Hence conditionally on $\mathbf{z}_1, \ldots, \mathbf{z}_n$ (or equivalently conditionally on $\mathbf{y}_1, \ldots, \mathbf{y}_n$) the conditional covariance matrix of $\mathbf{S}_8(\mathbf{0})$ is $\widehat{\mathbf{A}} = \frac{1}{4}\sum_i \mathbf{z}_i \mathbf{z}_i^T$. A conditional distribution-free test is

$$\text{Reject } H_0 : \boldsymbol{\theta} = \mathbf{0} \text{ when } \mathbf{S}_8^T(\mathbf{0})\widehat{\mathbf{A}}^{-1}\mathbf{S}_8(\mathbf{0}) \geq \chi_\alpha^2(2). \quad (6.4.7)$$

Theorem 6.4.3 shows that, conditionally on the data, the χ^2 approximation is appropriate. The next theorem shows that the approximation is appropriate unconditionally as well. For additional discussion of this test, see Brown and Hettmansperger (1989). We want to describe the asymptotically distribution-free version of the **Oja sign test**. Then we will show that, for elliptical models, the Oja sign test and Blumen's test are equivalent. It is left to the exercises to show that the Oja median is affine equivariant and the Oja sign test is affine invariant so they compete with Blumen's invariant test and with the L_2 methods (vector of means and Hotelling's test); see Exercise 6.8.16.

Since the Oja sign test is affine invariant, we will consider the behavior under spherical models, without loss of generality. The elliptical models can be reduced to spherical models by affine transformations. The next proposition shows that $\mathbf{z}_i$ has a useful limiting value.

Theorem 6.4.4 *Suppose that we sample from a spherical distribution centered at the origin. Let*

$$\mathbf{z}(t) = -\frac{2}{\pi}E(r)\begin{pmatrix}\cos t\pi \\ \sin t\pi\end{pmatrix};$$

then

$$\frac{1}{n^{3/2}}\mathbf{S}_8(\mathbf{0}) = \frac{1}{2\sqrt{n}}\sum_{i=1}^{n} s_i \mathbf{z}\left(\frac{i}{n}\right) + \mathbf{o}_p(1).$$

Proof. We sketch the argument. A more general result and a rigorous argument can be found in Brown *et al.* (1992). We begin by referring to formula (6.4.6). Recall that

$$\frac{1}{n}\sum_{i=1}^{n}\mathbf{y}_i^* = \frac{1}{n}\sum_{i=1}^{n}r_i\begin{pmatrix}-\sin\varphi_i \\ \cos\varphi_i\end{pmatrix}.$$

Consider the second component and let $\cong$ mean that the approximation is valid up to $o_p(1)$ terms. From the discussion of $\max_i |\pi i/(n+1) - \varphi_{(i)}|$ in Theorem 6.4.1, we have

$$\frac{1}{n}\sum_{i=1}^{[nt]} r_i \cos \varphi_i \cong \{Er\}\frac{1}{n}\sum_{i=1}^{[nt]} \cos \varphi_i$$

$$\cong \left\{\frac{Er}{\pi}\right\}\frac{\pi}{n}\sum_{i=1}^{[nt]} \cos \frac{\pi i}{n+1}$$

$$\cong \left\{\frac{Er}{\pi}\right\}\int_0^{\pi t} \cos u\, du = \left\{\frac{Er}{\pi}\right\} \sin \pi t.$$

Furthermore,

$$\frac{1}{n}\sum_{i=[nt]}^{n} r_i \cos \phi_i \cong \left\{\frac{Er}{\pi}\right\}\int_{\pi t}^{\pi} \cos u\, du = -\left\{\frac{Er}{n}\right\} \sin \pi t.$$

Hence the formula holds for the second component. The first component formula follows in a similar way.

This proposition is important since it shows that the Oja sign test is asymptotically equivalent to Blumen's test under elliptical models since they are both invariant under affine transformations. Hence, the efficiency results for Blumen's test carry over for spherical and elliptical models to the Oja sign test. Also recall that Blumen's test is locally most powerful invariant for the class of elliptical models so the Oja sign test should be quite good for elliptical models in general. The two tests are not equivalent for nonelliptical models. In Brown et al. (1992) the efficiency of the Oja sign test relative to Blumen's test was computed for a class of symmetric densities with contours of the form $|x_1|^m + |x_2|^m$. When $m = 2$ we have spherical densities, and when $m = 1$ we have Laplacian densities with independent marginals. Table 1 of Brown et al. (1992) shows that the Oja sign test is more efficient than Blumen's test except when $m = 2$ where, of course, the efficiency is 1. Hettmansperger, Nyblom, and Oja (1994) extend the Oja methods to dimensions higher than 2 in the one-sample case and Hettmansperger and Oja (1994) extend the methods to higher dimensions for the multi-sample problem.

6.4.2 Affine Invariant Rank Methods

Peters and Randles (1990b) extend the test of Blumen in two dimensions and the test of Randles in higher dimensions to a multivariate version of a signed-rank test, much like the extension by Hössjer and Croux in the case of spatial models; see Section 6.3.1. The test statistic is a modification of (6.4.4) and we give the formula for k dimensions:

$$\frac{3k}{n^2}\sum_{i=1}^{n}\sum_{j=1}^{n} \cos(\pi p_{ij}) \frac{R(D_i)R(D_j)}{n^2}, \qquad (6.4.8)$$

where D_i is the squared Mahalanobis distance of $\mathbf{x}_i$ from the mean vector, $D_i = (\mathbf{x}_i - \overline{\mathbf{x}})^T \mathbf{S}^{-1} (\mathbf{x}_i - \overline{\mathbf{x}})$, $\mathbf{S}$ the sample covariance matrix of the underlying distribution, and $R(D_i)$ is the rank of the distance among all n distances. We reject the null hypothesis $H_0 : \boldsymbol{\theta} = \mathbf{0}$ when the statistic exceeds a chi-square critical value with k degrees of freedom. The distribution theory is presented in the paper by Peters and Randles (1990b) and is based on the projection:

$$\frac{3k}{n^2} \sum_{i=1}^{n} \sum_{j=1}^{n} \cos(\pi p_{ij}) H(D_i) H(D_j), \tag{6.4.9}$$

where $H(\cdot)$ is the cdf of D_i. Peters and Randles also compute the efficiency of the signed-rank test relative to Hotelling's test for a power family of elliptical distributions. They compute the same results as in Table 6.3.4 for the spatial signed-rank test of Hössjer and Croux (1995). The signed-rank test can be much more efficient than Hotelling's test when the underlying distribution has heavy tails and the dimension is not too high.

In Brown and Hettmansperger (1987a), the idea of an **affine invariant rank vector** is introduced. The approach is similar to that of Möttönen and Oja (1995) for the spatial rank vector discussed earlier; see Section 6.3.1. The Oja criterion $D_8(\boldsymbol{\theta})$ with $m = 1$ in Section 6.4.1 is a multivariate extension of the univariate L_1 criterion function and we take its gradient to be the **centered rank vector**. Recall in the univariate case $D(\theta) = \sum |x_j - \theta|$ and the derivative $D'(\theta) = \sum \text{sgn}(\theta - x_j)$. Hence, $D'(x_i)$ is the centered rank of x_i. Likewise the vector centered rank of $\mathbf{x}_k$ is defined to be:

$$\mathbf{R}_n(\mathbf{x}_k) = \nabla D_8(\mathbf{x}_k) = \tfrac{1}{2} \sum \sum_{i<j} \text{sgn}\left\{ \det \begin{pmatrix} 1 & 1 & 1 \\ x_{k1} & x_{i1} & x_{j1} \\ x_{k2} & x_{i2} & x_{j2} \end{pmatrix} \right\} (\mathbf{x}_j^* - \mathbf{x}_i^*). \tag{6.4.10}$$

Again we use the idea of affine invariant vector rank to define a **one-sample signed rank statistic**. Let $\mathbf{R}_{2n}(\mathbf{x}_k)$ be the rank vector when $\mathbf{x}_k$ is ranked among the observation vectors $\mathbf{x}_1, \ldots, \mathbf{x}_n$ and their reflections $-\mathbf{x}_1, \ldots, -\mathbf{x}_n$. Then the test statistic is $\mathbf{S}_9(\mathbf{0}) = \sum \mathbf{R}_{2n}(\mathbf{x}_j)$. Now $\mathbf{R}_{2n}(-\mathbf{x}_j) = -\mathbf{R}_{2n}(\mathbf{x}_j)$ so that the conditional covariance matrix (conditioning on the observed data) is

$$\widehat{\mathbf{A}} = \sum_{j=1}^{n} \mathbf{R}_{2n}(\mathbf{x}_j) \mathbf{R}_{2n}^T(\mathbf{x}_j).$$

The approximately size α test of $H_0 : \boldsymbol{\theta} = \mathbf{0}$ is:

$$\text{Reject } H_0 \text{ if } \mathbf{S}_9^T(\mathbf{0}) \widehat{\mathbf{A}}^{-1} \mathbf{S}_9(\mathbf{0}) \geq \chi_\alpha^2(2). \tag{6.4.11}$$

In addition, the Hodges–Lehmann estimate of $\boldsymbol{\theta}$ based on $\mathbf{S}_9(\boldsymbol{\theta}) \doteq \mathbf{0}$ is the Oja median of the linked pairwise averages; see Brown and Hettmansperger (1987a) for details.

Example 6.4.1 *Mathematics and Statistics Exam Scores*

We now illustrate the one-sample signed-rank test (6.4.11) on a small data set. The data consists of 20 vectors, chosen at random from a larger data set published in Mardia, Kent, and Bibby (1979). Each vector consists of four components and records test scores in Mechanics, Vectors, Analysis and Statistics. The extension of (6.4.11) to four dimensions is obvious. We wish to test the hypothesis that there are no differences among the examination topics. This is a traditional hypothesis in repeated measures designs; see Jan and Randles (1996) for a thorough discussion of this problem. Similar to our findings above on efficiencies, they found that multivariate sign and signed-rank tests were often superior in robustness of level and efficiency.

Table 6.4.1 provides the original quadrivariate data along with the trivariate data that result when the Statistics score is subtracted from the other three. We suppose that the trivariate data are a sample of size 20 from a symmeteric distribution with center $\boldsymbol{\theta} = (\theta_1, \theta_2, \theta_3)^T$ and we wish to test $H_0 : \boldsymbol{\theta} = \mathbf{0}$ versus $H_A : \boldsymbol{\theta} \neq \mathbf{0}$. In Table 6.4.2 we have the means of the signed-rank vectors and the usual mean vector for Hotelling's T^2. We also provide the Hodges–Lehmann estimate for $\boldsymbol{\theta}$. Table 6.4.1 also contains the same material after the first person's Vectors test score of 70 is changed to 0. This simulates a data entry error.

The test statistic is then referred to a χ^2 distribution; see Table 6.4.2. For these data, the signed-rank methods show that there are significant differences among the typical test scores. The estimates point to Mechanics as the most difficult subject. Bootstrapping is not practical because of the required time. Finally, note that the signed-rank methods were not as affected by the data contamination as the traditional T^2 method. Computations of the tests and estimates were done using S-Plus functions developed by Möttönen (1997b) in the Department of Mathematical Sciences/Statistics, University of Oulu, Finland. See Example 6.6.2 for a discussion of the two-sample affine invariant test.

Table 6.4.1: Test score data: Mechanics (M), Vectors (V), Analysis (A), Statistics (S) and differences when Statistics is subtracted from the other three

M	V	A	S	M − S	V − S	A − S
59	70	62	56	3	14	6
52	64	63	54	−2	10	9
44	61	62	46	−2	15	16
44	56	61	36	8	20	25
30	69	52	45	−15	24	7
46	49	59	37	9	12	22
31	42	54	68	−37	−26	−14
42	60	49	33	9	27	16
46	52	41	40	6	12	1
49	49	48	39	10	10	9
17	53	43	51	−34	2	−8
37	56	28	45	−8	11	−17
40	43	21	61	−21	−18	−40
35	36	48	29	6	7	19
31	52	27	40	−9	12	−13
17	51	35	31	−14	20	4
49	50	23	9	40	41	14
8	42	26	40	−32	2	−14
15	38	28	17	−2	21	11
0	40	9	14	−14	26	−5

Table 6.4.2: Results for the original and contaminated test score data: mean of signed-rank vectors, usual mean vectors, the Hodges–Lehmann estimate of θ; results for the signed-rank test (6.4.11) and Hotelling's T^2 test

	$M - S$	$V - S$	$A - S$	Test Statistic	Asymp. p-value
Original data					
Mean of signed ranks	−83.35	130.38	9.95	14.07	0.0028
Mean vector	−4.95	12.10	2.40	13.47	0.0037
HL estimate	−3.05	14.06	4.96		
Contaminated data					
Mean of signed ranks	−125.31	95.94	71.72	10.09	0.0178
Mean vector	−4.95	8.60	2.40	6.95	0.0736
HL estimate	−3.90	12.69	4.64		

6.4.3 Summary

Hettmansperger, Möttönen and Oja (1997a; 1997b) extend the affine invariant one- and two-sample rank tests to dimensions greater than 2. Table 6.4.3 provides the efficiencies relative to Hotelling's test for a multivariate t distribution; see Möttönen et al. (1997). Note that the efficiency is quite high even for the multivariate normal distribution. Further, note that this efficiency is the same for all elliptical normal distributions as well since the test is affine invariant.

Jan and Randles (1995) extend the Randles (1989) affine invariant sign test to a signed-rank-like test. Essentially, they apply (6.4.4) to the pairwise sums

Table 6.4.3: Asymptotic relative efficiencies of the signed-rank test (6.4.11), the spatial rank test (6.3.21), and the sign test (6.4.7) relative to Hotelling's T^2 test for the multivariate t-distributions

Dimension	Test	Degrees of freedom							
		3	4	6	8	10	15	20	∞
1	Signed-rank	1.900	1.401	1.164	1.089	1.054	1.014	0.997	0.955
	Spatial	1.900	1.401	1.164	1.089	1.054	1.014	0.997	0.955
	Sign	1.621	1.125	0.879	0.798	0.757	0.710	0.690	0.637
2	Signed-rank	2.026	1.469	1.196	1.107	1.064	1.014	0.9222	0.937
	Spatial	1.953	1.435	1.187	1.108	1.071	1.029	1.011	0.967
	Sign	2.000	1.388	1.084	0.984	0.934	0.877	0.851	0.785
3	Signed-rank	2.112	1.515	1.221	1.124	1.076	1.021	0.997	0.934
	Spatial	1.994	1.453	1.200	1.119	1.081	1.038	1.019	0.973
	Sign	2.162	1.500	1.172	1.063	1.009	0.947	0.920	0.849
4	Signed-rank	2.173	1.550	1.241	1.139	1.088	1.030	1.004	0.937
	Spatial	2.018	1.467	1.208	1.127	1.087	1.044	1.025	0.978
	Sign	2.250	1.561	1.220	1.107	1.051	0.986	0.958	0.884
6	Signed-rank	2.256	1.598	1.270	1.162	1.108	1.045	1.018	0.947
	Spatial	2.050	1.484	1.219	1.136	1.095	1.051	1.031	0.984
	Sign	2.344	1.626	1.271	1.153	1.094	1.027	0.997	0.920
10	Signed-rank	2.346	1.650	1.304	1.189	1.132	1.066	1.037	0.961
	Spatial	2.093	1.503	1.229	1.144	1.103	1.058	1.038	0.989
	Sign	2.422	1.681	1.313	1.192	1.131	1.062	1.031	0.951

($\mathbf{x}_i + \mathbf{x}_j$) and hence the statistic is similar to the Walsh average form of the univariate Wilcoxon signed-rank test, (1.3.24). The test requires the estimation of a scaling parameter in order to have an asymptotically distribution-free test. This test has excellent efficiency properties relative to Hotelling's test. In particular, the efficiency increases with the dimension of a multivariate normal distribution. Some examples of efficiency as a function of dimension are: (dimension, efficiency) = (2, 0.966), (3, 0.974), (4, 0.978), (5, 0.981).

Many authors have worked on the problem of developing multidimensional sign tests under various invariance conditions. The sign statistics are important for defining medians, and further in defining the concept of centered rank. Oja and Nyblom (1989) propose a family of locally most powerful sign tests that are affine invariant and show that the Blumen (1958) test is optimal for elliptical alternatives. Using a different approach that involves data-based coordinate systems, Chaudhuri and Sengupta (1993) introduce a family of affine invariant sign tests. See also Dietz (1982) for a development of affine invariant sign and rank procedures based on rotations of the coordinate systems. Another interesting approach to the construction of a multivariate median and rank is based on the idea of data depth due to Liu (1990). In this case, the median is a point contained in a maximum number of triangles formed by the $\binom{n}{3}$ different choices of 3 data vectors. See also Liu and Singh (1993).

Hence, we conclude that if we are fairly certain that we have a spherical model, in a spatial statistics context, for example, then the spatial median and the angle test are quite good. If the model is likely to be elliptical with heavy tails then either Blumen's test or the Oja sign test and Oja median should be quite good. If we suspect that the model is nonelliptical then the methods of Oja are preferable. On the other hand, if invariance and equivariance considerations are not relevant then the componentwise methods should work quite well. Finally, departures from bivariate normality should be considered. The L_1-type methods are good when there is a heavy-tailed model. However, the efficiency can be improved by rank-type methods when the tailweight is more moderate and perhaps close to normality. Even for the bivariate normal model the rank methods lose very little efficiency when invariance is taken into account.

6.5 Robustness of Multivariate Estimates of Location

In this section we sketch some of the recent results on the influence and breakdown points for the estimators derived from the various estimating equations. Recall from Theorem 6.1.2 that the vector influence is proportional to the vector $\Psi(\mathbf{x})$. Typically $\Omega(\mathbf{x})$ is a projection and reduces the problem of finding the asymptotic distribution of the estimating function $(1/\sqrt{n})\mathbf{S}(\boldsymbol{\theta})$ to a central limit problem. To determine whether an estimator has bounded influence or not, it is only necessary to check that the norm of $\Omega(\mathbf{x})$ is bounded. Further, recall that the breakdown point is the smallest proportion of contamination needed to carry the estimator beyond all bounds. We now briefly consider the different invariance models.

6.5.1 Location and Scale Invariance: Componentwise Methods

In the case of component medians, the influence function is given by

$$\Omega^T(\mathbf{x}) \propto (\text{sgn}(x_{11}), \text{sgn}(x_{21})).$$

The norm is clearly bounded. Further, the breakdown point is 50% as it is in the univariate case. Likewise, for the Hodges–Lehmann component estimates $\Omega^T(\mathbf{x}) \propto (F_1(x_{11}) - \frac{1}{2}, F_2(x_{21}) - \frac{1}{2})$, where $F_i(\cdot)$ is the ith marginal cdf. Hence, the influence is bounded in this case as well. The breakdown point is 29%, the same as the univariate case. Note, however, that the componentwise methods are neither rotation nor affine invariant/equivariant.

6.5.2 Rotation Invariance: Spatial Methods

We assume in this subsection that the underlying distribution is spherical. For the spatial median, we have $\Omega(\mathbf{x}) = \mathbf{u}(\mathbf{x})$, the unit vector in the $\mathbf{x}$ direction. Hence, again we have bounded influence. Lopuhaä and Rousseeuw (1991) were the first to point out that the spatial median has 50% breakdown point. The proof is given in the following theorem. First note from Exercise 6.8.17 that the maximum breakdown point for any translation equivariant estimator is $[(n + 1)/2]/n$ and the spatial median is translation equivariant.

Theorem 6.5.1 *The spatial median $\widehat{\boldsymbol{\theta}}$ has breakdown point $\varepsilon^* = [(n + 1)/2]/n$ for every dimension.*

Proof. In view of the preceding remarks, we only need to show $\varepsilon^* \geq [(n + 1)/2]/n$. Let $\mathbf{X} = (\mathbf{x}_1, \ldots, \mathbf{x}_n)$ be a collection of n observations in k dimensions. Let $\mathbf{Y}_m = (\mathbf{y}_1, \ldots, \mathbf{y}_n)$ be formed from $\mathbf{X}$ by corrupting any m observations. Then $\widehat{\boldsymbol{\theta}}(\mathbf{Y}_m)$ minimizes $\sum \|\mathbf{y}_i - \boldsymbol{\theta}\|$. Assume, without loss of generality, that $\widehat{\boldsymbol{\theta}}(\mathbf{X}) = \mathbf{0}$. (Use translation equivariance.) We suggest that the reader follow the argument with a picture in two dimensions.

Let $M = \max_i \|\mathbf{x}_i\|$ and let $B(0, 2M)$ be the sphere of radius $2M$ centered at the origin. Suppose the number of corrupted observations $m \leq [(n - 1)/2]$. We will show that $\sup \|\widehat{\boldsymbol{\theta}}(\mathbf{Y}_m)\|$ over $\mathbf{Y}_m$ is finite. Hence, $\varepsilon^* \geq [(n - 1)/2 + 1]/n = [(n + 1)/2]/n$ and we will be finished.

Let $d_m = \inf\{\|\widehat{\boldsymbol{\theta}}(\mathbf{Y}_m) - \boldsymbol{\gamma}\| : \boldsymbol{\gamma} \text{ in } B(0, 2M)\}$, the distance of $\widehat{\boldsymbol{\theta}}(\mathbf{Y}_m)$ from $B(0, 2M)$. Then the distance of $\widehat{\boldsymbol{\theta}}(\mathbf{Y}_m)$ from the origin is $\|\widehat{\boldsymbol{\theta}}(\mathbf{Y}_m)\| \leq d_m + 2M$. Now

$$\|\mathbf{y}_j - \widehat{\boldsymbol{\theta}}(\mathbf{Y}_m)\| \geq \|\mathbf{y}_j\| - \|\widehat{\boldsymbol{\theta}}(\mathbf{Y}_m)\| \geq \|\mathbf{y}_j\| - (d_m + 2M). \tag{6.5.1}$$

Suppose the contamination has pushed $\widehat{\boldsymbol{\theta}}(\mathbf{Y}_m)$ far outside $B(0, 2M)$. In particular, suppose $d_m > 2M[(n + 1)/2]$. We will show this leads to a contradiction. We know that $\mathbf{X} \subset B(0, M)$ and if $\mathbf{x}_k$ is not contaminated,

$$\|\mathbf{x}_k - \widehat{\boldsymbol{\theta}}(\mathbf{Y}_m)\| \geq M + \|\mathbf{x}_k\| + d_m. \tag{6.5.2}$$

Next split the following sum over contaminated and not contaminated observations using (6.5.1) and (6.5.2).

$$\sum_{i=1}^{n} \|\mathbf{y}_i - \widehat{\boldsymbol{\theta}}(\mathbf{Y}_m)\| \geq \underbrace{\sum (\|\mathbf{y}_i\| - (d_m + 2M))}_{\text{contam}} + \underbrace{\sum (\|\mathbf{y}_i\| - d_m)}_{\text{not}}$$

$$= \sum \|\mathbf{y}_i\| - \left[\frac{n-1}{2}\right](d_m + 2M) + \left(n - \left[\frac{n-1}{2}\right]\right) d_m$$

$$= \sum \|\mathbf{y}_i\| - 2M\left(\left[\frac{n-1}{2}\right]\right) + d_m\left(n - 2\left(\left[\frac{n-1}{2}\right]\right)\right)$$

$$> \sum \|\mathbf{y}_i\| - 2M\left(\left[\frac{n-1}{2}\right]\right) + 2M\left(\left[\frac{n-1}{2}\right]\right)\left(n - 2\left(\left[\frac{n-1}{2}\right]\right)\right)$$

$$= \sum \|\mathbf{y}_i\| + 2M\left(\left[\frac{n-1}{2}\right]\right)\left(n - 1 - 2\left(\left[\frac{n-1}{2}\right]\right)\right)$$

$$\geq \sum \|\mathbf{y}_i\|.$$

But, recall that $\widehat{\boldsymbol{\theta}}(\mathbf{Y}_m)$ minimizes $\sum \|\mathbf{y}_i - \boldsymbol{\theta}\|$, hence we have a contradiction. Thus $d_m \leq 2M([(n-1)/2])$.

But then ε^* must be at least $\{[(n-1)/2] + 1\}/n$ and $[(n-1)/2] + 1 = [(n+1)/2]$ and the proof is complete.

6.5.3 Spatial R-Estimate of Hössjer and Croux

From (6.3.15), $\Omega(\mathbf{x}) = G(\|\mathbf{x}\|)\mathbf{u}(\|\mathbf{x}\|)$ where $G(\cdot)$ is the cdf of $\|\mathbf{x}\|$. Hence, the influence is bounded. Hössjer and Croux also show that the estimate has 29% breakdown point for all dimensions, the same as the univariate Hodges–Lehmann estimate.

6.5.4 The Spatial Hodges–Lehmann Estimate

This estimate is the spatial median of the pairwise averages: $\frac{1}{2}(\mathbf{x}_i + \mathbf{x}_j)$. It was first studied in detail by Chaudhuri (1992) and it is the estimate corresponding to the spatial signed rank statistic (6.3.16) of Möttönen and Oja (1995).

From (6.3.22) it is clear that the influence function is bounded. Further, since it is the spatial median of the pairwise averages, the argument that shows that the breakdown of the univariate Hodges–Lehmann estimate is $1 - (1/\sqrt{2}) \approx 0.29$ works in the multivariate case; see Exercise 1.13.10 in Chapter 1.

6.5.5 Affine Equivariance: The Oja Median

This estimator is affine equivariant and solves equation (6.4.5). From the projection representation of the statistic in Theorem 6.4.3, notice that the vector $\mathbf{z}(t)$ is bounded. It then follows that, for spherical models (with finite first moment), the influence function is bounded. See Niinimaa and Oja (1995) for a rigorous derivation of the influence function.

370 *Multivariate Models*

The breakdown properties of the Oja median are more interesting. In the next theorem we show that, even though the influence function is bounded, the estimate can be broken down with just two contaminated points. Niinimaa, Oja, and Tableman (1990) give the original proof. The proof given below is due to Gerald Sievers (personal communication).

Theorem 6.5.2 *The Oja median has breakdown point $2/n$.*

Proof. Suppose $n = 2r + 2$. Let $\mathbf{x}_1$ and $\mathbf{x}_n$ be the corrupted points. We can assume (via affine equivariance) that the axes are rotated and translated, resulting in a situation as described by Figure 6.5.1.

Further, $\widehat{\boldsymbol{\theta}}$ is somewhere on the horizontal axis; hence, $\widehat{\theta}_2 = 0$ and $\widehat{\theta}_1 > 0$. There are r good points in each of the regions U and L. The corrupted points $\mathbf{x}_1$ and $\mathbf{x}_n$ are above and below the regions U and L, respectively. We will work with the first component (horizontal) of $\mathbf{S}_8(\boldsymbol{\theta})$, say $S_8^{(1)}(\boldsymbol{\theta})$. We will show that if $\theta_1 > Q$ then $S_8^{(1)}(\boldsymbol{\theta}) > 0$ and if $\theta_1 < P$ then $S_8^{(1)}(\boldsymbol{\theta}) < 0$. Hence, $\widehat{\boldsymbol{\theta}}$ must be between P and Q. Finally, by maintaining the separation between $\mathbf{x}_1$ and $\mathbf{x}_n$ and letting the horizontal components go to $+\infty$, we will show that we can

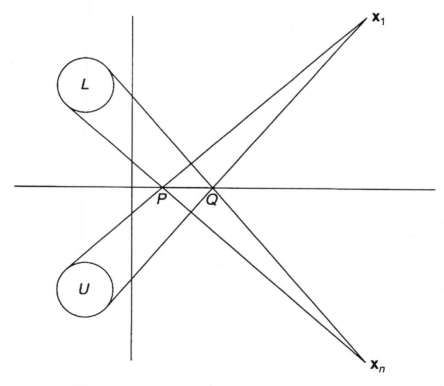

Figure 6.5.1: There are r data points in each of the regions labeled U and L. P and Q are the smallest and largest crossing points of lines from U and L to $\mathbf{x}_1$ and $\mathbf{x}_n$. Finally, $\mathbf{x}_1$ and $\mathbf{x}_n$ have vertical separation exceeding the separation of U and L.

shift the segment, PQ, as far to the right as desired. Hence, $\hat{\theta}$ will be shifted to the right and break down.

Recall the following: $2S_8(\theta) = \sum \sum_{i<j} \mathbf{u}_{ij}(\theta)$ and $\mathbf{u}_{ij}(\theta) = \text{sgn}\{(\mathbf{x}_j^* - \mathbf{x}_i^*)^T(\theta - \mathbf{x}_i)\}(\mathbf{x}_j^* - \mathbf{x}_i^*)$, where $\mathbf{x}^*$ means that the vector has been rotated 90 degrees in a counterclockwise direction. Then

$$\mathbf{u}_{ij}(\theta) = \begin{cases} (\mathbf{x}_j^* - \mathbf{x}_i^*) & \text{if } \mathbf{x}_i \to \mathbf{x}_j \to \theta \text{ is in a counterclockwise direction} \\ -(\mathbf{x}_j^* - \mathbf{x}_i^*) & \text{if } \mathbf{x}_i \to \mathbf{x}_j \to \theta \text{ is in a clockwise direction.} \end{cases}$$

Finally, note that

$$\mathbf{x}_j^* - \mathbf{x}_i^* = \begin{pmatrix} x_{2i} - x_{2j} \\ x_{1j} - x_{1i} \end{pmatrix}.$$

Hence the horizontal component (first component) of $\mathbf{u}_{ij}(\theta)$, say $u_{ij}^{(1)}(\theta)$, is determined by the vertical component (second component) of $\mathbf{x}_1$ and $\mathbf{x}_2$.

Now we consider the case when $\theta_1 > Q$, i.e. θ_1 is to the right of the largest crossing point.

1. Consider $\mathbf{x}_i, \mathbf{x}_j$, either in U or L. Let $(1) = \sum \sum u_{ij}^{(1)}(\theta)$, where the sums are over $\{i, j : \neq 1 \text{ or } n\}$. Then

$$|(1)| \leq M_1 = \binom{2r}{2} \max |x_{2j} - x_{2i}|,$$

where the max is taken over $\{i, j : \neq 1 \text{ or } n\}$, independently of θ.
2. Consider $\mathbf{x}_1$ versus $\mathbf{x}_j$ in U. We have $\mathbf{u}_{1j}(\theta) = \mathbf{x}_j^* - \mathbf{x}_1^*$.
   Let $(2) = \sum_{\mathbf{x}_j \in U}(x_{21} - x_{2j})$.
3. Consider $\mathbf{x}_1$ versus $\mathbf{x}_j$ in L. We have $\mathbf{u}_{1j}(\theta) = \mathbf{x}_j^* - \mathbf{x}_1^*$ and $(3) = \sum_{\mathbf{x}_j \in L}(x_{21} - x_{2j})$.
4. Consider $\mathbf{x}_n$ versus $\mathbf{x}_j$ in U. Then $\mathbf{u}_{jn}(\theta) = \mathbf{x}_j^* - \mathbf{x}_n^*$ and $(4) = \sum_{\mathbf{x}_j \in U}(x_{2j} - x_{2n})$.
5. Consider $\mathbf{x}_n$ versus $\mathbf{x}_j$ in L. Then $(5) = \sum_{\mathbf{x}_j \in L}(x_{2j} - x_{2n})$.
6. Consider $\mathbf{x}_1$ versus $\mathbf{x}_n$ and θ to the left of $\mathbf{x}_1$ and $\mathbf{x}_n$ since it is in the convex hull of the data. Now $\mathbf{u}_{1n}(\theta) = \mathbf{x}_1^* - \mathbf{x}_n^*$ and $u_{ij}^{(1)}(\theta) = (x_{2n} - x_{21})$.

We now have the following:

$$\sum_{i<j} u_{ij}^{(1)}(\theta) = (1) + (2) + (3) + (4) + (5) + (6)$$

$$\geq -M_1 + (2r - 1)(x_{21} - x_{2n}).$$

Now x_1, x_n must be chosen large enough so that $-M_1 + (2r - 1) \times (x_{21} - x_{2n}) > 0$. Then the first component of $S_8(\theta)$ will be positive for $\theta_1 > Q$. A similar argument shows that the horizontal component is negative for $\theta_1 < P$. Hence, $\hat{\theta}$ is trapped between P and Q and as x_{11} and x_{1n} increase P and Q also increase along the horizontal axis. Then $\hat{\theta}$ breaks down.

Niinimaa and Oja (1995) show that, in fact, the breakdown point of the Oja median depends on the dispersion of the contaminated data. When the dispersion of the contaminated data is less than the dispersion of the original data then the asymptotic breakdown point is positive. If, for example, the contaminated points are all at a single point, then the breakdown point is $\frac{1}{3}$.

6.6 Linear Model

We consider the bivariate linear model. As examples of the linear model, we will find bivariate estimates and tests for a general regression effect as well as shifts in the bivariate two-sample location model and multi-sample location models. We will focus primarily on compontentwise rank methods; however, we will discuss some other methods for the multi-sample location model in the examples of Section 6.6.1. Spatial and affine invariant/equivariant methods for the general linear model are currently under development in the research literature.

In Section 3.2, we present the notation for the univariate linear model. Here, we will think of the multivariate linear model as a series of concatenations of the univariate models. Hence, we introduce

$$\mathbf{Y}_{n\times 2} = \begin{pmatrix} Y_{11} & Y_{12} \\ \vdots & \vdots \\ Y_{n1} & Y_{n2} \end{pmatrix} = (\mathbf{Y}^{(1)}, \mathbf{Y}^{(2)}) = \begin{pmatrix} \mathbf{Y}_1^T \\ \vdots \\ \mathbf{Y}_n^T \end{pmatrix}. \tag{6.6.1}$$

A superscript indicates a column, a subscript a row, and, as usual in this chapter, T denotes transpose. Now the multivariate linear model is

$$\mathbf{Y} = \mathbf{1}\boldsymbol{\alpha}^T + \mathbf{X}\boldsymbol{\beta} + \boldsymbol{\epsilon}, \tag{6.6.2}$$

where $\mathbf{1}$ is an $n \times 1$ vector of ones, $\boldsymbol{\alpha}^T = (\alpha^{(1)}, \alpha^{(2)})$, $\mathbf{X}$ is an $n \times p$ full-rank, centered design matrix, $\boldsymbol{\beta}$ is a $p \times 2$ matrix of unknown regression constants, and $\boldsymbol{\epsilon}$ is an $n \times 2$ matrix of errors. The rows of $\boldsymbol{\epsilon}$, and hence, $\mathbf{Y}$, are independent and the rows of $\boldsymbol{\epsilon}$ are identically distributed with a continuous bivariate cdf $F(s, t)$. Model (6.6.2) is the concatenation of two univariate linear models: $\mathbf{Y}^{(i)} = \mathbf{1}\alpha^{(i)} + \mathbf{X}\boldsymbol{\beta}^{(i)} + \boldsymbol{\epsilon}^{(i)}$ for $i = 1, 2$. We have restricted attention to the bivariate case to simplify the presentation. In most cases the general multivariate results are obvious.

We rank within components or columns. Hence, the rank score of the ith item in the jth column is

$$a_{ij} = a(R_{ij}) = a(R(Y_{ij} - \mathbf{x}_i^T \boldsymbol{\beta}^{(j)})), \tag{6.6.3}$$

where R_{ij} is the rank of $Y_{ij} - \mathbf{x}_i^T \boldsymbol{\beta}^{(j)}$ when ranked among $Y_{1j} - \mathbf{x}_1^T \boldsymbol{\beta}^{(j)}, \ldots, Y_{nj} - \mathbf{x}_n^T \boldsymbol{\beta}^{(j)}$. The rank scores are generated by $a(i) = \varphi(i/n+1)$, $0 < \varphi(u) < 1$, $\int \varphi(u)du = 0$, and $\int \varphi^2(u)du = 1$; see Section 3.4. Let the score matrix $\mathbf{A}$ be defined as:

$$\mathbf{A} = \begin{pmatrix} a_{11} & a_{12} \\ \vdots & \vdots \\ a_{n1} & a_{n2} \end{pmatrix} = (\mathbf{a}^{(1)}, \mathbf{a}^{(2)}), \tag{6.6.4}$$

so that each column is the set of rank scores within the column.
The criterion function is

$$D(\boldsymbol{\beta}) = \sum_{i=1}^{n} \mathbf{a}_i^T \mathbf{r}_i, \qquad (6.6.5)$$

where $\mathbf{a}_i^T = (a_{i1} a_{i2}) = (a(R(Y_{i1} - \mathbf{x}_i^T \boldsymbol{\beta}^{(1)})), a(R(Y_{i2} - \mathbf{x}_i^T \boldsymbol{\beta}^{(2)})))$ and $\mathbf{r}_i^T = (Y_{i1} - \mathbf{x}_i^T \boldsymbol{\beta}^{(1)}, Y_{i2} - \mathbf{x}_i^T \boldsymbol{\beta}^{(2)})$. Note at once that this is an analog, using inner products, of the univariate criterion in Section 3.2.1. In fact, $D(\boldsymbol{\beta})$ is the sum of the corresponding univariate criterion functions. The matrix of the negatives of the partial derivatives is:

$$\mathbf{L}(\boldsymbol{\beta}) = \mathbf{X}^T \mathbf{A} = \begin{pmatrix} \sum x_{i1} a_{i1} & \sum x_{i1} a_{i2} \\ \vdots & \vdots \\ \sum x_{ip} a_{i1} & \sum x_{ip} a_{i2} \end{pmatrix} = \left(\sum a_{i1} \mathbf{x}_i, \sum a_{i2} \mathbf{x}_i \right); \qquad (6.6.6)$$

see Exercise 6.8.18 and equation (3.2.11). Again, note that the two columns in (6.6.6) are the estimating equations for the two concatenated univariate linear models and $\mathbf{x}_i$ is the ith row of $\mathbf{X}$ written as a column.

Hence, the componentwise multivariate R-estimator of $\boldsymbol{\beta}$ is $\widehat{\boldsymbol{\beta}}$ that minimizes (6.6.5) or solves $\mathbf{L}(\boldsymbol{\beta}) \doteq \mathbf{0}$. Further, $\mathbf{L}(\mathbf{0})$ is the basic quantity that we will use to test $H_0 : \boldsymbol{\beta} = \mathbf{0}$. We must statistically assess the size of $\mathbf{L}(\mathbf{0})$ and reject H_0 and claim the presence of a regression effect when $\mathbf{L}(\mathbf{0})$ is 'too large' or 'too far from the zero matrix'.

We first consider testing $H_0 : \boldsymbol{\beta} = \mathbf{0}$ since the distribution theory of the test statistic will be useful later for the asymptotic distribution theory of the estimate.

For the linear model we need some results on direct products; see Magnus and Neudecker (1988) for a complete discussion. We list here the results that we need:

1. Let $\mathbf{A}$ and $\mathbf{B}$ be $m \times n$ and $p \times q$ matrices. The $mp \times nq$ matrix $\mathbf{A} \otimes \mathbf{B}$ defined by

$$\mathbf{A} \otimes \mathbf{B} = \begin{bmatrix} a_{11} \mathbf{B} & \cdots & a_{1n} \mathbf{B} \\ \vdots & & \vdots \\ a_{m1} \mathbf{B} & \cdots & a_{mn} \mathbf{B} \end{bmatrix} \qquad (6.6.7)$$

is called the **direct product** or **Kronecker product** of $\mathbf{A}$ and $\mathbf{B}$.

2.
$$(\mathbf{A} \otimes \mathbf{B})^T = \mathbf{A}^T \otimes \mathbf{B}^T, \qquad (6.6.8)$$

$$(\mathbf{A} \otimes \mathbf{B})^{-1} = \mathbf{A}^{-1} \otimes \mathbf{B}^{-1}, \qquad (6.6.9)$$

$$(\mathbf{A} \otimes \mathbf{B})(\mathbf{C} \otimes \mathbf{D}) = (\mathbf{AC} \otimes \mathbf{BD}). \qquad (6.6.10)$$

3. Let $\mathbf{D}$ be an $m \times n$ matrix. Then $\mathbf{D}_{\text{col}}$ is the $mn \times 1$ vector formed by stacking the columns of $\mathbf{D}$. We then have

$$\text{tr}(\mathbf{ABCD}) = (\mathbf{D}_{\text{col}}^T)^T (\mathbf{C}^T \otimes \mathbf{A}) \mathbf{B}_{\text{col}} = \mathbf{D}_{\text{col}}^T (\mathbf{A} \otimes \mathbf{C}^T)(\mathbf{B}^T)_{\text{col}}. \qquad (6.6.11)$$

4.
$$(AB)_{col} = (B^T \otimes I)A_{col} = (I \otimes A)B_{col}. \tag{6.6.12}$$

These facts are used in the proofs of the theorems in the rest of this chapter.

6.6.1 Test for Regression Effect

As mentioned above, we will base the test of $H_0 : \beta = 0$ on the size of the random matrix $L(0)$. We deal with this random matrix by rolling out the matrix by columns. Note from (6.6.4) and (6.6.6) that $L(0) = X^T A = (X^T a^{(1)}, X^T a^{(2)})$. Then we define the vector

$$L_{col} = \begin{pmatrix} X^T a^{(1)} \\ X^T a^{(2)} \end{pmatrix} = \begin{pmatrix} X^T & 0 \\ 0 & X^T \end{pmatrix} \begin{pmatrix} a^{(1)} \\ a^{(2)} \end{pmatrix}. \tag{6.6.13}$$

Now from the discussion in Section 3.5.1, let the column variances and covariances be

$$\sigma^2_{a^{(i)}} = \frac{1}{n-1} \sum_{j=1}^{n} a_{ji}^2 \to \sigma_i^2 = \int \varphi^2(u)\,du = 1$$

$$\sigma^2_{a^{(1)}a^{(2)}} = \frac{1}{n-1} \sum_{j=1}^{n} a_{j1} a_{j2} \to \sigma_{12} = \int \varphi(F_1(s))\varphi(F_2(t))\,dF(s,t), \tag{6.6.14}$$

where $F_1(s)$ and $F_2(t)$ are the marginal cdfs of $F(s,t)$. Since the ranks are centered and using the same argument as in Theorem 3.5.1, $E(L_{col}) = 0$ and

$$V = \text{Cov}(L_{col}) = \begin{bmatrix} \sigma^2_{a^{(1)}} X^T X & \sigma_{a^{(1)}a^{(2)}} X^T X \\ \sigma_{a^{(1)}a^{(2)}} X^T X & \sigma^2_{a^{(2)}} X^T X \end{bmatrix}$$

$$= \left(\frac{1}{n-1} A^T A \right) \otimes (X^T X). \tag{6.6.15}$$

Further,

$$\frac{1}{n} V \to \begin{bmatrix} 1 & \sigma_{12} \\ \sigma_{12} & 1 \end{bmatrix} \otimes \Sigma, \tag{6.6.16}$$

where $n^{-1} X^T X \to \Sigma$ and Σ is positive definite.

The test statistic for $H_0 : \beta = 0$ is the quadratic form

$$A_R = L_{col}^T V^{-1} L_{col} = (n-1) L_{col}^T [(A^T A)^{-1} \otimes (X^T X)^{-1}] L_{col}, \tag{6.6.17}$$

where we use a basic formula for finding the inverse of a direct product; see (6.6.9). Before discussing the distribution theory we record one final result from traditional multivariate analysis:

$$A_R = (n-1)\text{tr}\{L^T (X^T X)^{-1} L (A^T A)^{-1}\}; \tag{6.6.18}$$

see Exercise 6.8.19. This result is useful in translating a quadratic form involving a direct product into a trace involving ordinary matrix products.

Expression (6.6.18) corresponds to the **Lawley–Hotelling trace statistic** based on ranks within the components. The following theorem summarizes the distribution theory needed to carry out the test.

Theorem 6.6.1 *Suppose $H_0 : \boldsymbol{\beta} = 0$ is true and the conditions in Section 3.4 hold. Then*

$$P_0(A_R \geq \chi_\alpha^2(2p)) \to \alpha \text{ as } n \to \infty$$

where $\chi_\alpha^2(2p)$ is the upper α percentile from a chi-square distribution with $2p$ degrees of freedom.

Proof. This theorem follows along the same lines as Theorem 3.5.2. Use a projection to establish that $(1/\sqrt{n})\mathbf{L}_{\text{col}}$ is asymptotically normally distributed and then A_R will be asymptotically chi-squared. The details are left as Exercise 6.8.20; however, the projection is provided below for use with the estimator.

$$\frac{1}{\sqrt{n}}\mathbf{L}_{\text{col}} = \frac{1}{\sqrt{n}}\begin{pmatrix} \mathbf{X}^T \boldsymbol{\varphi}^{(1)} \\ \mathbf{X}^T \boldsymbol{\varphi}^{(2)} \end{pmatrix} + o_p(1), \quad (6.6.19)$$

where $\boldsymbol{\varphi}^{(i)T} = (\varphi(F_i(\varepsilon_{1i})) \ldots \varphi(F_i(\varepsilon_{ni})))$, $i = 1, 2$, and F_1, F_2 are the marginal cdfs. Recall also that $a(i) = \varphi(i/(n+1))$, where $\varphi(\cdot)$ is the score generating function. The asymptotic covariance matrix is given in (6.6.16).

Example 6.6.1 *Multivariate Mann–Whitney–Wilcoxon Test*

We now specialize to the Wilcoxon score function $a(i) = \sqrt{12}((i/(n+1)) - 0.5)$ and consider the two-sample model. The test is a multivariate version of the Mann–Whitney–Wilcoxon test. Note that $\sum a(i) = 0$, $\sigma_a^2 = (1/(n-1))\sum a^2(i) = (n/(n+1)) \to 1$, and

$$\sigma_{a^{(1)}a^{(2)}} = \frac{12}{n-1}\sum_{i=1}^n \left(\frac{R_{i1}}{n+1} - \frac{1}{2}\right)\left(\frac{R_{i2}}{n+1} - \frac{1}{2}\right),$$

where $R_{11}, \ldots, R_{n1}$ are the ranks of the combined samples in the first component and similarly for $R_{12}, \ldots, R_{n2}$ for the second component. Note that $\sigma_{a^{(1)}a^{(2)}} = (n/(n+1))r_s$, where r_s is **Spearman's rank correlation coefficient**. Hence,

$$\frac{1}{n-1}\mathbf{A}^T\mathbf{A} = \begin{bmatrix} \frac{n}{n+1} & \sigma_{a^{(1)}a^{(2)}} \\ \sigma_{a^{(1)}a^{(2)}} & \frac{n}{n+1} \end{bmatrix} = \frac{n}{n+1}\begin{bmatrix} 1 & r_s \\ r_s & 1 \end{bmatrix} \to \begin{bmatrix} 1 & \sigma_{12} \\ \sigma_{12} & 1 \end{bmatrix},$$

where

$$\sigma_{12} = 12\iint (F_1(r) - \tfrac{1}{2})(F_2(s) - \tfrac{1}{2})dF(r,s)$$

depends on the underlying bivariate distribution.

Next, we must consider the design matrix $\mathbf{X}$ for the two-sample model. Recall (2.2.1) and (2.2.2) which cast the two-sample model as a linear model in

376   *Multivariate Models*

the univariate case. The design matrix (or vector in this case) is not centered. For convenience we modify $\mathbf{C}$ in (2.2.1) to have 1 in the first n_1 places and 0 elsewhere. Note that the mean of $\mathbf{C}$ is (n_1/n), and subtracting this from the elements of $\mathbf{C}$ yields the centered design

$$\mathbf{X} = \frac{1}{n}\begin{pmatrix} n_2 \\ \vdots \\ n_2 \\ -n_1 \\ \vdots \\ -n_1 \end{pmatrix},$$

where n_2 appears n_1 times. Then $\mathbf{X}^T\mathbf{X} = (n_1 n_2/n)$ and $(1/n)\mathbf{X}^T\mathbf{X} = (n_1 n_2/n^2) \to \lambda_1 \lambda_2$. We assume as usual that $0 < \lambda_i < 1, i = 1, 2$.

Now $\mathbf{L} = \mathbf{L}(0) = (l_1, l_2)$ and $l_i = \mathbf{X}^T \mathbf{a}^{(i)}$. It is easy to see that

$$l_i = \sum_{j=1}^{n_1} a_{ji} = \sqrt{12} \sum_{j=1}^{n_1} \left(\frac{R_{ji}}{n+1} - \frac{1}{2} \right), \; i = 1, 2.$$

So l_i is the centered and scaled sum of ranks of the first sample in the ith component.

Now $\mathbf{L}_{\text{col}} = (l_1, l_2)^T$ has an approximate bivariate normal distribution with covariance matrix

$$\text{Cov}(\mathbf{L}_{\text{col}}) = \frac{1}{n-1}(\mathbf{A}^T\mathbf{A}) \otimes (\mathbf{X}^T\mathbf{X}) = \frac{n_1 n_2}{n(n-1)} \mathbf{A}^T\mathbf{A} = \frac{n_1 n_2}{n+1} \begin{bmatrix} 1 & r_s \\ r_s & 1 \end{bmatrix}.$$

Note that σ_{12} is unknown but estimated by Spearman's rank correlation coefficient r_s (see above discussion). Hence, the test is based on A_R in (6.6.18). It is easy to invert $\text{Cov}(\mathbf{L}_{\text{col}})$ and we have (see Exercise 6.8.20)

$$A_R = \frac{n+1}{n_1 n_2 (1-r_s^2)} \{l_1^2 + l_2^2 - 2r_s l_1 l_2\} = \frac{1}{1-r_s^2} \{l_1^{*2} + l_2^{*2} - 2r_s l_1^* l_2^*\},$$

where l_1^* and l_2^* are the standardized MWW statistics. We reject $H_0 : \boldsymbol{\beta} = \mathbf{0}$ at approximately level α when $A_R \geq \chi_\alpha^2(2)$. The test statistic A_R is a quadratic form in the component MWW rank statistics and r_s provides the adjustment for the correlation between the components.

Example 6.6.2 *Brains of Mice Data*

In Table 6.6.1 we provide bivariate data on levels of certain biochemical components in the brains of mice. The treatment group received a drug which was hypothesized to alter these levels. The control group received a placebo.

The ranks of the combined treatment and control data for each component are given in the table, under component ranks. The Spearman rank correlation coefficient is $r_s = 0.149$, the standardized MWW statistics are $l_1^* = -2.74$ and $l_2^* = 2.17$. Hence $A_R = 14.31$ with the p-value 0.0008 based on a χ^2-distribution

Table 6.6.1: Levels of biochemical components in the brains of mice

Data				Component Ranks				Centered Affine Ranks			
Control		Treatment		Control		Treatment		Control		Treatment	
(1)	(2)	(1)	(2)	(1)	(2)	(1)	(2)	(1)	(2)	(1)	(2)
1.21	0.61	1.40	1.50	16	22	22	18.5	0.90	18.53	8.28	8.06
0.92	0.43	1.17	0.39	3	12	18	13	−9.02	3.55	2.06	−7.76
0.80	0.35	1.23	0.44	1	4.5	18	13	−8.81	−6.26	4.90	1.05
0.85	0.48	1.19	0.37	2	17	15	6	−9.40	9.37	4.15	−11.78
0.98	0.42	1.38	0.42	4	10.5	21	10.5	−6.80	0.74	9.79	−4.63
1.15	0.52	1.17	0.45	11	20	12.5	14.5	−0.53	15.51	0.55	5.96
1.10	0.50	1.31	0.41	9	18.5	20	9	−3.25	13.63	8.15	−7.05
1.02	0.53	1.30	0.47	6	21	19	16	−6.23	15.97	6.72	6.71
1.18	0.45	1.22	0.29	14	14.5	17	2	2.28	5.27	5.52	−16.78
1.09	0.40	1.00	0.30	7.5	8	5	3	−3.04	−4.56	−4.84	−15.02
		1.12	0.27			10	1			0.95	−18.31
		1.09	0.35			7.5	4.5			−2.03	−12.54

with 2 degrees of freedom. Figure 6.6.1(a,b) show plots of the bivariate data and the component ranks. The strong separation between treatment and control is clear from the plot. The treatment group contains an outlier which is brought in by the component ranking.

We have discussed the multivariate version of the MWW test based on the centered ranks of the combined data where the ranking mechanism is represented by the matrix $\mathbf{A}$. Given $\mathbf{A}$ and the design matrix $\mathbf{X}$, the test statistic A_R can be computed. Recall from Section 6.3.1 that Möttönen and Oja (1995) introduced the vector spatial rank $\mathbf{R}(\mathbf{x}_i) = \sum_j \mathbf{u}(\mathbf{x}_i - \mathbf{x}_j)$, where $\mathbf{u}(\mathbf{x}) = \|\mathbf{x}\|^{-1}\mathbf{x}$ is a unit vector in the direction of $\mathbf{x}$. In Section 6.4.2, an affine rank vector $\mathbf{R}(\mathbf{x}_i)$ is given by (6.4.10). Both spatial and affine rank vectors are centered. Let $\mathbf{R}(\mathbf{x}_i)$ be the ith row of $\mathbf{A}$. Note that in these two cases that the columns of $\mathbf{A}$ are not of length 1. Nevertheless, from (6.6.18), we have

$$A_R = \frac{n(n-1)}{n_1 n_2} [l_1 \; l_2](\mathbf{A}^T \mathbf{A})^{-1}[l_1 \; l_2]^T$$

$$= \frac{n(n-1)}{n_1 n_2} \frac{1}{1-r^2} \left\{ \frac{l_1^2}{\|\mathbf{a}^{(1)}\|^2} + \frac{l_2^2}{\|\mathbf{a}^{(2)}\|^2} - 2r \frac{l_1 l_2}{\|\mathbf{a}^{(1)}\| \|\mathbf{a}^{(2)}\|} \right\},$$

where r is the correlation coefficient between the two columns of $\mathbf{A}$; see Brown and Hettmansperger (1987b). Table 6.6.1 contains the affine rank vectors and the corresponding affine invariant MWW test is $A_R = 15.69$ with an approximate p-value of 0.0004 based on a $\chi^2(2)$ distribution. See Exercise 6.8.21. Peters and Randles (1990a) also develop affine invariant two-sample rank tests.

Example 6.6.3 *Multivariate Kruskal-Wallis Test*

In this example we develop the multivariate version of the Kruskal–Wallis statistic for use in a multivariate one-way layout; see Section 4.2.2 for the

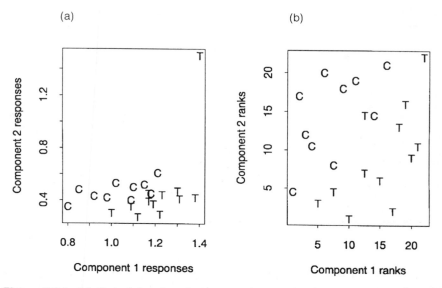

Figure 6.6.1: (a) Plot of the data for the brains of mice data; (b) Plot of the corresponding ranks of the brains of mice data.

univariate case. Suppose we have k samples from k independent distributions. The $n \times (k-1)$ design matrix is given by

$$\mathbf{C} = \begin{pmatrix} \mathbf{0}_{n_1} & \mathbf{0}_{n_1} & \cdots & \mathbf{0}_{n_1} \\ \mathbf{1}_{n_2} & \mathbf{0}_{n_2} & \cdots & \mathbf{0}_{n_2} \\ \mathbf{0}_{n_3} & \mathbf{1}_{n_3} & \cdots & \mathbf{0}_{n_3} \\ \vdots & \vdots & \vdots & \vdots \\ \mathbf{0}_{n_k} & \mathbf{0}_{n_k} & \cdots & \mathbf{1}_{n_k} \end{pmatrix}$$

and the column means are $\bar{\mathbf{c}}' = (\lambda_2, \ldots, \lambda_k)$, where $\lambda_i = (n_i/n)$. The centered design is $\mathbf{X} = \mathbf{C} - \mathbf{1}\bar{\mathbf{c}}'$ and has full column rank $k - 1$.

In this design the first of the k populations is taken as the reference population with location $(\alpha_1, \alpha_2)^T$. The ith row of the $\boldsymbol{\beta}$ matrix is the vector of shift parameters for the $(i+1)$th population relative to the first population. We wish to test $H_0 : \boldsymbol{\beta} = \mathbf{0}$ that all populations have the same (unknown) location vector.

The matrix $\mathbf{A} = (\mathbf{a}^{(1)}, \mathbf{a}^{(2)})$ has the centered and scaled Wilcoxon scores of the previous example. Hence, $\mathbf{a}^{(1)}$ is the vector of rank scores for the combined k samples in the first component. Since the rank scores are centered, we have

$$\mathbf{X}^T \mathbf{a}^{(i)} = (\mathbf{C} - \mathbf{1}\bar{\mathbf{c}}^T)^T \mathbf{a}^{(i)} = \mathbf{C}^T \mathbf{a}^{(i)}$$

and the second version is easier to compute. Now $\mathbf{L}(\mathbf{0}) = (\mathbf{L}^{(1)}, \mathbf{L}^{(2)})$ and the

hth component of column i is

$$l_{hi} = \sqrt{12} \sum_{j \in S_h} \left(\frac{R_{ji}}{n+1} - \frac{1}{2} \right)$$

$$= \frac{\sqrt{12}}{n+1} n_h \left(\overline{R}_{hi} - \frac{n+1}{2} \right),$$

where S_h is the index set corresponding to the hth sample and $\overline{R}_{hi}$ is the average rank of the hth sample in the ith component.

As in the previous example, we replace $(1/(n-1))\mathbf{A}^T\mathbf{A}$ by its limit with 1 on the main diagonal and σ_{12} off the diagonal. Then let $((\sigma^{ij}))$ be the inverse matrix. This is easy to compute and will be useful below. The test statistic is then, from (6.6.17),

$$A_R \approx (\mathbf{L}^{(1)^T}, \mathbf{L}^{(2)^T}) \begin{pmatrix} \sigma^{11}(\mathbf{X}^T\mathbf{X})^{-1} & \sigma^{12}(\mathbf{X}^T\mathbf{X})^{-1} \\ \sigma^{12}(\mathbf{X}^T\mathbf{X})^{-1} & \sigma^{22}(\mathbf{X}^T\mathbf{X})^{-1} \end{pmatrix} \begin{pmatrix} \mathbf{L}^{(1)} \\ \mathbf{L}^{(2)} \end{pmatrix}$$

$$= \sigma^{11}\mathbf{L}^{(1)^T}(\mathbf{X}^T\mathbf{X})^{-1}\mathbf{L}^{(1)} + 2\sigma^{12}\mathbf{L}^{(1)^T}(\mathbf{X}^T\mathbf{X})^{-1}\mathbf{L}^{(2)} + \sigma^{22}\mathbf{L}^{(2)^T}(\mathbf{X}^T\mathbf{X})^{-1}\mathbf{L}^{(2)}.$$

The $\approx$ indicates that the right-hand side contains asymptotic quantities which must be estimated in practice. Now

$$\mathbf{L}^{(1)^T}(\mathbf{X}^T\mathbf{X})^{-1}\mathbf{L}^{(1)} = \sum_{j=1}^{k} n_j^{-1} l_{j1}^2 = \frac{12}{(n+1)^2} \sum_{j=1}^{k} n_j \left(\overline{R}_{j1} - \frac{n+1}{2} \right) = \frac{n}{n+1} H_1,$$

where H_1 is the Kruskal–Wallis statistic computed on the first component. Similarly,

$$\mathbf{L}^{(1)^T}(\mathbf{X}^T\mathbf{X})^{-1}\mathbf{L}^{(2)} = \frac{12}{(n+1)^2} \sum_{j=1}^{k} n_j \left(\overline{R}_{j1} - \frac{n+1}{2} \right) \left(\overline{R}_{j2} - \frac{n+1}{2} \right) = \frac{n}{n+1} H_{12}$$

and H_{12} is a cross-component statistic. Using Spearman's rank correlation coefficient r_s to estimate σ_{12}, we obtain

$$A_R = \frac{1}{1-r_s^2} \{H_1 - 2r_s H_{12} + H_2\}.$$

The test rejects the null hypothesis of equal location vectors at approximately level α when $A_R \geq \chi_\alpha^2(2(k-1))$.

In order to compute the test, first compute componentwise rankings. We can display the means of the rankings in a $2 \times k$ table as follows:

	Treatment			
	1	2	$\cdots$	k
Component 1	$\overline{R}_{11}$	$\overline{R}_{21}$	$\cdots$	$\overline{R}_{1k}$
Component 2	$\overline{R}_{12}$	$\overline{R}_{22}$	$\cdots$	$\overline{R}_{2k}$

Then use Minitab or some other package to find the two Kruskal–Wallis statistics. To compute H_{12} use either the formula above or use

$$H_{12} = \sum_{j=1}^{k} \left(1 - \frac{n_j}{n}\right) Z_{j1} Z_{j2}, \tag{6.6.20}$$

where $Z_{ji} = (\bar{R}_{ji} - (n+1)/2)/\sqrt{\text{Var}\bar{R}_{ji}}$ and $\text{Var}\bar{R}_{ji} = (n - n_j)(n+1)/n_j$; see Exercise 6.8.22. The package Minitab lists Z_{ji} in its output.

Example 6.6.3 shows that in the general regression problem with Wilcoxon scores, if we wish to test $H_0: \boldsymbol{\beta} = \mathbf{0}$, the test statistic (6.6.17) can be written as

$$A_R = \frac{1}{1 - \widehat{\sigma_{12}}^2} \{\mathbf{L}^{(1)^T}(\mathbf{X}^T\mathbf{X})^{-1}\mathbf{L}^{(1)} - 2\widehat{\sigma_{12}}\mathbf{L}^{(1)^T}(\mathbf{X}^T\mathbf{X})^{-1}\mathbf{L}^{(2)} + \mathbf{L}^{(2)^T}(\mathbf{X}^T\mathbf{X})^{-1}\mathbf{L}^{(2)}\}, \tag{6.6.21}$$

where the estimate of σ_{12} can be taken to be r_s or $(n/(n+1))r_s$ and r_s is Spearman's rank correlation coefficient, and

$$l_{hi} = \frac{\sqrt{12}}{n+1} \sum_{j=1}^{n} \left(R_{ji} - \frac{n+1}{2}\right) x_{jh} = \sum_{j=1}^{n} a(R(Y_{ji})) x_{jh}.$$

Then reject $H_0: \boldsymbol{\beta} = \mathbf{0}$ when $A_R \geq \chi_\alpha^2(2p)$.

6.6.2 The Estimate of the Regression Effect

In the introduction to this section, we pointed out that the R-estimate $\widehat{\boldsymbol{\beta}}$ solves $\mathbf{L}(\boldsymbol{\beta}) \doteq \mathbf{0}$, (6.6.6). Recall the representation of the R-estimate in the univariate case given in Corollary 3.5.7. This immediately extends to the multivariate case as

$$\sqrt{n}(\widehat{\boldsymbol{\beta}} - \boldsymbol{\beta}_0) = \left(\frac{1}{n}\mathbf{X}^T\mathbf{X}\right)^{-1} \frac{1}{\sqrt{n}} (\tau_1 \mathbf{X}^T \boldsymbol{\varphi}^{(1)}, \tau_2 \mathbf{X}^T \boldsymbol{\varphi}^{(2)}) + \mathbf{o}_p(1), \tag{6.6.22}$$

where $\boldsymbol{\varphi}^{(i)^T} = (\varphi(F_i(\varepsilon_{1i})), \ldots, \varphi(F_i(\varepsilon_{ni})))$, $i = 1, 2$. Further, τ_i is given by (3.4.4) and we define the matrix $\boldsymbol{\tau}$ by $\boldsymbol{\tau} = \text{diag}\{\tau_1, \tau_2\}$. To investigate the asymptotic distribution of the random matrix (6.6.22), we again roll it out by columns. We need only consider the linear approximation on the right-hand side.

Theorem 6.6.2 *Assume the regularity conditions in Section 3.4. Then, if $\boldsymbol{\beta}$ is the true matrix,*

$$\sqrt{n}(\widehat{\boldsymbol{\beta}}_{\text{col}} - \boldsymbol{\beta}_{\text{col}}) \xrightarrow{D} N_{2p}\left(0, \left(\boldsymbol{\tau}\begin{pmatrix} 1 & \sigma_{12} \\ \sigma_{12} & 1 \end{pmatrix}\boldsymbol{\tau}\right) \otimes \boldsymbol{\Sigma}^{-1}\right),$$

where

$$\sigma_{12} = \iint \varphi(F_1(s))\varphi(F_2(t))dF(s,t), \quad \boldsymbol{\tau} = \text{diag}\{\tau_1, \tau_2\},$$

τ_i is given by (3.4.4), and $(1/n)\mathbf{X}'\mathbf{X} \to \boldsymbol{\Sigma}$ positive definite.

Proof. We will sketch the argument based on (6.6.1), (6.6.13), and Theorem 3.5.2. Consider, with Σ^{-1} replaced by $((1/n)\mathbf{X'X})^{-1}$,

$$\frac{1}{\sqrt{n}}\begin{pmatrix}\tau_1\Sigma^{-1}\mathbf{X}^T\varphi^{(1)}\\ \tau_2\Sigma^{-1}\mathbf{X}^T\varphi^{(2)}\end{pmatrix}=\begin{pmatrix}\tau_1\Sigma^{-1} & 0\\ 0 & \tau_2\Sigma^{-1}\end{pmatrix}\begin{pmatrix}\frac{1}{\sqrt{n}}\mathbf{X}^T\varphi^{(1)}\\ \frac{1}{\sqrt{n}}\mathbf{X}^T\varphi^{(2)}\end{pmatrix}.$$

The multivariate central limit theorem establishes the asymptotic multivariate normality. From the discussion after (6.6.1), we have $E\varphi^{(i)}\varphi^{(i)^T}=\mathbf{I}, i=1,2$, and $E\varphi^{(1)}\varphi^{(2)^T}=\sigma_{12}\mathbf{I}$. Hence, the covariance matrix of the above vector is:

$$\begin{pmatrix}\tau_1\Sigma^{-1} & 0\\ 0 & \tau_2\Sigma^{-1}\end{pmatrix}\begin{pmatrix}\Sigma & \sigma_{12}\Sigma\\ \sigma_{12}\Sigma & \Sigma\end{pmatrix}\begin{pmatrix}\tau_1\Sigma^{-1} & 0\\ 0 & \tau_2\Sigma^{-1}\end{pmatrix}$$

$$=\begin{pmatrix}\tau_1^2\Sigma^{-1} & \tau_1\tau_2\sigma_{12}\Sigma^{-1}\\ \tau_1\tau_2\sigma_{12}\Sigma^{-1} & \tau_2^2\Sigma^{-1}\end{pmatrix}$$

$$=\begin{pmatrix}\tau_1^2 & \tau_1\tau_2\sigma_{12}\\ \tau_1\tau_2\sigma_{12} & \tau_2^2\end{pmatrix}\otimes\Sigma^{-1}$$

$$=\left(\boldsymbol{\tau}\begin{pmatrix}1 & \sigma_{12}\\ \sigma_{12} & 1\end{pmatrix}\boldsymbol{\tau}\right)\otimes\Sigma^{-1},$$

and this is the asymptotic covariance matrix for $\sqrt{n}(\widehat{\boldsymbol{\beta}}_{col}-\boldsymbol{\beta}_{col})$.

We remind the reader that when we use the Wilcoxon score $\varphi(u)=\sqrt{12}(u-\frac{1}{2})$, then $\tau_i^{-1}=\sqrt{12}\int f_i^2(x)dx, f_i$ the marginal pdf $i=1,2$, and $\widehat{\sigma}_{12}=(n/(n+1))r_s$, where r_s is Spearman's rank correlation coefficient. See Section 3.7.1 for a discussion of the estimation of τ_i.

6.6.3 Tests of General Hypotheses

Recall the model (6.6.1) and let the matrix $\mathbf{M}$ be $r\times p$ of rank r and the matrix $\mathbf{K}$ be $2\times s$ of rank s. The matrices $\mathbf{M}$ and $\mathbf{K}$ are fully specified by the researcher. We consider a test of $H_0:\mathbf{M}\boldsymbol{\beta}\mathbf{K}=\mathbf{0}$. For example, when $\mathbf{K}=\mathbf{I}_2$, and $\mathbf{M}=(\mathbf{O}\ \mathbf{I}_r)$, where $\mathbf{O}$ denotes the $r\times(p-r)$ matrix of zeros, we have $H_0:\mathbf{M}\boldsymbol{\beta}=\mathbf{0}$ and this null hypothesis specifies that the last r parameters in both components are 0. This is the usual subhypothesis in the linear model applied to both components. Alternatively we might let $\mathbf{M}=\mathbf{I}_p, p\times p$, and

$$\mathbf{K}=\begin{pmatrix}1\\ -1\end{pmatrix}.$$

Then $H_0:\boldsymbol{\beta}\mathbf{K}=\mathbf{0}$ and we test the null hypothesis that the parameters of the two concatenated linear models are equal: $\beta_{i1}=\beta_{i2}$ for $i=1,\ldots,p$. This could be appropriate for a pre-post test model. Thus, we generalize (3.6.1) to the

382 *Multivariate Models*

multivariate linear model. The development will proceed in steps beginning with $H_0 : \mathbf{M}\boldsymbol{\beta} = \mathbf{0}$, i.e. $\mathbf{K} = \mathbf{I}_2$.

Theorem 6.6.3 *Under $H_0 : \mathbf{M}\boldsymbol{\beta} = \mathbf{0}$,*

$$\sqrt{n}(\mathbf{M}\widehat{\boldsymbol{\beta}})_{\text{col}} \xrightarrow{D} N_{2r}\left(\mathbf{0}, \left(\boldsymbol{\tau}\begin{pmatrix} 1 & \sigma_{12} \\ \sigma_{12} & 1 \end{pmatrix}\boldsymbol{\tau}\right) \otimes [\mathbf{M}\boldsymbol{\Sigma}^{-1}\mathbf{M}^T]\right);$$

here $\boldsymbol{\tau}, \sigma_{12}, \boldsymbol{\Sigma}$ are given in Theorem 6.6.2. Let $\mathbf{V}$ denote the asymptotic covariance matrix; then

$$n(\mathbf{M}\widehat{\boldsymbol{\beta}})_{\text{col}}^T \widehat{\mathbf{V}}^{-1}(\mathbf{M}\widehat{\boldsymbol{\beta}})_{\text{col}} = \widehat{\boldsymbol{\beta}}_{\text{col}}^T \left\{ \left[\widehat{\boldsymbol{\tau}}\frac{1}{n-1}\mathbf{A}^T\mathbf{A}\widehat{\boldsymbol{\tau}}\right]^{-1} \otimes \mathbf{M}^T(\mathbf{M}(\mathbf{X}^T\mathbf{X})^{-1}\mathbf{M}^T)^{-1}\mathbf{M} \right\} \widehat{\boldsymbol{\beta}}_{\text{col}}$$

$$= \text{tr}\left\{ (\mathbf{M}\widehat{\boldsymbol{\beta}})^T (\mathbf{M}(\mathbf{X}^T\mathbf{X})^{-1}\mathbf{M}^T)^{-1}(\mathbf{M}\widehat{\boldsymbol{\beta}}) \cdot \left[\widehat{\boldsymbol{\tau}}\frac{1}{n-1}\mathbf{A}^T\mathbf{A}\widehat{\boldsymbol{\tau}}\right]^{-1} \right\}$$

(6.6.23)

is asymptotically $\chi^2(2r)$. Note that we have estimated unknown parameters in the asymptotic covariance matrix $\mathbf{V}$.

Proof. First note that

$$(\mathbf{M}\widehat{\boldsymbol{\beta}})_{\text{col}} = \begin{pmatrix} \mathbf{M} & \mathbf{0} \\ \mathbf{0} & \mathbf{M} \end{pmatrix} \widehat{\boldsymbol{\beta}}_{\text{col}}$$

Using Theorem 6.6.2, the asymptotic covariance is, with $\boldsymbol{\tau} = \text{diag}\{\tau_1, \tau_2\}$,

$$\mathbf{V} = \begin{pmatrix} \mathbf{M} & \mathbf{0} \\ \mathbf{0} & \mathbf{M} \end{pmatrix} \left(\left(\boldsymbol{\tau}\begin{pmatrix} 1 & \sigma_{12} \\ \sigma_{12} & 1 \end{pmatrix}\boldsymbol{\tau} \right) \otimes \boldsymbol{\Sigma}^{-1} \right) \begin{pmatrix} \mathbf{M}^T & \mathbf{0} \\ \mathbf{0} & \mathbf{M}^T \end{pmatrix}$$

$$= \begin{pmatrix} \mathbf{M} & \mathbf{0} \\ \mathbf{0} & \mathbf{M} \end{pmatrix} \begin{pmatrix} \tau_1 \boldsymbol{\Sigma}^{-1} & \tau_1\sigma_{12}\boldsymbol{\Sigma}^{-1} \\ \tau_2\sigma_{12}\boldsymbol{\Sigma}^{-1} & \tau_2\boldsymbol{\Sigma}^{-1} \end{pmatrix} \begin{pmatrix} \mathbf{M}^T & \mathbf{0} \\ \mathbf{0} & \mathbf{M}^T \end{pmatrix}$$

$$= \begin{pmatrix} \tau_1 \mathbf{M}\boldsymbol{\Sigma}^{-1}\mathbf{M}^T & \tau_1\sigma_{12}\mathbf{M}\boldsymbol{\Sigma}^{-1}\mathbf{M}^T \\ \tau_2\sigma_{12}\mathbf{M}\boldsymbol{\Sigma}^{-1}\mathbf{M}^T & \tau_2\mathbf{M}\boldsymbol{\Sigma}^{-1}\mathbf{M}^T \end{pmatrix}$$

$$= \left(\boldsymbol{\tau}\begin{pmatrix} 1 & \sigma_{12} \\ \sigma_{12} & 1 \end{pmatrix}\boldsymbol{\tau} \right) \otimes \mathbf{M}\boldsymbol{\Sigma}^{-1}\mathbf{M}^T.$$

Hence, by the same argument,

$$\widehat{\boldsymbol{\beta}}_{col}^T \begin{pmatrix} \mathbf{M}^T & 0 \\ 0 & \mathbf{M}^T \end{pmatrix} \mathbf{V}^{-1} \begin{pmatrix} \mathbf{M} & 0 \\ 0 & \mathbf{M} \end{pmatrix} \widehat{\boldsymbol{\beta}}_{col}$$

$$= \widehat{\boldsymbol{\beta}}_{col}^T \left\{ \left[\boldsymbol{\tau} \begin{pmatrix} 1 & \sigma_{12} \\ \sigma_{12} & 1 \end{pmatrix} \boldsymbol{\tau} \right]^{-1} \otimes \mathbf{M}^T (\mathbf{M}\boldsymbol{\Sigma}^{-1}\mathbf{M}^T)^{-1} \mathbf{M} \right\} \widehat{\boldsymbol{\beta}}_{col}$$

$$= \mathrm{tr} \left\{ (\mathbf{M}\widehat{\boldsymbol{\beta}})^T (\mathbf{M}\boldsymbol{\Sigma}^{-1}\mathbf{M}^T)^{-1} (\mathbf{M}\widehat{\boldsymbol{\beta}}) \left[\boldsymbol{\tau} \begin{pmatrix} 1 & \sigma_{12} \\ \sigma_{12} & 1 \end{pmatrix} \boldsymbol{\tau} \right]^{-1} \right\}.$$

Denote the test statistic, (6.6.23) as

$$Q_{MVR} = \mathrm{tr} \left\{ (\mathbf{M}\widehat{\boldsymbol{\beta}})^T (\mathbf{M}(\mathbf{X}^T\mathbf{X})^{-1}\mathbf{M}^T)^{-1} (\mathbf{M}\widehat{\boldsymbol{\beta}}) \left[\widehat{\boldsymbol{\tau}} \frac{1}{n-1} \mathbf{A}^T \mathbf{A} \widehat{\boldsymbol{\tau}} \right]^{-1} \right\}. \quad (6.6.24)$$

Then the corresponding level α asymptotic decision rule is:

Reject $H_0: \mathbf{M}\boldsymbol{\beta} = 0$ in favor of $H_A: \mathbf{M}\boldsymbol{\beta} \neq 0$ if $Q_{MVR} \geq \chi_\alpha(2r)$. (6.6.25)

The next theorem describes the test when only $\mathbf{K}$ is involved. After that we put the two results together for the general statement.

Theorem 6.6.4 *Under $H_0: \boldsymbol{\beta}\mathbf{K} = 0$, where $\mathbf{K}$ is a $2 \times s$ matrix, $\sqrt{n}(\widehat{\boldsymbol{\beta}}\mathbf{K})_{col}$ is asymptotically*

$$N_{ps} \left(0, \left[\mathbf{K}^T \boldsymbol{\tau} \begin{pmatrix} 1 & \sigma_{12} \\ \sigma_{12} & 1 \end{pmatrix} \boldsymbol{\tau} \mathbf{K} \right] \otimes \boldsymbol{\Sigma}^{-1} \right),$$

where $\boldsymbol{\tau}, \sigma_{12}$, and $\boldsymbol{\Sigma}$ are given in Theorem 6.6.2. Let $\mathbf{V}$ denote the asymptotic covariance matrix. Then

$$n(\widehat{\boldsymbol{\beta}}\mathbf{K})_{col}^T \widehat{\mathbf{V}}^{-1} (\widehat{\boldsymbol{\beta}}\mathbf{K})_{col} = \mathrm{tr} \left\{ (\widehat{\boldsymbol{\beta}}\mathbf{K})^T (\mathbf{X}^T\mathbf{X})(\widehat{\boldsymbol{\beta}}\mathbf{K}) \left[\mathbf{K}\widehat{\boldsymbol{\tau}} \frac{1}{n-1} \mathbf{A}^T \mathbf{A} \widehat{\boldsymbol{\tau}} \mathbf{K} \right]^{-1} \right\}$$

is asymptotically $\chi^2(ps)$.

Proof. First note that $(\widehat{\boldsymbol{\beta}}^T\mathbf{K})_{col} = (\mathbf{K}^T \otimes \mathbf{I})\widehat{\boldsymbol{\beta}}^T_{col}$. Then from Theorem 6.6.2, the asymptotic covariance matrix of $\sqrt{n}\widehat{\boldsymbol{\beta}}^T_{col}$ is

$$A_{sy}\mathrm{Cov}(\sqrt{n}\,\widehat{\boldsymbol{\beta}}^T_{col}) = \left(\boldsymbol{\tau} \begin{bmatrix} 1 & \sigma_{12} \\ \sigma_{12} & 1 \end{bmatrix} \boldsymbol{\tau} \right) \otimes \boldsymbol{\Sigma}^{-1}.$$

384 Multivariate Models

Hence, the asymptotic covariance matrix of $\sqrt{n}(\widehat{\boldsymbol{\beta}}\mathbf{K})_{col}$ is,

$$A_{sy}Cov(\sqrt{n}(\widehat{\boldsymbol{\beta}}\mathbf{K})_{col}) = (\mathbf{K}^T \otimes \mathbf{I})\left(\left(\boldsymbol{\tau}\begin{bmatrix} 1 & \sigma_{12} \\ \sigma_{12} & 1 \end{bmatrix}\boldsymbol{\tau}\right) \otimes \boldsymbol{\Sigma}^{-1}\right)(\mathbf{K}^T \otimes \mathbf{I})^T$$

$$= \left(\mathbf{K}^T\boldsymbol{\tau}\begin{bmatrix} 1 & \sigma_{12} \\ \sigma_{12} & 1 \end{bmatrix}\boldsymbol{\tau} \otimes \boldsymbol{\Sigma}^{-1}\right)(\mathbf{K} \otimes \mathbf{I})$$

$$= \left(\mathbf{K}^T\boldsymbol{\tau}\begin{bmatrix} 1 & \sigma_{12} \\ \sigma_{12} & 1 \end{bmatrix}\boldsymbol{\tau}\mathbf{K}\right) \otimes \boldsymbol{\Sigma}^{-1},$$

which is the desired result. The asymptotic normality and chi-square distribution follow from Theorem 6.6.2.

The previous two theorems can be combined to yield the general case.

Theorem 6.6.5 *Under* $H_0 : \mathbf{M}\boldsymbol{\beta}\mathbf{K} = \mathbf{0}$,

$$\sqrt{n}(\mathbf{M}\widehat{\boldsymbol{\beta}}\mathbf{K})_{col} \xrightarrow{D} N_{rs}\left(\mathbf{0}, \left[\mathbf{K}^T\boldsymbol{\tau}\begin{pmatrix} 1 & \sigma_{12} \\ \sigma_{12} & 1 \end{pmatrix}\boldsymbol{\tau}\mathbf{K}\right] \otimes [\mathbf{M}\boldsymbol{\Sigma}^{-1}\mathbf{M}^T]\right).$$

If $\mathbf{V}$ *is the asymptotic covariance matrix with estimate* $\widehat{\mathbf{V}}$, *then*

$$n(\mathbf{M}\widehat{\boldsymbol{\beta}}\mathbf{K})_{col}^T \widehat{\mathbf{V}}^{-1}(\mathbf{M}\widehat{\boldsymbol{\beta}}\mathbf{K})_{col}$$

$$= (\mathbf{M}\widehat{\boldsymbol{\beta}}\mathbf{K})_{col}^T \left\{[\mathbf{K}^T\widehat{\boldsymbol{\tau}}\frac{1}{n-1}\mathbf{A}^T\mathbf{A}\widehat{\boldsymbol{\tau}}\mathbf{K}]^{-1} \otimes [\mathbf{M}(\mathbf{X}^T\mathbf{X})^{-1}\mathbf{M}^T]^{-1}\right\}(\mathbf{M}\widehat{\boldsymbol{\beta}}\mathbf{K})_{col}$$

$$= \operatorname{tr}\left\{(\mathbf{M}\widehat{\boldsymbol{\beta}}\mathbf{K})^T[\mathbf{M}(\mathbf{X}^T\mathbf{X})^{-1}\mathbf{M}^T]^{-1}(\mathbf{M}\widehat{\boldsymbol{\beta}}\mathbf{K})\left[\mathbf{K}^T\widehat{\boldsymbol{\tau}}\frac{1}{n-1}\mathbf{A}^T\mathbf{A}\widehat{\boldsymbol{\tau}}\mathbf{K}\right]^{-1}\right\} \quad (6.6.26)$$

has an asymptotic $\chi^2(rs)$ *distribution.*

The last theorem provides great flexibility in composing and testing hypotheses in the multivariate linear model. We must estimate the matrix $\boldsymbol{\beta}$ along with the other parameters familiar in the linear model. However, once we have these estimates it is a simple series of matrix multiplications and the trace operation to yield the test statistic.

Denote the test statistic (6.6.26) as

$$Q_{MVRK} = \operatorname{tr}\left\{(\mathbf{M}\widehat{\boldsymbol{\beta}}\mathbf{K})^T[\mathbf{M}(\mathbf{X}^T\mathbf{X})^{-1}\mathbf{M}^T]^{-1}(\mathbf{M}\widehat{\boldsymbol{\beta}}\mathbf{K})\left[\mathbf{K}^T\widehat{\boldsymbol{\tau}}\frac{1}{n-1}\mathbf{A}^T\mathbf{A}\widehat{\boldsymbol{\tau}}\mathbf{K}\right]^{-1}\right\}.$$

(6.6.27)

Then the corresponding level α asymptotic decision rule is:

Reject $H_0 : \mathbf{M}\boldsymbol{\beta}\mathbf{K} = \mathbf{0}$ in favor of $H_A : \mathbf{M}\boldsymbol{\beta}\mathbf{K} \neq \mathbf{0}$ if $Q_{MVRK} \geq \chi_\alpha(rs)$.

(6.6.28)

The test statistics Q_{MVR} and Q_{MVRK} are extensions to the multivariate linear model of the quadratic form test statistic $F_{\varphi,Q}$, (3.6.14). The score or aligned test and the drop in dispersion test are also available. Davis and McKean (1993) develop these in detail and provide the rigorous development of the asymptotic theory. See also Puri and Sen (1985) for a development of rank methods in the multivariate linear model.

In traditional analysis, based on the least-squares estimate of the matrix of regression coefficients, there are several tests of the hypothesis $H_0 : \mathbf{M\beta K} = \mathbf{0}$. The test statistic Q_{MVRK}, (6.6.26), is an analog of the Lawley (1938) and Hotelling (1951) trace criterion. This traditional test statistic is given by

$$Q_{LH} = \mathrm{tr}\left\{(\mathbf{M\widehat{\beta}}_{LS}\mathbf{K})^T[\mathbf{M}(\mathbf{X}^T\mathbf{X})^{-1}\mathbf{M}^T]^{-1}\mathbf{M\widehat{\beta}}_{LS}\mathbf{K}(\mathbf{K}^T\widehat{\mathbf{\Lambda}}\mathbf{K})^{-1}\right\}, \qquad (6.6.29)$$

where $\widehat{\boldsymbol{\beta}}_{LS} = (\mathbf{X'X})^{-1}\mathbf{X'Y}$ is the least-squares estimate of the matrix of regression coefficients $\boldsymbol{\beta}$ and $\widehat{\boldsymbol{\Lambda}}$ is the usual estimate of $\boldsymbol{\Lambda}$, the covariance matrix of the matrix of errors $\boldsymbol{\epsilon}$, given by

$$\widehat{\boldsymbol{\Lambda}} = (\mathbf{Y} - \mathbf{X}\widehat{\boldsymbol{\beta}}_{LS})^T(\mathbf{Y} - \mathbf{X}\widehat{\boldsymbol{\beta}}_{LS})/(n - p - 1). \qquad (6.6.30)$$

Under the above assumptions and the assumption that $\boldsymbol{\Lambda}$ is positive definite and assuming $H_0 : \mathbf{M\beta K} = \mathbf{0}$ is true, Q_{LH} has an asymptotic χ^2 distribution with rs degrees of freedom. This type of hypothesis arises in profile analysis; see Chinchilli and Sen (1982) for this application.

In order to illustrate these tests, we complete this section with an example.

Example 6.6.4 *Tablet Potency Data*

The data are the results from a pharmaceutical experiment on the effects of four factors on five measurements of a tablet. There are $n = 34$ data cases. The five responses are the potency of the tablet at the end of 2 weeks (POT2); the potency of the tablet at the end of 4 weeks (POT4); measures of the tablet's purity (RSDCU) and hardness (HARD); and water content (H_2O); hence, we have a five-dimensional response rather than the bivariate responses discussed so far. This means that the degrees of freedom are $5r$ rather than $2r$ in Theorem 6.6.3. The factors are: SAI, the amount of intragranular steric acid, which was set at the three levels, $-1, 0$, and 1; SAE, the amount of extragranular steric acid, which was set at the three levels, $-1, 0$, and 1; ADS, the amount of cross carmellose sodium, which was set at the three levels, $-1, 0$, and 1; and TYPE of steric acid, which was set at two levels, -1 and 1. The initial potency of the compound, POT0, served as a covariate. The data are displayed in Table 6.6.2. It was used as an example in the article by Davis and McKean (1993) and much of our discussion below is taken from that article.

This data set was treated as a univariate model for the response POT2 in Chapter 3; see Examples 3.3.3 and 3.9.2. As our full model we choose the same model as described in expression (3.3.1). It includes: the linear effects of the four factors; six simple two-way interactions between the factors; the three quadratic terms of the factors SAI, SAE, and ADS; and the covariate – a total of 15 terms. The need for the quadratic terms was discussed in the diagnostic analysis of this model for the response POT2; see Example 3.9.2. Hence, $\mathbf{Y}$ is 34×5, $\mathbf{X}$ is 34×14, and $\boldsymbol{\beta}$ is 14×5.

Table 6.6.2: Responses and levels of the factors for the potency data

Responses					Factors				Covariate
POT2	POT4	RSDCU	HARD	H_2O	SAE	SAI	ADS	TYPE	POT0
7.94	3.15	1.20	8.50	0.188	1	1	1	1	9.38
8.13	3.00	0.90	6.80	0.250	1	1	1	−1	9.67
8.11	2.70	2.00	9.50	0.107	1	1	−1	1	9.91
7.96	4.05	2.30	6.00	0.125	1	1	−1	−1	9.77
7.83	1.90	0.50	9.80	0.142	−1	1	1	1	9.50
7.91	2.30	0.90	6.60	0.229	−1	1	1	−1	9.35
7.82	1.40	1.10	8.43	0.112	−1	1	−1	1	9.58
7.42	2.60	2.60	8.50	0.093	−1	1	−1	−1	9.69
8.06	2.00	1.90	6.17	0.207	1	−1	1	1	9.62
8.51	2.80	1.70	7.20	0.184	1	−1	1	−1	9.89
7.88	3.35	4.70	9.30	0.107	1	−1	−1	1	9.80
7.58	3.05	4.00	8.10	0.102	1	−1	−1	−1	9.73
8.14	1.20	0.80	7.17	0.202	−1	−1	1	1	9.51
8.06	2.95	2.50	7.80	0.027	−1	−1	1	−1	9.82
7.31	1.85	2.10	8.70	0.116	−1	−1	−1	1	9.20
8.66	4.10	3.60	6.40	0.114	−1	−1	−1	−1	9.53
8.16	3.95	2.00	8.00	0.183	0	0	0	1	9.67
8.02	2.85	1.10	6.61	0.139	0	0	0	−1	9.41
8.03	3.20	3.60	9.80	0.171	0	1	0	1	9.62
7.93	3.20	6.10	7.33	0.152	0	1	0	−1	9.49
7.84	3.95	2.00	7.70	0.165	0	−1	0	1	9.96
7.59	1.15	2.10	7.03	0.149	0	−1	0	−1	9.79
8.28	3.95	0.70	8.40	0.195	1	0	0	1	9.46
7.75	3.35	2.20	6.37	0.168	1	0	0	−1	9.78
7.95	3.85	7.20	9.30	0.158	−1	0	0	1	9.48
8.69	2.80	1.30	6.57	0.169	−1	0	0	−1	9.46
8.38	3.50	1.70	8.00	0.249	0	0	1	1	9.73
8.15	2.00	2.30	6.80	0.189	0	0	1	−1	9.67
8.12	3.85	2.50	7.90	0.116	0	0	−1	1	9.84
7.72	3.50	2.20	5.60	0.110	0	0	−1	−1	9.84
7.96	3.55	1.80	7.85	0.135	0	0	0	1	9.50
8.20	2.75	0.60	7.20	0.161	0	0	0	−1	9.78
8.10	3.30	0.97	8.73	0.152	0	0	0	1	9.71
8.16	3.90	2.40	7.50	0.155	0	0	0	−1	9.57

Table 6.6.3 displays the results for the test statistic Q_{MVR}, (6.6.24), for the usual ANOVA hypotheses of interest: main effects, interaction effects broken down as simple two-way and quadratic, and covariate. Also listed are the hypothesis matrices **M** for each effect, where the notation $O_{t \times u}$ represents a

Table 6.6.3: Tests of the effects for the potency data

Effects	**M**-matrix	df	Q_{MVR}	p-value	Q_{LH}	p-value
Main	$[I_4 \ O_{4 \times 10}]$	20	179.6	0.00	91.8	0.00
Higher Order	$[O_{9 \times 4} \ I_9 \ O_{9 \times 1}]$	45	102.1	0.00	70.7	0.01
Interaction	$[O_{6 \times 4} \ I_6 \ O_{6 \times 4}]$	30	70.2	0.00	52.2	0.01
Quadratic	$[O_{3 \times 10} \ I_3 \ O_{3 \times 1}]$	15	34.5	0.00	18.7	0.23
Covariate	$[O_{1 \times 13} \ 1]$	5	3.88	0.57	4.34	0.50

$t \times u$ matrix of 0s and $\mathbf{I}_t$ is the $t \times t$ identity matrix. Also given for comparison purposes are the results of the traditional Lawley–Hotelling test, based on the statistic (6.6.29) with $\mathbf{K} = \mathbf{I}_5$. For example, $\mathbf{M} = [\mathbf{I}_4 \; \mathbf{O}_{4 \times 10}]$ yields a test of the hypothesis:

$$H_0 : \beta_{11} = \cdots = \beta_{41} = 0, \beta_{12} = \cdots = \beta_{42} = 0, \ldots, \beta_{15} = \cdots = \beta_{45} = 0;$$

that is, the linear terms vanish in all five components. Note that $\mathbf{M}$ is 4×14 so $r = 4$ and hence we have $4 \times 5 = 20$ degrees of freedom in Theorem 6.6.3. The other hypothesis matrices are developed similarly. The robust analysis indicates that all effects are significant except the covariate effect. In particular, the quadratic effect is significant for the robust analysis but not for the Lawley–Hotelling test. This confirms the discussion on least squares and robust residual plots for this data set given in Example 3.9.2.

Are the effects of the factors different on potencies of the tablet after 2 weeks, POT2, or 4 weeks, POT4? This question can be answered by evaluating the statistic Q_{MVRK}, (6.6.27), for hypotheses of the form $\mathbf{M}\boldsymbol{\beta}\mathbf{K}$, for the matrices $\mathbf{M}$ given in Table 6.6.3 and the 5×1 matrix $\mathbf{K}$ where $\mathbf{K}' = [1 \; -1 \; 0 \; 0 \; 0]$. For example, $\beta_{11}, \ldots, \beta_{41}$ are the linear effects of SAE, SAI, ADS, and TYPE on PO2 and $\beta_{12}, \ldots, \beta_{42}$ are the linear effects on PO4. We may want to test the hypothesis

$$H_0 : \beta_{11} = \beta_{12}, \ldots, \beta_{41} = \beta_{42}.$$

The $\mathbf{M}$ matrix picks the appropriate βs within a component and the $\mathbf{K}$ matrix compares the results across components. From Table 6.6.3, choose $\mathbf{M} = [\mathbf{I}_4 \; \mathbf{O}_{4 \times 10}]$. Then

$$\mathbf{M}\boldsymbol{\beta} = \begin{bmatrix} \beta_{11} & \beta_{12} & \cdots & \beta_{15} \\ \vdots & \vdots & & \vdots \\ \beta_{41} & \beta_{42} & \cdots & \beta_{45} \end{bmatrix}.$$

Next choose $\mathbf{K}^T = [1 \; -1 \; 0 \; 0 \; 0]$ so that

$$\mathbf{M}\boldsymbol{\beta}\mathbf{K} = \begin{bmatrix} \beta_{11} - \beta_{12} \\ \beta_{21} - \beta_{22} \\ \vdots \\ \beta_{41} - \beta_{42} \end{bmatrix}.$$

Then the null hypothesis is $H_0 : \mathbf{M}\boldsymbol{\beta}\mathbf{K} = \mathbf{0}$. In this example $r = 4$ and $s = 1$, so the test has $rs = 4$ degrees of freedom. The test is illustrated in column 3 of Table 6.6.4. Other comparisons are also given. Once again, for comparison purposes the results for the Lawley–Hotelling test based on the test statistic, (6.6.29), are given also. The robust and traditional analyses seem to agree on the contrasts. Although there is some overall difference the factors behave somewhat the same on the responses.

Suppose we have the linear model (6.6.2) along with a matrix of scores that sum to zero. The criterion function and the matrix of partial derivatives are given by (6.6.5) and (6.6.6). Then the test statistic for a general regression

Table 6.6.4: Contrast analyses between responses POT2 and POT4

	All terms except mean	Covariate terms	Main effect terms	Higher order terms	Interaction terms	Quadratic terms
df	14	1	4	9	6	3
Q_{MVRK}	21.93	3.00	5.07	12.20	8.22	4.77
p-value	0.08	0.08	0.28	0.20	0.22	0.19
Q_{LH}	22.28	2.67	6.36	11.73	6.99	5.48
p-value	0.07	0.10	0.17	0.23	0.32	0.14

effect is given by (6.6.17) or (6.6.18). Special cases yield the two-sample and k-sample tests discussed in Examples 6.6.1 and 6.6.3. The componentwise rank case uses chi-square critical values. The computation of the tests requires the score matrix **A** along with the design matrix **X**. For example, we could use the L_1 norm componentwise and produce multivariate sign tests that extend Mood's test to the multivariate model.

This approach can be extended to the spatial rank and affine rank cases; recall the discussion in Example 6.3.2. In the spatial case the criterion function is $D(\alpha, \beta) = \sum \|\mathbf{y}_i^T - \alpha^T - \mathbf{x}_i'\beta\|$, (6.1.4). Let $\mathbf{u}(\mathbf{x}) = \|\mathbf{x}\|^{-1}\mathbf{x}$ and $\mathbf{r}_i^T = \alpha^T - \mathbf{x}_i'\beta$; then $D(\alpha, \beta) = \sum \mathbf{u}^T(\mathbf{r}_i)\mathbf{r}_i$ and, hence,

$$\mathbf{A} = \begin{pmatrix} \mathbf{u}^T(\mathbf{r}_1) \\ \vdots \\ \mathbf{u}^T(\mathbf{r}_n) \end{pmatrix}.$$

Further, let $\mathbf{R}_c(\mathbf{r}_i) = \sum_j \mathbf{u}(\mathbf{r}_i - \mathbf{r}_j)$ be the centered spatial rank vector. Then the criterion function is $D(\alpha, \beta) = \sum \mathbf{R}_c^T(\mathbf{r}_i)\mathbf{r}_i$ and

$$\mathbf{A}^* = \begin{pmatrix} \mathbf{R}^T(\mathbf{r}_1) \\ \vdots \\ \mathbf{R}^T(\mathbf{r}_n) \end{pmatrix}.$$

The tests then can be carried out using the chi-square critical values. See Brown and Hettmansperger (1987b) and Möttönen and Oja (1995) for details. For the details in the affine invariant sign or rank vector cases see Brown and Hettmansperger (1987b), Hettmansperger, Nyblom, and Oja (1994), and Hettmansperger, Möttönen, and Oja (1997a; 1997b).

Rao (1988) and Bai *et al.* (1990) consider a different formulation of a linear model. Suppose, for $i = 1, \ldots, n$, that $\mathbf{Y}_i = \mathbf{X}_i\beta + \epsilon_i$, where $\mathbf{Y}_i$ is a 2×1 vector, $\mathbf{X}_i$ is a $q \times 2$ matrix of known values, and β is a 2×1 vector of unknown parameters. Further, $\epsilon_1, \ldots, \epsilon_n$ is an iid set of random vectors from a distribution with median vector **0**. The criterion function is $\sum \|\mathbf{Y}_i - \mathbf{X}_i\beta\|$, the spatial criterion function. Estimates, tests, and the asymptotic theory are developed in the above references.

6.7 Experimental Designs

Recall that in Chapter 4 we developed rank-based procedures for experimental designs based on the general R-estimation and testing theory of Chapter 3. Analogously in the multivariate case, rank-based procedures for experimental designs can be based on the R-estimation and testing theory of the previous section. In this short section we show how this development can proceed. In particular, we use the cell median model (the basic model of Chapter 4), and show how the test (6.6.28) can be used to test general linear hypotheses involving contrasts in these cell medians. This allows the testing of MANOVA-type hypotheses as well as, for instance, profile analyses for multivariate data.

Suppose we have k groups and within the jth group, $j = 1, \ldots, k$, we have a sample of size n_j. For each subject a d-dimensional vector of variables has been recorded. Let y_{ijl} denote the response for the ith subject in group j for the lth variable and let $\mathbf{y}_{ij} = (y_{ij1}, \ldots, y_{ij2})^T$ denote the vector of responses for this subject. Consider the model,

$$\mathbf{y}_{ij} = \boldsymbol{\mu}_j + \mathbf{e}_{ij}, \quad j = 1, \ldots, k; \quad i = 1, \ldots, n_k,$$

where the $\mathbf{e}_{ij}$ are independent and identically distributed. Let $n = \sum n_j$ denote the total sample size. Let $\mathbf{Y}_{n \times d}$ denote the matrix of responses (the $\mathbf{y}_{ij}$s are stacked sequentially by group) and let $\boldsymbol{\epsilon}$ be the corresponding $n \times d$ matrix of $\mathbf{e}_{ij}$. Let $\boldsymbol{\Gamma} = (\boldsymbol{\mu}_1, \ldots, \boldsymbol{\mu}_k)^T$ be the $k \times d$ matrix of parameters. We can then write the model as

$$\mathbf{Y} = \mathbf{W}\boldsymbol{\Gamma} + \boldsymbol{\epsilon}, \tag{6.7.1}$$

where $\mathbf{W}$ is the incidence matrix in expression (4.2.5). This is our **full model** and it is the multivariate analog of the basic model of Chapter 4, (4.2.1). If $\boldsymbol{\mu}_j$ is the vector of medians then this is the **multivariate medians model**. On the other hand, if $\boldsymbol{\mu}_j$ is the vector of means then this is the **multivariate means model**.

We are interested in the following general hypotheses:

$$H_0 : \mathbf{M}\boldsymbol{\Gamma}\mathbf{K} = \mathbf{O} \text{ versus } H_A : \mathbf{M}\boldsymbol{\Gamma}\mathbf{K} \neq \mathbf{O}, \tag{6.7.2}$$

where $\mathbf{M}$ is an $r \times k$ contrast matrix (the rows of $\mathbf{M}$ sum to zero) of rank r and $\mathbf{K}$ is a $d \times s$ matrix of rank s.

In order to use the theory of Section 6.6 we need to transform model (6.7.1) into a model of the form (6.6.2). Consider the $k \times k$ elementary column matrix $\mathbf{E}$ which replaces the first column of a matrix by the sum of all columns of the matrix, i.e.

$$[\mathbf{c}_1 \; \mathbf{c}_2 \; \cdots \; \mathbf{c}_k]\mathbf{E} = \left[\sum_{i=1}^{k} \mathbf{c}_i \; \mathbf{c}_2 \; \cdots \; \mathbf{c}_k\right], \tag{6.7.3}$$

for any matrix $[\mathbf{c}_1 \; \mathbf{c}_2 \; \cdots \; \mathbf{c}_k]$. Note that $\mathbf{E}$ is nonsingular. Hence we can write model (6.7.1) as

$$\mathbf{Y} = \mathbf{W}\boldsymbol{\Gamma} + \boldsymbol{\epsilon} = \mathbf{W}\mathbf{E}\mathbf{E}^{-1}\boldsymbol{\Gamma} + \boldsymbol{\epsilon} = [\mathbf{1} \; \mathbf{W}_1]\begin{bmatrix}\boldsymbol{\alpha}^T \\ \boldsymbol{\beta}\end{bmatrix} + \boldsymbol{\epsilon}, \tag{6.7.4}$$

390 *Multivariate Models*

where $\mathbf{W}_1$ is the last $k-1$ columns of $\mathbf{W}$ and $\mathbf{E}^{-1}\boldsymbol{\Gamma} = [\boldsymbol{\alpha}\ \boldsymbol{\beta}^T]^T$. This is a model of the form (6.6.2). Since $\mathbf{M}$ is a contrast matrix, its rows sum to zero. Hence the hypothesis simplifies to:

$$\mathbf{M}\boldsymbol{\Gamma}\mathbf{K} = \mathbf{M}\mathbf{E}\mathbf{E}^{-1}\boldsymbol{\Gamma}\mathbf{K} = [\mathbf{0}\ \mathbf{M}_1]\begin{bmatrix}\boldsymbol{\alpha}^T\\ \boldsymbol{\beta}\end{bmatrix}\mathbf{K} = \mathbf{M}_1\boldsymbol{\beta}\mathbf{K}. \tag{6.7.5}$$

Therefore the hypotheses (6.7.2) can be tested by the procedure (6.6.28) based on the fit of model (6.7.4).

Most of the interesting hypotheses in MANOVA can be written in the form (6.7.2) for some specified contrast matrix $\mathbf{M}$. Therefore, based on the theory developed in Section 6.6, a robust rank-based methodology can be developed for MANOVA-type models. This methodology is demonstrated in Example 6.7.1, which follows, and Exercise 6.8.23.

For the multivariate setting, Davis and McKean (1993) developed an analog of Theorem 3.5.11 which gives the joint asymptotic distribution of $[\widehat{\boldsymbol{\alpha}}\ \widehat{\boldsymbol{\beta}}^T]^T$. They further developed a test of the hypothesis $H_0: \mathbf{M}\boldsymbol{\Gamma}\mathbf{K} = \mathbf{O}$, where $\mathbf{M}$ is any full row rank matrix, not necessarily a contrast matrix. Hence, this provides a robust rank-based analysis for any multivariate linear model.

Example 6.7.1 *Paspalum Grass*

This data set, discussed by Seber (1984, p. 460), concerns the effect on growth of paspalum grass due to a fungal infection. The experiment was a 4×2 two-way design. Half of the 48 pots of paspalum grass in the experiment were inoculated with a fungal infection and half were left as controls. The second factor was the temperature (14, 18, 22, 26°C) at which the inoculation was applied. The design was balanced so that six plants were used for each combination of treatment and temperature. After a specified amount of time, the following three measurements were made on each plant:

$y_1 = $ the fresh weight of the roots of the plant (g)

$y_2 = $ the maximum root length of the plant (mm)

$y_3 = $ the fresh weight of the tops of the plant (g).

The data are displayed in Table 6.7.1.

As a full model we fit model (6.7.1). Based on the residual analysis found in Exercise 6.8.24, though, the fit clearly shows heteroscedasticity and suggests the log transformation. The subsequent analysis is based on the transformed data. Table 6.7.2 displays the estimates of model (6.7.1) based on the Wilcoxon score function and least squares. Note the fits are very similar. The estimates of the vector $\boldsymbol{\tau}$ and the matrix $\mathbf{A}^T\mathbf{A}$ are also displayed.

The hypotheses of interest concern the average main effects and interaction. For model (6.7.1), matrices for treatment effects, temperature effects and interaction are given by

Table 6.7.1: Responses for the paspalum grass data, Example 6.7.1

Temperature	Control			Inoculated		
14°C	2.2	23.5	1.7	2.3	23.5	2.0
	3.0	27.0	2.3	3.0	21.0	2.7
	3.3	24.5	3.2	2.3	22.0	1.8
	2.2	20.5	1.5	2.5	22.5	2.4
	2.0	19.0	2.0	2.4	21.5	1.1
	3.5	23.5	2.9	2.7	25.0	2.6
18°C	21.8	41.5	23.0	10.1	43.5	14.2
	11.0	32.5	15.4	7.6	27.0	14.7
	16.4	46.5	22.8	19.7	32.5	21.4
	13.1	31.0	21.5	4.3	28.5	9.7
	15.4	41.5	20.8	5.2	33.5	12.2
	14.5	46.0	20.3	3.9	24.5	8.2
22°C	13.6	29.5	30.8	10.0	21.0	23.6
	6.2	23.5	14.6	12.3	49.0	28.1
	16.7	58.5	36.0	4.9	28.5	13.3
	12.2	40.5	23.9	9.6	27.0	24.6
	8.7	37.0	20.3	6.5	29.0	19.3
	12.3	41.5	27.7	13.6	30.5	31.5
26°C	3.0	24.0	10.2	4.2	25.5	13.3
	5.3	26.5	15.6	2.2	23.5	8.5
	3.1	24.5	14.7	2.8	19.5	11.8
	4.8	34.0	20.5	1.3	21.5	7.8
	3.4	22.5	14.3	4.2	28.5	15.1
	7.4	32.0	23.2	3.0	25.0	11.8

Table 6.7.2: Estimates based on the Wilcoxon and LS fits for the paspalum grass data, Example 6.7.1. **V** is the variance–covariance matrix of vector of random errors ϵ

Parameter	Wilcoxon Fit Components			LS Fit Components		
	(1)	(2)	(3)	(1)	(2)	(3)
μ_{11}	1.04	3.14	0.82	0.97	3.12	0.78
μ_{21}	2.74	3.70	3.05	2.71	3.67	3.02
μ_{31}	2.47	3.63	3.25	2.40	3.61	3.20
μ_{41}	1.49	3.29	2.79	1.45	3.29	2.76
μ_{12}	0.94	3.12	0.77	0.92	3.12	0.70
μ_{22}	1.95	3.43	2.58	1.96	3.43	2.55
μ_{32}	2.26	3.36	3.19	2.19	3.39	3.11
μ_{42}	1.09	3.18	2.45	1.01	3.17	2.41
τ or σ	0.376	0.188	0.333	0.370	0.197	0.292
$\mathbf{A}^T\mathbf{A}$ or **V**	1.04	0.62	0.92	0.14	0.04	0.09
	0.62	1.04	0.57	0.04	0.04	0.03
	0.92	0.57	1.04	0.09	0.03	0.09

$$\mathbf{M}_{\text{Treat.}} = [1 \quad 1 \quad 1 \quad 1 \quad -1 \quad -1 \quad -1 \quad -1]$$

$$\mathbf{M}_{\text{Temp.}} = \begin{bmatrix} 1 & -1 & 0 & 0 & 1 & -1 & 0 & 0 \\ 1 & 0 & -1 & 0 & 1 & 0 & -1 & 0 \\ 1 & 0 & 0 & -1 & 1 & 0 & 0 & -1 \end{bmatrix}$$

$$\mathbf{M}_{\text{Treat.} \times \text{Temp.}} = \begin{bmatrix} 1 & -1 & 0 & 0 & -1 & 1 & 0 & 0 \\ 0 & 1 & -1 & 0 & 0 & -1 & 1 & 0 \\ 0 & 0 & 1 & -1 & 0 & 0 & -1 & 1 \end{bmatrix}.$$

Take the matrix $\mathbf{K}$ to be $\mathbf{I}_3$. Then the hypotheses of interest can be expressed as $\mathbf{M\Gamma K} = \mathbf{O}$ for the above M-matrices. Using the summary statistics in Table 6.7.2 and the elemenatry column matrix $\mathbf{E}$, as defined above expression (6.7.3), we obtained the test statistics Q_{MVRK}, (6.6.27), based on the Wilcoxon fit. For comparison we also obtain the LS test statistics Q_{LH}, (6.6.29). The values of these statistics for the hypotheses of interest are summarized in Table 6.7.3. The test for interaction is not significant while both main effects, treatment and temperature, are significant. The results are quite similar for the traditional test also. We also tabulated the marginal test statistics, F_φ. The results for each component are the similar to the multivariate result.

Table 6.7.3: Test statistics Q_{MVRK} and Q_{LH} based on the Wilcoxon and LS fits, respectively, for the paspalum grass data, Example 6.7.1. Marginal F-tests are also given. The numerator degrees of freedom are given. Note that the denominator degrees of freedom for the marginal F-tests is 40

	Wilcoxon					LS				
	MVAR			Marginal F_φ		MVAR			Marginal F_{LS}	
Effect	df	Q_{MVRK}	df	(1)	(2)	(3)	df	Q_{LH}	df	(1)	(2)	(3)
Treat.	3	14.9	1	9.19	7.07	11.6	3	12.2	1	11.4	6.72	8.66
Temp.	9	819	3	32.5	13.4	61.4	9	980	3	45.2	13.4	162
Treat. x Temp.	9	11.2	3	2.27	1.49	1.35	9	7.98	3	2.01	0.79	1.36

6.8 Exercises

6.8.1 *Show that the vector of sample means of the components is affine equivariant. See Definition 6.1.1.*

6.8.2 *Compute the gradients of the three criterion functions (6.1.3)–(6.1.5).*

6.8.3 *Show that in the univariate case* $\mathbf{S}_2(\boldsymbol{\theta}) = \mathbf{S}_3(\boldsymbol{\theta})$, *(6.1.7) and (6.1.8).*

6.8.4 *Establish (6.2.6).*

6.8.5 *Construct an example in the bivariate case for which the mean vector rotates into the new mean vector but the vector of componentwise medians does not rotate into the new vector of medians.*

6.8.6 Students were given a math aptitude and reading comprehension test before starting an intensive study skills workshop. At the end of the program they were given the test again. The following data represents the change in the math and reading tests for the five students in the program.

Math	Reading
11	7
20	40
−10	−4
10	12
16	5

We would like to test the hypothesis $H_0 : \boldsymbol{\theta} = \mathbf{0}$ versus $H_A : \boldsymbol{\theta} \neq \mathbf{0}$. Following the discussion at the beginning of Section 6.2.2, find the sign change distribution of the componentwise sign test and find the conditional p-value.

6.8.7 Prove Theorem 6.2.2.

6.8.8 Using the projection method discussed in Chapter 2, derive the projection of the statistic given in (6.2.13).

6.8.9 Apply Lemma 6.2.1 and show that (6.2.18) provides the bounds on the testing efficiency of the Wilcoxon test relative to Hotelling's test in the case of a bivariate normal distribution.

6.8.10 Prove Theorem 6.3.1.

6.8.11 Show that (6.3.10) can be generalized to k dimensions.

6.8.12 Consider the spatial L_1 methods.
(a) Show that the efficiency of the spatial L_1 methods relative to the L_2 methods with a k-variate spherical model is given by

$$e_k(\text{spatial } L_1, L_2) = \left(\frac{k-1}{k}\right)^2 E(r^2)[E(r^{-1})]^2$$

(b) Next assume that the k-variate spherical model is normal. Show that $Er^{-1} = \Gamma[(k-1)/2]/\sqrt{2}\Gamma(k/2)$ with $\Gamma(1/2) = \sqrt{\pi}$.

6.8.13 Verify by direct calculation that the efficiency of the Hössjer and Croux spatial rank test relative to Hotelling's test is 0.985 for the bivariate spherical normal model.

6.8.14 Verify (6.3.16).

6.8.15 Complete the proof of Theorem 6.4.3 by establishing the third formula for $S_8(\mathbf{0})$.

6.8.16 Show that the Oja median and Oja sign test are affine equivariant and affine invariant, respectively. See Section 6.4.1.

6.8.17 Show that the maximum breakdown point for a translation equivariant estimator is $(n+1)/(2n)$. An estimator is translation equivariant if $T(\mathbf{X} + a\mathbf{1}) = T(\mathbf{X}) + a\mathbf{1}$, for every real a. Note that $\mathbf{1}$ is the vector of all ones.

6.8.18 Verify (6.6.6).

6.8.19 Show that (6.6.18) can be derived from (6.6.17).

6.8.20 Fill in the details of the proof of Theorem 6.6.1.

6.8.21 Show that $A_R = 15.69$ in Example 6.6.2, using Table 6.6.1.

6.8.22 Verify formula (6.6.20).

6.8.23 Consider model (6.7.1) for a repeated measures design in which the responses are recorded on the same variable over time, i.e. y_{ijl} is response for the ith subject in group j at time period l. In this model the vector $\boldsymbol{\mu}_j$ is the **profile vector** for the jth group and the plot of μ_{ij} versus i is called the profile plot for group j. Let $\widehat{\boldsymbol{\mu}}_j$ denote the estimate of $\boldsymbol{\mu}_j$ based on the R fit of model (6.7.1). The plot of $\widehat{\mu}_{ij}$ versus j is called the sample profile plot of group j. These group plots are overlaid and are called the **sample profiles**. A hypothesis of interest is whether or not the population profiles are parallel.

(a) Let $\mathbf{A}_{t-1}$ be the $(t-1) \times t$ matrix given by

$$\mathbf{A}_{t-1} = \begin{bmatrix} 1 & -1 & 0 & \cdots & 0 \\ 0 & 1 & -1 & \cdots & 0 \\ \vdots & \vdots & \vdots & & \vdots \\ 0 & 0 & 0 & \cdots & -1 \end{bmatrix}.$$

Show that parallel profiles are equivalent to the null hypothesis H_0 defined by:

$$H_0: \mathbf{A}_{k-1}\boldsymbol{\Gamma}\mathbf{A}_{d-1} = \mathbf{O} \text{ versus } H_A: \mathbf{A}_{k-1}\boldsymbol{\Gamma}\mathbf{A}_{d-1} \neq \mathbf{O}, \quad (6.8.1)$$

where $\boldsymbol{\Gamma}$ is defined in model (6.7.1). Hence show that a test of parallel profiles can be based on the test (6.6.28).

(b) The data in Table 6.8.1 are the times (in seconds) it took three different species (A, B, and C) of rats to run a maze at four different times (I, II, III, and IV). Each row contains the scores of a single rat. Compare the sample profile plots based on Wilcoxon and LS estimates.

Table 6.8.1: Data for Exercise 6.8.23

	Group A					Group B					Group C			
	Times					Times					Times			
Rat	I	II	III	IV	Rat	I	II	III	IV	Rat	I	II	III	IV
1	47	53	51	28	6	44	57	46	27	11	45	33	30	18
2	35	66	38	39	7	47	29	21	30	12	30	50	21	25
3	43	40	34	40	8	28	76	29	39	13	33	32	32	24
4	49	60	44	32	9	57	63	60	15	14	44	62	38	22
5	41	61	38	32	10	34	62	41	27	15	40	42	33	24

(c) Test the hypotheses (6.8.1) using the procedure (6.6.28) based on Wilcoxon scores. Repeat using the LS test procedure (6.6.29).

(d) Repeat items (b) and (c) if the 13th rat at time period 2 took 80 seconds to run the maze instead of 32. Note that p-value of the LS procedure changes from 0.77 to 0.15 while the p-value of the Wilcoxon procedure changes from 0.95 to 0.85.

6.8.24 Consider the data of Example 6.7.1.
(a) Using the Wilcoxon scores, fit model (6.7.4) to the original data displayed in Table 6.7.1. Obtain the marginal residual plots which show heteroscedasticity. Reason that the log transformation is appropriate. Show that the residual plots based on the transformed data remove much of the heteroscedasticity. For both the transformed and original data obtain the internal Wilcoxon studentized residuals. Identify the outliers.
(b) In order to see the effect of the transformation, obtain the Wilcoxon and LS analyses of Example 6.7.1 based on the original data. Discuss your findings.

Appendix A
Asymptotic Results

A.1 Central Limit Theorems

The following version of the Lindeberg–Feller central limit theorem will be useful. A proof of it can be found in Arnold (1981).

Theorem A.1.1 *Consider the sequence of independent random variables* $W_{1n}, \ldots, W_{nn}$, *for* $n = 1, 2, \ldots$. *Suppose* $E(W_{in}) = 0$, $\text{Var}(W_{in}) = \sigma_{in}^2 < \infty$,

$$\max_{1 \leq i \leq n} \sigma_{in}^2 \to 0, \text{ as } n \to \infty, \tag{A.1.1}$$

$$\sum_{i=1}^{n} \sigma_{in}^2 \to \sigma^2, \ 0 < \sigma^2 < \infty, \text{ as } n \to \infty, \tag{A.1.2}$$

and

$$\lim_{n \to \infty} \sum_{i=1}^{n} E(W_{in}^2 I_\varepsilon(|W_{in}|)) = 0, \tag{A.1.3}$$

for all $\varepsilon > 0$, *where* $I_a(|x|)$ *is 0 or 1 when* $|x| > a$ *or* $|x| \leq a$, *respectively. Then*

$$\sum_{i=1}^{n} W_{in} \xrightarrow{D} N(0, \sigma^2).$$

A useful corollary to this theorem is given next; see Hájek and Šidák (1967, p. 153).

Corollary A.1.2 *Suppose that the sequence of random variables* $X_1, \ldots, X_n$ *are iid with* $E(X_i) = 0$ *and* $\text{Var}(X_i) = \sigma^2 < \infty$. *Suppose the sequence of constants* $a_{1n}, \ldots, a_{nn}$ *are such that*

$$\sum_{i=1}^{n} a_{in}^2 \to \sigma_a^2, \text{ as } n \to \infty, \ 0 < \sigma_a^2 < \infty, \tag{A.1.4}$$

$$\max_{1 \leq i \leq n} |a_{in}| \to 0, \text{ as } n \to \infty. \tag{A.1.5}$$

Then

$$\sum_{i=1}^{n} a_{in} X_i \xrightarrow{D} N(0, \sigma^2 \sigma_a^2).$$

Proof. Take W_{in} of Theorem A.1.1 to be $a_{in}X_i$. Then the mean of W_{in} is 0 and its variance is $\sigma_{in}^2 = a_{in}^2\sigma^2$. By (A.1.5), $\max \sigma_{in}^2 \to 0$, and by (A.1.4), $\sum \sigma_{in}^2 \to \sigma^2\sigma_a^2$. Hence we need only show that condition (A.1.3) is true. For $i = 1, \ldots, n$, define

$$W_{in}^* = \max_{1 \leq j \leq n} |a_{jn}||X_i|.$$

Then $|W_{in}^*| \geq |W_{in}|$; hence, $I_\varepsilon(|W_{in}|) \leq I_\varepsilon(|W_{in}^*|)$, for $\varepsilon > 0$. Therefore,

$$\sum_{i=1}^n E[W_{in}^2 I_\varepsilon(|W_{in}|)] \leq \sum_{i=1}^n E[W_{in}^2 I_\varepsilon(|W_{in}^*|)] = \left\{\sum_{i=1}^n a_{in}^2\right\} E[X_1^2 I_\varepsilon(|W_{1n}^*|)]. \quad \text{(A.1.6)}$$

Note that the sum in braces converges to $\sigma^2 \sigma_a^2$. Because $X_1^2 I_\varepsilon(|W_{1n}^*|)$ converges to 0 pointwise and is bounded above by the integrable function X_1^2, it then follows from Lebesgue's dominated convergence theorem that the right-hand side of (A.1.6) converges to 0. Thus condition (A.1.3) of Theorem A.1.1 is true and we are finished.

Note that the simple central limit theorem follows from this corollary by taking $a_{in} = n^{-1/2}$, so that (A.1.4) and (A.1.5) hold.

A.2 Simple Linear Rank Statistics

In the next two subsections, we present the asymptotic distribution theory for a simple linear rank statistic under the null and local alternative hypotheses. This theory is used in Chapters 1 and 2 for location models, and in Section A.3 will be useful in establishing asymptotic linearity and quadraticity results for Chapters 3 and 5. The theory for a simple linear rank statistic is presented in detail in Chapters 5 and 6 of the book by Hájek and Šidák (1967); hence, here we will only present a heuristic development with appropriate references to Hájek and Šidák. Also, Chapter 8 of Randles and Wolfe (1979) presents a detailed development of the null asymptotic theory of a simple linear rank statistic.

In this section we assume that the sequence of random variables $Y_1, \ldots, Y_n$ are iid with common density function $f(y)$ which follows assumption (E.1), (3.4.1). Let $x_1, \ldots, x_n$ denote a sequence of centered, ($\bar{x} = 0$) regression coefficients and assume that they follow assumptions (D.2), (3.4.7), and (D.3), (3.4.8). For this one-dimensional case, these assumptions simplify to:

$$\frac{\max x_i^2}{\sum_{i=1}^n x_i^2} \to 0 \quad \text{(A.2.1)}$$

$$\frac{1}{n}\sum_{i=1}^n x_i^2 \to \sigma_x^2, \sigma_x^2 > 0, \quad \text{(A.2.2)}$$

for some constant σ_x^2. It follows from these assumptions that $\max_i |x_i|/\sqrt{n} \to 0$, a fact that we will find useful. Assume that the score function $\varphi(u)$ is defined on the interval $(0, 1)$ and that it satisfies (S.1), (3.4.10); in particular,

$$\int_0^1 \varphi(u)\, du = 0 \text{ and } \int_0^1 \varphi^2(u)\, du = 1. \qquad (A.2.3)$$

Consider then the linear rank statistics

$$S = \sum_{i=1}^n x_i a(R(Y_i)), \qquad (A.2.4)$$

where the scores are generated as $a(i) = \varphi(i/(n+1))$.

A.2.1 Null Asymptotic Distribution Theory

It follows immediately that the mean and variance of S are given by

$$E(S) = 0 \text{ and } \mathrm{Var}(S) = \sum_{i=1}^n x_i^2 \left\{ \frac{1}{n-1} \sum_{i=1}^n a^2(i) \right\} \doteq \sum_{i=1}^n x_i^2, \qquad (A.2.5)$$

where the approximation is due to the fact that the quantity in braces is a Riemann sum of $\int_0^1 \varphi^2(u)\, du = 1$.

Note that we can write S as

$$S = \sum_{i=1}^n x_i \varphi\left(\frac{n}{n+1} F_n(Y_i) \right), \qquad (A.2.6)$$

where F_n is the empirical distribution function of $Y_1, \ldots, Y_n$. This suggests the approximation

$$T = \sum_{i=1}^n x_i \varphi(F(Y_i)). \qquad (A.2.7)$$

We have immediately from (A.2.3) that the mean and variance of T are

$$E(T) = 0 \text{ and } \mathrm{Var}(T) = \sum_{i=1}^n x_i^2. \qquad (A.2.8)$$

Furthermore, by assumptions (A.2.1) and (A.2.2), we can apply Corollary A.1.2 to show that

$$\frac{1}{\sqrt{n}} T \text{ is asymptotically distributed as } N(0, \sigma_x^2). \qquad (A.2.9)$$

Because the means of S and T are the same, it will follow that S has the same asymptotic distribution as T provided the second moment of their difference goes to 0. But this follows:

$$E\left[\left(\frac{1}{\sqrt{n}}S - \frac{1}{\sqrt{n}}T\right)^2\right] = \frac{1}{n}E\left[\left(\sum_{i=1}^{n} x_i\left(\varphi\left(\frac{n}{n+1}F_n(Y_i)\right) - \varphi(F(Y_i))\right)\right)^2\right]$$

$$\leq \frac{n}{n-1}\left\{\frac{1}{n}\sum_{i=1}^{n}x_i^2\right\}E\left[\left(\varphi\left(\frac{n}{n+1}F_n(Y_1)\right) - \varphi(F(Y_1))\right)^2\right]$$

$$\to \sigma_x^2 \cdot 0,$$

where the inequality and the derivation of the limit are in Hájek and Šidák (1967, p. 160). This results in the following theorem,

Theorem A.2.1 *Under the above assumptions,*

$$\frac{1}{\sqrt{n}}(T-S) \xrightarrow{P} 0, \tag{A.2.10}$$

and

$$\frac{1}{\sqrt{n}}S \xrightarrow{D} N(0, \sigma_x^2). \tag{A.2.11}$$

Hence we have established the null asymptotic distribution theory of a simple linear rank statistic.

A.2.2 Local Asymptotic Distribution Theory

We first need the definition of contiguity between two sequences of densities.

Definition A.2.1 *A sequence of densities* $\{q_n\}$ *is* **contiguous** *to another sequence of densities* $\{p_n\}$ *if, for any sequence of events* $\{A_n\}$,

$$\int_{A_n} p_n \to 0 \Rightarrow \int_{A_n} q_n \to 0.$$

This concept is discussed in some detail in Hájek and Šidák (1967).

The following fact follows immediately from this definition. Suppose the sequence of densities $\{q_n\}$ is contiguous to the sequence of densities $\{p_n\}$. Let $\{X_n\}$ be a sequence of random variables. If $X_n \xrightarrow{P} 0$ under p_n then $X_n \xrightarrow{P} 0$ under q_n.

Then according to **LeCam's first lemma**, if $\log(q_n/p_n)$ is asymptotically $N(-\sigma^2/2, \sigma^2)$ under p_n, then q_n is contiguous to p_n. Further, by **LeCam's third lemma**, if $(S_n, \log(q_n/p_n))$ is asymptotically bivariate normal $(\mu_1, \mu_2, \sigma_1^2, \sigma_2^2, \rho\sigma_1\sigma_2)$ with $\mu_2 = -\sigma_2^2/2$ under p_n, then S_n is asymptotically $N(\mu_1 + \rho\sigma_1\sigma_2, \sigma_1^2)$ under q_n; see Hájek and Šidák (1967, pp. 202–209).

In this section, we assume that the random variables $Y_1, \ldots, Y_n$ and the regression coefficients $x_1, \ldots, x_n$ follow the same assumptions that we made in the previous section; see expressions (A.2.1) and (A.2.2). We denote the likelihood function of $Y_1, \ldots, Y_n$ by

$$p_y = \Pi_{i=1}^n f(y_i). \tag{A.2.12}$$

In the previous section we derived the asymptotic distribution of S under p_y. In this section we are further concerned with the likelihood function

$$q_d = \Pi_{i=1}^n f(y_i + d_i), \tag{A.2.13}$$

for a sequence of constants $d_1, \ldots, d_n$ satisfying the conditions

$$\sum_{i=1}^n d_i = 0 \tag{A.2.14}$$

$$\sum_{i=1}^n d_i^2 \to \sigma_d^2 > 0, \text{ as } n \to \infty \tag{A.2.15}$$

$$\max_{1 \leq i \leq n} d_i^2 \to 0, \text{ as } n \to \infty \tag{A.2.16}$$

$$\frac{1}{\sqrt{n}} \sum_{i=1}^n x_i d_i \to \sigma_{xd}, \text{ as } n \to \infty. \tag{A.2.17}$$

In applications (for example, power in simple linear models) we take $d_i = -x_i\beta/\sqrt{n}$. For x_is following assumptions (A.2.1) and (A.2.2), the above assumptions would hold for these d_is.

In this section, we establish the asymptotic distribution of S under q_d. Consider the log of the ratio of the likehood functions q_d and p_y given by

$$l(\eta) = \sum_{i=1}^n \log \frac{f(Y_i + \eta d_i)}{f(Y_i)}. \tag{A.2.18}$$

Expanding $l(\eta)$ about 0 and evaluating the resulting expression at $\eta = 1$ results in

$$l = \sum_{i=1}^n d_i \frac{f'(Y_i)}{f(Y_i)} + \frac{1}{2} \sum_{i=1}^n d_i^2 \frac{f(Y_i)f''(Y_i) - (f'(Y_i))^2}{f^2(Y_i)} + o_p(1), \tag{A.2.19}$$

provided that the third derivative of the log ratio, evaluated at 0, is square integrable. Under p_y, the middle term converges in probability to $-I(f)\sigma_d^2/2$, provided that the second derivative of the log ratio, evaluated at 0, is square integrable.

Hence, under p_y and some further regularity conditions, we can write

$$l = \sum_{i=1}^n d_i \frac{f'(Y_i)}{f(Y_i)} - \frac{I(f)\sigma_d^2}{2} + o_p(1). \tag{A.2.20}$$

The random variables in the first term, $f'(Y_i)/f(Y_i)$ are iid with mean 0 and variance $I(f)$. Because the sequence $d_1, \ldots, d_n$ satisfies (A.2.14)–(A.2.16), we can use Corollary A.1.2 to show that, under p_y, l converges in distribution to

$N(-I(f)\sigma_d^2/2, I(f)\sigma_d^2)$. By the definition of contiguity (A.2.1) and LeCam's first lemma, we have the result that

the densities $q_d = \Pi_{i=1}^n f(y_i + d_i)$ are contiguous to $p_y = \Pi_{i=1}^n f(y_i)$; (A.2.21)

see also Hájek and Šidák (1967, p. 204).

We next establish the following key result:

Theorem A.2.2 For T given by (A.2.7) and under p_y and the assumptions (3.4.1), (A.2.1), (A.2.2), and (A.2.14)–(A.2.17),

$$\begin{pmatrix} \frac{1}{\sqrt{n}} T \\ l \end{pmatrix} \xrightarrow{D} N_2 \left(\begin{pmatrix} 0 \\ -\frac{I(f)\sigma_d^2}{2} \end{pmatrix}, \begin{bmatrix} \sigma_x^2 & \sigma_{xd}\gamma_f \\ \sigma_{xd}\gamma_f & I(f)\sigma_d^2 \end{bmatrix} \right). \quad (A.2.22)$$

Proof. Consider the random vector $\mathbf{V} = (T/\sqrt{n}, l)'$, where T is defined in expression (A.2.7). To show that $\mathbf{V}$ is asymptotically normal under p_n it suffices to show that for $\mathbf{t} \in \mathcal{R}^2$, $\mathbf{t} \neq \mathbf{0}$, $\mathbf{t}'\mathbf{V}$ is asymptotically univariate normal. By the above discussion, for the second component of $\mathbf{V}$, we need only be concerned with the first term in expression (A.2.19); hence, for $\mathbf{t} = (t_1, t_2)'$, define the random variables W_{in} by

$$\sum_{i=1}^n \left[\frac{1}{\sqrt{n}} x_i t_1 \varphi(F(Y_i)) + t_2 d_i \frac{f'(Y_i)}{f(Y_i)} \right] = \sum_{i=1}^n W_{in}. \quad (A.2.23)$$

We want to apply Theorem A.1.1. The random variables W_{in} are independent and have mean 0. After some simplification, we can show that the variance of W_{in} is

$$\sigma_{in}^2 = \frac{1}{n} x_i^2 t_1^2 + t_2^2 d_i^2 I(f) - 2t_1 t_2 d_i \frac{x_i}{\sqrt{n}} \gamma_f, \quad (A.2.24)$$

where γ_f is given by

$$\gamma_f = \int_0^1 \varphi(u) \left(-\frac{f'(F^{-1}(u))}{f(F^{-1}(u))} \right) du. \quad (A.2.25)$$

Note, by assumptions (A.2.1), (A.2.2), and (A.2.15)–(A.2.17), that

$$\sum_{i=1}^n \sigma_{in}^2 \to t_1^2 \sigma_x^2 + t_2^2 \sigma_d^2 I(f) - 2t_1 t_2 \gamma_f \sigma_{xd} > 0, \quad (A.2.26)$$

and that

$$\max_{1 \leq i \leq n} \sigma_{in}^2 \leq \max_{1 \leq i \leq n} \frac{1}{n} x_i^2 t_1^2 + t_2^2 I(f) \max_{1 \leq i \leq n} d_i^2$$

$$+ 2|t_1 t_2| \gamma_f \max_{1 \leq i \leq n} \frac{1}{\sqrt{n}} |x_i| \max_{1 \leq i \leq n} |d_i| \to 0; \quad (A.2.27)$$

hence conditions (A.1.2) and (A.1.1) are true. Thus to obtain the result we need to show

$$\lim_{n \to \infty} \sum_{i=1}^{n} E[W_{in}^2 I_\varepsilon(|W_{in}|)] = 0, \qquad (A.2.28)$$

for $\varepsilon > 0$. But $|W_{in}| \leq W_{in}^*$, where

$$W_{in}^* = |t_1| \max_{1 \leq j \leq n} \frac{1}{\sqrt{n}} |x_j| |\varphi(F(Y_i))| + |t_2| \max_{1 \leq j \leq n} |d_j| \left|\frac{f'(Y_i)}{f(Y_i)}\right|.$$

Hence,

$$\sum_{i=1}^{n} E[W_{in}^2 I_\varepsilon(|W_{in}|)] \leq \sum_{i=1}^{n} E[W_{in}^2 I_\varepsilon(W_{in}^*)] = \sum_{i=1}^{n} E\left[\left(\frac{1}{n} t_1^2 x_i^2 \varphi^2(F(Y_1))\right.\right.$$

$$+ t_2^2 d_i^2 \left(\frac{f'(Y_1)}{f(Y_1)}\right)^2 + 2 t_1 t_2 \frac{1}{\sqrt{n}} x_i d_i \varphi(F(Y_1)) \left(-\frac{f'(Y_1)}{f(Y_1)}\right)\right) I_\varepsilon(W_{1n}^*)\right]$$

$$= t_1^2 \left\{\sum_{i=1}^{n} \frac{1}{n} x_i^2\right\} E[\varphi^2(F(Y_1)) I_\varepsilon(W_{1n}^*)]$$

$$+ t_2^2 \left\{\sum_{i=1}^{n} d_i^2\right\} E\left[\left(\frac{f'(Y_i)}{f(Y_i)}\right)^2 I_\varepsilon(W_{1n}^*)\right]$$

$$+ 2 t_1 t_2 \left\{\frac{1}{\sqrt{n}} \sum_{i=1}^{n} x_i d_i\right\} E\left[\varphi(F(Y_1)) \left(-\frac{f'(Y_1)}{f(Y_1)}\right)^2 I_\varepsilon(W_{1n}^*)\right].$$
$$(A.2.29)$$

Because $I_\varepsilon(W_{1n}^*) \to 0$ pointwise and each of the other random variables in the expectations of (A.2.29) are absolutely integrable, the Lebesgue dominated convergence theorem implies that each of these expectations converges to 0. The desired limit in expression (A.2.28) then follows from assumptions (A.2.1), (A.2.2), and (A.2.15)–(A.2.17). Hence **V** is asymptotically bivariate normal. We can obtain its asymptotic variance-covariance matrix from expression (A.2.26), which completes the proof.

Based on Theorem A.2.2, an application of LeCam's third lemma leads to the asymptotic distribution of $T/\sqrt{n}$ under local alternatives which we state in the following theorem.

Theorem A.2.3 *Under the sequence of densities $q_d = \Pi_{i=1}^n f(y_i + d_i)$, and assumptions (3.4.1), (A.2.1), (A.2.2), and (A.2.14)–(A.2.17),*

$$\frac{1}{\sqrt{n}} T \xrightarrow{D} N(\sigma_{xd}\gamma_f, \sigma_x^2), \qquad (A.2.30)$$

$$\frac{1}{\sqrt{n}} S \xrightarrow{D} N(\sigma_{xd}\gamma_f, \sigma_x^2). \qquad (A.2.31)$$

The result for $S/\sqrt{n}$ follows because $(T - S)/\sqrt{n} \to 0$ in probability under the densities p_y; hence, due to the contiguity cited in expression (A.2.21), $(T - S)/\sqrt{n} \to 0$ also under the densities q_d. A proof of the asymptotic power lemma, Theorem 2.4.17, follows from this result.

We now investigate the relationship between S and the shifted process given by

$$S_d = \sum_{i=1}^{n} x_i a(R(Y_i + d_i)). \qquad (A.2.32)$$

Consider the analogous process,

$$T_d = \sum_{i=1}^{n} x_i \varphi(F(Y_i + d_i)). \qquad (A.2.33)$$

We next establish the connection between T and T_d; see also Theorem 1.3.1.

Theorem A.2.4 *Under the likelihoods q_d and p_y, we have the following identity:*

$$P_{q_d}\left[\frac{1}{\sqrt{n}} T \leq t\right] = P_{p_y}\left[\frac{1}{\sqrt{n}} T_d \leq t\right]. \qquad (A.2.34)$$

Proof. The proof follows from the following string of equalities.

$$\begin{aligned}
P_{q_d}\left[\frac{1}{\sqrt{n}} T \leq t\right] &= P_{q_d}\left[\frac{1}{\sqrt{n}} \sum_{i=1}^{n} x_i \varphi(F(Y_i)) \leq t\right] \\
&= P_{q_d}\left[\frac{1}{\sqrt{n}} \sum_{i=1}^{n} x_i \varphi(F((Y_i - d_i) + d_i)) \leq t\right] \\
&= P_{p_y}\left[\frac{1}{\sqrt{n}} \sum_{i=1}^{n} x_i \varphi(F(Z_i + d_i)) \leq t\right] \\
&= P_{p_y}\left[\frac{1}{\sqrt{n}} T_d \leq t\right],
\end{aligned} \qquad (A.2.35)$$

where in the third line the sequence of random variables $Z_1, \ldots, Z_n$ follows the likelihood p_y.

We next establish an asymptotic relationship between T and T_d.

Theorem A.2.5 *Under p_y and assumptions (3.4.1), (A.2.1), (A.2.2), and (A.2.14)–(A.2.17),*

$$\left[\frac{T - [T_d - E_{p_y}(T_d)]}{\sqrt{n}}\right] \xrightarrow{P} 0.$$

Proof. Since $E(T) = 0$, it suffices to show that $\text{Var}[(T - T_d)/\sqrt{n}] \to 0$. We have,

$$\text{Var}\left[\frac{T - T_d}{\sqrt{n}}\right] = \frac{1}{n}\sum_{i=1}^{n} x_i^2 \text{Var}[\varphi(F(Y_i)) - \varphi(F(Y_i + d_i))]$$

$$\leq \frac{1}{n}\sum_{i=1}^{n} x_i^2 E[\varphi(F(Y_i)) - \varphi(F(Y_i + d_i))]^2$$

$$= \frac{1}{n}\sum_{i=1}^{n} x_i^2 \int_{-\infty}^{\infty} [\varphi(F(y)) - \varphi(F(y + d_i))]^2 f(y)\, dy$$

$$\leq \left(\frac{1}{n}\sum_{i=1}^{n} x_i^2\right)\left(\int_{-\infty}^{\infty} \max_{1 \leq i \leq n}[\varphi(F(y)) - \varphi(F(y + d_i))]^2 f(y)\, dy\right).$$

The first factor in the last expression converges to σ_x^2; hence, it suffices to show that the $\overline{\lim}$ of the second factor is 0. Fix y. Let $\varepsilon > 0$ be given. Then since $\varphi(u)$ is continuous a.e. we can assume it is continuous at $F(y)$. Hence there exists a $\delta_1 > 0$ such that $|\varphi(z) - \varphi(F(y))| < \varepsilon$ for $|z - F(y)| < \delta_1$. By the uniform continuity of F, choose $\delta_2 > 0$ such that $|F(t) - F(s)| < \delta_1$ for $|s - t| < \delta_2$. By (A.2.16) choose N_0 so that for $n > N_0$ implies

$$\max_{1 \leq i \leq n}\{|d_i|\} < \delta_2.$$

Thus for $n > N_0$,

$$|F(y) - F(y + d_i)| < \delta_1, \quad \text{for } i = 1, \ldots, n,$$

and, hence,

$$|\varphi(F(y)) - \varphi(F(y + d_i))| < \varepsilon, \quad \text{for } i = 1, \ldots, n.$$

Thus for $n > N_0$,

$$\max_{1 \leq i \leq n}[\varphi(F(y)) - \varphi(F(y + d_i))]^2 < \varepsilon^2.$$

Therefore,

$$\overline{\lim}\left(\int_{-\infty}^{\infty} \max_{1 \leq i \leq n}[\varphi(F(y)) - \varphi(F(y + d_i))]^2 f(y)\, dy\right) \leq \varepsilon^2,$$

and we are finished.

The next result yields the asymptotic mean of T_d.

Theorem A.2.6 *Under p_y and the assumptions (3.4.1), (A.2.1), (A.2.2), and (A.2.14)–(A.2.17),*

$$E_{p_y}\left[\frac{1}{\sqrt{n}} T_d\right] \to \gamma_f \sigma_{xd}.$$

Proof. By Theorem A.2.3,

$$\frac{\frac{1}{\sqrt{n}}T - \gamma_f \sigma_{xd}}{\sigma_x} \xrightarrow{D} N(0, 1), \text{ under } q_d.$$

Hence by Theorem A.2.4,

$$\frac{\frac{1}{\sqrt{n}}T_d - \gamma_f \sigma_{xd}}{\sigma_x} \xrightarrow{D} N(0, 1), \text{ under } p_y. \tag{A.2.36}$$

By (A.2.9),

$$\frac{\frac{1}{\sqrt{n}}T}{\sigma_x} \xrightarrow{D} N(0, 1), \text{ under } p_y;$$

hence by Theorem A.2.5, we must have

$$\frac{\frac{1}{\sqrt{n}}T_d - E\left[\frac{1}{\sqrt{n}}T_d\right]}{\sigma_x} \xrightarrow{D} N(0, 1), \text{ under } p_y. \tag{A.2.37}$$

The conclusion follows from (A.2.36) and (A.2.37).

By the last two theorems we have, under p_y,

$$\frac{1}{\sqrt{n}}T_d = \frac{1}{\sqrt{n}}T + \gamma_f \sigma_{xd} + o_p(1).$$

We need to express these results for the random variables S, (A.2.4), and S_d, (A.2.32). Because the densities q_d are contiguous to p_y and $(T - S)/\sqrt{n} \to 0$ in probability under p_y, it follows that $(T - S)/\sqrt{n} \to 0$ in probability under q_d. By a change of variable this means $(T_d - S_d)/\sqrt{n} \to 0$ in probability under p_y. This discussion leads to the following two results which we state in a theorem.

Theorem A.2.7 *Under p_y and the assumptions (3.4.1), (A.2.1), (A.2.2), and (A.2.14)–(A.2.17),*

$$\frac{1}{\sqrt{n}}S_d = \frac{1}{\sqrt{n}}S + \gamma_f \sigma_{xd} + o_p(1) \tag{A.2.38}$$

$$\frac{1}{\sqrt{n}}S_d = \frac{1}{\sqrt{n}}T + \gamma_f \sigma_{xd} + o_p(1). \tag{A.2.39}$$

Next we relate Theorem A.2.7 to (2.5.27), the asymptotic linearity of the general scores statistic in the two-sample problem. Recall in the two-sample problem that $c_i = 0$ for $1 \leq i \leq n_1$ and $c_i = 1$ for $n_1 + 1 \leq i \leq n_1 + n_2 = n$, (2.2.1). Hence, $x_i = c_i - \bar{c} = -n_2/n$ for $1 \leq i \leq n_1$ and $x_i = n_1/n$ for $n_1 + 1 \leq i \leq n$. Defining $d_i = -\delta x_i/\sqrt{n}$, it is easy to check that conditions (A.2.14)–(A.2.17) hold with $\sigma_{xd} = -\lambda_1 \lambda_2 \delta$. Further, (A.2.32) becomes

$S_\varphi(\delta/\sqrt{n}) = \sum x_i a(R(Y_i - \delta x_i/\sqrt{n}))$ and (A.2.4) becomes $S_\varphi(0) = \sum x_i a(R(Y_i))$, where $a(i) = \varphi(i/(n+1))$, $\int \varphi = 0$ and $\int \varphi^2 = 1$. Hence (A.2.38) becomes

$$\frac{1}{\sqrt{n}} S_\varphi(\delta/\sqrt{n}) = \frac{1}{\sqrt{n}} S_\varphi(0) - \lambda_1 \lambda_2 \gamma_f \delta + o_p(1).$$

Finally, using the usual partition argument, Theorem 1.5.6, and the monotonicity of $S_\varphi(\delta/\sqrt{n})$, we have:

Theorem A.2.8 *Assuming Finite Fisher information, nondecreasing and square integrable $\varphi(u)$, and $n_i/n \to \lambda_i$, $0 < \lambda_i < 1$, $i = 1, 2$,*

$$P_{p_x}\left(\sup_{\sqrt{n}|\delta| \leq c} \left| \frac{1}{\sqrt{n}} S_\varphi\left(\frac{\delta}{\sqrt{n}}\right) - \frac{1}{\sqrt{n}} S_\varphi(0) + \lambda_1 \lambda_2 \gamma_f \delta \right| \geq \varepsilon \right) \to 0, \quad \text{(A.2.40)}$$

for all $\varepsilon > 0$ and for all $c > 0$.

This theorem establishes (2.5.27). As a final note from (A.2.11), $n^{-1/2} S_\varphi(0)$ is asymptotically $N(0, \sigma_x^2)$, where $\sigma_x^2 = \sigma^2(0) = \lim n^{-1} \sum x_i^2 = \lambda_1 \lambda_2$. Hence to determine the efficacy using this approach, we have

$$c_\varphi = \frac{\lambda_1 \lambda_2 \gamma_f}{\sigma(0)} = \sqrt{\lambda_1 \lambda_2} \tau_\varphi^{-1}; \quad \text{(A.2.41)}$$

see (2.5.28).

A.2.3 Signed-Rank Statistics

In this section we develop the asymptotic local behavior for the general signed-rank statistics defined in Section 1.8. Assume that $X_1, \ldots X_n$ are a random sample having distribution function $H(x)$ with density $h(x)$ which is symmetric about 0. Recall that general signed-rank statistics are given by

$$T_{\varphi^+} = \sum a^+(R(|X_i|)) \operatorname{sgn}(X_i), \quad \text{(A.2.42)}$$

where the scores are generated as $a^+(i) = \varphi^+(i/(n+1))$ for a nonnegative and square-integrable function $\varphi^+(u)$ which is standardized such that $\int (\varphi^+(u))^2 \, du = 1$.

The null asymptotic distribution of T_{φ^+} was derived in Section 1.8, so here we will be concerned with its behavior under local alternatives. Also the derivations here are similar to those for simple linear rank statistics (Section A.2.2); hence, we will be brief.

Note that we can write T_{φ^+} as

$$T_{\varphi^+} = \sum \varphi^+\left(\frac{n}{n+1} H_n^+(|X_i|)\right) \operatorname{sgn}(X_i),$$

where H_n^+ denotes the empirical distribution function of $|X_1|, \ldots, |X_n|$. This suggests the approximation

$$T_{\varphi^+}^* = \sum \varphi^+(H^+(|X_i|)) \operatorname{sgn}(X_i), \quad \text{(A.2.43)}$$

where $H^+(x)$ is the distribution function of $|X_i|$.

Denote the likelihood of the sample $X_1, \ldots X_n$ by

$$p_x = \Pi_{i=1}^n h(x_i). \tag{A.2.44}$$

A result that we will need is

$$\frac{1}{\sqrt{n}}\left(T_{\varphi^+} - T_{\varphi^+}^*\right) \xrightarrow{P} 0, \text{ under } p_x. \tag{A.2.45}$$

This result is shown in Hájek and Šidák (1967, p. 167).

For the sequence of local alternatives $b/\sqrt{n}$, with $b \in \mathcal{R}$ (here we are taking $d_i = -b/\sqrt{n}$), we denote the likelihood by

$$q_b = \Pi_{i=1}^n h\left(x_i - \frac{b}{\sqrt{n}}\right). \tag{A.2.46}$$

For $b \in \mathcal{R}$, consider the log of the likelihoods given by

$$l(\eta) = \sum_{i=1}^n \log \frac{h\left(X_i - \eta \frac{b}{\sqrt{n}}\right)}{h(X_i)}. \tag{A.2.47}$$

If we expand $l(\eta)$ about 0 and evaluate it at $\eta = 1$, similar to the expansion (A.2.19), we obtain

$$l = -\frac{b}{\sqrt{n}} \sum_{i=1}^n \frac{h'(X_i)}{h(X_i)} + \frac{b^2}{2n} \sum_{i=1}^n \frac{h(X_i)h''(X_i) - (h'(X_i))^2}{h^2(X_i)} + o_p(1), \tag{A.2.48}$$

provided that the third derivative of the log ratio, evaluated at 0, is square integrable. Under p_x, the middle term converges in probability to $-I(h)b^2/2$, provided that the second derivative of the log ratio, evaluated at 0, is square integrable. An application of Theorem A.1.1 shows that l converges in distribution to $N(-[I(h)b^2/2], I(h)b^2)$. Hence, by LeCam's first lemma,

the densities $q_b = \Pi_{i=1}^n h\left(x_i - \frac{b}{\sqrt{n}}\right)$ are contiguous to $p_x = \Pi_{i=1}^n h(x_i);$

$$\tag{A.2.49}$$

Similar to Section A.2.2, by using Theorem A.1.1 we can derive the asymptotic distribution of the random vector $(T_{\varphi^+}^*/\sqrt{n}, l)$, which we record as the following theorem:

Theorem A.2.9 *Under p_x and some regularity conditions on h,*

$$\begin{pmatrix} \frac{1}{\sqrt{n}} T_{\varphi^+}^* \\ l \end{pmatrix} \xrightarrow{\mathcal{D}} N_2\left(\begin{pmatrix} 0 \\ -\frac{I(h)b^2}{2} \end{pmatrix}, \begin{bmatrix} 1 & b\gamma_h \\ b\gamma_h & I(h)b^2 \end{bmatrix}\right), \tag{A.2.50}$$

where $\gamma_h = 1/\tau_{\varphi^+}$ and τ_{φ^+} is given in expression (1.8.24).

By this theorem and LeCam's third lemma, we have

$$\frac{1}{\sqrt{n}} T_{\varphi^+}^* \xrightarrow{\mathcal{D}} N(b\gamma_h, 1), \text{ under } q_b. \tag{A.2.51}$$

By the result on contiguity, (A.2.49), the test statistic $T_{\varphi^+}/\sqrt{n}$ has the same distribution under q_b. A proof of the asymptotic power lemma, Theorem 1.8.1, follows from this result.

Next consider a shifted version of $T_{\varphi^+}^*$ given by

$$T_{b\varphi^+}^* = \sum_{i=1}^n \varphi^+\left(H^+\left(\left|X_i + \frac{b}{\sqrt{n}}\right|\right)\right) \operatorname{sgn}\left(X_i + \frac{b}{\sqrt{n}}\right). \tag{A.2.52}$$

The following identity is readily established:

$$P_{q_b}[T_{\varphi^+}^* \leq t] = P_{p_x}[T_{b\varphi^+}^* \leq t]; \tag{A.2.53}$$

see also Theorem 1.3.1. We need the following theorem:

Theorem A.2.10 *Under p_x,*

$$\left[\frac{T_{\varphi^+}^* - [T_{b\varphi^+}^* - E_{p_x}(T_{b\varphi^+}^*)]}{\sqrt{n}}\right] \xrightarrow{P} 0.$$

Proof. As in Theorem A.2.5, it suffices to show that $\operatorname{Var}[(T_{\varphi^+}^* - T_{b\varphi^+}^*)/\sqrt{n}] \to 0$. But this variance reduces to

$$\operatorname{Var}\left[\frac{T_{\varphi^+}^* - T_{b\varphi^+}^*}{\sqrt{n}}\right] = \int_{-\infty}^{\infty} \left\{\varphi^+\left(H^+(|x|)\right)\operatorname{sgn}(x) - \varphi^+\left(H^+\left(\left|x + \frac{b}{\sqrt{n}}\right|\right)\right)\operatorname{sgn}\left(x + \frac{b}{\sqrt{n}}\right)\right\}^2 h(x)\,dx.$$

Since $\varphi^+(u)$ is square integrable, the quantity in braces is dominated by an integrable function. Since it converges pointwise to 0, a.e., an application of the Lebesgue dominated convergence theorem establishes the result.

Using the above results, we can proceed as we did for Theorem A.2.6 to show that under p_x,

$$E_{p_x}\left[\frac{1}{\sqrt{n}} T_{b\varphi^+}^*\right] \to b\gamma_h. \tag{A.2.54}$$

Hence,

$$\frac{1}{\sqrt{n}} T_{b\varphi^+}^* = \frac{1}{\sqrt{n}} T_{\varphi^+}^* + b\gamma_h + o_p(1). \tag{A.2.55}$$

A similar result holds for the signed-rank statistic.

For the results needed in Chapter 1, however, it is convenient to change the notation to:

$$T_{\varphi^+}(b) = \sum_{i=1}^n a^+(R|X_i - b|)\operatorname{sgn}(X_i - b). \tag{A.2.56}$$

The above results imply that

$$\frac{1}{\sqrt{n}} T_{\varphi^+}(\theta) = \frac{1}{\sqrt{n}} T_{\varphi^+}(0) - \theta\gamma_h + o_p(1), \tag{A.2.57}$$

for $\sqrt{n}|\theta| \leq B$, for $B > 0$.

The general signed-rank statistics found in Chapter 1 are based on norms. In this case, since the scores are nondecreasing, we can strengthen our results to include uniformity; that is.

Theorem A.2.11 *Assuming Finite Fisher information, nondecreasing and square integrable $\varphi^+(u)$,*

$$P_{P_x}\left[\sup_{\sqrt{n}|\theta| \leq B}\left|\frac{1}{\sqrt{n}}T_{\varphi^+}(\theta) - \frac{1}{\sqrt{n}}T_{\varphi^+}(0) + \theta\gamma_h\right| \geq \varepsilon\right] \to 0, \qquad (A.2.58)$$

for all $\varepsilon > 0$ and all $B > 0$.

A proof can be obtained by the usual partitioning type of argument on the interval $[-B, B]$; see the proof of Theorem 1.5.6. Hence, since $\int(\varphi^+(u))^2\,du = 1$, the efficacy is given by $c_{\varphi^+} = \gamma_h$; see (1.8.21).

A.3 Results for Rank-Based Analysis of Linear Models

In this section we consider the linear model defined by (3.2.3) in Chapter 3. The distribution of the errors satisfies assumption (E.1), (3.4.1). The design matrix satisfies conditions (D.2), (3.4.7), and (D.3), (3.4.8). We shall assume without loss of generality that the true vector of parameters is $\mathbf{0}$.

It will be easier to work with the following transformation of the design matrix and parameters. We consider $\boldsymbol{\beta}$ such that $\sqrt{n}\boldsymbol{\beta} = O(1)$. Note that we will suppress the notation indicating that $\boldsymbol{\beta}$ depends on n. Let

$$\boldsymbol{\Delta} = (\mathbf{X}'\mathbf{X})^{1/2}\boldsymbol{\beta}, \qquad (A.3.1)$$

$$\mathbf{C} = \mathbf{X}(\mathbf{X}'\mathbf{X})^{-1/2}, \qquad (A.3.2)$$

$$d_i = -\mathbf{c}'_i\boldsymbol{\Delta}, \qquad (A.3.3)$$

where $\mathbf{c}_i$ is the ith row of $\mathbf{C}$, and note that $\boldsymbol{\Delta} = O(1)$ because $n^{-1}\mathbf{X}'\mathbf{X} \to \boldsymbol{\Sigma} > 0$ and $\sqrt{n}\boldsymbol{\beta} = O(1)$. Then $\mathbf{C}'\mathbf{C} = \mathbf{I}_p$ and $\mathbf{H}_\mathbf{C} = \mathbf{H}_\mathbf{X}$, where $\mathbf{H}_\mathbf{C}$ is the projection matrix onto the column space of $\mathbf{C}$. Note that since $\mathbf{X}$ is centered, $\mathbf{C}$ is also. Also $\|\mathbf{c}_i\|^2 = h_{nii}^2$, where h_{nii}^2 is the ith diagonal entry of $\mathbf{H}_\mathbf{X}$. It is straightforward to show that $\mathbf{c}'_i\boldsymbol{\Delta} = \mathbf{x}'_i\boldsymbol{\beta}$. Using conditions (D.2) and (D.3), the following conditions are readily established:

$$\bar{d} = 0 \qquad (A.3.4)$$

$$\sum_{i=1}^n d_i^2 \leq \sum_{i=1}^n \|\mathbf{c}_i\|^2\|\boldsymbol{\Delta}\|^2 = p\|\boldsymbol{\Delta}\|^2, \text{ for all } n \qquad (A.3.5)$$

$$\max_{1 \leq i \leq n} d_i^2 \leq \|\boldsymbol{\Delta}\|^2 \max_{1 \leq i \leq n} \|\mathbf{c}_i\|^2 \qquad (A.3.6)$$

$$= \|\boldsymbol{\Delta}\|^2 \max_{1 \leq i \leq n} h_{nii}^2 \to 0 \text{ as } n \to \infty,$$

since $\|\boldsymbol{\Delta}\|$ is bounded.

For $j = 1, \ldots, p$ define

$$S_{nj}(\Delta) = \sum_{i=1}^{n} c_{ij} a(R(Y_i - \mathbf{c}_i'\Delta)), \qquad (A.3.7)$$

where the scores are generated by a function φ which satisfies (S.1), (3.4.10). We now show that the theory established in Section A.2 for simple linear rank statistics holds for S_{nj}, for each j.

Fix j then the regression coefficients x_i of Section A.2 are given by $x_i = \sqrt{n} c_{ij}$. Note from (A.3.2) that $\sum x_i^2/n = \sum c_{ij}^2 = 1$; hence, condition (A.2.2) is true. Further, by (A.3 6),

$$\frac{\max_{1 \leq i \leq n} x_i^2}{\sum_{i=1}^{n} x_i^2} = \max_{1 \leq i \leq n} c_{ij}^2 \to 0;$$

hence, condition (A.2.1) is true.

For the sequence $d_i = -\mathbf{c}_i'\Delta$, conditions (A.3.4)–(A.3.6) imply conditions (A.2.14)–(A.2.16) (the upper bound in condition (A.3.6) was actually all that was needed in the proofs of Section A.2). Finally, for (A.2.17), because $\mathbf{C}$ is orthogonal, σ_{xd} is given by

$$\sigma_{xd} = \frac{1}{\sqrt{n}} \sum_{i=1}^{n} x_i d_i = -\sum_{i=1}^{n} c_{ij} \mathbf{c}_i'\Delta = -\sum_{k=1}^{p} \left\{ \sum_{i=1}^{n} c_{ij} c_{ik} \right\} \Delta_k = -\Delta_j. \qquad (A.3.8)$$

Thus by Theorem A.2.7, for $j = 1, \ldots, p$, we have the results

$$S_{nj}(\Delta) = S_{nj}(0) - \gamma_f \Delta_j + o_p(1) \qquad (A.3.9)$$

$$S_{nj}(\Delta) = T_{nj}(0) - \gamma_f \Delta_j + o_p(1), \qquad (A.3.10)$$

where

$$T_{nj}(0) = \sum_{i=1}^{n} c_{ij} \varphi(F(Y_i)). \qquad (A.3.11)$$

Let $\mathbf{S}_n(\Delta)' = (S_{n1}(\Delta), \ldots, S_{np}(\Delta))$. Because componentwise convergence in probability implies that the corresponding vector converges, we have shown that the following theorem is true:

Theorem A.3.1 *Under the above assumptions, for $\varepsilon > 0$ and for all Δ*

$$\lim_{n \to \infty} P(\|\mathbf{S}_n(\Delta) - (\mathbf{S}_n(0) - \gamma \Delta)\| \geq \varepsilon) = 0. \qquad (A.3.12)$$

The conditions we want are asymptotic linearity and quadraticity. **Asymptotic linearity** is the condition

$$\lim_{n \to \infty} P\left(\sup_{\|\Delta\| \leq c} \|\mathbf{S}_n(\Delta) - (\mathbf{S}_n(0) - \gamma \Delta)\| \geq \varepsilon \right) = 0, \qquad (A.3.13)$$

for arbitrary $c > 0$ and $\varepsilon > 0$. This result was first shown by Jurečková (1971) under more stringent conditions on the design matrix.

Consider the dispersion function discussed in Chapter 2. In terms of the above notation,

$$D_n(\Delta) = \sum_{i=1}^{n} a(R(Y_i - \mathbf{c}_i'\Delta))(Y_i - \mathbf{c}_i'\Delta). \quad (A.3.14)$$

An approximation of $D_n(\Delta)$ is the quadratic function

$$Q_n(\Delta) = \gamma\Delta'\Delta/2 - \Delta'\mathbf{S}_n(0) + D_n(0). \quad (A.3.15)$$

Using Jurečková's conditions, Jaeckel (1972) extended the result (A.3.13) to **asymptotic quadraticity**, which is given by

$$\lim_{n \to \infty} P\left(\sup_{\|\Delta\| \leq c} |D_n(\Delta) - Q_n(\Delta)| \geq \varepsilon \right) = 0, \quad (A.3.16)$$

for arbitrary $c > 0$ and $\varepsilon > 0$. Our main result of this section shows that (A.3.12), (A.3.13), and (A.3.16) are equivalent. The proof proceeds as in Heiler and Willers (1988) who established their results based on convex function theory. Before proceeding with the proof, for the reader's convenience, we present some notes on convex functions.

A.3.1 Convex Functions

Let f be a real-valued function defined on $\mathcal{R}^p$. Recall the definition of a convex function:

Definition A.3.1 *The function f is* **convex** *if*

$$f(\lambda \mathbf{x} + (1-\lambda)\mathbf{y}) \leq \lambda f(\mathbf{x}) + (1-\lambda)f(\mathbf{y}), \quad (A.3.17)$$

for $0 < \lambda < 1$. Further, a convex function f is called **proper** *if it is defined on an open set $\mathbf{C} \in \mathcal{R}^p$ and is everywhere finite on $\mathbf{C}$.*

The convex functions of interest here are proper with $\mathbf{C} = \mathbf{R}^p$.

The proof of the following theorem can be found in Rockafellar (1970, pp. 82 and 246).

Theorem A.3.2 *Suppose f is convex and proper on an open subset $\mathbf{C}$ of $\mathcal{R}^p$. Then f is continuous on $\mathbf{C}$ and is differentiable almost everywhere on $\mathbf{C}$.*

We will find it useful to define a subgradient:

Definition A.3.2 *The vector $D(\mathbf{x}_0)$ is called a* **subgradient** *of f at $\mathbf{x}_0$ if*

$$f(\mathbf{x}) - f(\mathbf{x}_0) \geq D(\mathbf{x}_0)'(\mathbf{x} - \mathbf{x}_0), \text{ for all } \mathbf{x} \in \mathbf{C}. \quad (A.3.18)$$

As Rockafellar (1970, p. 217 shows), a proper convex function which is defined on an open set $\mathbf{C}$ has a subgradient at each point in $\mathbf{C}$. Furthermore, at the points of differentiability, the subgradient is unique and agrees with the gradient. This is a theorem proved in Rockafellar (1970, p. 242), which we next state.

Theorem A.3.3 *Let f be convex. If f is differentiable at x_0 then $\nabla f(x_0)$, the gradient of f at x_0, is the unique subgradient of f at x_0.*

Hence, combining Theorems A.3.2 and A.3.3, we see that for proper convex functions the subgradient is the gradient almost everywhere; hence if f is a proper convex function we have

$$f(x) - f(x_0) \geq \nabla f(x_0)'(x - x_0), \text{ a.e. } x \in C. \tag{A.3.19}$$

The next theorem can be found in Rockafellar (1970, p. 90).

Theorem A.3.4 *Let the sequence of convex functions $\{f_n\}$ be proper on C and suppose the sequence converges for all $x \in C^*$ where C^* is dense in C. Then the functions f_n converge on the whole set C to a proper and convex function f and, furthermore, the convergence is uniform on each compact subset of C.*

The following theorem is a modification by Heiler and Willers (1988) of a theorem found in Rockafellar (1970, p. 248).

Theorem A.3.5 *Suppose in addition to the assumptions of the last theorem the limit function f is differentiable, then*

$$\lim_{n \to \infty} \nabla f_n(x) = \nabla f(x), \text{ for all } x \in C. \tag{A.3.20}$$

Furthermore the convergence is uniform on each compact subset of C.

The following result is proved in Heiler and Willers (1988).

Theorem A.3.6 *Suppose the hypotheses of Theorem A.3.4 hold. Assume also that the limit function f is differentiable. Then*

$$\lim_{n \to \infty} \nabla f_n(x) = \nabla f(x), \text{ for all } x \in C^*, \tag{A.3.21}$$

and

$$\lim_{n \to \infty} f_n(x_0) = f(x_0), \text{ for at least one } x_0 \in C^*, \tag{A.3.22}$$

where C^* is dense in C, imply that

$$\lim_{n \to \infty} f_n(x) = f(x), \text{ for all } x \in C \tag{A.3.23}$$

and the convergence is uniform on each compact subset of C.

A.3.2 Asymptotic Linearity and Quadraticity

We now proceed with the proof by Heiler and Willers (1988) of the equivalence of (A.3.12), (A.3.13), and (A.3.16).

Theorem A.3.7 *Under model (3.2.3) and assumptions (3.4.7), (3.4.8), and (3.4.1), the expressions (A.3.12), (A.3.13), and (A.3.16) are equivalent.*

Proof.

(A.3.12) $\Rightarrow$ (A.3.16). Both functions $D_n(\Delta)$ and $Q_n(\Delta)$ are proper convex functions for $\Delta \in \mathcal{R}^p$. Their gradients are given by,

$$\nabla Q_n(\Delta) = \gamma\Delta - \mathbf{S}_n(\mathbf{0}) \tag{A.3.24}$$

$$\nabla D_n(\Delta) = -\mathbf{S}_n(\Delta), \text{ a.e. } \Delta \in \mathcal{R}^p. \tag{A.3.25}$$

By Theorem A.3.2 the gradient of D exists almost everwhere. Where the derivative of $D_n(\Delta)$ is not defined, we will use the subgradient of $D_n(\Delta)$, (A.3.2), which, in the case of proper convex functions, exists everywhere and which agrees uniquely with the gradient at points where $D(\Delta)$ is differentiable; see Theorem A.3.3 and the surrounding discussion. Combining these results, we have

$$\nabla(D_n(\Delta) - Q_n(\Delta)) = -[\mathbf{S}_n(\Delta) - \mathbf{S}_n(\mathbf{0}) + \gamma\Delta]. \tag{A.3.26}$$

Let N denote the set of positive integers. Let $\Delta^{(1)}, \Delta^{(2)}, \ldots$ be a listing of the vectors in p-space with rational components. By (A.3.12), the right-hand side of (A.3.26) goes to 0 in probability for $\Delta^{(1)}$. Hence, for every infinite index set $N^* \subset N$ there exists another infinite index set $N_1^{**} \subset N^*$ such that

$$[\mathbf{S}_n(\Delta^{(1)}) - \mathbf{S}_n(\mathbf{0}) + \gamma\Delta^{(1)}] \xrightarrow{\text{a.s.}} 0, \tag{A.3.27}$$

for $n \in N_1^{**}$. Since the right-hand side of (A.3.26) goes to 0 in probability for $\Delta^{(2)}$ and N_1^{**} is an infinite index set, there exists another infinite index set $N_2^{**} \subset N_1^{**}$ such that

$$[\mathbf{S}_n(\Delta^{(i)}) - \mathbf{S}_n(\mathbf{0}) + \gamma\Delta^{(i)}] \xrightarrow{\text{a.s.}} 0, \tag{A.3.28}$$

for $n \in N_2^{**}$ and $i \leq 2$. We continue and, hence, obtain a sequence of nested infinite index sets $N_1^{**} \supset N_2^{**} \supset \cdots \supset N_i^{**} \supset \cdots$ such that

$$[\mathbf{S}_n(\Delta^{(j)}) - \mathbf{S}_n(\mathbf{0}) + \gamma\Delta^{(j)}] \xrightarrow{\text{a.s.}} 0, \tag{A.3.29}$$

for $n \in N_i^{**} \supset N_{i+1}^{**} \supset \cdots$ and $j \leq i$. Let $\tilde{N}$ be a diagonal infinite index set of the sequence $N_1^{**} \supset N_2^{**} \supset \cdots \supset N_i^{**} \supset \cdots$. Then

$$[\mathbf{S}_n(\Delta) - \mathbf{S}_n(\mathbf{0}) + \gamma\Delta] \xrightarrow{\text{a.s.}} 0, \tag{A.3.30}$$

for $n \in \tilde{N}$ and for all rational Δ.

Define the convex function $H_n(\Delta) = D_n(\Delta) - D_n(\mathbf{0}) + \Delta'\mathbf{S}_n(\mathbf{0})$. Then

$$D_n(\Delta) - Q_n(\Delta) = H_n(\Delta) - \gamma\Delta'\Delta/2 \tag{A.3.31}$$

$$\nabla(D_n(\Delta) - Q_n(\Delta)) = \nabla H_n(\Delta) - \gamma\Delta. \tag{A.3.32}$$

Hence, by (A.3.30), we have

$$\nabla H_n(\Delta) \xrightarrow{\text{a.s.}} \gamma\Delta = \nabla\gamma\Delta'\Delta/2, \tag{A.3.33}$$

for $n \in \tilde{N}$ and for all rational Δ. Also note

$$H_n(0) = 0 = \gamma\Delta'\Delta/2|_{\Delta=0}. \qquad (A.3.34)$$

Since H_n is convex and (A.3.33) and (A.3.34) hold, we have by Theorem A.3.6 that $\{H_n(\Delta)\}_{n \in \tilde{N}}$ converges to $\gamma\Delta'\Delta/2$ a.s., uniformly on each compact subset of $\mathcal{R}^p$. That is by (A.3.31), $D_n(\Delta) - Q_n(\Delta) \to 0$ a.s., uniformly on each compact subset of $\mathcal{R}^p$. Since N^* is arbitrary, we can conclude (see Theorem 4 of Tucker, 1967 p. 103) that $D_n(\Delta) - Q_n(\Delta) \xrightarrow{P} 0$ uniformly on each compact subset of $\mathcal{R}^p$.

(A.3.16) $\Rightarrow$ (A.3.13). Let $c > 0$ be given and let $C = \{\Delta : \|\Delta\| \leq c\}$. By (A.3.16) we know that $D_n(\Delta) - Q_n(\Delta) \xrightarrow{P} 0$ on C. Using the same diagonal argument as above, for any infinite index set $N^* \subset N$ there exists an infinite index set $\tilde{N} \subset N^*$ such that $D_n(\Delta) - Q_n(\Delta) \xrightarrow{\text{a.s.}} 0$ for $n \in \tilde{N}$ and for all rational Δ. As in the last part, introduce the function H_n as

$$D_n(\Delta) - Q_n(\Delta) = H_n(\Delta) - \gamma\Delta'\Delta/2. \qquad (A.3.35)$$

Hence,

$$H_n(\Delta) \xrightarrow{\text{a.s.}} \gamma\Delta'\Delta/2, \qquad (A.3.36)$$

for $n \in \tilde{N}$ and for all rational Δ. By (A.3.36) and the fact that the function $\gamma\Delta'\Delta/2$ is differentiable we have by Theorem A.3.5,

$$\nabla H_n(\Delta) \xrightarrow{\text{a.s.}} \gamma\Delta, \qquad (A.3.37)$$

for $n \in \tilde{N}$ and uniformly on C. This leads to the following string of convergences,

$$\nabla(D_n(\Delta) - Q_n(\Delta)) \xrightarrow{\text{a.s.}} 0$$

$$S_n(\Delta) - (S_n(0) - \gamma\Delta) \xrightarrow{\text{a.s.}} 0, \qquad (A.3.38)$$

where both convergences are for $n \in \tilde{N}$ and uniformly on C. Since N^* was arbitrary we can conclude that

$$S_n(\Delta) - (S_n(0) - \gamma\Delta) \xrightarrow{P} 0 \qquad (A.3.39)$$

uniformly on C. Hence (A.3.13) holds.

(A.3.13) $\Rightarrow$ (A.3.12). This is trivial.

These are the results we wanted. For convenience we summarize **asymptotic linearity** and **asymptotic quadraticity** in the following theorem:

Theorem A.3.8 *Under model (3.2.3) and the assumptions (3.4.7), (3.4.8), and (3.4.1),*

$$\lim_{n \to \infty} P\left(\sup_{\|\Delta\| \leq c} \|\mathbf{S}_n(\Delta) - (\mathbf{S}_n(0) - \gamma \Delta)\| \geq \varepsilon \right) = 0, \quad (A.3.40)$$

$$\lim_{n \to \infty} P\left(\sup_{\|\Delta\| \leq c} |D_n(\Delta) - Q_n(\Delta)| \geq \varepsilon \right) = 0, \quad (A.3.41)$$

for all $\varepsilon > 0$ and all $c > 0$.

Proof. This follows from Theorems A.3.1 and A.3.7.

A.3.3 Asymptotic Distance Between $\hat{\boldsymbol{\beta}}$ and $\tilde{\boldsymbol{\beta}}$

This section contains a proof of Theorem 3.5.5. It shows that the R-estimate in Chapter 3 is close to the value which minimizes the quadratic approximation to the dispersion function. The proof is due to Jaeckel (1972). For convenience, we have restated the theorem.

Theorem A.3.9 *Under the model (3.2.3), (E.1), (D.1), (D.2) and (S.1) in Section 3.4,*

$$\sqrt{n}(\hat{\boldsymbol{\beta}} - \tilde{\boldsymbol{\beta}}) \xrightarrow{P} \mathbf{0}.$$

Proof. Choose $\varepsilon > 0$ and $\delta > 0$. Since $\sqrt{n}\tilde{\boldsymbol{\beta}}$ converges in distribution, there exists a c_0 such that

$$P\left[\|\tilde{\boldsymbol{\beta}}\| \geq c_0/\sqrt{n} \right] < \delta/2, \quad (A.3.42)$$

for n sufficiently large. Let

$$T = \min\left\{ Q(\mathbf{Y} - \mathbf{X}\boldsymbol{\beta}) : \|\boldsymbol{\beta} - \tilde{\boldsymbol{\beta}}\| = \varepsilon/\sqrt{n} \right\} - Q(\mathbf{Y} - \mathbf{X}\tilde{\boldsymbol{\beta}}). \quad (A.3.43)$$

Since $\tilde{\boldsymbol{\beta}}$ is the unique minimizer of Q, $T > 0$; hence, by asymptotic quadraticity we have

$$P\left[\max_{\|\boldsymbol{\beta}\| < (c_0+\varepsilon)/\sqrt{n}} |D(\mathbf{Y} - \mathbf{X}\boldsymbol{\beta}) - Q(\mathbf{Y} - \mathbf{X}\boldsymbol{\beta})| \geq T/2 \right] \leq \delta/2, \quad (A.3.44)$$

for sufficiently large n. By (A.3.42) and (A.3.44) we can assert with probability greater than $1 - \delta$ that for sufficiently large n, $|Q(\mathbf{Y} - \mathbf{X}\boldsymbol{\beta}) - D(\mathbf{Y} - \mathbf{X}\boldsymbol{\beta})| < T/2$ and $\|\tilde{\boldsymbol{\beta}}\| < c_0/\sqrt{n}$. This implies with probability greater than $1 - \delta$ that for sufficiently large n,

$$D(\mathbf{Y} - \mathbf{X}\tilde{\boldsymbol{\beta}}) < Q(\mathbf{Y} - \mathbf{X}\tilde{\boldsymbol{\beta}}) + T/2 \text{ and } \|\tilde{\boldsymbol{\beta}}\| < c_0/\sqrt{n}. \quad (A.3.45)$$

Next suppose $\boldsymbol{\beta}$ is arbitrary and on the ring $\|\boldsymbol{\beta} - \tilde{\boldsymbol{\beta}}\| = \varepsilon/\sqrt{n}$. For $\|\tilde{\boldsymbol{\beta}}\| < c_0/\sqrt{n}$ it then follows that $\|\boldsymbol{\beta}\| \leq (c_0 + \varepsilon)/\sqrt{n}$. Arguing as above, we have with probability greater than $1 - \delta$ that $D(\mathbf{Y} - \mathbf{X}\boldsymbol{\beta}) > Q(\mathbf{Y} - \mathbf{X}\boldsymbol{\beta}) - T/2$, for sufficiently large n. From this, (A.3.43), and (A.3.45) we obtain the following string of inequalities:

$$D(\mathbf{Y} - \mathbf{X}\boldsymbol{\beta}) > Q(\mathbf{Y} - \mathbf{X}\boldsymbol{\beta}) - T/2$$

$$\geq \min\left\{Q(\mathbf{Y} - \mathbf{X}\boldsymbol{\beta}) : \|\boldsymbol{\beta} - \widetilde{\boldsymbol{\beta}}\| = \varepsilon/\sqrt{n}\right\} - T/2$$

$$= T + Q(\mathbf{Y} - \mathbf{X}\widetilde{\boldsymbol{\beta}}) - T/2$$

$$= T/2 + Q(\mathbf{Y} - \mathbf{X}\widetilde{\boldsymbol{\beta}}) > D(\mathbf{Y} - \mathbf{X}\widetilde{\boldsymbol{\beta}}). \qquad (A.3.46)$$

Thus, $D(\mathbf{Y} - \mathbf{X}\boldsymbol{\beta}) > D(\mathbf{Y} - \mathbf{X}\widetilde{\boldsymbol{\beta}})$, for $\|\boldsymbol{\beta} - \widetilde{\boldsymbol{\beta}}\| = \varepsilon/\sqrt{n}$. Since D is convex, we must also have $D(\mathbf{Y} - \mathbf{X}\boldsymbol{\beta}) > D(\mathbf{Y} - \mathbf{X}\widetilde{\boldsymbol{\beta}})$, for $\|\boldsymbol{\beta} - \widetilde{\boldsymbol{\beta}}\| \geq \varepsilon/\sqrt{n}$. But $D(\mathbf{Y} - \mathbf{X}\widetilde{\boldsymbol{\beta}}) \geq \min D(\mathbf{Y} - \mathbf{X}\boldsymbol{\beta}) = D(\mathbf{Y} - \mathbf{X}\widehat{\boldsymbol{\beta}})$. Hence $\widehat{\boldsymbol{\beta}}$ must lie inside the disk $\|\boldsymbol{\beta} - \widetilde{\boldsymbol{\beta}}\| = \varepsilon/\sqrt{n}$ with probability of at least $1 - 2\delta$; that is, $P[\|\widehat{\boldsymbol{\beta}} - \widetilde{\boldsymbol{\beta}}\| < \varepsilon/\sqrt{n}] > 1 - 2\delta$. This yields the result.

A.3.4 Consistency of the Test Statistic F_φ

This section contains a proof of the consistency of the test statistic F_φ (Theorem 3.6.5). We begin with a lemma.

Lemma A.3.10 *Let $a > 0$ be given and let $t_n = \min_{\sqrt{n}\|\boldsymbol{\beta} - \widetilde{\boldsymbol{\beta}}\| = a}(Q(\boldsymbol{\beta}) - Q(\widetilde{\boldsymbol{\beta}}))$. Then $t_n = (2\tau)^{-1}a^2 \lambda_{n,1}$, where $\lambda_{n,1}$ is the minimum eigenvalue of $n^{-1}\mathbf{X}'\mathbf{X}$.*

Proof. After some computation, we have

$$Q(\boldsymbol{\beta}) - Q(\widetilde{\boldsymbol{\beta}}) = (2\tau)^{-1}\sqrt{n}(\boldsymbol{\beta} - \widetilde{\boldsymbol{\beta}})' n^{-1}\mathbf{X}'\mathbf{X}\sqrt{n}(\boldsymbol{\beta} - \widetilde{\boldsymbol{\beta}}).$$

Let $0 < \lambda_{n,1} \leq \cdots \leq \lambda_{n,p}$ be the eigenvalues of $n^{-1}\mathbf{X}'\mathbf{X}$ and let $\boldsymbol{\gamma}_{n,1}, \ldots, \boldsymbol{\gamma}_{n,p}$ be a corresponding set of orthonormal eigenvectors. The spectral decomposition of $n^{-1}\mathbf{X}'\mathbf{X}$ is $n^{-1}\mathbf{X}'\mathbf{X} = \sum_{i=1}^{p} \lambda_{n,i} \boldsymbol{\gamma}_{n,i} \boldsymbol{\gamma}'_{n,i}$. From this we can show for any vector $\boldsymbol{\delta}$ that $\boldsymbol{\delta}' n^{-1}\mathbf{X}'\mathbf{X}\boldsymbol{\delta} \geq \lambda_{n,1}\|\boldsymbol{\delta}\|^2$ and that, further, the minimum is achieved over all vectors of unit length when $\boldsymbol{\delta} = \boldsymbol{\gamma}_{n,1}$. It then follows that

$$\min_{\|\boldsymbol{\delta}\| = a} \boldsymbol{\delta}' n^{-1}\mathbf{X}'\mathbf{X}\boldsymbol{\delta} = \lambda_{n,1} a^2,$$

which yields the conclusion.

Note that by (D.2) of Section 3.4, $\lambda_{n,1} \to \lambda_1$, for some $\lambda_1 > 0$. The following is a restatement and a proof of Theorem 3.6.5.

Theorem A.3.11 *Suppose conditions (E.1), (D.1), (D.2), and (S.1) of Section 3.4 hold. The test statistic F_φ is consistent for the hypotheses (3.2.5).*

Proof. By the above discussion we need only show that (3.6.23) is true. Let $\varepsilon > 0$ be given. Let $c_0 = (2\tau)^{-1}\chi^2_{\alpha,q}$. By Lemma A.3.10, choose $a > 0$ so large that $(2\tau)^{-1}a^2\lambda_1 > 3c_0 + \varepsilon$. Next choose n_0 so large that $(2\tau)^{-1}a^2\lambda_{n,1} > 3c_0$, for $n \geq n_0$. Since $\sqrt{n}\|\boldsymbol{\beta} - \boldsymbol{\beta}_0\|$ is bounded in probability, there exits a $c > 0$ and an n_1 such that for $n \geq n_1$,

$$P_{\boldsymbol{\beta}_0}(C_{1,n}) \geq 1 - \varepsilon/2, \qquad (A.3.47)$$

418 *Asymptotic Results*

where we define the event $C_{1,n} = \{\sqrt{n}\|\widetilde{\boldsymbol{\beta}} - \boldsymbol{\beta}_0\| < c\}$. Since $t > 0$ by asymptotic quadraticity, (Theorem A.3.8), there exits an n_2 such that for $n > n_2$,

$$P_{\boldsymbol{\beta}_0}(C_{2,n}) \geq 1 - \varepsilon/2, \qquad (A.3.48)$$

where $C_{2,n} = \{\max_{\sqrt{n}\|\boldsymbol{\beta}-\boldsymbol{\beta}_0\| \leq c+a} |Q(\boldsymbol{\beta}) - D(\boldsymbol{\beta})| < (t/3)\}$. For the remainder of the proof assume that $n \geq \max\{n_0, n_1, n_2\} = n^*$. Next suppose $\boldsymbol{\beta}$ is such that $\sqrt{n}\|\boldsymbol{\beta} - \widetilde{\boldsymbol{\beta}}\| = a$. Then on $C_{1,n}$ it follows that $\sqrt{n}\|\boldsymbol{\beta} - \boldsymbol{\beta}\| \leq c + a$. Hence on both $C_{1,n}$ and $C_{2,n}$ we have

$$\begin{aligned} D(\boldsymbol{\beta}) &> Q(\boldsymbol{\beta}) - t/3 \\ &\geq Q(\widetilde{\boldsymbol{\beta}}) + t - t/3 \\ &= Q(\widetilde{\boldsymbol{\beta}}) + 2t/3 \\ &> D(\widetilde{\boldsymbol{\beta}}) + t/3. \end{aligned}$$

Therefore, for all $\boldsymbol{\beta}$ such that $\sqrt{n}\|\boldsymbol{\beta} - \widetilde{\boldsymbol{\beta}}\| = a$, $D(\boldsymbol{\beta}) - D(\widetilde{\boldsymbol{\beta}}) > t/3 \geq c_0$. But D is convex; hence on $C_{1,n} \cap C_{2,n}$, for all $\boldsymbol{\beta}$ such that $\sqrt{n}\|\boldsymbol{\beta} - \widetilde{\boldsymbol{\beta}}\| \geq a$, $D(\boldsymbol{\beta}) - D(\widetilde{\boldsymbol{\beta}}) > (t/3) > c_0$.

Finally, choose n_3 such that for $n \geq n_3$, $\delta > (c+a)/\sqrt{n}$, where δ is the positive distance between $\boldsymbol{\beta}_0$ and $\mathcal{R}^r$. Now assume that $n \geq \max\{n^*, n_3\}$ and $C_{1,n} \cap C_{2,n}$ is true. Recall that the reduced model R-estimate is $\widehat{\boldsymbol{\beta}}_r = (\widehat{\boldsymbol{\beta}}_{r,1}', \mathbf{0}')'$ where $\widehat{\boldsymbol{\beta}}_{r,1}$ lies in $\mathcal{R}^r$; hence,

$$\sqrt{n}\|\widehat{\boldsymbol{\beta}}_r - \widetilde{\boldsymbol{\beta}}\| \geq \sqrt{n}\|\widehat{\boldsymbol{\beta}}_r - \boldsymbol{\beta}_0\| - \sqrt{n}\|\widetilde{\boldsymbol{\beta}} - \boldsymbol{\beta}_0\| \geq \sqrt{n}\delta - c > a.$$

Thus on $C_{1,n} \cap C_{2,n}$, $D(\widehat{\boldsymbol{\beta}}_r) - D(\widetilde{\boldsymbol{\beta}}) > c_0$. Thus for n sufficiently large we have,

$$P[D(\widehat{\boldsymbol{\beta}}_r) - D(\widetilde{\boldsymbol{\beta}}) > (2\tau)^{-1}\chi^2_{\alpha,q}] \geq 1 - \varepsilon.$$

Because ε was arbitrary (3.6.23) is true and consistency of F_φ follows.

A.3.5 Proof of Lemma 3.5.8

The following lemma was used to establish the asymptotic linearity for the sign process for linear models in Chapter 3. The proof of this lemma was first given by Jurečková (1971) for general scores. We restate the lemma and give its proof for sign scores.

Lemma A.3.12 *Assume conditions (E.1), (E.2), (S.1), (D.1) and (D.2) of Section 3.4. For any $\varepsilon > 0$ and for any $a \in \mathcal{R}$,*

$$\lim_{n \to \infty} P[|S_1(\mathbf{Y} - an^{-1/2} - \mathbf{X}\widehat{\boldsymbol{\beta}}_R) - S_1(\mathbf{Y} - an^{-1/2})| \geq \varepsilon\sqrt{n}] = 0.$$

Proof. Let a be arbitrary but fixed and let $c > |a|$. After matching notation, Theorem A.4.5 leads to the result

$$\max_{\|(\mathbf{X}'\mathbf{X})^{1/2}\boldsymbol{\beta}\| \leq c} \left| \frac{1}{\sqrt{n}} S_1(\mathbf{Y} - an^{-1/2} - \mathbf{X}\boldsymbol{\beta}) - \frac{1}{\sqrt{n}} S_1(\mathbf{Y}) + (2f(0))a \right| = o_p(1).$$

$$(A.3.49)$$

Obviously the above result holds for $\boldsymbol{\beta} = \mathbf{0}$. Hence, for any $\varepsilon > 0$,

$$P\left[\max_{\|(\mathbf{X}'\mathbf{X})^{1/2}\boldsymbol{\beta}\| \leq c} \left|\frac{1}{\sqrt{n}} S_1(\mathbf{Y} - an^{-1/2} - \mathbf{X}\boldsymbol{\beta}) - \frac{1}{\sqrt{n}} S_1(\mathbf{Y} - an^{-1/2})\right| \geq \varepsilon\right]$$

$$\leq P\left[\max_{\|(\mathbf{X}'\mathbf{X})^{1/2}\boldsymbol{\beta}\| \leq c} \left|\frac{1}{\sqrt{n}} S_1(\mathbf{Y} - an^{-1/2} - \mathbf{X}\boldsymbol{\beta}) - \frac{1}{\sqrt{n}} S_1(\mathbf{Y}) + (2f(0))a\right| \geq \frac{\varepsilon}{2}\right]$$

$$+ P\left[\left|\frac{1}{\sqrt{n}} S_1(\mathbf{Y} - an^{-1/2}) - \frac{1}{\sqrt{n}} S_1(\mathbf{Y}) + (2f(0))a\right| \geq \frac{\varepsilon}{2}\right]$$

By (A.3.49), for n sufficiently large, the two terms on the right-hand side are arbitrarily small. The desired result follows from this since $(\mathbf{X}'\mathbf{X})^{1/2}\widehat{\boldsymbol{\beta}}$ is bounded in probability.

A.4 Asymptotic Linearity for the L_1 Analysis

In this section we obtain a linearity result for the L_1 analysis of a linear model. Recall from Section 3.6 that the L_1 estimates are equivalent to the R-estimates when the rank scores are generated by the sign function; hence, the distribution theory for the L_1 estimates is derived in Section 3.4. The linearity result derived below offers another way to obtain this result. More importantly though, we need the linearity result for the proof of Lemma 3.5.9 of Section 3.5. As we next show, this result is a corollary to the linearity results derived in the previous section.

We will assume the same linear model and use the same notation as in Section 3.2. Recall that the L_1 estimate of $\boldsymbol{\beta}$ minimizes the dispersion function,

$$D_1(\alpha, \boldsymbol{\beta}) = \sum_{i=1}^{n} |Y_i - \alpha - \mathbf{x}_i\boldsymbol{\beta}|.$$

The corresponding gradient function is the $(p+1) \times 1$ vector whose components are

$$\nabla_j D_1 = \begin{cases} -\sum_{i=1}^{n} \text{sgn}(Y_i - \alpha - \mathbf{x}_i\boldsymbol{\beta}) & \text{if } j = 0 \\ -\sum_{i=1}^{n} x_{ij}\text{sgn}(Y_i - \alpha - \mathbf{x}_i\boldsymbol{\beta}) & \text{if } j = 1, \ldots, p, \end{cases}$$

where $j = 0$ denotes he partial of D_1 with respect to α. The parameter α will denote the location functional $\text{med}(Y_i - \mathbf{x}_i\boldsymbol{\beta})$, i.e. the median of the errors. Without loss of generality we will assume that the true parameters are 0.

We first consider the **simple linear model**. Consider, then, the notation of Section A.3 see (A.3.1)–(A.3.7). We will derive the analog of Theorem A.3.8

for the processes

$$U_0(\alpha, \Delta) = \sum_{i=1}^{n} \text{sgn}\left(Y_i - \frac{\alpha}{\sqrt{n}} - \Delta c_i\right) \qquad (A.4.1)$$

$$U_1(\alpha, \Delta) = \sum_{i=1}^{n} c_i \text{sgn}\left(Y_i - \frac{\alpha}{\sqrt{n}} - \Delta c_i\right). \qquad (A.4.2)$$

Let $p_d = \prod_{i=1}^{n} f_0(y_i)$ denote the likelihood for the iid observations $Y_1, \ldots, Y_n$ and let $q_d = \prod_{i=1}^{n} f_0(y_i + \alpha/\sqrt{n} + \Delta c_i)$ denote the likelihood of the variables $Y_i - (\alpha/\sqrt{n}) - \Delta c_i$. We assume throughout that $f(0) > 0$. Similar to Section A.2.2, the sequence of densities q_d is contiguous to the sequence p_d. Note that the processes U_0 and U_1 are already sums of independent variables; hence, projections are unnecessary.

We first work with the process U_1.

Lemma A.4.1 *Under the above assumptions and as $n \to \infty$,*

$$E_0(U_1(\alpha, \Delta)) \to -2\Delta f_0(0).$$

Proof. After some simplification we obtain

$$E_0(U_1(\alpha, \Delta)) = 2 \sum_{i=1}^{n} c_i [F_0(0) - F_0(\alpha/\sqrt{n} + \Delta c_i)]$$

$$= 2 \sum_{i=1}^{n} c_i(-\Delta c_i - \alpha/\sqrt{n}) f_0(\xi_{in}),$$

where, by the mean value theorem, ξ_{in} is between 0 and $|\alpha/\sqrt{n} + \Delta c_i|$. Since the c_is are centered, we further obtain

$$E_0(U_1(\alpha, \Delta)) = -2\Delta \sum_{i=1}^{n} c_i^2 [f_0(\xi_{in}) - f_0(0)] - 2\Delta \sum_{i=1}^{n} c_i^2 f_0(0).$$

By assumptions of Section A.2.2, it follows that $\max_i |\alpha/\sqrt{n} + \Delta c_i| \to 0$ as $n \to \infty$. Recall that f_0 is assumed to be continuous and positive at 0. Futhermore, $\sum_{i=1}^{n} c_i^2 = 1$. Hence, the desired result easily follows.

This leads us to our main result for $U_1(\alpha, \Delta)$:

Theorem A.4.2 *Under the above assumptions, for all α and Δ,*

$$U_1(\alpha, \Delta) - [U_1(0, 0) - \Delta 2 f_0(0)] \xrightarrow{P} 0,$$

as $n \to \infty$.

Because the c_is are centered it follows that $E_{p_d}(U_1(0, 0)) = 0$. Thus by the last lemma, we need only show that $\text{Var}(U_1(\alpha, \Delta) - U_1(0, 0)) \to 0$. By considering the variance of the sign of a random variable, simplification leads to the bound

$$\text{Var}(U_1(\alpha, \Delta) - U_1(0, 0)) \leq 4 \sum_{i=1}^{n} c_i^2 |F_0(\alpha/\sqrt{n} + \Delta c_i) - F_0(0)|.$$

By our assumptions, $\max_i |\Delta c_i + \alpha/\sqrt{n}| \to 0$ as $n \to \infty$. From this and the continuity of F_0 at 0, it follows that $\text{Var}(U_1(\alpha, \Delta) - U_1(0, 0)) \to 0$.

We need analogous results for the process $U_0(\alpha, \Delta)$.

Lemma A.4.3 *Under the above assumptions,*

$$E_0[U_0(\alpha, \Delta)] \to -2\alpha f_0(0),$$

as $n \to \infty$.

Proof. Upon simplifying, and applying the mean value theorem,

$$E_0[U_0(\alpha, \Delta)] = \frac{2}{\sqrt{n}} \sum_{i=1}^{n} \left[F_0(0) - F_0\left(\frac{\alpha}{\sqrt{n}} + c_i \Delta\right) \right]$$

$$= \frac{-2}{\sqrt{n}} \sum_{i=1}^{n} \left[\frac{\alpha}{\sqrt{n}} + c_i \Delta \right] f_0(\xi_{in})$$

$$= \frac{-2\alpha}{n} \sum_{i=1}^{n} [f_0(\xi_{in}) - f_0(0)] - 2\alpha f_0(0)$$

where we have used the fact that the c_is are centered. Note that $|\xi_{in}|$ is between 0 and $|\alpha/\sqrt{n} + c_i \Delta|$ and that $\max |\alpha/\sqrt{n} + c_i \Delta| \to 0$ as $n \to \infty$. By the continuity of f_0 at 0, the desired result follows.

Theorem A.4.4 *Under the above assumptions, for all α and Δ,*

$$U_0(\alpha, \Delta) - [U_0(0, 0) - 2\alpha f_0(0)] \xrightarrow{P} 0,$$

as $n \to \infty$.

Because med Y_i is 0, $E_0[U_0(0, 0)] = 0$. Hence, by the last lemma it suffices to show that $\text{Var}(U_0(\alpha, \Delta) - U_0(0, 0)) \to 0$. But

$$\text{Var}(U_0(\alpha, \Delta) - U_0(0, 0)) \leq \frac{4}{n} \sum_{i=1}^{n} \left| F_0\left(\frac{\alpha}{\sqrt{n}} + c_i \Delta\right) - F_0(0) \right|.$$

Because $\max |\alpha/\sqrt{n} + c_i \Delta| \to 0$ and F_0 is continuous at 0, $\text{Var}(U_0(\alpha, \Delta) - U_0(0, 0)) \to 0$.

Next consider the multiple regression model as discussed in Section A.3. The only difference in notation is that here we have the intercept parameter included. Let $\mathbf{\Delta} = (\alpha, \Delta_1, \ldots, \Delta_p)'$ denote the vector of all regression parameters. Take $\mathbf{X} = [\mathbf{1}_n : \mathbf{X}_c]$, where $\mathbf{X}_c$ denotes a centered design matrix and, as in (A.3.2), take $\mathbf{C} = \mathbf{X}(\mathbf{X}'\mathbf{X})^{-1/2}$. Note that the first column of $\mathbf{C}$ is $(1/\sqrt{n})\mathbf{1}_n$.

Let $\mathbf{U}(\boldsymbol{\Delta}) = (U_0(\boldsymbol{\Delta}), \ldots, U_p(\boldsymbol{\Delta}))'$ denote the vector of processes. Similar to the discussion prior to Theorem A.3.1, the last two theorems imply that

$$\mathbf{U}(\boldsymbol{\Delta}) - [\mathbf{U}(0) - 2f_0(0)\boldsymbol{\Delta}] \xrightarrow{P} 0,$$

for all real $\boldsymbol{\Delta}$ in $\mathcal{R}^{p+1}$.

As in Section A.3, we define the approximation quadratic to D_1 as

$$Q_{1n}(\boldsymbol{\Delta}) = (2f_0(0))\boldsymbol{\Delta}'\boldsymbol{\Delta}/2 - \boldsymbol{\Delta}'\mathbf{U}(0) + D_1(0).$$

The asymptotic linearity of U and the asymptotic quadraticity of D_1 then follow as in the previous section. We state the result for reference:

Theorem A.4.5 *Under conditions (3.4.1), (3.4.3), (3.4.7), and (3.4.8),*

$$\lim_{n \to \infty} P\left(\max_{\|\boldsymbol{\Delta}\| \leq c} \|\mathbf{U}(\boldsymbol{\Delta}) - (\mathbf{U}(0) - (2f_0(0))\boldsymbol{\Delta})\| \geq \varepsilon \right) = 0, \qquad (A.4.3)$$

$$\lim_{n \to \infty} P\left(\max_{\|\boldsymbol{\Delta}\| \leq c} |D_1(\boldsymbol{\Delta}) - Q_1(\boldsymbol{\Delta})| \geq \varepsilon \right) = 0, \qquad (A.4.4)$$

for all $\varepsilon > 0$ and all $c > 0$.

A.5 Influence Functions

In this section we derive the influence functions found in Chapters 1–3 and 5. Discussions of the influence function can be found in Staudte and Sheather (1990), Hampel *et al.* (1986), and Huber (1981). For the influence functions of Chapter 3, we will find the Gâteux derivative of a functional; see Fernholz (1983) and Huber (1981) for rigorous discussions of functionals and derivatives.

Definition A.5.1 *Let T be a statistical functional defined on a space of distribution functions and let H denote a distribution function in the domain of T. We say T is **Gateux differentiable** at H if for any distribution function W, such that the distribution functions $\{(1-s)H + sW\}$ lie in the domain of T, the following limit exists:*

$$\lim_{s \to 0} \frac{T[(1-s)H + sW] - T[H]}{s} = \int \psi_H dW, \qquad (A.5.1)$$

for some function ψ_H.

Note that, by taking W to be H in the above definition, we have

$$\int \psi_H dH = 0. \qquad (A.5.2)$$

The usual definition of the influence function is obtained by taking the distribution function W to be a point mass distribution. Denote the point mass distribution function at t by $\Delta_t(x)$. Letting $W(x) = \Delta_t(x)$, the Gâteux derivative of $T(H)$ is

$$\lim_{s \to 0} \frac{T[(1-s)H + s\Delta_s(x)] - T[H]}{s} = \psi_H(t). \quad \text{(A.5.3)}$$

The function $\psi_H(t)$ is called the **influence function** of $T(H)$. Note that this is the derivative of the functional $T[(1-s)H + s\Delta_t(x)]$ at $s = 0$. It measures the rate of change of the functional $T(H)$ at H in the direction of Δ_t. A functional is said to be **robust** when this derivative is bounded.

A.5.1 Influence Function for Estimates Based on Signed-Rank Statistics

In this section we derive the influence function for the one-sample location estimate $\hat{\theta}_{\varphi^+}$, (1.8.5), discussed in Chapter 1. We will assume that we are sampling from a symmetric density $h(x)$ with distribution function $H(x)$, as in Section 1.8. As in Chapter 2, we will assume that the one-sample score function $\varphi^+(u)$ is defined by

$$\varphi^+(u) = \varphi\left(\frac{u+1}{2}\right), \quad \text{(A.5.4)}$$

where $\varphi(u)$ is a nondecreasing, differentiable function defined on the interval $(0, 1)$ satisfying

$$\varphi(1-u) = -\varphi(u). \quad \text{(A.5.5)}$$

Recall from Chapter 2 that this assumption is appropriate for scores for samples from symmetrical distributions. For convenience we extend $\varphi^+(u)$ to the interval $(-1, 0)$ by

$$\varphi^+(-u) = -\varphi^+(u). \quad \text{(A.5.6)}$$

Our functional $T(H)$ is defined implicitly by equation (1.8.5). Using the symmetry of $h(x)$, (A.5.5), and (A.5.6) we can write the defining equation for $\theta = T(H)$ as

$$0 = \int_{-\infty}^{\infty} \varphi^+(H(x) - H(2\theta - x))h(x)\,dx$$

$$0 = \int_{-\infty}^{\infty} \varphi(1 - H(2\theta - x))h(x)\,dx. \quad \text{(A.5.7)}$$

For the derivation, we will proceed as discussed above; see the discussion around expression (A.5.3). Consider the contaminated distribution of $H(x)$ given by

$$H_{t,\varepsilon}(x) = (1-\varepsilon)H(x) + \varepsilon\Delta_t(x), \quad \text{(A.5.8)}$$

where $0 < \varepsilon < 1$ is the proportion of contamination and $\Delta_t(x)$ is the distribution function for a point mass at t. By (A.5.3) the influence function is the derivative of the functional at $\varepsilon = 0$. To obtain this derivative we implicitly differentiate the defining equation (A.5.7) at $H_{t,\varepsilon}(x)$; i.e. at

$$0 = (1-\varepsilon)\int_{-\infty}^{\infty} \varphi(1-(1-\varepsilon)H(2\theta-x)-\varepsilon\Delta_t(2\theta-x))h(x)\,dx$$

$$= \varepsilon\int_{-\infty}^{\infty} \varphi(1-(1-\varepsilon)H(2\theta-x)-\varepsilon\Delta_t(2\theta-x))\,d\Delta_t(x).$$

Let $\dot\theta$ denote the derivative of the functional. Implicitly differentiating this equation and then setting $\varepsilon = 0$ and without loss of generality $\theta = 0$, we obtain

$$0 = -\int_{-\infty}^{\infty} \varphi(H(x))h(x)\,dx + \int_{-\infty}^{\infty} \varphi'(H(x))H(-x)h(x)\,dx$$

$$= -2\dot\theta\int_{-\infty}^{\infty} \varphi'(H(x))h^2(x)\,dx - \int_{-\infty}^{\infty} \varphi'(H(x))\Delta_t(-x)h(x)\,dx + \varphi(H(t)).$$

Label the four integrals in the above equation as $I_1, \ldots, I_4$. Since $\int \varphi(u)\,du = 0$, $I_1 = 0$. For I_2 we obtain

$$I_2 = \int_{-\infty}^{\infty} \varphi'(H(x))h(x)\,dx - \int_{-\infty}^{\infty} \varphi'(H(x))H(x)h(x)\,dx$$

$$= \int_0^1 \varphi'(u)\,du - \int_0^1 \varphi'(u)u\,du = -\varphi(0).$$

Next I_4 reduces to

$$-\int_{-\infty}^{-t} \varphi'(H(x))h(x)\,dx = -\int_0^{H(-t)} \varphi'(u)\,du = \varphi(H(t)) + \varphi(0).$$

Combining these results and solving for $\dot\theta$ leads to the influence function which we can write in either of the following two ways:

$$\Omega(t, \widehat{\theta}_{\varphi^+}) = \frac{\varphi(H(t))}{\int_{-\infty}^{\infty} \varphi'(H(x))h^2(x)\,dx}$$

$$= \frac{\varphi^+(2H(t)-1)}{4\int_0^{\infty} \varphi^{+\prime}(2H(x)-1)h^2(x)\,dx}. \tag{A.5.9}$$

A.5.2 Influence Functions for Chapter 3

In this subsection, we derive the influence functions which were presented in Chapter 3. Much of this work was developed in Witt (1989) and Witt, Naranjo, and McKean (1995). The correlation model of Section 3.11 is the underlying model for the influence functions derived here. Recall that the joint distribution function of $\mathbf{x}$ and Y is H, the distribution functions of $\mathbf{x}$, Y and e are M, G and F, respectively, and Σ is the variance-covariance matrix of $\mathbf{x}$.

Let $\widehat{\boldsymbol{\beta}}_\varphi$ denote the R-estimate of $\boldsymbol{\beta}$ for a specified score function $\varphi(u)$. In this section we are interested in deriving the influence functions of this R-estimate and of the corresponding R test statistic for the general linear hypotheses. We will obtain these influence functions by using the definition of the Gâteux derivative of a functional, (A.5.1). The influence functions are then obtained by taking W to be the point mass distribution function $\Delta_{(\mathbf{x}_0, y_0)}$; see expression (A.5.3). If T is Gâteux differentiable at H then by setting $W = \Delta_{(\mathbf{x}_0, y_0)}$ we see that the influence function of T is given by

$$\Omega(\mathbf{x}_0, y_0; T) = \int \psi_H d\Delta_{(\mathbf{x}_0, y_0)} = \psi_H(\mathbf{x}_0, y_0). \quad (A.5.10)$$

As a simple example, we will obtain the influence function of the statistic $D(\mathbf{0}) = \sum a(R(Y_i))Y_i$. Since G is the distribution function of Y, the corresponding functional is $T[G] = \int \varphi(G(y))y \, dG(y)$. Hence for a given distribution function W,

$$T[(1-s)G + sW] = (1-s)\int \varphi[(1-s)G(y) + sW(y)]y \, dG(y)$$

$$+ s \int \varphi[(1-s)G(y) + sW(y)]y \, dW(y).$$

Taking the partial derivative of the right-hand side with respect to s, setting $s = 0$, and substituting Δ_{y_0} for W leads to the influence function

$$\Omega(y_0; D(\mathbf{0})) = -\int \varphi(G(y))y \, dG(y) - \int \varphi'(G(y))G(y)y \, dG(y)$$

$$+ \int_{y_0}^{\infty} \varphi'(G(y))y \, dG(y) + \varphi(G(y_0))y_0. \quad (A.5.11)$$

Note that this is not bounded in the Y-space and, hence, the statistic $D(\mathbf{0})$ is not robust. Thus, as noted in Section 3.11, the coefficient of multiple determination R_1, (3.11.16), is not robust. A similar development establishes the influence function for the denominator of LS coefficient of multiple determination R^2, showing too that it is not bounded. Hence, R^2 is not a robust statistic.

Another example is the the influence function of the least-squares estimate of $\boldsymbol{\beta}$ which is given by

$$\Omega(\mathbf{x}_0, y_0; \widehat{\boldsymbol{\beta}}_{LS}) = \sigma^{-1} y_0 \Sigma^{-1} \mathbf{x}_0. \quad (A.5.12)$$

The influence function of the least-squares estimate is, thus, unbounded in both the Y- and $\mathbf{x}$-spaces.

Influence Function of $\widehat{\boldsymbol{\beta}}_\varphi$

Recall that H is the joint distribution function of $\mathbf{x}$ and Y. Let the $p \times 1$ vector $T(H)$ denote the functional corresponding to $\widehat{\boldsymbol{\beta}}_\varphi$. Assume without loss of generality that the true $\boldsymbol{\beta} = \mathbf{0}$, $\alpha = 0$, and that $E\mathbf{x} = \mathbf{0}$. Hence, the distribution function of Y is $F(y)$ and Y and $\mathbf{x}$ are independent, i.e. $H(\mathbf{x}, y) = M(\mathbf{x})F(y)$.

426 Asymptotic Results

Recall that the R-estimate satisfies the equations

$$\sum_{i=1}^{n} x_i a(R(Y_i - x_i'\beta)) \doteq 0.$$

Let $\widehat{G}_n^*$ denote the empirical distribution function of $Y_i - x_i'\beta$. Then we can rewrite the above equations as

$$n \sum_{i=1}^{n} x_i \varphi\left(\frac{n}{n+1}\widehat{G}_n^*(Y_i - x_i'\beta)\right)\frac{1}{n} \doteq 0.$$

Let G^* denote the distribution function of $Y - x'T(H)$. Then the functional $T(H)$ satisfies

$$\int \varphi(G^*(y - x'T(H)))x \, dH(x, y) = 0. \tag{A.5.13}$$

We can show that

$$G^*(t) = \iint_{u \leq v'T(H)+t} dH(v, u). \tag{A.5.14}$$

Let $H_s = (1 - s)H + sW$ for an arbitrary distribution function W. Then the functional $T(H)$ evaluated at H_s satisfies the equation

$$(1 - s)\int \varphi(G_s^*(y - x'T(H_s)))x \, dH(x, y) + s \int \varphi(G_s^*(y - x'T(H_s)))x \, dW(x, y) = 0,$$

where G_s^* is the distribution function of $Y - x'T(H_s)$. We will obtain $\partial T/\partial s$ by implicit differentiation. Then upon substituting Δ_{x_0, y_0} for W the influence function is given by $(\partial T/\partial s)|_{s=0}$, which we will denote by $\dot{T}$. Implicit differentiation leads to

$$0 = -\int \varphi(G_s^*(y - x'T(H_s)))x \, dH(x, y)$$

$$- (1 - s)\int \varphi'(G_s^*(y - x'T(H_s)))\frac{\partial G_s^*}{\partial s} x \, dH(x, y)$$

$$+ \int \varphi(G_s^*(y - x'T(H_s)))x \, dW(x, y) + s\mathbf{B}_1, \tag{A.5.15}$$

where $\mathbf{B}_1$ is irrelevant since we will be setting s to 0. We first obtain the partial derivative of G_s^* with respect to s. By (A.5.14) and the independence between Y and $\mathbf{x}$ at H, we have

$$G_s^*(y - x'T(H_s)) = \iint_{u \leq y - T(H_s)'(x-v)} dH_s(v, u)$$

$$= (1 - s)\int F[y - T(H_s)'(x - v)]dM(v)$$

$$+ s \iint_{u \leq y - T(H_s)'(x-v)} dW(v, u).$$

Thus,

$$\frac{\partial G_s^*(y - \mathbf{x}'T(H_s))}{\partial s} = -\int F[y - T(H_s)'(\mathbf{x} - \mathbf{v})]dM(\mathbf{v})$$

$$+ (1 - s)\int F'[y - T(H_s)'(\mathbf{x} - \mathbf{v})](\mathbf{v} - \mathbf{x})'\frac{\partial T}{\partial s}dM(\mathbf{v})$$

$$+ \int\int_{u \leq y - T(H_s)'(\mathbf{x} - \mathbf{v})} dW(\mathbf{v}, u) + s\mathbf{B}_2,$$

where $\mathbf{B}_2$ is irrelevant since we are setting s to 0. Therefore using the independence between Y and $\mathbf{x}$ at H, $T(H) = \mathbf{0}$, and $E\mathbf{x} = \mathbf{0}$, we obtain

$$\frac{\partial G_s^*(y - \mathbf{x}'T(H_s))}{\partial s}\bigg|_{s=0} = -F(y) - f(y)\mathbf{x}'\dot{T} + W_Y(y), \quad (A.5.16)$$

where W_Y denotes the marginal (second variable) distribution function of W.

Upon evaluating expression (A.5.15) at $s = 0$ and substituting into it expression (A.5.16), we have

$$\mathbf{0} = -\int \mathbf{x}\varphi(F(y))dH(\mathbf{x}, y) + \int \mathbf{x}\varphi'(F(y))[-F(y) - f(y)\mathbf{x}'\dot{T} + W_Y(y)]dH(\mathbf{x}, y)$$

$$+ \int \mathbf{x}\varphi(F(y))dW(\mathbf{x}, y)$$

$$= -\int \varphi'(F(y))f(y)\mathbf{x}\mathbf{x}'\dot{T}dH(\mathbf{x}, y) + \int \mathbf{x}\varphi(F(y))dW(\mathbf{x}, y).$$

Substituting $\Delta_{\mathbf{x}_0, y_0}$ for W, we obtain

$$\mathbf{0} = -\tau\Sigma\dot{T} + \mathbf{x}_0\varphi(F(y_0)).$$

Solving this last expression for $\dot{T}$, the influence function of $\widehat{\boldsymbol{\beta}}_\varphi$ is given by

$$\Omega(\mathbf{x}_0, y_0; \widehat{\boldsymbol{\beta}}_\varphi) = \tau\Sigma^{-1}\varphi(F(y_0))\mathbf{x}_0. \quad (A.5.17)$$

Hence the influence function of $\widehat{\boldsymbol{\beta}}_\varphi$ is bounded in the Y-space but not in the x-space. The estimate is thus bias robust. In Chapter 5 we presented R-estimates whose influence functions are bounded in both spaces; see Theorems 5.7.4 and 5.8.6. Note that the asymptotic representation of $\widehat{\boldsymbol{\beta}}_\varphi$ in Corollary 3.5.24 can be written in terms of this influence function as

$$\sqrt{n}\widehat{\boldsymbol{\beta}}_\varphi = n^{-1/2}\sum_{i=1}^{n}\Omega(\mathbf{x}_i, Y_i; \widehat{\boldsymbol{\beta}}_\varphi) + o_p(1).$$

Influence Function of F_φ

Rewrite the correlation model as

$$Y = \alpha + \mathbf{x}_1'\boldsymbol{\beta}_1 + \mathbf{x}_2'\boldsymbol{\beta}_2 + e$$

and consider testing the general linear hypotheses

$$H_0 : \boldsymbol{\beta}_2 = \mathbf{0} \text{ versus } H_A : \boldsymbol{\beta}_2 \neq \mathbf{0}, \tag{A.5.18}$$

where $\boldsymbol{\beta}_1$ and $\boldsymbol{\beta}_2$ are $q \times 1$ and $(p-q) \times 1$ vectors of parameters, respectively. Let $\widehat{\boldsymbol{\beta}}_{1,\varphi}$ denote the reduced model estimate. Recall that the R test based upon the drop in dispersion is given by

$$F_\varphi = \frac{RD/q}{\widehat{\tau}/2},$$

where $RD = D(\widehat{\boldsymbol{\beta}}_{1,\varphi}) - D(\widehat{\boldsymbol{\beta}}_\varphi)$ is the reduction in dispersion. In this section we want to derive the influence function of the test statistic.

Let $RD(H)$ denote the functional for the statistic RD. Then

$$RD(H) = D_1(H) - D_2(H),$$

where $D_1(H)$ and $D_2(H)$ are the reduced and full model functionals given by

$$D_1(H) = \int \varphi[G^*(y - \mathbf{x}_1' T_1(H))](y - \mathbf{x}_1' T_1(H))dH(\mathbf{x}, y)$$

$$D_2(H) = \int \varphi[G^*(y - \mathbf{x}' T(H))](y - \mathbf{x}' T(H))dH(\mathbf{x}, y), \tag{A.5.19}$$

and $T_1(H)$ and $T_2(H)$ denote the reduced and full model functionals for $\boldsymbol{\beta}_1$ and $\boldsymbol{\beta}$, respectively. Let $\boldsymbol{\beta}_r = (\boldsymbol{\beta}_1', \mathbf{0}')'$ denote the true vector of parameters under H_0. Then the random variables $Y - \mathbf{x}'\boldsymbol{\beta}_r$ and $\mathbf{x}$ are independent. Next write Σ as

$$\Sigma = \begin{bmatrix} \Sigma_{11} & \Sigma_{12} \\ \Sigma_{21} & \Sigma_{22} \end{bmatrix}.$$

It will be convenient to define the matrices Σ_r and Σ_r^+ as

$$\Sigma_r = \begin{bmatrix} \Sigma_{11} & 0 \\ 0 & 0 \end{bmatrix} \text{ and } \Sigma_r^+ = \begin{bmatrix} \Sigma_{11}^{-1} & 0 \\ 0 & 0 \end{bmatrix}.$$

As above, let $H_s = (1-s)H + sW$. We begin with a lemma.

Lemma A.5.1 *Under the correlation model,*

(a) $RD(0) = 0$

(b) $\left.\dfrac{\partial RD(H_s)}{\partial s}\right|_{s=0} = 0$

(c) $\left.\dfrac{\partial^2 RD(H_s)}{\partial s^2}\right|_{s=0} = \tau \varphi^2 [F(y - \mathbf{x}'\boldsymbol{\beta}_r)] \mathbf{x}' [\Sigma^{-1} - \Sigma^+] \mathbf{x}. \tag{A.5.20}$

Proof. (a) is immediate. For (b), it follows from (A.5.19) that

$$\frac{\partial D_2(H_s)}{\partial s} = -\int \varphi[G_s^*(y - \mathbf{x}'T(H_s))](y - \mathbf{x}'T(H_s))dH$$

$$+ (1-s) \int \varphi'[G_s^*(y - \mathbf{x}'T(H_s))](y - \mathbf{x}'T(H_s)) \frac{\partial G_s^*}{\partial s} dH$$

$$+ (1-s) \int \varphi[G_s^*(y - \mathbf{x}'T(H_s))] \left(-\mathbf{x}' \frac{\partial T}{\partial s}\right) dH$$

$$+ \int \varphi[G_s^*(y - \mathbf{x}'T(H_s))](y - \mathbf{x}'T(H_s))dW(y) + s\mathbf{B}, \qquad (A.5.21)$$

where **B** is irrelevant because we are setting s to 0. Evaluating this at $s = 0$ and using the independence of $Y - \mathbf{x}'\boldsymbol{\beta}_r$ and $\mathbf{x}$, and $E(\mathbf{x}) = 0$, we obtain after some simplification

$$\left.\frac{\partial D_2(H_s)}{\partial s}\right|_{s=0} = -\int \varphi[F(y - \mathbf{x}'\boldsymbol{\beta}_r)](y - \mathbf{x}'\boldsymbol{\beta}_r)dH$$

$$- \int \varphi'[F(y - \mathbf{x}'\boldsymbol{\beta}_r)]F(y - \mathbf{x}'\boldsymbol{\beta}_r)(y - \mathbf{x}'\boldsymbol{\beta}_r)dH$$

$$+ \int \varphi'[F(y - \mathbf{x}'\boldsymbol{\beta}_r)]W_Y(y - \mathbf{x}'\boldsymbol{\beta}_r)(y - \mathbf{x}'\boldsymbol{\beta}_r)dH$$

$$+ \varphi[F(y_0 - \mathbf{x}_0'\boldsymbol{\beta}_r)](y_0 - \mathbf{x}_0'\boldsymbol{\beta}_r).$$

Differentiating as above and using $\mathbf{x}'\boldsymbol{\beta}_r = \mathbf{x}_1'\boldsymbol{\beta}_1$, we get the same expression for $(\partial D_1/\partial s)|_{s=0}$. Hence, (b) is true. Taking the second partial derivatives of $D_1(H)$ and $D_2(H)$ with respect to s, the result for (c) can be obtained. This is a tedious derivation and details of it can be found in Witt (1989) and Witt et al. (1995).

Since F_φ is nonnegative, there is no loss of generality in deriving the influence function of $\sqrt{qF_\varphi}$. Letting $Q^2 = 2\tau^{-1}RD$, we have

$$\Omega(\mathbf{x}_0, y_0; \sqrt{qF_\varphi}) = \lim_{s \to 0} \frac{Q[(1-s)H + s\Delta_{\mathbf{x}_0, y_0}] - Q[H]}{s}.$$

But $Q[H] = 0$ by part (a) of Lemma A.5.1. Hence we can rewrite the above limit as

$$\Omega(\mathbf{x}_0, y_0; \sqrt{qF_\varphi}) = \left[\lim_{s \to 0} \frac{Q^2[(1-s)H + s\Delta_{\mathbf{x}_0, y_0}]}{s^2}\right]^{1/2}.$$

Using parts (a) and (b) of Lemma A.5.1, we can apply L'Hospital's rule twice to evaluate this limit. Thus

$$\Omega(\mathbf{x}_0, y_0; \sqrt{qF_\varphi}) = \left[\lim_{s \to 0} \frac{1}{2} \frac{\partial^2 Q^2}{\partial s^2}\right]^{1/2}$$

$$= \left[2\tau^{-1} \frac{\partial^2 RD}{\partial s^2}\right]^{1/2}$$

$$= |\varphi[F(y - \mathbf{x}'\boldsymbol{\beta}_r)]|\sqrt{\mathbf{x}'[\Sigma^{-1} - \Sigma^+]\mathbf{x}}. \quad (A.5.22)$$

Hence, the influence function of the rank-based test statistic F_φ is bounded in the Y-space as long as the score function is bounded. It can be shown that the influence function of the least-squares test statistic is not bounded in Y-space. It is clear from the above argument that the coefficient of multiple determination R_2 is also robust. Hence, for R fits, R_2 is the preferred coefficient of determination.

However, F_φ is not bounded in the x-space. In Chapter 5 we present statistics whose influence functions are bounded in both spaces; however, they are less efficient.

The asymptotic distribution of qF_φ was derived in Section 3.6; however, we can use the above result on the influence function to display it immediately. If we expand Q^2 into a von Mises expansion at H, we have

$$Q^2(H_s) = Q^2(H) + \frac{\partial Q^2}{\partial s}\bigg|_{s=0} + \frac{1}{2}\frac{\partial^2 Q^2}{\partial s^2}\bigg|_{s=0} + R$$

$$= \left[\int \varphi(F(y - \mathbf{x}'\boldsymbol{\beta}_r))\mathbf{x}' d\Delta_{\mathbf{x}_0, y_0}(\mathbf{x}, y)\right][\Sigma^{-1} - \Sigma^+]$$

$$\cdot \left[\int \varphi(F(y - \mathbf{x}'\boldsymbol{\beta}_r))\mathbf{x} d\Delta_{\mathbf{x}_0, y_0}(\mathbf{x}, y)\right] + R. \quad (A.5.23)$$

Upon substituting the empirical distribution function for $\Delta_{\mathbf{x}_0, y_0}$ in expression (A.5.23), we have at the sample

$$nQ^2(H_s) = \left[\frac{1}{\sqrt{n}}\sum_{i=1}^n \mathbf{x}'_i\varphi\left(\frac{1}{n}R(Y_i - \mathbf{x}'_i\boldsymbol{\beta}_r)\right)\right][\Sigma^{-1} - \Sigma^+]$$

$$\times \left[\frac{1}{\sqrt{n}}\sum_{i=1}^n \mathbf{x}_i\varphi\left(\frac{1}{n}R(Y_i - \mathbf{x}'_i\boldsymbol{\beta}_r)\right)\right] + o_p(1).$$

This expression is equivalent to expression (3.6.11) which yields the asymptotic distribution of the test statistic in Section 3.6.

A.5.3 Influence Function of $\widehat{\boldsymbol{\beta}}_{HBR}$ of Chapter 5

The influence function of the high-breakdown estimator $\widehat{\boldsymbol{\beta}}_{HBR}$ is discussed in Section 5.8.3. In this section, we restate Theorem A.5.24 and then derive a proof of it.

Theorem A.5.3 *The influence function for the estimate $\widehat{\boldsymbol{\beta}}_{HBR}$ is given by*

$$\Omega(\mathbf{x}_0, y_0, \widehat{\boldsymbol{\beta}}_{HBR}) = \mathbf{C}_H^{-1} \frac{1}{2} \iint (\mathbf{x}_0 - \mathbf{x}_1) b(\mathbf{x}_1, \mathbf{x}_0, y_1, y_0) \operatorname{sgn}\{y_0 - y_1\} \, dF(y_1) dM(\mathbf{x}_1),$$
(A.5.24)

where $\mathbf{C}_H$ is given by expression (5.8.26).

Proof. Let $\Delta_0(\mathbf{x}, y)$ denote the distribution function of the point mass at the point $(\mathbf{x}_0, y_0)$ and consider the contaminated distribution $H_t = (1-t)H + t\Delta_0$ for $0 < t < 1$. Let $\boldsymbol{\beta}(H_t)$ denote the functional at H_t. Then $\boldsymbol{\beta}(H_t)$ satisfies

$$0 = \iint \mathbf{x}_1 b(\mathbf{x}_1, \mathbf{x}_2, y_1, y_2) [I(y_2 - y_1 < (\mathbf{x}_2 - \mathbf{x}_1)' \boldsymbol{\beta}(H_t)) - \tfrac{1}{2}] dH_t(\mathbf{x}_1, y_1) dH_t(\mathbf{x}_2, y_2).$$
(A.5.25)

We next implicitly differentiate (A.5.25) with respect to t to obtain the derivative of the functional. The value of this derivative at $t = 0$ is the influence function. Without loss of generality, we can assume that the true parameter $\boldsymbol{\beta} = 0$. Under this assumption $\mathbf{x}$ and y are independent. Substituting the value of H_t into (A.5.25) and expanding, we obtain the four terms:

$$0 = (1-t)^2 \iiint \mathbf{x}_1 \left[\int_{-\infty}^{y_1 + (\mathbf{x}_2 - \mathbf{x}_1)' \boldsymbol{\beta}(H_t)} b(\mathbf{x}_1, \mathbf{x}_2, y_1, y_2) dF(y_2) - \tfrac{1}{2} \right]$$
$$\times dM(\mathbf{x}_2) dM(\mathbf{x}_1) dF(y_1)$$

$$+ (1-t)t \iiiint \mathbf{x}_1 b(\mathbf{x}_1, \mathbf{x}_2, y_1, y_2)[I(y_2 - y_1 < (\mathbf{x}_2 - \mathbf{x}_1)' \boldsymbol{\beta}(H)) - \tfrac{1}{2}]$$
$$\times dM(\mathbf{x}_2) dF(y_2) d\Delta_0(\mathbf{x}_1, y_1)$$

$$+ (1-t)t \iiiint \mathbf{x}_1 b(\mathbf{x}_1, \mathbf{x}_2, y_1, y_2)[I(y_2 - y_1 < (\mathbf{x}_2 - \mathbf{x}_1)' \boldsymbol{\beta}(H)) - \tfrac{1}{2}]$$
$$\times d\Delta_0(\mathbf{x}_2, y_2) dM(\mathbf{x}_1) dF(y_1)$$

$$+ t^2 \iiiint \mathbf{x}_1 b(\mathbf{x}_1, \mathbf{x}_2, y_1, y_2)[I(y_2 - y_1 < (\mathbf{x}_2 - \mathbf{x}_1)' \boldsymbol{\beta}(H)) - \tfrac{1}{2}]$$
$$\times d\Delta_0(\mathbf{x}_2, y_2) d\delta_0(\mathbf{x}_1, y_1).$$

Let $\dot{\boldsymbol{\beta}}$ denote the derivative of the functional evaluted at 0. Proceeding to implicitly differentiate this equation and evaluating the derivative at 0, we get, after some derivation,

$$0 = \iiint \mathbf{x}_1 b(\mathbf{x}_1, \mathbf{x}_2, y_1, y_1) f^2(y_1)(\mathbf{x}_2 - \mathbf{x}_1)' \, dy_1 dM(\mathbf{x}_1) dM(\mathbf{x}_2) \dot{\boldsymbol{\beta}}$$

$$+ \iint \mathbf{x}_0 b(\mathbf{x}_0, \mathbf{x}_2, y_0, y_2)[I(y_2 < y_0) - \tfrac{1}{2}] dF(y_2) dM(\mathbf{x}_2)$$

$$+ \iint \mathbf{x}_1 b(\mathbf{x}_1, \mathbf{x}_0, y_1, y_0)[I(y_0 < y_1) - \tfrac{1}{2}] dF(y_1) dM(\mathbf{x}_1).$$

Once again using the symmetry in the **x** arguments and y arguments of the function b, we can simplify this expression to

$$0 = -\left\{\tfrac{1}{2}\int\int\int (\mathbf{x}_2 - \mathbf{x}_1)b(\mathbf{x}_1, \mathbf{x}_2, y_1, y_1)(\mathbf{x}_2 - \mathbf{x}_1)'f^2(y_1)\,dy_1 dM(\mathbf{x}_1)dM(\mathbf{x}_2)\right\}\dot{\boldsymbol{\beta}}$$
$$+ \int\int (\mathbf{x}_0 - \mathbf{x}_1)b(\mathbf{x}_1, \mathbf{x}_0, y_1, y_0)[I(y_1 < y_0) - \tfrac{1}{2}]dF(y_1)dM(\mathbf{x}_1).$$

Using the relationship between the indicator function and the sign function and the definition of $\mathbf{C}_H$, (5.8.26), we can rewrite this last expression as

$$0 = -\mathbf{C}_H \dot{\boldsymbol{\beta}} + \tfrac{1}{2}\int\int (\mathbf{x}_0 - \mathbf{x}_1)b(\mathbf{x}_1, \mathbf{x}_0, y_1, y_0)\operatorname{sgn}\{y_0 - y_1\}\,dF(y_1)dM(\mathbf{x}_1).$$

Solving for $\dot{\boldsymbol{\beta}}$ leads to the desired result.

A.6 Asymptotic Theory for Chapter 5

In this section we derive the results that are needed in Section 5.8.2 of Chapter 5. These results were first derived by Chang (1995). Our development is taken from Chang et al. (1997). The main goal is to prove Theorem 5.8.2 which we restate here:

Theorem A.6.1 *Under assumptions (E.1), (3.4.1), and (H.1)–H.4), (5.8.13)–(5.8.16),*

$$\sqrt{n}(\widehat{\boldsymbol{\beta}}_{HBR} - \boldsymbol{\beta}) \xrightarrow{d} N(\mathbf{0},\ (\tfrac{1}{4})\mathbf{C}^{-1}\Sigma_H \mathbf{C}^{-1}).$$

Besides the notation of Chapter 5, we need:

1. $W_{ij}(\Delta) = (\tfrac{1}{2})[\operatorname{sgn}(z_j - z_i) - \operatorname{sgn}(y_j - y_i)]$,
   where $z_j = y_j - \mathbf{x}_j'\Delta/\sqrt{n}$. (A.6.1)

2. $t_{ij}(\Delta) = (\mathbf{x}_j - \mathbf{x}_i)'\Delta/\sqrt{n}$. (A.6.2)

3. $B_{ij}(t) = E[b_{ij}I(0 < y_i - y_j < t)]$. (A.6.3)

4. $\gamma_{ij} = B'_{ij}(0)/E(b_{ij})$. (A.6.4)

5. $\mathbf{C}_n = \sum_{i<j}\gamma_{ij}b_{ij}(\mathbf{x}_j - \mathbf{x}_i)(\mathbf{x}_j - \mathbf{x}_i)'$. (A.6.5)

6. $\mathbf{R}(\Delta) = n^{-3/2}\left[\sum_{i<j} b_{ij}(\mathbf{x}_j - \mathbf{x}_i)W_{ij}(\Delta) + \mathbf{C}_n\Delta/\sqrt{n}\right]$. (A.6.6)

Without loss of generality we will assume that the true $\boldsymbol{\beta}_0$ is **0**. We begin with the following lemma.

Asymptotic Theory for Chapter 5 433

Lemma A.6.2 *Under assumptions (E.1), (3.4.1), and (H.1), (5.8.16),*

$$B'_{ij}(t) = \int_{-\infty}^{\infty} \cdots \int_{-\infty}^{\infty} b(x_i, x_j, y_j + t, y_j, \widehat{\boldsymbol{\beta}}_0) f(y_j + t) f(y_j) \prod_{k \neq i,j} f(y_k) \, dy_1 \cdots dy_n$$

is continuous in t.

Proof. This result follows from (3.4.1), (5.8.16) and an application of Leibniz's rule on differentiation of definite integrals.

Let $\boldsymbol{\Delta}$ be arbitrary but fixed. Denote $W_{ij}(\boldsymbol{\Delta})$ by W_{ij}, suppressing dependence on $\boldsymbol{\Delta}$.

Lemma A.6.3 *Under assumptions (E.1), (3.4.1), and (H.4), (5.8.16), there exist constants $|\xi_{ij}| < |t_{ij}|$ such that $E(b_{ij}W_{ij}) = -t_{ij}\, B'_{ij}(\xi_{ij})$.*

Proof. Since $W_{ij} = 1, -1$, or 0 according as $t_{ij} < y_j - y_i < 0$, $0 < y_j - y_i < t_{ij}$, or otherwise, we have

$$E_{\boldsymbol{\beta}_0}(b_{ij}W_{ij}) = \int_{t_{ij} < y_j - y_i < 0} b_{ij} f_{\mathbf{Y}}(\mathbf{y})\, d\mathbf{y} - \int_{0 < y_j - y_i < t_{ij}} b_{ij} f_{\mathbf{Y}}(\mathbf{y})\, d\mathbf{y}.$$

When $t_{ij} > 0$, $E(b_{ij}W_{ij}) = -B_{ij}(t_{ij}) = B_{ij}(0) - B_{ij}(t_{ij}) = -t_{ij}\, B'_{ij}(\xi_{ij})$ by Lemma A.6.2 and the mean value theorem. The same result holds for $t_{ij} < 0$, which proves the lemma.

Lemma A.6.4 *Under assumptions (H.3), (5.8.15), and (H.4), (5.8.16), we have*

$$b_{ij} = g_{ij}(\widehat{\boldsymbol{\beta}}_0) = g_{ij}(0) + [\nabla g_{ij}(\boldsymbol{\xi})]' \widehat{\boldsymbol{\beta}}_0 = g_{ij}(0) + O_p(1/\sqrt{n}),$$

uniformly over all i and j, where $\|\boldsymbol{\xi}\| \leq \|\widehat{\boldsymbol{\beta}}_0\|$.

Proof. This follows from a multivariate mean value theorem (see Apostol, 1974, p. 355), and by (5.8.15) and (5.8.16).

Lemma A.6.5 *Under assumptions (5.8.13)–(5.8.16), (3.4.1), (3.4.7), and (3.4.8),*

(i) $E(g_{ij}(0)g_{ik}(0)W_{ij}W_{ik}) \longrightarrow 0$, as $n \to \infty$

(ii) $E(g_{ij}(0)W_{ij}) \longrightarrow 0$, as $n \to \infty$

uniformly over i and j.

Proof. Without loss of generality, let $t_{ij} > 0$ and $t_{ik} > 0$, where the indices i, j and k are all different. Then

$$E(g_{ij}(0)g_{ik}(0)W_{ij}W_{ik}) = E[g_{ij}g_{ik}\mathbf{I}(0 < y_j - y_i < t_{ij})\, \mathbf{I}(0 < y_k - y_i < t_{ik})]$$

$$= \left| \int_{-\infty}^{\infty} \int_{y_i}^{y_i + t_{ik}} \int_{y_i}^{y_i + t_{ij}} g_{ij}g_{ik}\, f_i f_j f_k \, dy_j dy_k dy_i \right|.$$

Assumptions (3.4.7) and (3.4.8) imply $(1/n)\max_i(x_{ik} - \bar{x}_k)^2 \to 0$ for all k, or equivalently $(1/\sqrt{n})\max_i|x_{ik} - \bar{x}_k| \to 0$ for all k, which implies that $t_{ij} \to 0$. Since the integrand is bounded, this proves (i).

Similarly, $E(g_{ij}(0)W_{ij}) = \int_{-\infty}^{\infty} \int_{y_i}^{y_i+t_{ij}} g_{ij}f_if_j dy_j dy_i \to 0$, which proves (ii).

Lemma A.6.6 *Under assumptions (5.8.13)–(5.8.16), (3.4.1), (3.4.7), and (3.4.8),*

(i) $\text{Cov}(b_{12}W_{12}, b_{34}W_{34}) = o(n^{-1})$,

(ii) $\text{Cov}(b_{12}W_{12}, b_{34}) = o(n^{-1})$,

(iii) $\text{Cov}(b_{12}W_{12}, b_{13}W_{13}) = o(1)$,

(iv) $\text{Cov}(b_{12}W_{12}, b_{13}) = o(1)$.

Proof. To prove (i), recall that $b_{12} = g_{12}(0) + [\nabla g_{12}(\xi)]' \cdot \hat{\boldsymbol{\beta}}_0$. Thus

$$\text{Cov}(b_{12}W_{12}, b_{34}W_{34}) = \text{Cov}(g_{12}(0)W_{12}, g_{34}(0)W_{34})$$
$$+ 2\,\text{Cov}([\nabla g_{12}(\xi)]' \cdot \hat{\boldsymbol{\beta}}_0\, W_{12}, g_{34}(0)W_{34})$$
$$+ \text{Cov}([\nabla g_{12}(\xi)]' \cdot \hat{\boldsymbol{\beta}}_0\, W_{12}, [\nabla g_{34}(\xi)]' \cdot \hat{\boldsymbol{\beta}}_0\, W_{34}).$$

Let I_1, I_2 and I_3 denote the three terms on the right-hand side. I_1 is 0, by independence. Now,

$$I_2 = 2E\{[\nabla g_{12}(\xi)]' \cdot \hat{\boldsymbol{\beta}}_0\, W_{12}\, g_{34}(0)W_{34}\} - 2E\{[\nabla g_{12}(\xi)]' \cdot \hat{\boldsymbol{\beta}}_0\, W_{12}\} E\{g_{34}(0)W_{34}\}$$
$$= I_{21} - I_{22}.$$

Write the first term above as

$$I_{21} = 2(1/n)E\{[\nabla g_{12}(\xi)]' \cdot \hat{\boldsymbol{\beta}}_0\, g_{34}(0)(\sqrt{n}W_{12})(\sqrt{n}W_{34})\}.$$

The term $[\nabla g_{12}(\xi)]' \cdot \hat{\boldsymbol{\beta}}_0 = b_{12} - g_{12}(0)$ is bounded and of magnitude $o_p(1)$. If we can show that $\sqrt{n}W_{12}$ is integrable, then it follows using standard arguments that $I_{21} = o(1/n)$. Let F^* denote the cdf of $y_2 - y_1$ and f^* denote its pdf. Using the mean value theorem,

$$E[\sqrt{n}W_{12}(\Delta)] = \sqrt{n}(1/2)E[\text{sgn}(y_2 - y_1 - (\mathbf{x}_2 - \mathbf{x}_1)'\Delta/\sqrt{n}) - \text{sgn}(y_2 - y_1)]$$
$$= \sqrt{n}(1/2)[2F^*(-(\mathbf{x}_2 - \mathbf{x}_1)'\Delta/\sqrt{n}) - 2F^*(0)]$$
$$= -\sqrt{n}f^*(\xi^*)(\mathbf{x}_2 - \mathbf{x}_1)'\Delta/\sqrt{n} \leqslant f^*(\xi^*)|(\mathbf{x}_2 - \mathbf{x}_1)'\Delta|,$$

for $|\xi*| < |(\mathbf{x}_2 - \mathbf{x}_1)'\Delta/\sqrt{n}|$. The right-hand side of the inequality in expression (A.6.7) is bounded. This proves that $I_{21} = o(1/n)$. Similarly,

$$I_{22} = 2(1/n)E\{[\nabla g_{12}(\xi)]' \cdot \hat{\boldsymbol{\beta}}_0\,(\sqrt{n}W_{12})\} E\{g_{34}(0)(\sqrt{n}W_{34})\} = o(1/n),$$

which proves $I_2 = 0$.

The term I_3 can be shown to be $o(n^{-1})$ similarly, which proves (i). The proof of (ii) is analogous to (i). To prove (iii), note that

$$\text{Cov}(b_{12}W_{12}, b_{13}W_{13}) = \text{Cov}(g_{12}(0)W_{12}, g_{13}(0)W_{13})$$
$$+ 2\,\text{Cov}([\nabla g_{12}(\xi)]' \cdot \hat{\boldsymbol{\beta}}_0\, W_{12}, g_{13}(0)W_{13})$$
$$+ \text{Cov}([\nabla g_{12}(\xi)]' \cdot \hat{\boldsymbol{\beta}}_0\, W_{12}, [\nabla g_{13}(\xi)]' \cdot \hat{\boldsymbol{\beta}}_0\, W_{13}).$$

The first term is $o(1)$ by Lemma A.6.5. The second and third terms are clearly $o(1)$. This proves (iii). Result (iv) is analogously proved.

We are now ready to state and prove asymptotic linearity. Consider the negative gradient function

$$S(\boldsymbol{\beta}) = -\nabla D(\boldsymbol{\beta}) = \sum\sum_{i<j} b_{ij}\text{sgn}(z_j - z_i)(\mathbf{x}_j - \mathbf{x}_i). \tag{A.6.7}$$

Theorem A.6.7 *Under assumptions (5.8.13)–(5.8.16), (3.4.1), (3.4.7), and (3.4.8),*

$$\sup_{\|\sqrt{n}\boldsymbol{\beta}\|\leqslant C} n^{-3/2}[S(\boldsymbol{\beta}) - S(\mathbf{0}) + 2\,\mathbf{C}_n\boldsymbol{\beta}] \xrightarrow{P} 0.$$

Proof. Write $\mathbf{R}(\boldsymbol{\Delta}) = [S(n^{-1/2}\boldsymbol{\Delta}) - S(\mathbf{0}) + 2n^{-1/2}\,\mathbf{C}_n\boldsymbol{\Delta}]$. We will show that

$$\sup_{\|\boldsymbol{\Delta}\|\leqslant C} \mathbf{R}(\boldsymbol{\Delta}) = 2 \sup_{\|\boldsymbol{\Delta}\|\leqslant C}\left\{n^{-3/2}\sum\sum_{i<j} b_{ij}(\mathbf{x}_j - \mathbf{x}_i)\,W_{ij}(\boldsymbol{\Delta}) + n^{-1/2}\mathbf{C}_n\boldsymbol{\Delta}\right\} \xrightarrow{P} 0.$$

It will suffice to show that each component converges to 0. Consider the kth component

$$R_k(\boldsymbol{\Delta}) = 2\left[n^{-3/2}\sum\sum_{i<j} b_{ij}(x_{jk} - x_{ik})\,W_{ij}(\boldsymbol{\Delta}) + \sum\sum_{i<j} \gamma_{ij}b_{ij}(x_{jk} - x_{ik})\,t_{ij}\right]$$

$$= 2\,n^{-3/2}\sum\sum_{i<j}(x_{jk} - x_{ik})(b_{ij}W_{ij} + \gamma_{ij}t_{ij}b_{ij}).$$

We will show that $E(R_k(\boldsymbol{\Delta})) \to 0$ and $\text{Var}(R_k(\boldsymbol{\Delta})) \to 0$. By Lemma A.6.3 and the definition of γ_{ij},

$$E(R_k) = 2n^{-3/2}\sum\sum_{i<j}(x_{jk} - x_{ik})[E(b_{ij}W_{ij}) + \gamma_{ij}t_{ij}E(b_{ij})]$$

$$= 2n^{-3/2}\sum\sum_{i<j}(x_{jk} - x_{ik})t_{ij}[B'_{ij}(0) - B'_{ij}(\xi_{ij})]$$

$$\leqslant 2n^{-3/2}\left[\sum\sum_{i<j}(x_{jk} - x_{ik})^2\right]^{1/2}\left[\sum\sum_{i<j}t_{ij}^2\right]^{1/2}\sup_{i,j}|B'_{ij}(0) - B'_{ij}(\xi_{ij})|$$

$$= 2\left[(1/n^2)\sum\sum_{i<j}(x_{jk} - x_{ik})^2\right]^{1/2}\left[(1/n)\sum\sum_{i<j}t_{ij}^2\right]^{1/2}\sup_{i,j}|B'_{ij}(0) - B'_{ij}(\xi_{ij})| \to 0,$$

since $(1/n)\sum\sum_{i<j} t_{ij}^2 = (1/n)\boldsymbol{\Delta}'\mathbf{X}'\mathbf{X}\boldsymbol{\Delta} = O(1)$ and $\sup_{i,j} |B'_{ij}(0) - B'_{ij}(\xi_{ij})| \to 0$ by Lemma A.6.2.

Next, we will show that $\text{Var}(R_k) \to 0$.

$$\text{Var}(R_k) = \text{Var}\left[2n^{-3/2} \sum\sum_{i<j}(x_{jk} - x_{ik})(b_{ij}W_{ij} + \gamma_{ij}t_{ij}b_{ij})\right]$$

$$= \text{Var}\left[2n^{-3/2} \sum_{i=1}^{n}\sum_{j=1}^{n}(x_{jk} - \bar{x}_k)(b_{ij}W_{ij} + \gamma_{ij}t_{ij}b_{ij})\right]$$

$$= 4n^{-3} \sum_{i=1}^{n}\sum_{j=1}^{n}(x_{jk} - \bar{x}_k)^2 \text{Var}(b_{ij}W_{ij} + \gamma_{ij}t_{ij}b_{ij})$$

$$+ 4n^{-3} \sum\sum\sum \sum_{(i,j) \neq (l,m)} (x_{jk} - \bar{x}_k)(x_{mk} - \bar{x}_k)$$

$$\times \text{Cov}(b_{ij}W_{ij} + \gamma_{ij}t_{ij}b_{ij}, b_{lm}W_{lm} + \gamma_{lm}t_{lm}b_{lm}).$$

The double-sum term above goes to 0, since there there n^2 bounded terms in the double sum, multiplied by n^{-3}. There are two types of covariance term in the quadruple sum, covariance terms with all four indices different, e.g. $((i,j), (l,m)) = ((1,2), (3,4))$, and covariance terms with one index of the first pair equal to one index of the second pair, e.g. $((i,j), (l,m)) = ((1,2), (1,3))$. Since there can be n^4 terms with all four indices different, it will suffice to show that each covariance term is $o(n^{-1})$. This immediately follows from Lemma A.6.6. Finally, since there can be n^3 covariance terms with one shared index, it suffices to show that each term is $o(1)$. Again, this immediately follows from Lemma A.6.6. Hence, we have established the desired result.

Next define the approximating quadratic process,

$$Q(\boldsymbol{\beta}) = D(0) - \sum\sum_{i<j} b_{ij}\text{sgn}(y_j - y_i)(\mathbf{x}_j - \mathbf{x}_i)'\boldsymbol{\beta} + \boldsymbol{\beta}'\mathbf{C}_n\boldsymbol{\beta}. \tag{A.6.8}$$

Let

$$D^*(\boldsymbol{\Delta}) = n^{-1}D(n^{-1/2}\boldsymbol{\Delta}) \tag{A.6.9}$$

and

$$Q^*(\boldsymbol{\Delta}) = n^{-1}Q(n^{-1/2}\boldsymbol{\Delta}). \tag{A.6.10}$$

Note that minimizing $D^*(\boldsymbol{\Delta})$ and $Q^*(\boldsymbol{\Delta})$ is equivalent to minimizing $D(n^{-1/2}\boldsymbol{\Delta})$ and $Q(n^{-1/2}\boldsymbol{\Delta})$, respectively.

The next result is asymptotic quadraticity.

Theorem A.6.8 *Under assumptions (5.8.13)–(5.8.16), (3.4.1), (3.4.7), and (3.4.8), for a fixed constant C and for any $\varepsilon > 0$,*

$$P\left(\sup_{\|\boldsymbol{\Delta}\|<C} |Q^*(\boldsymbol{\Delta}) - D^*(\boldsymbol{\Delta})| \geq \varepsilon\right) \to 0. \tag{A.6.11}$$

Proof. Since

$$(\partial Q^*/\partial \boldsymbol{\Delta}) - (\partial D^*/\partial \boldsymbol{\Delta}) = 2n^{-3/2}\left[\sum\sum_{i<j} b_{ij}(\mathbf{x}_j - \mathbf{x}_i)W_{ij} + \mathbf{C}(n^{-1/2}\boldsymbol{\Delta})\right] = R(\boldsymbol{\Delta}),$$

it follows from Theorem A.6.7 that for $\varepsilon > 0$ and $C > 0$,

$$P\left(\sup_{\|\boldsymbol{\Delta}\|<C} \left\|\frac{\partial Q^*}{\partial \boldsymbol{\Delta}} - \frac{\partial D^*}{\partial \boldsymbol{\Delta}}\right\| \geq \varepsilon/C\right) \to 0.$$

For $0 \leq t \leq 1$, let $\boldsymbol{\Delta}_t = t\boldsymbol{\Delta}$. Then

$$\left|\frac{d}{dt}[Q^*(\boldsymbol{\Delta}_t) - D^*(\boldsymbol{\Delta}_t)]\right| = \left|\sum_{k=1}^{p}\Delta_k\left(\frac{\partial Q^*}{\partial \Delta_{tk}} - \frac{\partial D^*}{\partial \Delta_{tk}}\right)\right|$$

$$\leq \|\boldsymbol{\Delta}\| \sup_{\|\boldsymbol{\Delta}\|<C}\left\|\frac{\partial Q^*}{\partial \boldsymbol{\Delta}} - \frac{\partial D^*}{\partial \boldsymbol{\Delta}}\right\| < \|\boldsymbol{\Delta}\|(\varepsilon/C) < \varepsilon$$

with probability approaching 1. Now, let $h(t) = Q^*(\boldsymbol{\Delta}_t) - D^*(\boldsymbol{\Delta}_t)$. By the previous result, we have $|h'(t)| < \varepsilon$ with high probability. Thus

$$|h(1)| = |h(1) - h(0)| = \left|\int_0^1 h'(t)\,dt\right| \leq \int_0^1 |h'(t)|\,dt < \varepsilon,$$

with probability approaching one. This proves the theorem.

The next theorem states asymptotic normality of $\mathbf{S}(\mathbf{0})$.

Theorem A.6.9 *Under assumptions (5.8.13)–(5.8.16), (3.4.1), (3.4.7), and (3.4.8),*

$$n^{-3/2}\mathbf{S}(\mathbf{0}) \xrightarrow{D} N(\mathbf{0},\, \Sigma_H). \tag{A.6.12}$$

Proof. Let $\mathbf{S}_P$ denote the projection of $\mathbf{S}^*(\mathbf{0}) = n^{-3/2}\mathbf{S}(\mathbf{0})$ onto the space of linear combinations of independent random variables. Then

$$\mathbf{S}_P = \sum_{k=1}^{n} E[\mathbf{S}^*(\mathbf{0})|y_k] = \sum_{k=1}^{n} E\left[n^{-3/2}\sum\sum_{i<j}(\mathbf{x}_j - \mathbf{x}_i)b_{ij}\mathrm{sgn}(y_j - y_i)|y_k\right]$$

$$= \sum_{k=1}^{n} n^{-3/2}\left[\sum_{i=1}^{k-1}(\mathbf{x}_k - \mathbf{x}_i)E[b_{ik}\mathrm{sgn}(y_k - y_i)|y_k] + \sum_{j=k+1}^{n}(\mathbf{x}_j - \mathbf{x}_k)E[b_{kj}\mathrm{sgn}(y_j - y_k)|y_k]\right]$$

$$= n^{-3/2}\sum_{k=1}^{n}\sum_{j=1}^{n}(\mathbf{x}_j - \mathbf{x}_k)E[b_{kj}\mathrm{sgn}(y_j - y_k)|y_k]$$

$$= (1/\sqrt{n})\sum_{k=1}^{n}\mathbf{U}_k,$$

438 *Asymptotic Results*

where $\mathbf{U}_k$ is defined in expression (5.8.12) of Chapter 5. By assumption (D.3), (3.4.8), and a multivariate extension of the Lindeberg–Feller theorem (Rao, 1973), it follows that $\mathbf{S}_P \sim AN(\mathbf{0}, \Sigma_\mathbf{H})$. If we show that $E \parallel \mathbf{S}_P - \mathbf{S}^*(\mathbf{0}) \parallel^2 \to 0$, then it follows from the projection theorem (Theorem 2.4.6) that $\mathbf{S}^*(\mathbf{0})$ has the same asymptotic distribution as $\mathbf{S}_P$, and the proof will be done. Equivalently, we may show that $E(S_{P,r} - S_r^*(0))^2 \to 0$ for each component $r = 1, \ldots, p$. Since for each r we have $E(S_{P,r} - S_r^*(0)) = 0$, then

$$E(S_{P,r} - S_r^*(0))^2 = \text{Var}(S_{P,r} - S_r^*(0))$$

$$= \text{Var}\left[n^{-3/2} \sum_{k=1}^{n} \sum_{j=1}^{n} (x_{jr} - x_{kr}) \right.$$

$$\left. \times \{ E[b_{kj}\text{sgn}(y_j - y_k)|y_k] - b_{kj}\text{sgn}(y_j - y_k) \} \right]$$

$$\equiv \text{Var}\left[n^{-3/2} \sum_{k=1}^{n} \sum_{j=1}^{n} T(y_j, y_k) \right]$$

$$= n^{-3} \sum_{k=1}^{n} \sum_{j=1}^{n} \text{Var}(T(y_j, y_k))$$

$$+ n^{-3} \sum_{k} \sum_{j} \sum_{l} \sum_{m} \text{Cov}[T(y_j, y_k), T(y_l, y_m)],$$

where the quadruple sum is taken over $(j, k) \neq (l, m)$. The double-sum term goes to 0 since there are n^2 bounded terms divided by n^3. There are two types of covariance term in the quadruple sum: terms with four different indices, and terms with three different indices (i.e. one shared index). Covariance terms with four different indices are zero (this can be shown by writing out the covariance in terms of expectations, and using symmetry to show that each covariance term is zero). Thus we only need to consider covariance terms with three different indices and show that the sum goes to 0. Letting k be the shared index (without loss of generality), and noting that $ET(y_j, y_k) = 0$ for all j, k, we have

$$n^{-3} \sum_{k} \sum_{j \neq k} \sum_{l \neq k, j} \text{Cov}[T(y_j, y_k), T(y_l, y_k)]$$

$$= n^{-3} \sum_{k} \sum_{j \neq k} \sum_{l \neq k, j} E\{T(y_j, y_k) \cdot T(y_l, y_k)\}$$

$$= n^{-3} \sum_{k} \sum_{j \neq k} \sum_{l \neq k, j} E\{[E(b_{kj}\text{sgn}(y_j - y_k)|y_k) - b_{kj}\text{sgn}(y_j - y_k)]$$

$$\cdot [E(b_{kl}\text{sgn}(y_l - y_k)|y_k) - b_{kl}\text{sgn}(y_l - y_k)]\}$$

$$= n^{-3} \sum_{k} \sum_{j \neq k} \sum_{l \neq k, j} E\{[E(g_{kj}(0)\,\text{sgn}(y_j - y_k)|y_k) - g_{kj}(0)\,\text{sgn}(y_j - y_k)]$$

$$\cdot [E(g_{kl}(0)\,\text{sgn}(y_l - y_k)|y_k) - g_{kl}(0)\,\text{sgn}(y_l - y_k)]\} + o_p(1),$$

where the last equality follows from the relation $b_{kj} = g_{kj}(0) + 0_p(1/\sqrt{n})$. Expanding the product, each term in the triple sum may be written as

$$E\{[E(g_{kj}(0)\,\text{sgn}(y_j - y_k)|y_k)]^2\} + E\{g_{kj}(0)\,\text{sgn}(y_j - y_k)g_{kl}(0)\,\text{sgn}(y_l - y_k)\}$$

$$- 2\,E\{g_{kj}(0)\,\text{sgn}(y_j - y_k)[E(g_{kl}(0)\,\text{sgn}(y_l - y_k)|y_k)]\}$$

$$= (1 + 1 - 2)\{[E(g_{kj}(0)\,\text{sgn}(y_j - y_k)|y_k)]^2\} = 0,$$

where the first equality follows by taking conditional expectations with respect to k inside appropriate terms.

A similar method applies to terms where k is not the shared index. The theorem is proved.

Proof of Theorem A.6.1. Let $\tilde{\beta}$ denote the value which minimizes $Q(\beta)$. Then $\tilde{\beta}$ is the solution to

$$0 = S(0) - 2C_n\beta,$$

so that $\sqrt{n}\tilde{\beta} = (\frac{1}{2})n^2 C_n^{-1}[n^{-3/2}S(0)] \sim AN(0, (\frac{1}{4})\,C^{-1}\Sigma C^{-1})$, by Theorem A.6.9 and assumption (D.2), (3.4.7). It remains to show that $\sqrt{n}(\tilde{\beta} - \hat{\beta}) = o_p(1)$. This follows from Theorem A.6.8 and convexity of $D(\beta)$, using standard arguments as in Jaeckel (1972).

Bibliography

Adichi, J. N. (1978) Rank tests of sub-hypotheses in the general regression model, *Annals of Statistics*, 6, 1012–1026.

Afifi, A. A. and Azen, S. P. (1972) *Statistical Analysis: A Computer Oriented Approach*, New York: Academic Press.

Akritas, M. G. (1990) The rank transform method in some two-factor designs, *Journal of the American Statistical Association*, 85, 73–78.

Akritas, M. G. (1991) Limitations of the rank transform procedure: A study of repeated measures designs. Part I, *Journal of the American Statistical Association*, 86, 457–460.

Akritas, M. G. (1993) Limitations of the rank transform procedure: A study of repeated measures designs. Part II, *Statistics and Probability Letters*, 17, 149–156.

Akritas, M. G. and Arnold, S. F. (1994) Fully nonparametric hypotheses for factorial designs I: Multivariate repeated measures designs, *Journal of the American Statistical Association*, 89, 336–343.

Akritas, M. G., Arnold, S. F., and Brunner, E. (1997) Nonparametric hypotheses and rank statistics for unbalanced factorial designs, *Journal of the American Statistical Association*, 92, 258–265.

Ammann, L. P. (1993) Robust singular value decompositions: A new approach to projection pursuit, *Journal of the American Statistical Association*, 88, 505–514.

Ansari, A. R. and Bradley, R. A. (1960) Rank-sum tests for dispersion, *Annals of Mathematical Statistics*, 31, 1174–1189.

Apostol, T. M. (1974) *Mathematical Analysis*, 2nd Edition, Reading, MA: Addison-Wesley.

Arnold, S. F. (1980) Asymptotic validity of F-tests for the ordinary linear model and the multiple correlation model, *Journal of the American Statistical Association*, 75, 890–894.

Arnold, S. F. (1981) *The Theory of Linear Models and Multivariate Analysis*, New York: John Wiley and Sons.

Aubuchon, J. C. and Hettmansperger, T. P. (1984) A note on the estimation of the integral of $f^2(x)$, *Journal of Statistical Inference and Planning*, 9, 321–331.

Aubuchon, J. C. and Hettmansperger, T. P. (1989) Rank-based inference for linear models: Asymmetric errors, *Statistics and Probability Letters*, 8, 97–107.

Babu, G. J. and Koti, K. M. (1996) Sign test for ranked-set sampling, *Communications in Statistics. Theory and Methods*, 25(7), 1617–1630.

Bahadur, R. R. (1967) Rates of convergence of estimates and test statistics, *Annals of Mathematical Statistics*, 31, 276–295.

Bai, Z. D., Chen, X. R., Miao, B. Q., and Rao, C. R. (1990) Asymptotic theory of least distance estimate in multivariate linear models, *Statistics*, 21, 503–519.

Bassett, G. and Koenker, R. (1978) Asymptotic theory of least absolute error regression, *Journal of the American Statistical Association*, 73, 618–622.

Bedall, F. K. and Zimmerman, H. (1979) Algorithm AS143, the median-center, *Applied Statistics*, 28, 325–328.

Belsley, D. A., Kuh, K., and Welsch, R. E. (1980) *Regression Diagnostics*, New York: John Wiley and Sons.

Bickel, P. J. (1964) On some alternative estimates for shift in the p-variate one sample problem, *Annals of Mathematical Statistics*, 35, 1079–1090.

Bickel, P. J. (1965) On some asymptotically nonparametric competitors of Hotelling's T^2, *Annals of Mathematical Statistics*, 36, 160–173.

Bickel, P. J. (1974) Edgeworth expansions in nonparametric statistics, *Annals of Statistics*, 2, 1–20.

Bickel, P. J. (1976) Another look at robustness: A review of reviews and some new developments (reply to discussant), *Scandinavian Journal of Statistics*, 3, 167.

Bickel, P. J. and Lehmann, E. L. (1975) Descriptive statistics for nonparametric model, II. Location, *Annals of Statistics*, 3, 1045–1069.

Blair, R. C., Sawilowsky, S. S., and Higgins, J. J. (1987) Limitations of the rank transform statistic in tests for interaction, *Communications in Statistics. Simulation and Computation*, 16, 1133–1145.

Bloomfield, B. and Steiger, W. L. (1983) *Least Absolute Deviations*, Boston: Birkhäuser.

Blumen, I. (1958) A new bivariate sign test, *Journal of the American Statistical Association*, 53, 448–456.

Bohn, L. L. and Wolfe, D. A. (1992) Nonparametric two-sample procedures for ranked-set samples data, *Journal of the American Statistical Association*, 87, 552–561.

Boos, D. D. (1982) A test for asymmetry associated with the Hodges–Lehmann estimator, *Journal of the American Statistical Association*, 77, 647–651.

Bose, A. and Chaudhuri, P. (1993) On the dispersion of multivariate median, *Annals of the Institute of Statistical Mathematics*, 45, 541–550.

Box, G. E. P. and Cox, D. R. (1964) An analysis of transformations, *Journal of the Royal Statistical Society, Series B*, 26, 211–252.

Brown, B. M. (1983) Statistical uses of the spatial median, *Journal of the Royal Statistical Society, Series B*, 45, 25–30.

Brown, B. M. (1985) Multiparameter linearization theorems, *Journal of the Royal Statistical Society, Series B*, 47, 323–331.

Brown, B. M. and Hettmansperger, T. P. (1987a) Affine invariant rank methods in the bivariate location model, *Journal of the Royal Statistical Society, Series B*, 49, 301–310.

Brown, B. M. and Hettmansperger, T. P. (1987b) Invariant tests in bivariate models and the L_1 criterion function, in: Y. Dodge, ed., *Statistical Data Analysis Based on the L_1 Norm and Related Methods*, Amsterdam: North Holland, 333–344.

Brown, B. M. and Hettmansperger, T. P. (1989) An affine invariant version of the sign test, *Journal of the Royal Statistical Society, Series B*, 51, 117–125.

Brown, B. M. and Hettmansperger, T. P. (1994) Regular redescending rank estimates, *Journal of the American Statistical Association*, 89, 538–542.

Brown, B. M., Hettmansperger, T. P., Nyblom, J., and Oja, H., (1992) On certain bivariate sign tests and medians, *Journal of the American Statistical Association*, 87, 127–135.

Brunner, E. and Neumann, N. (1986) Rank tests in 2×2 designs, *Statistica Neerlandica*, 40, 251–272.

Brunner, E. and Puri, M. L. (1996) Nonparametric methods in design and analysis of experiments, in: S. Ghosh and C. R. Rao, eds, *Handbook of Statistics*, Vol. 13, Amsterdam: Elsevier Science, 631–703.

Carmer, S. G. and Swanson, M. R. (1973) An evaluation of ten pairwise multiple comparison procedures by Monte Carlo methods, *Journal of the American Statistical Association*, 68, 66–74.

Chang, W. H. (1995) High break-down rank-based estimates for linear models, Unpublished Ph.D. thesis, Western Michigan University, Kalamazoo.

Chang, W. H., McKean, J. W., Naranjo, J. D. and Sheather, S. J. (1997) High breakdown rank regression. Submitted.

Chaudhuri, P. (1992) Multivariate location estimation using extension of R-estimates through U-statistics type approach, *Annals of Statistics*, 20, 897–916.

Chaudhuri, P. and Sengupta, D. (1993) Sign tests in multidimensional inference based on the geometry of the data cloud, *Journal of the American Statistical Association*, 88, 1363–1370.

Chernoff, H. and Savage, I. R. (1958) Asymptotic normality and efficiency of certain nonparametric test statistics, *Annals of Mathematical Statistics*, 39, 972–994.

Chiang, C.-Y. and Puri, M. L. (1984) Rank procedures for testing subhypotheses in linear regression, *Annals of the Institute of Statistical Mathematics*, 36, 35–50.

Chinchilli, V. M. and Sen, P. K. (1982) Multivariate linear rank statistics for profile analysis, *Journal of Multivariate Analysis*, 12, 219–229.

Choi, K. and Marden, J. (1997) An approach to multivariate rank tests in multivariate analysis of variance, *Journal of the American Statistical Association*. To appear.

Coakley, C. W. and Hettmansperger, T. P. (1992) Breakdown bounds and expected test resistance, *Journal of Nonparametric Statistics*, 1, 267–276.

Conover, W. J. and Iman, R. L. (1981) Rank transform as a bridge between parametric and nonparametric statistics, *American Statistician*, 35, 124–133.

Conover, W. J., Johnson, M. E., and Johnson, M. M. (1981) A comparative study of tests for homogeneity of variances, with applications to the outer continental shelf bidding data, *Technometrics*, 23, 351–361.

Cook, R. D., Hawkins, D. M. and Weisberg, S. (1992) Comparison of model misspecification diagnostics using residuals from least mean of squares and least median of squares fits, *Journal of the American Statistical Association*, 87, 419–424.

Cook, R. D. and Weisberg, S. (1982) *Residuals and Influence in Regression*, New York: Chapman & Hall.

Cook, R. D. and Weisberg, S. (1994) *An Introduction to Regression Graphics*, New York: John Wiley and Sons.

Croux, C., Rousseeuw, P. J., and Hössjer, O. (1994) Generalized S-estimators, *Journal of the American Statistical Association*, 89, 1271–1281.

Cushney, A. R. and Peebles, A. R. (1905) The action of optical isomers, II. Hyoscines, *Journal of Physiology*, 32, 501–510.

Davis, J. B. and McKean, J. W. (1993) Rank based methods for multivariate linear models, *Journal of the American Statistical Association*, 88, 245–251.

Dietz, E. J. (1982) Bivariate nonparametric tests for the one-sample location problem, *Journal of the American Statistical Association*, 77, 163–169.

Dixon, S. L. and McKean, J. W. (1996) Rank-based analysis of the heteroscedastic linear model, *Journal of the American Statistical Association*, 91, 699–712.

Dongarra, J. J., Bunch, J. R., Moler, C. B., and Stewart, G. W. (1979) *Linpack Users' Guide*, Philadelphia: SIAM.

Donoho, D. L. and Huber, P. J. (1983) The notion of breakdown point, in: P. J. Bickel, K. A. Doksum, J. L. Hodges Jr., eds, *A Festschrift for Erich L. Lehmann*, Belmont, CA: Wadsworth, 157–184.

Dowell, M. and and Jarratt, P. (1971) A modified regula falsi method for computing the root of an equation, *BIT*, 11, 168–171.

DuBois, C., ed. (1960) *Lowie's Selected Papers in Anthropology*, Berkeley: University of California Press.

Draper, D. (1988) Rank-based robust analysis of linear models. I. Exposition and review, *Statistical Science*, 3, 239–257.

Draper, N. R. and Smith, H. (1966) *Applied Regression Analysis*, New York: John Wiley and Sons.

Ducharme, G. R. and Milasevic, P. (1987) Spatial median and directional data, *Biometrika*, 74, 212–215.

Dwass, M. (1960) Some k-sample rank order tests, in: I. Olkin, *et al.*, eds, *Contributions to Probability and Statistics in Honor of Harold Hotelling*, Stanford, CA: Stanford University Press.

Efron, B. (1979) Bootstrap methods: another look at the jackknife, *Annals of Statistics*, 7, 1–26.

Efron B. and Tibshirani, R. J. (1993) *An Introduction to the Bootstrap*, New York: Chapman & Hall.

Eubank, R. L., LaRiccia, V. N. and Rosenstein, R. B. (1992) Testing symmetry about an unknown median, via linear rank procedures, *Journal of Nonparametric Statistics*, 1, 301–311.

Fernholz, L. T. (1983) *Von Mises Calculus for Statistical Functionals*, Lecture Notes in Statistics 19, New York: Springer-Verlag.

Fisher, N. I. (1987) *Statistical Analysis for Spherical Data*, Cambridge: Cambridge University Press.

Fisher, N. I. (1993) *Statistical Analysis for Circular Data*, Cambridge: Cambridge University Press.

Fix, E. and Hodges, J. L., Jr. (1955) Significance probabilities of the Wilcoxon test, *Annals of Mathematical Statistics*, 26, 301–312.

Fligner, M. A. (1981) Comment, *American Statistician*, 35, 131–132.

Fligner, M. A. and Hettmansperger, T. P. (1979) On the use of conditional asymptotic normality, *Journal of the Royal Statistical Society, Series B*, 41, 178–183.

Fligner, M. A. and Killeen, T. J. (1976) Distribution-free two-sample test for scale, *Journal of the American Statistical Association*, 71, 210–213.

Fligner, M. A. and Policello, G. E. (1981) Robust rank procedures for the Behrens-Fisher problem, *Journal of the American Statistical Association*, 76, 162–168.

Fligner, M. A. and Rust, S. W. (1982) A modification of Mood's median test for the generalized Behrens–Fisher problem, *Biometrika*, 69, 221–226.

Fraser, D. A. S. (1957) *Nonparametric Methods in Statistics*, New York: John Wiley and Sons.

Gastwirth, J. L. (1968) The first median test: A two-sided version of the control median test, *Journal of the American Statistical Association*, 63, 692–706.

Gastwirth, J. L. (1971) On the sign test for symmetry, *Journal of the American Statistical Association*, 66, 821–823.

Gastwirth, J. L. and Wolff, S. S. (1986), An elementary method for obtaining lower bounds on the asymptotic power of rank tests, *Annals of Mathematical Statistics*, 39, 2128–2130.

George, K. J., McKean, J. W., Schucany, W. R., and Sheather, S. J. (1995) A comparison of confidence intervals from R-estimators in regression, *Journal of Statistical Computation and Simulation*, 53, 13–22.

Ghosh, M. and Sen, P. K. (1971) On a class of rank order tests for regression with partially formed stochastic predictors, *Annals of Mathematical Statistics*, 42, 650–661.

Gower, J. C. (1974) The mediancenter, *Applied Statistics*, 32, 466–470.

Graybill, F. A. (1976) *Theory and Application of the Linear Model*, North Scituate, MA: Duxbury.

Graybill, F. A. (1983) *Matrices with Applications in Statistics*, 2nd edn, Belmont, CA: Wadsworth.

Graybill, F. A. and Iyer, H. K. (1994) *Regression Analysis: Concepts and Applications*, Belmont, CA: Duxbury.

Hadi, A. S. and Simonoff, J. S. (1993) Procedures for the identification of multiple outliers in linear models, *Journal of the American Statistical Association*, 88, 1264–1272.

Hájek, J. and Šidák, Z. (1967) *Theory of Rank Tests*, New York: Academic Press.

Hald, A. (1952) *Statistical Theory with Engineering Applications*, New York: John Wiley and Sons.

Hampel, F. R. (1974) The influence curve and its role in robust estimation, *Journal of the American Statistical Association*, 69, 383–393.

Hampel, F. R., Ronchetti, E. M., Rousseeuw, P. J., and Stahel, W. J. (1986) *Robust Statistics, the Approach Based on Influence Functions*, New York: John Wiley and Sons.

Hardy, G. H., Littlewood, J. E., and Pólya, G. (1952) *Inequalities*, 2nd Edition, Cambridge: Cambridge University Press.

Hawkins, D. M., Bradu, D., and Kass, G. V. (1984) Location of several outliers in multiple regression data using elemental sets, *Technometrics*, 26, 197–208

He, X., Simpson, D. G., and Portnoy, S. L. (1990) Breakdown robustness of tests, *Journal of the American Statistical Association*, 85, 446–452.

Heiler, S. and Willers, R. (1988) Asymptotic normality of R-estimation in the linear model, *Statistics*, 19, 173–184.

Hendy, M. F. and Charles, J. A. (1970) The production techniques, silver content and circulation history of the twelfth-century Byzantine, *Archaeometry*, 12, 13–21.

Hettmansperger, T. P. (1984a), *Statistical Inference Based on Ranks*, New York: John Wiley and Sons.

Hettmansperger, T. P. (1984b), Two-sample inference based on one-sample sign statistics, *Applied Statistics*, 33, 45–51.

Hettmansperger, T. P. (1995) The rank-set sample sign test, *Journal of Nonparametric Statistics*, 4, 263–270.

Hettmansperger, T. P. and Malin, J. S. (1975) A modified Mood's test for location with no shape assumptions on the underlying distributions, *Biometrika*, 62, 527–529.

Hettmansperger, T. P. and McKean, J. W. (1978) Statistical inference based on ranks, *Psychometrika*, 43, 69–79.

Hettmansperger, T. P. and McKean, J. W. (1983) A geometric interpretation of inferences based on ranks in the linear model, *Journal of the American Statistical Association*, 78, 885–893.

Hettmansperger, T. P. McKean, J. W., and Sheather, S. J. (1997) Rank-based analyses of linear models, in: S. Ghosh and C. R. Rao, eds, *Handbook of Statistics*, Vol. 15, Amsterdam: Elsevier Science, 145–173.

Hettmansperger, T. P., Möttönen, J., and Oja, H. (1997a), Affine invariant multivariate one-sample signed-rank tests, *Journal of the American Statistical Association*. To appear.

Hettmansperger, T. P., Möttönen, J., and Oja, H. (1997b), Affine invariant multivariate two-sample rank tests, *Statistica Sinica*. To appear.

Hettmansperger, T. P. and Oja, H. (1994) Affine invariant multivariate multi-sample sign tests, *Journal of the Royal Statistical Society, Series B*, 56, 235–249.

Hettmansperger, T. P. Nyblom, J., and Oja, H. (1994) Affine invariant multivariate one-sample sign tests, *Journal of the Royal Statistical Society, Series B*, 56, 221–234.

Hettmansperger, T. P. and Sheather, S. J. (1986) Confidence intervals based on interpolated order statistics, *Statistics and Probability Letters*, 4, 75–79.

Hocking, R. R. (1985) *The Analysis of Linear Models*, Monterey, California: Brooks/Cole.

Hodges, J. L., Jr. (1967) Efficiency in normal samples and tolerance of extreme values for some estimates of location, in: L. LeCam and J. Neyman, eds, *Proceedings of the Fifth Berkeley Symposium on Mathematical Statistics and Probability*, Vol. 1, Berkeley: University of California Press, 163–186.

Hodges, J. L., Jr. and Lehmann, E. L. (1956) The efficiency of some nonparametric competitors of the t-test, *Annals of Mathematical Statistics*, 27, 324–335.

Hodges, J. L., Jr. and Lehmann, E. L. (1961) Comparison of the normal scores and Wilcoxon tests, in: *Proceedings of the Fourth Berkeley Symposium on Mathematical Statistics and Probability*, Vol. 1, Berkeley: University of California Press, 307–317.

Hodges, J. L., Jr. and Lehmann, E. L. (1962) Rank methods for combination of independent experiments in analysis of variance, *Annals of Mathematical Statistics* 33, 482–497.

Hodges, J. L., Jr. and Lehmann, E. L. (1963) Estimates of location based on rank tests, *Annals of Mathematical Statistics*, 34, 598–611.

Hogg, R. V. (1974) Adaptive robust procedures: A partial review and some suggestions for future applications and theory, *Journal of the American Statistical Association*, 69, 909–923.

Hora, S. C. and Conover, W. J. (1984) The F-statistic in the two-way layout with rank-score transformed data, *Journal of the American Statistical Association*, 79, 688–673.

Hössjer, O. (1994) Rank-based estimates in the linear model with high breakdown point, *Journal of the American Statistical Association*, 89, 149–158.

Hössjer, O. and Croux, C. (1995) Generalizing univariate signed rank statistics for testing and estimating a multivariate location parameter, *Journal of Nonparametric Statistics*, 4, 293–308.

Hotelling, H. (1951), A generalized T-test and measure of multivariate dispersion, in: *Proceedings of the Second Berkeley Symposium on Mathematical Statistics and Probability*, Berkeley: University of California Press, 23–41.

Høyland, A. (1965) Robustness of the Hodges–Lehmann estimates for shift, *Annals of Mathematical Statistics*, 36, 174–197.

Hsu, J. C. (1996) *Multiple Comparisons*, London: Chapman & Hall.

Huber, P. J. (1981) *Robust Statistics*, New York: John Wiley and Sons.

Huitema, B. E. (1980) *The Analysis of Covariates and Alternatives*, New York: John Wiley and Sons.

Iman, R. L. (1974) A power study of the rank transform for the two-way classification model when interaction may be present, *Canadian Journal of Statistics*, 2, 227–239.

International Mathematical and Statistical Libraries, Inc. (1987) *User's Manual: Stat/Library*, Houston, TX: Author.

Jaeckel, L. A. (1972) Estimating regression coefficients by minimizing the dispersion of the residuals, *Annals of Mathematical Statistics*, 43, 1449–1458.

Jan, S. L. and Randles, R. H. (1995) A multivariate signed sum test for the one-sample location problem, *Journal of Nonparametric Statistics*, 4, 49–63.

Jan, S. L. and Randles, R. H. (1996) Interdirection tests for simple repeated measures designs, *Journal of the American Statistical Association*, 91, 1611–1618.

Johnson, G. D., Nussbaum, B. D., Patil, G. P., and Ross, N. P. (1996) Designing cost-effective environmental sampling using concomitant information. *Chance*, 9, 4–16.

Jonckheere, A. R. (1954) A distribution-free k-sample test against ordered alternatives, *Biometrika*, 41, 133–145.

Jurečková, J. (1969) Asymptotic linearity of rank statistics in regression parameters, *Annals of Mathematical Statistics*, 40, 1449–1458.

Jurečková, J. (1971) Nonparametric estimate of regression coefficients, *Annals of Mathematical Statistics*, 42, 1328–1338.

Kahaner, D., Moler, C., and Nash, S. (1989) *Numerical Methods and Software*, Englewood Cliffs, NJ: Prentice Hall.

Kalbfleisch, J. D. and Prentice, R. L. (1980) *The Statistical Analysis of Failure Time Data*, New York: John Wiley and Sons.

Kapenga, J. A., McKean, J. W., and Vidmar, T. J. (1988) *RGLM: Users Manual*, Amer. Statist. Assoc. Short Course on Robust Statistical Procedures for the Analysis of Linear and Nonlinear Models, New Orleans.

Kepner, J. C. and Robinson, D. H. (1988) Nonparametric methods for detecting treatment effects in repeated measures designs, *Journal of the American Statistical Association*, 83, 456–461.

Killeen, T. J., Hettmansperger, T. P., and Sievers, G. L. (1972) An elementary theorem on the probability of large deviations, *Annals of Mathematical Statistics*, 43, 181–192.

Klotz, J. (1962) Nonparametric tests for scale, *Annals of Mathematical Statistics*, 33, 498–512.

Koul, H. L. (1992) *Weighted Empiricals and Linear Models*, Hayward, CA: Institute of Mathematical Statistics.

Koul, H. L., Sievers, G. L., and McKean, J. W. (1987) An estimator of the scale parameter for the rank analysis of linear models under general score functions, *Scandinavian Journal of Statistics*, 14, 131–141.

Kramer, C. Y. (1956) Extension of multiple range tests to group means with unequal numbers of replications, *Biometrics*, 12, 307–310.

Kruskal, W. H. and Wallis, W. A. (1952) Use of ranks in one criterion variance analysis, *Journal of the American Statistical Association*, 57, 583–621.

Larsen, R. J. and Stroup, D. F. (1976) *Statistics in the Real World*, New York: Macmillan.

Lawless, J. F. (1982) *Statistical Models and Methods for Lifetime Data*, New York: John Wiley and Sons.

Lawley, D. N. (1938), A generalization of Fisher's z-test, *Biometrika*, 30, 180–187.

Lehmann, E. L. (1975) *Nonparametrics: Statistical Methods Based on Ranks*, San Francisco: Holden-Day.

Li, H. (1991) Rank procedures for the logistic model, Unpublished Ph.D. thesis, Western Michigan University, Kalamazoo.

Liu, R. Y. (1990) On a notion of data depth based on simplices, *Annals of Statistics*, 18, 405–414.

Liu, R. Y. and Singh, K. (1993) A quality index based on data depth and multivariate rank tests, *Journal of the American Statistical Association*, 88, 405–414.

Lopuhaä, H. P. and Rousseeuw, P. J. (1991) Breakdown properties of affine equivariant estimators of multivariate location and covariance matrices, *Annals of Statistics*, 19, 229–248.

Magnus, J. R. and Neudecker, H. (1988) *Matrix Differential Calculus with Applications in Statistics and Econometrics*, New York: John Wiley and Sons.

Mann, H. B. (1945) Nonparametric tests against trend, *Econometrica*, 13, 245–259.

Mann, H. B. and Whitney, D. R. (1947) On a test of whether one of two random variables is stochastically larger than the other, *Annals of Mathematical Statistics*, 18, 50–60.

Mardia, K. V. (1972) *Statistics of Directional Data*, London: Academic Press.

Mardia, K. V., Kent, J. T., and Bibby, J. M. (1979) *Multivariate Analysis*, Orlando, FL: Academic Press.

Maritz, J. S. (1981) *Distribution-Free Statistical Methods*, London: Chapman & Hall.

Maritz, J. S. and Jarrett, R. G. (1978) A note on estimating the variance of the sample median, *Journal of the American Statistical Association*, 73, 194–196.

Maritz, J. S., Wu, M., and Staudte, R. G., Jr. (1977) A location estimator based on a U-statistic, *Annals of Statistics*, 5, 779–786.

Marsaglia, G. and Bray, T. A. (1964) A convenient method for generating normal variables, *SIAM Review*, 6, 260–264.

Mason, R. L., Gunst, R. F., and Hess, J. L. (1989) *Statistical Design and Analysis of Experiments*, New York: John Wiley and Sons.

Mathisen, H. C. (1943) A method of testing the hypothesis that two samples are from the same population, *Annals of Mathematical Statistics*, 14, 188–194.

McIntyre, G. A. (1952) A method of unbiased selective sampling, using ranked sets, *Australian Journal of Agricultural Research*, 3, 385–390.

McKean, J. W. and Hettmansperger, T. P. (1976) Tests of hypotheses of the general linear model based on ranks, *Communications in Statistics. Theory and Methods*, 5, 693–709.

McKean, J. W. and Hettmansperger, T. P. (1978) A robust analysis of the general linear model based on one step R-estimates, *Biometrika*, 65, 571–579.

McKean, J. W., Naranjo, J. D., and Sheather, S. J. (1996a), Diagnostics to detect differences in robust fits of linear models, *Computational Statistics*, 11, 223–243.

McKean, J. W., Naranjo, J. D., and Sheather, S. J. (1996b), An efficient and high breakdown procedure for model criticism, *Communications in Statistics. Theory and Methods*, 25, 2575–2595.

McKean, J. W. and Ryan, T. A., Jr. (1977) An algorithm for obtaining confidence intervals and point estimates based on ranks in the two sample location problem, *Transactions of Mathematical Software*, 3, 183–185.

McKean, J. W. and Schrader, R. (1980) The geometry of robust procedures in linear models, *Journal of the Royal Statistical Society, Series B*, 42, 366–371.

McKean, J. W. and Schrader, R. M. (1984) A comparison of methods for studentizing the sample median, *Communications in Statistics. Simulation and Computation*, 6, 751–773.

McKean, J. W. and Sheather, S. J. (1991) Small sample properties of robust analyses of linear models based on R-estimates: A survey, in: W. Stahel and S. Weisberg, eds., *Directions in Robust Statistics and Diagnostics*, Part II, New York: Springer-Verlag, 1–20.

McKean, J. W., Sheather, S. J., and Hettmansperger, T. P. (1990) Regression diagnostics for rank-based methods, *Journal of the American Statistical Association*, 85, 1018–1028.

McKean, J. W., Sheather, S. J., and Hettmansperger, T. P. (1991) Regression diagnostics for rank-based methods II, in: W. Stahel and S. Weisberg, eds., *Directions in Robust Statistics and Diagnostics*, Part II, Yew York: Springer-Verlag, 21–31.

McKean, J. W., Sheather, S. J., and Hettmansperger, T. P. (1993) The use and interpretation of residuals based on robust estimation, *Journal of the American Statistical Association*, 88, 1254–1263.

McKean, J. W., Sheather, S. J., and Hettmansperger, T. P. (1994) Robust and high breakdown fits of polynomial models, *Technometrics*, 36, 409–415.

McKean, J. W. and Sievers, G. L. (1987) Coefficients of determination for least absolute deviation analysis, *Statistics and Probability Letters*, 5, 49–54.

McKean, J. W. and Sievers, G. L. (1989) Rank scores suitable for the analysis of linear models under asymmetric error distributions, *Technometrics*, 31, 207–218.

McKean, J. W. and Vidmar, T. J. (1992) Using procedures based on ranks: cautions and recommendations, *American Statistical Association 1992 Proceedings of the Biopharmaceutical Section*, 280–289.

McKean, J. W. and Vidmar, T. J. (1994) A comparison of two rank-based methods for the analysis of linear models, *American Statistician*, 48, 220–229.

McKean, J. W., Vidmar, T.J., and Sievers, G. L. (1989) A robust two stage multiple comparison procedure with application to a random drug screen, *Biometrics 45*, 1281–1297.

Merchant, J. A., Halprin, G. M., Hudson, A. R., Kilburn, K. H., McKenzie, W.N., Jr., Hurst, D. J., and Bermazohn, P. (1975) Responses to cotton dust, *Archives of Environmental Health*, 30, 222–229.

Mielke, P. W. (1972) Asymptotic behavior of two-sample tests based on the powers of ranks for detecting scale and location alternatives, *Journal of the American Statistical Association*, 67, 850–854.

Milasevic, P. and Ducharme, G. R. (1987) Uniqueness of the spatial median, *Annals of Statistics*, 15, 1332–1333.

Miller, R. G. (1981) *Simultaneous Statistical Inference*, New York: Springer-Verlag.

Mood, A. M. (1950) *Introduction to the Theory of Statistics*, New York: McGraw-Hill.

Mood, A. M. (1954) On the asymptotic efficiency of certain nonparametric two-sample tests, *Annals of Mathematical Statistics*, 25, 514–533.

Morrison, D. F. (1983) *Applied Linear Statistical Models*, Englewood Cliffs, NJ: Prentice Hall.

Möttönen, J. (1997a), SAS/IML macros for spatial sign and rank tests, Mathematics Department, University of Oulu, Finland.

Möttönen, J. (1997b), SAS/IML macros for affine invariant multivariate sign and rank tests, Mathematics Department, University of Oulu, Finland.

Möttönen, J. and Oja, H. (1995) Multivariate spatial sign and rank methods. *Journal of Nonparametric Statistics*, 5, 201–213.

Möttönen, J., Oja, H., and Tienari, J. (1997) On efficiency of multivariate spatial sign and rank methods, *Annals of Statistics*. In press.

Möttönen, J., Hettmansperger, T. P., Oja, H., and Tienari, J. (1997) On the efficiency of multivariate affine invariant rank methods, *Journal of Multivariate Analysis*. In press.

Naranjo, J. D. and Hettmansperger, T. P. (1994) Bounded-influence rank regression, *Journal of the Royal Statistical Society, Series B*, 56(1), 209–220.

Naranjo, J. D., McKean, J. W., Sheather, S. J., and Hettmansperger, T. P. (1994) The use and interpretation of rank-based residuals, *Journal of Nonparametric Statistics*, 3, 323–341.

Nelson, W. (1982) *Applied Lifetime Data Analysis*, New York: John Wiley and Sons.

Neter, J., Kutner, M. H., Nachtsheim, C. J., and Wasserman, W. (1996) *Applied Linear Statistical Models*, 4th Edition., Chicago: Irwin.

Niinimaa, A. and Oja, H. (1995) On the influence functions of certain bivariate medians, *Journal of the Royal Statistical Society, Series B*, 57, 565–574.

Niinimaa, A., Oja, H., and Nyblom, J. (1992) Algorithm AS277: The Oja bivariate median, *Applied Statistics*, 41, 611–617.

Niinimaa, A., Oja, H., and Tableman, M. (1990) The finite-sample breakdown point of the Oja bivariate median, *Statistics and Probability Letters*, 10, 325–328.

Noether, G. E. (1955) On a theorem of Pitman, *Annals of Mathematical Statistics*, 26, 64–68.

Noether, G. E. (1987) Sample size determination for some common nonparametric tests, *Journal of the American Statistical Association*, 82, 645–647.

Numerical Algorithms Group, Inc. (1983) *Library Manual Mark 15*, Oxford: Numerical Algorithms Group.

Nyblom, J. (1992) Note on interpolated order statistics, *Statistics and Probability Letters*, 14, 129–131.

Oja, H. (1983) Descriptive statistics for multivariate distributions, *Statistics and Probability Letters*, 1, 327–333.

Oja, H. and Nyblom, J. (1989) Bivariate sign tests, *Journal of the American Statistical Association*, 84, 249–259.

Olshen, R. A. (1967) Sign and Wilcoxon test for linearity, *Annals of Mathematical Statistics*, 38, 1759–1769.

Osborne, M. R. (1985) *Finite Algorithms in Optimization and Data Analysis*, Chichester: John Wiley and Sons.

Peters, D. and Randles, R. H. (1990a), Multivariate rank tests in the two-sample location problem, *Communications in Statistics. Theory and Methods*, 15(11), 4225–4238.

Peters, D. and Randles, R. H. (1990b), A multivariate signed-rank test for the one-sample location problem, *Journal of the American Statistical Association*, 85, 552–557.

Pitman, E. J. G. (1948) Notes on nonparametric statistical inference, Unpublished notes.

Policello, G. E., II, and Hettmansperger, T. P. (1976) Adaptive robust procedures for the one-sample location model, *Journal of the American Statistical Association*, 71, 624–633.

Puri, M. L. (1968) Multisample scale problem: Unknown location parameters. *Annals of the Institute of Statistical Mathematics*, 40, 619–632.

Puri, M. L. and Sen, P. K. (1971) *Nonparametric Methods in Multivariate Analysis*, New York: John Wiley and Sons.

Puri, M. L. and Sen, P. K. (1985) *Nonparametric Methods in General Linear Models*, New York: John Wiley and Sons.

Randles, R. H. (1989) A distribution-free multivariate sign test based on interdirections, *Journal of the American Statistical Association*, 84, 1045–1050.

Randles, R. H. and Wolfe, D. A. (1979) *Introduction to the Theory of Nonparametric Statistics*, New York: John Wiley and Sons.

Randles, R. H., Fligner, M. A., Policello, G. E., and Wolfe, D. A. (1980) An asymptotically distribution-free test for symmetry versus asymmetry, *Journal of the American Statistical Association*, 75, 168–172.

Rao, C. R. (1948) Tests of significance in multivariate analysis, *Biometrika*, 35, 58–79.

Rao, C. R. (1973) *Linear Statistical Inference and Its Applications*, 2nd Edition, New York: John Wiley and Sons.

Rao, C. R. (1988) Methodology based on L_1-norm in statistical inference, *Sankhyā, Series A*, 50, 289–313.

Rockafellar, R. T. (1970) *Convex Analysis*, Princeton, NJ: Princeton University Press.

Rousseeuw, P. J. (1984) Least median squares regression, *Journal of the American Statistical Association*, 79, 871–880.

Rousseeuw, P. J. and Leroy, A. M. (1987) *Robust Regression and Outlier Detection*, New York: John Wiley and Sons.

Rousseeuw, P. J. and van Zomeren, B. C. (1990) Unmasking multivariate outliers and leverage points, *Journal of the American Statistical Association*, 85, 633–648.

Rousseeuw. P. J. and van Zomeren, B. C. (1991) Robust distances: Simulations and cutoff values, in: W. Stahel and S. Weisberg, eds., *Directions in Robust Statistics and Diagnostics*, Part II, New York: Springer-Verlag, 195–203.

Savage, I. R. (1956) Contributions to the theory of rank order statistics – the two sample case, *Annals of Mathematical Statistics*, 27, 590–615.

Sawilowsky, S. S. (1990) Nonparametric tests of interaction in experimental design, *Review of Educational Research*, 60, 91–126.

Sawilowsky, S. S., Blair, R. C., and Higgins, J. J. (1989) An investigation of the Type I error and power properties of the rank transform procedure in factorial ANOVA, *Journal of Educational Statistics*, 14, 255–267.

Scheffé, H. (1959) *The Analysis of Variance*, New York: John Wiley and Sons.

Schrader, R. M. and McKean, J. W. (1977) Robust analysis of variance, *Communications in Statistics. Theory and Methods*, 6, 879–894.

Schrader, R. M. and McKean, J. W. (1987) Small sample properties of least absolute values analysis of variance, in: Y. Dodge, ed., *Statistical Data Analysis Based on the L_1 Norm and Related Methods*, Amsterdam: North Holland, 307–321.

Schuster, E. F. (1975) Estimating the distribution function of a symmetric distribution, *Biometrika*, 62, 631–635.

Schuster, E. F. (1987) Identifying the closest symmetric distribution or density function, *Annals of Mathematical Statistics*, 15, 865–874.

Schuster, E. F. and Barker, R. C. (1987) Using the bootstrap in testing symmetry versus asymmetry, *Communications in Statistics. Simulation and Computation*, 16, 19–84.

Searle, S. R. (1971) *Linear Models*, New York: John Wiley and Sons.

Seber, G. A. F. (1984), *Multivariate Observations*, New York: Wiley and Sons.

Sheather, S. J. (1987) Assessing the accuracy of the sample median: Estimated standard errors versus interpolated confidence intervals. in: Y. Dodge, ed., *Statistical Data Analysis Based on the L_1 Norm and Related Methods*, Amsterdam: North-Holland, 203–216.

Sheather, S. J., McKean, J. W., and Hettmansperger, T. P. (1997) Finite sample stability properties of the least median of squares estimator, *Journal of Statistical Computation and Simulation*, 58, 371–383.

Shirley, E. A. C. (1981) A distribution-free method for analysis of covariance based on rank data, *Applied Statistics*, 30, 158–162.

Siegel, S. and Tukey, J. W. (1960) A nonparametric sum of ranks procedure for relative spread in unpaired samples, *Journal of the American Statistical Association*, 55, 429–444.

Sievers, G. L. (1983) A weighted dispersion function for estimation in linear models, *Communications in Statistics. Theory and Methods*, 12(10), 1161–1179.

Simonoff, J. S. and Hawkins, D. M. (1993), Algorithm AS 282: High breakdown regression and multivariate estimation, *Applied Statistics*, 42, 423–432.

Simpson, D. G. Ruppert, D., and Carroll, R. J. (1992) On one-step GM-estimates and stability of inferences in linear regression, *Journal of the American Statistical Association*, 87, 439–450.

Small, C. G. (1990) A survey of multidimensional medians, *International Statistical Review*, 58, 263–277.

Speed, F. M., Hocking, R. R., and Hackney, O. P. (1978) Methods of analysis with unbalanced data, *Journal of the American Statistical Association*, 73, 105–112.

Steel, R. G. D. (1960) A rank sum test for comparing all pairs of treatments, *Technometrics*, 2, 197–207.

Stefanski, L. A., Carroll, R. J., and Ruppert, D. (1986) Optimally bounded score functions for generalized linear models with applications to logistic regression, *Biometrika*, 73, 413–424.

Stewart, G. W. (1973) *Introduction to Matrix Computations*, New York: Academic Press.

Stromberg, A. J. (1993) Computing the exact least median of squares estimate and stability diagnostics in multiple linear regression, *SIAM Journal of Scientific Computing*, 14, 1289–1299.

Student (1908) The probable error of a mean, *Biometrika*, 6, 1–25.

Tableman, M. (1990) Bounded-influence rank regression: A one-step estimator based on Wilcoxon scores, *Journal of the American Statistical Association*, 85, 508–513.

Terpstra, T. J. (1952) The asymptotic normality and consistency for Kendall's test against trend, when ties are present, *Indagationes Mathematicae*, 14, 327–333.

Thompson, G. L. (1991a), A note on the rank transform for interaction, *Biometrika*, 78, 697–701.

Thompson, G. L. (1991b), A unified approach to rank tests for multivariate and repeated measure designs, *Journal of the American Statistical Association*, 86, 410–419.

Thompson, G. L. (1993) Correction note to: A note on the rank transform for interactions, (78, 697–701), *Biometrika*, 80, 211.

Thompson, G. L. and Ammann, L. P. (1989) Efficacies of rank-transform statistics in two-way models with no interaction, *Journal of the American Statistical Association*, 85, 519–528.

Tierney, L. (1990) *XLISP-STAT*, New York: John Wiley and Sons.

Tucker, H. G. (1967) *A Graduate Course in Probability*, New York: Academic Press.

Tukey, J. W. (1960), A survey of sampling from contaminated distributions, in: I. Olkin *et al.*, eds, *Contributions to Probability and Statistics in Honor of Harold Hotelling*, Stanford, CA: Stanford University Press.

Vidmar, T. J. and McKean, J. W., (1996) A Monte Carlo study of robust and least squares response surface methods, *Journal of Statistical Computation and Simulation*, 54, 1–18.

Vidmar, T. J., McKean, J. W., and Hettmansperger, T. P. (1992) Robust procedures for drug combination problems with quantal responses, *Applied Statistics*, 41, 299–315.

Wang, M. H. (1996) Statistical graphics: Applications to the R and GR methods in linear models, Unpublished Ph.D. thesis, Western Michigan University, Kalamazoo.

Welch, B. L. (1937) The significance of the difference between two means when the population variances are unequal, *Biometrika*, 29, 350–362.

Wilcoxon, F. (1945) Individual comparisons by ranking methods. *Biometrics*, 1, 80–83.

Wilks, S. S. (1960), Multidimensional statistical scatter, in: I. Olkin *et al.*, eds, *Contributions to Probability and Statistics in Honor of Harold Hotelling*, Stanford, CA: Stanford University Press, 486–503.

Witt, L. D. (1989) Coefficients of multiple determination based on rank estimates, Unpublished Ph.D. thesis, Western Michigan University, Kalamazoo.

Witt, L. D., McKean, J. W., and Naranjo, J. D. (1994) Robust measures of association in the correlation model, *Statistics and Probability Letters*, 20, 295–306.

Witt, L. D., Naranjo, J. D., and McKean, J. W. (1995) Influence functions for rank-based procedures in the linear model, *Journal of Nonparametric Statistics*, 5, 339–358.

Ylvisaker, D. (1977) Test resistance, *Journal of the American Statistical Association*, 72, 551–556.

Author Index

Adichi, J. N. 174
Afifi, A. A. 259
Akritas, M. G. 267–8, 347
Ammann, L. P. 203, 268, 289
Ansari, A. R. 123, 126
Apostol, T. M. 433
Arnold, S. F. 216–18, 267, 347, 397
Aubuchon, J. C. 181
Azen, S. P. 259

Babu, G. J. 55
Bedall, F. K. 344
Bahadur, R. R. 108
Bai, Z. D. 345, 388
Barker, R. C. 51
Bassett, G. 187
Belsley, D. A. 189, 202–4
Bibby, J. M. 364
Bickel, P. J. 2, 84, 244, 333, 342, 346
Blair, R. C. 269
Bloomfield, B. 187
Blumen, I. 357, 367
Bohn, L. L. 117
Boos, D. D. 52
Bose, A. 345 356
Box, G. E. P. 276
Bradley, R. A. 123, 126
Bradu, D. 289
Bray, T. A. 269
Brown, B. M. 41, 331, 344, 349, 357, 361, 362, 363, 364, 377, 388
Brunner, E. 233, 267

Carmer, S. G. 247
Carroll, R. J. 203, 234, 289
Chang, W. H. 313, 315, 317, 322, 432
Charles, J. A. 63
Chaudhuri, P. 345, 355–6, 367, 369
Chernoff, H. 47, 98
Chiang, C.-Y. 174
Chinchilli, V. M. 385

Choi, K. 356
Coakley, C. W. 33
Conover, W. J. 121–2, 127, 268
Cook, R. D. 189, 198–9, 201–3, 305
Cox, D. R. 276
Croux, C. 316, 322, 350–3, 355, 363–4, 369
Cushney, A. R. 12

Davis, J. B. 385, 390
Dietz, E. J. 367
Dixon, S. L. 190, 230
Dongarra, J. J. 185
Donoho, D. L. 29
Dowell, M. 187
Draper, D. 193
Draper, N. R. 225–6, 233, 305
DuBois, C. 13
Ducharme, G. R. 344, 360
Dwass, M. 250

Efron, B. 27–8
Eubank, R. L. 52

Fernholz, L. T. 422
Fisher, N. I. 350
Fix, E. 84
Fligner, M. A. 120–2, 127, 130, 132, 135–6, 268
Fraser, D. A. S. 43

Gastwirth, J. L. 47, 49, 50, 105, 108
George, K. J. 182
Ghosh, M. 224
Gosset, W., *see* Student 12
Gower, J. C. 344
Graybill, F. A. 158, 178, 231, 341
Gunst, R. F. 294

Hackney, O. P. 255
Hadi, A. S. 289, 324

Hájek, J. 80, 114, 122, 125, 397–8, 400, 402, 408
Hald, A. 225
Hampel, F. R. 30, 313, 422
Hardy, G. H. 9
Hawkins, D. M. 289
He, X. 33
Heiler, S. 412–13
Hendy, M. F. 63
Hess, J. L. 294
Hettmansperger, T. P. 26, 33, 34, 41, 52, 56–9, 62, 78, 108, 121, 127, 131, 145, 169, 172, 174–5, 180–1, 185, 190, 201, 233, 280, 288, 295, 298–9, 312–13, 322, 361–2, 364, 366, 377, 388
Higgins, J. J. 43, 269
Hocking, R. R. 234, 255
Hodges, J. L. 1, 10, 29, 36, 38, 74, 84, 174, 355
Hogg, R. V. 43
Hora, S. C. 268
Hössjer, O. 313, 316, 319, 350–2, 353, 355, 363, 364, 369
Hotelling, H. 385
Høyland, A. 4
Hsu, J. C. 246, 253
Huber, P. J. 29, 48, 158, 422
Huitema, B. E. 259

Iman, R. L. 268
Iyer, H. K. 231

Jaeckel, L. A. 162–3, 412, 416, 439
Jan, S. L. 345, 347, 365–6
Jarratt, P. 187
Jarrett, R. G. 28
Johnson, G. D. 52
Johnson, M. E. 121
Johnson, M. M. 121
Jonckherre, A. R. 277
Jurečková, J. 162, 165, 180, 412, 418

Kahaner, D. 269
Kalbfleisch, J. D. 114, 207–9
Kapenga, J. A. 184, 186–7, 324
Kass, G. V. 289
Kent, J. T. 364

Kepner, J. C. 347
Killeen, T. J. 108, 120, 122
Klotz, J. 123, 125
Koenker, R. 187
Koti, K. M. 55
Koul, H. L. 92, 180, 182–3, 288
Kramer, C. Y. 249
Kruskal, W. H. 240
Kuh, K. 189, 202, 204

LaRiccia, V. N. 52
Larsen, R. J. 64
Lawless, J. F. 115, 212, 214
Lawley, D. N. 385
Lehmann, E. L. 1, 2, 10, 36, 38, 74, 174, 251, 253, 355
Leroy, A. M. 152, 289, 316, 318, 324
Li, H. 233
Littlewood, J. E. 9
Liu, R. Y. 367
Lopuhaä, H. P. 368

Magnus, J. R. 373
Malin, J. S. 131
Mann, H. B. 66, 74
Marden, J. 356
Mardia, K. V. 350, 364
Maritz, J. S. 28, 41, 111
Marsaglia, G. 269
Mason, R. L. 294
Mathisen, H. C. 101, 105
McIntyre, G. A. 53
McKean, J. W. 28, 75, 76, 92, 100, 145, 150, 169, 172–5, 178, 180–7, 189, 190, 193, 201, 212, 214, 215, 220, 230, 233, 237, 247, 253, 255, 256, 258, 267, 269, 287, 288, 294, 295, 298, 299, 301, 303, 308, 313, 322, 326, 385, 390, 424
Merchant, J. A. 335
Mielke, P. W. 210
Milasevic, P. 344, 360
Miller, R. G. 246, 247, 250, 251, 253
Moler, C. B. 269
Mood, A. M. 101, 103, 123, 126
Morrison, D. F. 204
Möttönen, J. 344, 347, 350, 353–5, 364–6, 369, 377, 388

Naranjo, J. D. 192–3, 215, 280, 287, 293, 298, 301, 308, 312, 313, 424
Nash, S. 269
Nelson, W. 115, 212, 255
Neter, J. 262
Neudecker, H. 373
Neumann, N. 268
Niinimaa, A. 360, 369, 370, 371
Noether, G. E. 25, 158
Nyblom, J. 59, 359, 360, 363, 367, 388

Oja, H. 344, 347, 350, 353–5, 359, 360, 363, 364, 366, 367, 369, 370, 371, 377, 388
Olshen, R. A. 36
Osborne, M. R. 184

Peebles, A. R. 12
Peters, D. 363, 364, 377
Pitman, E. J. G. 25
Policello, G. E., II 43, 130, 132, 135–6
Pólya, G. 9
Portnoy, S. L. 33
Prentice, R. L. 114, 207–9
Puri, M. L. 121, 127, 174, 233, 344, 385

Randles, R. H. 52, 345, 347, 358, 359, 363–6, 377, 398
Rao, C. R. 173, 187, 265, 345, 388, 438
Robinson, D. H. 347
Rockafellar, R. T. 412, 413
Rosenstein, R. B. 52
Rousseeuw, P. J. 152, 203, 289, 305, 316, 318, 324, 368
Ruppert, D. 203, 234, 289
Rust, S. W. 135, 136
Ryan, T. A., Jr. 187

Savage, I. R. 47, 98, 112, 114
Sawilowsky, S. S. 254, 269
Scheffé, H. 242
Schrader, R. M. 28, 145, 150, 189, 288
Schuster, E. F. 50, 51

Searle, S. R. 172
Seber, G. A. F. 390
Sen, P. K. 174, 224, 344, 385
Sengupta, D. 367
Sheather, S. J. 26, 27, 59, 92, 173, 178, 184, 190, 201, 288, 295, 298, 299, 301, 313, 322, 422
Shirley, E. A. C. 270, 273
Šidák, Z. 80, 114, 122, 125, 397, 398, 400, 402, 408
Siegel, S. 123
Sievers, G. L. 75, 92, 108, 182, 183, 193, 212, 214, 220, 237, 247, 255, 256, 258, 280, 370
Simonoff, J. S. 289, 324
Simpson, D. G. 33, 203, 289
Singh, K. 367
Small, C. G. 333
Smith, H. 193, 225–6, 305
Speed, F. M. 255
Staudte, R. G., Jr. 41, 422
Steel, R. G. D. 250
Stefanski, L. A. 234
Steiger, W. L. 187
Stewart, G. W. 185
Stromberg, A. J. 324
Stroup, D. F. 64
Student 12
Swanson, M. R. 247

Tableman, M. 279, 370
Terpstra, T. J. 277
Thompson, G. L. 268
Tibshirani, R. J. 27
Tienari, J. 355
Tierney, L. 212
Tucker, H. G. 415
Tukey, J. W. 23, 123

van Zomeren, B. C. 203, 289
Vidmar, T. J. 75, 184, 186, 187, 233, 237, 247, 267, 269

Wallis, W. A. 240
Wang, M. H. 212
Weisberg, S. 189, 198, 199, 201–3, 305
Welch, B. L. 133
Welsch, R. E. 189, 202, 204

Whitney, D. R. 74
Wilcoxon, F. 1, 75
Wilks, S. S. 360
Willers, R. 412, 413
Witt, L. D. 215, 220, 224, 287, 308, 312, 424, 429
Wolfe, D. A. 117, 398
Wolff, S. S. 47
Wu, M. 41

Ylvisaker, D. 32

Zimmerman, H. 344

Subject Index

accelerated failure time models, 207
added variable plot, 193
affine, 331
    transformation, 331
affine invariant rank methods, 364
    one-sample signed-rank statistic, 364
affine invariant sign test, 353
aligned rank test, 174
analysis of covariance models, *see* experimental designs
angle sign test, 344
    efficiency relative to Hotelling's T^2, 348, 355
anti-ranks, 9, 34, 44
Argmin, 4
asymptotic linearity, 20
    L_1, 22
    general signed-rank scores, 410
    linear model, 162, 411
    L_1, 422
    two-sample scores, 98
asymptotic power, 24
asymptotic power lemma, 24
asymptotic relative efficiency, 24

Behrens–Fisher problem, 128
    Mann–Whitney–Wilcoxon, 128
    modified, 133
    Mathisen's test, 131
    Welch t-test, 134
Blumen's bivariate sign test, 357
    efficiency relative to Hotelling's T^2, 357
bootstrap, 27
    test for symmetry, 49
bounded in probability, 22
bounded influence, *see* GR-estimates
breakdown, *see* general signed-rank scores, Mann–Whitney–Wilcoxon
    acceptance breakdown, 31
    asymptotic value, 29
    estimation, 29
    GR-estimates, 313
    HBR-estimates, 317
    L_1 two-sample, 109
    rejection breakdown, 31
    expected rejection breakdown, 33

central limit theorem
    Lindeberg–Feller, 397
componentwise estimates, 334
    breakdown, 368
    efficiency, 338, 341
    Hodges–Lehmann estimate efficiency relative to the mean vector, 343
    influence function, 368
componentwise estimating equations, 334
componentwise tests
    sign tests, 340
    Wilcoxon test efficiency relative to Hotelling's T^2, 343
confidence interval, 6
    comparison of two samples, 60
    efficiency, 25
    estimate of standard error, 25
    interpolated confidence intervals, 56
    shift parameter, 62
consistent test, 18
    rank-based tests in linear model reduction in dispersion, 175
contaminated normal distribution, 23, 38
contiguity, 400
convex, 412
coordinate-free model, 146
correlation model, 215, 308
    GR coefficient of multiple determination, 309

correlation model (*contd*)
  Huber's condition, 215
  R coefficient of multiple
    determination, 218
    properties of, 220
  traditional coefficient of multiple
    determination, 218

delete i model, 202
difference in means, 73
direct product, 373
dispersion function
  linear model, 148
    functional representation, 179
  one-sample, 4
  quadratic approximation, 162
  two-sample, 71

efficacy, 19
efficiency, 23
  bivariate, 332
  L_1 versus L_2, 23
elliptical model, 329
equivariance
  scale, 91
  translation, 91
equivariant estimator, 108
  bivariate, 330
examples, data
  correlation model
    Hald data, 225
  diagnostics
    cloud data, 193, 200
    free fatty acid data, 204
  experimental design
    Box–Cox data, 273
    LDL cholesterol in quail, 236, 241, 243, 245, 251
    lifetime of motors, 255
    marketing data, 262
    pigs and diets, 265
    Poland China pigs, 242
    rat data, 270
    snake data, 259
  linear model
    baseball salaries data, 152
    Hawkins data, 289
    potency data, 155, 194
    quadratic data, 324
    stars data, 289, 324
    synthetic rubber data, 294
    telephone data, 151
    wood data, 305
  log-linear model
    insulating fluid data, 212
  multivariate experimental design
    paspalum grass data, 390
  multivariate linear model
    tablet potency data, 385
  multivariate location
    brains of mice data, 376
    cork borings data, 345, 354
    cotton dust data, 335
    Mathematics and Statistics exam scores, 364
  one sample
    Cushney–Peebles data, 12, 59
    Darwin data, 137
    Shoshoni rectangles data, 13
  proportional hazards
    lifetime of insulation fluid, 115
  two samples
    Hendy and Charles coin data, 63, 76, 92
    quail data, 75, 100, 105, 108
experimental designs
  analysis of covariance models, 258
    covariates, 258
  contrasts, 234
    estimation, 242
    hypothesis testing, 241
  incidence matrix, 235
  means model, 234
  medians model, 234, 243
  multiple comparison procedures, 246
    Bonferroni, 246
    experiment error rate, 246
    family error rate, 246
    pairwise confidence intervals, 251
    pairwise tests, joint rankings, 249
    pairwise tests, separate rankings, 250
    protected LSD procedure, 247
    Tukey, 247
    Tukey–Kramer, 249, 255

experimental designs (*contd*)
  multivariate
    means model, 389
    medians model, 389
  one-way design, 235
  pseudo-observations, 244
  two-way model, 254
    additive model, 254
    interaction, 254
    main effect, 254
    profile plots, 254
extreme value distribution, 113

Fisher information, 46, 88
Friedman test statistic, 276
full model, 4

Gâteux derivative, 422
general rank scores, 159,
    see regression model
  piecewise linear, 100
  two-sample scores, 94
    asymptotic linearity, 98
    estimate of shift, 95, 98
    gradient function, 95
    gradient rank test, 95
    normal scores, 99
    null distribution, 96
    pseudo-norm, 93
general signed-rank scores, 42, 407
  asymptotic breakdown, 48
  confidence interval, 44
  derived from two-sample general
    rank scores, 101
  efficacy, 46
  functional, 43
  gradient function, 42
  influence function, 423
  influence function of the estimate,
    48
  linearity, 410
  local asymptotic distribution
    theory, 400, 403
  optimal score function, 45
  Pitman regular, 45
  test statistic
    asymptotic power lemma, 46
    null distribution, 44
$GF(2m_1, 2m_2)$, 209

GR-estimates, 279, *see* correlation
    model,
    HBR-estimates
  asymptotic distribution, 286
  asymptotic linearity, 286
  breakdown, 313
  comparison with R-estimates,
    301
  curvature detection, 298
  gradient, 280, 283
    asymptotic null distribution,
      283
  influence function, 312
  intercept, 288
    joint asymptotic distribution,
      288
  internal GR-studentized residuals,
    300
  pseudo-norm, 279
  reduction in dispersion, 287
    null asymptotic distribution,
      287
  weight matrix, 280
gradient process, 5
gradient test, 5, 148, 174
  consistency, 18

hazard function, 111
HBR-estimates, 314,
    see GR-estimates
  asymptotic distribution, 316,
    432
  breakdown, 317
  dispersion function, 314
    quadratic approximation,
      436
  gradient, 314
    asymptotic null distribution,
      316, 437
  influence functions, 319
  intercept, 323
  internal HBR-studentized
    residuals, 323
  linearity, 435
  pseudo-norm, 314
high breakdown estimates,
    see HBR-estimates
Hotelling's T^2, 330
Huber's condition, 158

influence function, 30, 422,
    see R-estimates in linear
    model, rank-based tests in
    linear model
interdirections, 359
interpolated confidence intervals, 56
invariant test statistic, 330

Kendall's τ, 227
Kronecker product, 373
Kruskal–Wallis, 240,
    see experimental designs,
    multivariate linear model

lack of fit, 214
Lawley–Hotelling trace statistic, 375
least squares, see norm
LeCam's lemmas, 400
Lehmann alternatives, 111
linear model, 145, see R-estimates in
    linear model, rank-based tests
    in linear model,
    GR-estimates, correlation
    model, experimental design,
    HBR-estimates, multivariate
    linear model
  approximation of R-residual, 197
  $CFITS_{D,i}$, 302
  external R-studentized residual, 201
  general linear hypotheses, 146
  internal HBR-studentized residuals, 323
  internal R-studentized residuals, 199
  least squares, 148
    reduction in sums of squares, 150
  L_1 estimates, 188
    reduction in dispersion, 189
  model misspecification, 190, 293
    departure from orthogonality, 192, 293
  R pseudo-norm, 146
  $RDBETAS$, 203
  $RDCOOK$, 202
  $RDFFIT$, 202
  $TDBETAS_R$, 302

linear rank statistics, see regression model
linearity, see asymptotic linearity
location functional, 2
  center of symmetry, 4
  in two-sample model, 70
location model
  bivariate location model, 329
  one-sample, 2
  symmetric, 34
  two-sample, 70
location parameter, 2
log-linear models, 207

Mann's test for trend, 66
Mann–Whitney–Wilcoxon, 73,
    see multivariate linear model
  asymptotic linearity, 89
  confidence interval, 74, 86, 92
  efficacy, 89
  estimate of shift, 74, 91
    approximate distribution, 91
    breakdown, 109
    influence function, 110
    relative efficiency to difference
      in means, 91
  gradient, 73
  Hodges–Lehmann estimate, 74
  Pitman regular, 88
  τ, 89
    estimate of, 92
  test, 74
    consistency, 85
    efficiency relative to the t-test, 89
    general asymptotic distribution, 84
    null asymptotic distribution, 84
    null distribution theory, 78
    power function, 87
    projection, 81
    sample size determination, 90
    unbiased, 87
masking effect, 305
Mathisen's two-sample test, 105
  confidence interval, 107
mean
  breakdown, 29

mean (contd)
    influence function, 31
        two-sample location, 110
    location functional, 3
mean shift model, 201
median, 7
    asymptotic distribution, 7
    bootstrap, 27
    breakdown, 29
    confidence interval, 8
    estimate of standard error, 26
    influence function, 30
    location functional, 2
    standard deviation of, 7
minimum distance, 5
minimum volume ellipsoid, 289, 314
Mood's median test, 103
    confidence interval, 103
    efficacy, 105
    estimate of shift, 104
        influence function, 110
multiple comparison procedures,
        see experimental designs
multivariate linear model, 372
    estimating equations, 373
    Kruskal–Wallis, 377
    Mann–Whitney–Wilcoxon, 375
    means model, 389
    medians model, 389
    profile analysis, 394
    R-estimates, 380
    test for regression effect, 374
    tests of general hypotheses, 381
MVE, see minimum volume
        ellipsoid
MWW, see Mann–Whitney–
        Wilcoxon

Noether's condition, 158
norm, 4, see pseudo-norm
    L_1 norm, 7, see Mathisen's two-
            sample test
        asymptotic linearity, 22
        confidence interval, 8
        dispersion, 7
        efficacy, 22
        estimating equation, 7
        gradient, 7
        Pitman regularity, 21

L_2 norm, 8
    t-test, 8
    efficacy, 22
    influence function, 425
    Pitman regularity, 22
    estimate induced by, 4
weighted L_1 norm, 9
    gradient function, 10
normal scores, 43
    breakdown, 49
    efficiency relative to L_2, 47

Oja criterion function, 360
Oja median, 360
    breakdown point, 370
Oja sign test, 361
one-sample location model,
        see norm
optimal score function
    one-sample, 46
    two-sample, 97
ordered alternative, 276
orthogonal transformation, 330
outlier, 197

paired design model, 136
    compared with completely
            randomized design, 136
Pitman regular, 19, 331
power function, 17
    asymptotic power lemma, 24
profile analysis, see multivariate
        linear model
projection theorem, 81
proportional hazards models, 111
    linear model, 208
    log exponential model, 113
    log-rank test, 113
    Mann-Whitney–Wilcoxon, 112
pseudo-median, 3, 11
pseudo-norm, 71, see linear model,
        see GR-estimates
    confidence interval, 72
    gradient, 72
    gradient test, 72
    HBR-estimates, 314
    L_1 pseudo-norm, 102, 103
    L_2 pseudo-norm, 73
    Mann–Whitney–Wilcoxon, 73

pseudo-norm (contd)
  reduction in dispersion, 72
pure error dispersion, 212

QR-decomposition, 185
quadratic approximation
  L_1, 422
  dispersion function linear model, 163, 413
  quadraticity, 162

R-estimates, see HBR-estimates
  comparison with GR-estimates, 301
R-estimates in linear model, 147, see correlation model
  asymptotic distribution, 163
  influence function, 164, 427
  intercept, 164
    joint asymptotic distribution, 166, 168, 170
  internal R-studentized residuals, 199
  k-step, 185
  Newton-type algorithm, 184
  R normal equations, 148
Randles' sign test, 359
rank scores, see general rank scores
rank transform, 267
rank-based tests in linear model, see multivariate linear model
  aligned rank test, 174
  efficiency relative to least squares, 178
  gradient test, 151
  influence function, 175
  reduction in dispersion, 150
    F_φ, 150, 151
    asymptotic distribution, local alternatives, 176
    consistency, 175, 417
    influence function, 427
    null asymptotic distribution, 170
  scores test, 151, 160
    null asymptotic distribution, 174
  Wald-type test, 151

  null asymptotic distribution, 173
ranked-set sampling, 52
  Mann–Whitney–Wilcoxon, 117
reduced model, 6
reduction in dispersion, 6
regression model, 145, 398
  linear rank statistics, 399
    local asymptotic distribution theory, 400, 403
    null distribution theory, 399
  through the origin, 230
resolving test, 17
rglm, 185
RSS, 52

Sample size determination
  one-sample, 25
  two-sample, 90
scale model, 118
  general scores, 118
  Mood test, 123
  Siegel–Tukey test, 142
scale problem
  Fligner–Killeen test, 122
  unknown locations, 121
score test, 5, 151, 174, 283
scores, see general signed-rank scores, general rank scores
selection of predictors, 226
shift parameter, 70
  estimate of, 72
sign test, 7
  consistency, 18
  distribution-free, 7
  nonparametric, 7
  null distribution, 7
  Pitman regularity, 21
  ranked-set sampling, 52
  rejection breakdown, 31
signed-rank scores, see general signed-rank scores
signed-rank statistics, see general signed-rank scores
spatial median, 344
  breakdown point, 368
spatial rank methods
  ranking lengths, 350

spatial rank methods (contd)
  rotational invariant rank vectors, 353
  spatial signed rank statistic, 353
spatial ranks
  spatial signed rank tests efficiency, 353
spatial sign test, see angle sign test
Spearman's r_s, 227
  multivariate linear model, 375
stepwise model building, 226
stochastic ordering, 2, 77
symmetry
  diagonal symmetry, 329
  test for, 49

t-test
  one-sample, 8
    Pitman regularity, 22
    rejection breakdown, 32
  two-sample, 73
τ_{φ^+}, 46
τ_φ, 97, 157
  estimate of, 181
    consistency, 183

two-sample location model, see pseudo-norm
two-sample scale model, see scale model

unbiased test, 17

Wald-type tests, 6, 151, 173
Walsh averages, 16
Wilcoxon
  efficacy, 36
  efficiency relative to L_2, 37
  Hodges–Lehmann
    approximate distribution, 11, 37
    confidence interval, 11
    estimate of location, 10, 37
  Pitman regular, 36
  pseudo-median, 3, 11
    breakdown, 28, 49
    influence function, 40
  signed-rank test, 11
    null distribution, 34
    rejection breakdown, 41
Wilks's generalized variance, 332